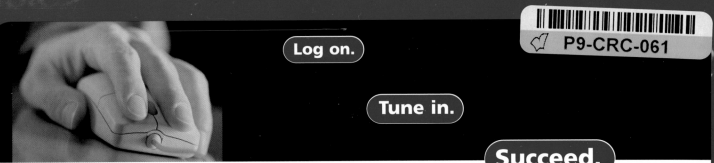

Your steps to success.

STEP 1: Register

All you need to get started is a valid email address and the access code below. To register, simply:

1. Go to www.aw-bc.com/envscience
2. Click the appropriate book cover.
   Cover must match the textbook edition being used for your class.
3. Click **"Register"** under **"First-Time User?"**
4. Leave **"No, I Am a New User"** selected.
5. Using a coin, scratch off the silver coating below to reveal your access code.
   Do not use a knife or other sharp object, which can damage the code.
6. Enter your access code in lowercase or uppercase, without the dashes.
7. Follow the on-screen instructions to complete registration.
   During registration, you will establish a personal login name and password to use for logging into the website. You will also be sent a registration confirmation email that contains your login name and password.

Your Access Code is:

*

Note: If there is no silver foil covering the access code, it may already have been redeemed, and therefore may no longer be valid. In that case, you can purchase access online using a major credit card. To do so, go to www.aw-bc.com/envscience, click the cover of your textbook, click **"Buy Now"**, and follow the on-screen instructions.

STEP 2: Log in

1. Go to www.aw-bc.com/envscience and click the appropriate book cover.
2. Under **"Established User?"** enter the login name and password that you created during registration. *If unsure of this information, refer to your registration confirmation email.*
3. Click **"Log In"**.

STEP 3: (Optional) Join a class

Instructors have the option of creating an online class for you to use with this website. If your instructor decides to do this, you'll need to complete the following steps using the Class ID your instructor provides you. By "joining a class," you enable your instructor to view the scored results of your work on the website in his or her online gradebook.

To join a class:

1. Log into the website. For instructions, see "STEP 2: Log in."
2. Click **"Join a Class"** near the top right.
3. Enter your instructor's **"Class ID"** and then click **"Next"**.
4. At the Confirm Class page you will see your instructor's name and class information. If this information is correct, click **"Next"**.
5. Click **"Enter Class Now"** from the Class Confirmation page.

- *To confirm your enrollment in the class, check for your instructor and class name at the top right of the page. You will be sent a class enrollment confirmation email.*
- *As you complete activities on the website from now through the class end date, your results will post to your instructor's gradebook, in addition to appearing in your personal view of the Results Reporter.*

To log into the class later, follow the instructions under "STEP 2: Log in."

Got technical questions?

Visit http://247.aw.com. Email technical support is available 24/7.

SITE REQUIREMENTS

For the latest updates on Site Requirements, go to www.aw-bc.com/envscience, choose your text cover, and click Site Reqs.

WINDOWS
Windows 2000/XP
64 MB RAM
1024 x768 screen resolution, thousands of colors
Browser: Internet Explorer 6.0; Netscape 7.0
Plug-Ins: Latest versions of Flash and QuickTime
Internet Connection: 56k modem minimum

MACINTOSH
OS 10
64 MB RAM
1024 x768 screen resolution, thousands of colors
Browsers: Internet Explorer 6.0; Netscape 7; Safari 1.2*
Plug-Ins: Latest versions of Flash and QuickTime
Internet Connection: 56k modem minimum

* The gradebook and class manager features of this website have been partially tested on Mac OS X 10.3.2 with Safari 1.2. A future update will fully support this configuration.

Register and log in

Join a Class

Important: Please read the Subscription and End-User License Agreement, accessible from the book website's login page, before using the Environmental Science Place website. By using the website, you indicate that you have read, understood, and accepted the terms of this agreement.

ESSENTIAL ENVIRONMENT

THE SCIENCE BEHIND THE STORIES

SECOND EDITION

JAY WITHGOTT

SCOTT BRENNAN

PEARSON

Prentice
Hall

Upper Saddle River, NJ 07458

Editor-in-Chief: Beth Wilbur
Senior Acquisitions Editor: Chalon Bridges
Project Editor: Nora Lally-Graves
Director of Development: Deborah Gale
Managing Editor: Michael Early
Production Supervisor: Lori Newman
Production Management: Holly Henjum, GGS Book Services
Copyeditor: Sally Peyrefitte
Compositor: GGS Book Services
Art Development Editor: Russell Chun
Design Manager: Mark Ong

Interior Designer: Gary Hespenheide
Cover Designer: Yvo Riezebos
Illustrators: Dragonfly Media Group
Photo Production Manager: Travis Amos
Photo Researcher: Kristin Piljay
Director, Image Resource Center: Melinda Patelli
Image Rights and Permissions Manager: Zina Arabia
Manufacturing Buyer: Stacy Wong
Executive Marketing Manager: Lauren Harp
Text printer: Quebecor World Dubuque
Cover printer: Phoenix Color Corporation

Cover Photo Credit: A scientist takes an ice sample. Scientists aboard the Louis S. St.-Laurent, an ice-breaker research ship, master summer ice floes to study waters of the Arctic Ocean in the Canada Basin. National Geographic/Getty Images.

Photo credits continue following the glossary.

Printed using soy-based ink. Paper is recycled, containing at least 20% post-consumer waste.

Library of Congress Cataloging-in-Publication Data
Withgott, Jay.
  Essential environment : the science behind the stories. — 2nd ed. / Jay Withgott, Scott Brennan.
    p. cm.
  Brennan's name appears first on the previous ed.
  Abridged ed. of: Environment. 2nd ed. c2007.
  Include bibliographical references and index.
  ISBN 0-8053-0640-4 (alk. paper)
    1. Environmental sciences.   I. Brennan, Scott R.   II. Withgott, Jay. Environment.   III. Title.
GE105.B74 2007b
363.7—dc22   2006021828
  ISBN **0-8053-0640-4** (Student edition)
  ISBN **0-8053-9367-6** (Professional copy)

                                                    10 9 8 7 6 5 4 3

www.prenhall.com

About the Authors

Jay H. Withgott is a science and environmental writer with a background in scientific research and teaching. He holds degrees from Yale University, the University of Arkansas, and the University of Arizona. As a researcher, he has published scientific papers on topics in ecology, evolution, animal behavior, and conservation biology in a variety of journals including *Proceedings of the National Academy of Sciences, Proceedings of the Royal Society of London B, Evolution*, and *Animal Behavior*. He has taught university-level laboratory courses in ecology, ornithology, vertebrate diversity, anatomy, and general biology.

As a science writer, Jay has authored articles for a variety of journals and magazines including *Science, New Scientist, BioScience, Smithsonian, Conservation in Practice*, and *Natural History*. He combines his scientific expertise with his past experience as a reporter and editor for daily newspapers to make science accessible and engaging for general audiences.

Jay lives with his wife, biologist Susan Masta, in Portland, Oregon, and takes every opportunity he can to explore the diverse landscapes of Oregon and the American West.

Scott Brennan has taught environmental science, ecology, resource policy, and journalism at Western Washington University and at Walla Walla Community College. He has also worked as a journalist, photographer, and consultant.

Scott has cultivated his expertise in environmental science and public policy by serving as Campaign Director of Alaskans for Responsible Mining, as Executive Conservation Fellow of the National Parks Conservation Association in Washington, D.C., and as a consultant to the U.S. Department of Defense Environmental Security Office at the Pentagon.

When not at work, Scott is likely to be found exploring the Chugach Mountains and the Bristol Bay drainages in southwest Alaska. He lives with his wife, Angela, and their dogs Raven and Hatcher, in south-central Alaska's Chester Creek Watershed.

5 Species Interactions and Community Ecology

Aggregation of zebra mussels

Upon successfully completing this chapter, you will be able to:

▶ Compare and contrast the major types of species interactions

▶ Characterize feeding relationships and energy flow, using them to construct trophic levels and food webs

▶ Distinguish characteristics of a keystone species

▶ Characterize the process of succession

▶ Perceive and predict the potential impacts of invasive species in communities

▶ Explain the goals and methods of ecological restoration

▶ Describe and illustrate the terrestrial biomes of the world

people and places?

INVESTIGATE it! on the Withgott/Brennan Companion Website provides an additional **120 case studies** beyond those presented in the text. Browse by topic or geographic region to access 100 recent articles from **The New York Times** and 20 **abc NEWS** clips that explore environmental issues that are in the news today.

Do you understand the **science** behind

The Science behind the Story highlights how scientists develop hypotheses, test predictions, and analyze and interpret data. Each *Science behind the Story* carefully walks you through the scientific process—not only *what* scientists have discovered, but *how* they discovered it. These engaging accounts help you understand "how we know what we know" about environmental issues.

The Science behind the Story

Inferring Zebra Mussels' Impacts on Fish Communities

When zebra mussels appeared in the Great Lakes, people feared for sport fisheries and estimated that fish population declines could cost billions of dollars. The mussels would deplete the phytoplankton and zooplankton that fish depended on, people reasoned, and many fewer fish would survive. However, food webs are complicated systems, and disentangling them to infer the effects of any one species is fraught with difficulty. Thus, even after 15 years, scientists had little solid evidence of widespread harm to fish populations.

So, aquatic biologist David Strayer of the Institute of Ecosystem Studies in Millbrook, New York, joined Kathyrn Hattala and Andrew Kahnle of New York State's Department of Environmental Conservation (DEC). They mined datasets on fish populations in the Hudson River, which zebra mussels had invaded in 1991.

Strayer and others had already been studying effects of zebra mussels on aspects of the community for years. Their data showed that since the species' introduction to the Hudson:

► Biomass of phytoplankton fell 80%.
► Biomass of small zooplankton fell 76%.
► Biomass of large zooplankton fell 52%.

Zebra mussels increased filter-feeding in the community 30-fold, thereby depleting the phytoplankton and small zooplankton, and leaving all sizes of zooplankton with less phytoplankton to eat. Overall, the zooplankton and invertebrate animals of the open water that are eaten by open-water fish declined by 70%.

However, Strayer's work had also found that *benthic*, or bottom-dwelling, invertebrates in shallow water (especially in the nearshore, or *littoral*, zone) had increased by 10%, and likely much more, because the mussels' shells provide habitat structure and their feces provide nutrients.

These contrasting trends in the benthic shallows and the open deep water led Strayer's team to hypothesize that zebra mussels would harm open-water fish but would help littoral-feeding fish. They predicted that after zebra mussel introduction, larvae and juveniles of six common open-water fish species would decline in number, decline in growth rate, and shift downriver toward saltier water, where mussels are absent. Conversely, they predicted that larvae and juveniles of 10 littoral fish species would increase in number, increase in growth rate, and shift upriver to regions of greatest zebra mussel density.

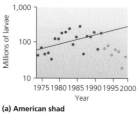

(a) American shad

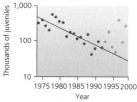

(b) Tessellated darter

Larvae of American shad **(a)**, an open-water fish, had been increasing in abundance before zebra mussels were introduced (red points and trend line). After zebra mussel introduction, shad larvae decreased in abundance (orange points). Juveniles of the tessellated darter **(b)**, a littoral zone fish, had been decreasing in abundance before zebra mussels were introduced (red points and trend line). After zebra mussel introduction, they increased in abundance (orange points). *Source:* Strayer, D., et al. 2004. Effects of an invasive bivalve (*Dreissena polymorpha*) on fish in the Hudson River estuary. *Can. J. Fish. Aquat. Sci.* 61: 924–941.

To test their predictions, the researchers analyzed data from

eat them; fish that eat phytoplankton and zooplankton; larger fish that eat the smaller fish; and lampreys that parasitize the fish. The food web would include a number of native mussels and clams and, since 1988, the zebra mussel that is outcompeting them. It would include diving ducks that used to feed on native bivalves and now are preying on zebra mussels. This food web would also show that crayfish and other bottom-dwelling invertebrates feed from the refuse of zebra mussels. Finally, the food web would include underwater plants and macroscopic algae, whose

the news stories?

We are placing a greater burden on the planet's systems each year. The ongoing rise in human population amplifies nearly all of our environmental impacts—and our consumption of resources has risen even faster than our population growth. The rise in affluence has been a positive development for humanity, and our conversion of the planet's natural capital has made life more pleasant for us so far. However, rising per capita consumption amplifies the demands we make on our environment. Moreover, affluence and consumption have not risen equally for all the world's citizens. Today the 20 wealthiest nations boast 40 times the income of the 20 poorest nations—twice the gap that existed four decades earlier. The ecological footprint of the average citizen of a developed nation such as the United States is considerably larger than that of the average resident of a developing country (Figure 1.15). Within the United States, the rich-

est fifth of people claim nearly half the income, whereas the poorest fifth receive only 5%.

The most comprehensive scientific assessment of the present condition of the world's ecological systems and their ability to continue supporting our civilization was completed in 2005. In that year, over 2,000 of the world's leading environmental scientists from nearly 100 nations completed the **Millennium Ecosystem Assessment.** The four main findings of this exhaustive project are summarized in Table 1.1. The Assessment makes clear that our degradation of the world's environmental systems is having negative impacts on all of us, but that with care and diligence we can still turn many of these trends around.

Environmental issues change quickly, so Withgott/Brennan uses the most current data available.

Sustainability need not require great sacrifice of us. We will naturally always desire to enhance our quality of life, and there are many ways we can do so while also encouraging a more sustainable lifestyle. Indeed, this is the goal of **sustainable development**, the use of resources for economic advancement in a manner that satisfies people's current needs without compromising the future availability of resources.

Sustainability depends largely on the ability of the current human population to limit its environmental impact. Doing so will require us to make an ethical commitment and also to apply knowledge we gain from the sciences. Science can help us devise ways to limit our impact and maintain the functioning of the environmental

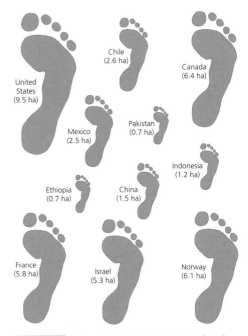

FIGURE 1.15 Citizens of some nations have larger ecological footprints than citizens of others. U.S. residents consume more resources—and thus use more land—than residents of any other nation. Shown are ecological footprints for average citizens of several developed and developing nations, as of 2001. Data from Global Footprint Network, 2005.

Table 1.1	Main Findings of the Millennium Ecosystem Assessment

▶ Over the past 50 years, humans have changed ecosystems more rapidly and extensively than in any comparable period of time in human history, largely to meet rapidly growing demands for food, freshwater, timber, fiber, and fuel. These changes have resulted in a substantial and largely irreversible loss in the diversity of life on Earth.

▶ The changes made to ecosystems have contributed to substantial net gains in human well-being and economic development, but the costs of achieving these gains have been growing. The costs include the degradation of ecosystems and the services they provide for us, and the exacerbation of poverty for some groups of people.

▶ Ecosystem degradation could grow significantly worse during the first half of this century.

▶ The challenge of reversing the degradation of ecosystems while meeting increasing demands for their services can be partially overcome. However, doing so will require that we significantly change many policies, institutions, and practices.

Adapted from *Millennium Ecosystem Assessment, Synthesis Report*, 2005.

References are clearly cited so you can trace the source of the information presented.

Do you know how to interpret graphs

Calculating Ecological Footprints activities at the end of each chapter let you work with numbers to evaluate the impact of actions—including your own—on local and global scales.

CALCULATING ECOLOGICAL FOOTPRINTS

In 2004, coffee consumption in the United States topped 2.7 billion pounds (out of 14.8 billion pounds produced globally). Most coffee is produced on large tropical plantations, where coffee is the only tree species and is grown in full sun. However, approximately 2% of coffee is produced in small groves where coffee trees and other species are intermingled. These *shade-grown* coffee forests maintain greater habitat diversity for tropical rainforest wildlife. Given the information above, estimate the coffee consumption rates in the table.

	Population	Pounds of coffee per day	Pounds of coffee per year
You (or the average American)	1	0.025	9
Your class			
Your hometown			
Your state			
United States			

Data from O'Brien, T. G. and M. F. Kinnaird. 2003. Caffeine and conservation. *Science* 300: 587; and International Coffee Organization.

1. What percentage of global coffee production is consumed in the United States? If only shade-grown coffee were consumed in the United States, how much would shade-grown production need to increase to meet that demand?
2. How much extra would you be willing to pay for a pound of shade-grown coffee, if you knew that your money would help to prevent habitat loss or extinction for animals such as Sumatran tigers, rhinoceroses, and the many songbirds that migrate between Latin America and North America each year?
3. If everyone in the United States were willing to pay as much extra per pound for shade-grown coffee as you are, how much additional money would that provide for conservation of biodiversity in the tropics each year?

GRAPHit! exercises on the Withgott/Brennan Companion Website help you to better understand how to work with and interpret graphs.

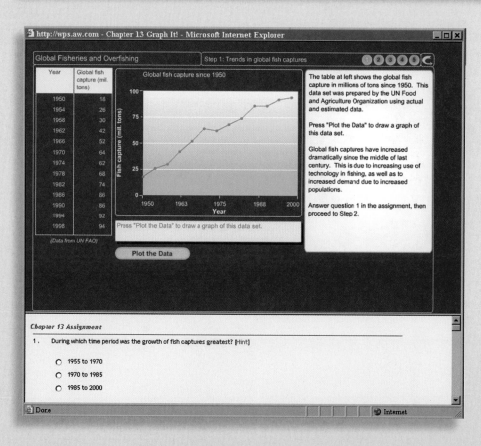

and **data**?

Bar Chart

A bar chart is most often used when one of the variables represents categories rather than numerical values (Figure A.3; see Figure 9.6a, ▸ p. 202). Bar charts allow us to visualize how a variable differs quantitatively among categories.

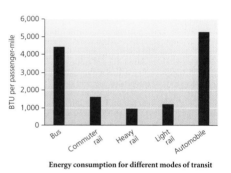

Energy consumption for different modes of transit

FIGURE A.3

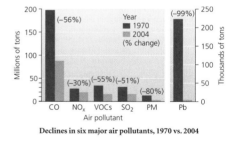

Declines in six major air pollutants, 1970 vs. 2004

FIGURE A.4

It is often instructive to graph two or more variables together to reveal patterns and relationships (Figure A.4; see Figure 12.6a, ▸ p. 278). Many of the bar charts you will see in this book illustrate several types of information at once in this manner.

An *appendix on graphs* (see pages A-1 through A-3 near the end of the book) guides you through the types of graphs you will see throughout the book.

Are you prepared to make **informed**

The *Viewpoints* feature in each chapter presents two opposing views on an environmental issue related to the chapter's central theme, allowing you to consider multiple sides of the story.

Reach your own conclusion on *Viewpoints* questions by accessing the *Viewpoints* link on the companion website. There you'll find questions to consider when exploring the issues further, and links to websites that support each opinion.

VIEWPOINTS

Genetically Modified Foods

Proponents of GM foods say these products can alleviate hunger and malnutrition while posing no known threats to human health or the environment. Opponents say that these foods help only the large corporations that sell them and that they do pose risks to human health, wild organisms, and ecosystems. **What do you think? Should we encourage the continued development of GM foods? If so, what, if any, restrictions should we put on their dissemination?**

A Global Experiment without Controls

Genetic engineering, specifically transgenesis, gives us the unprecedented capacity to move DNA. In so doing, this technology breaches boundaries established through millions of years of evolution. As such, we should expect fundamental alterations in ecosystems with the release of transgenic crops, fish, insects, microbes, and so forth into uncontrolled areas. These alterations are similar in nature to those caused by the introduction of exotic species into new environments. Both processes are unpredictable and could have serious consequences.

Yet because of political and short-term economic imperatives, releases of transgenic organisms have continued unabated for at least a decade. Science has barely started to imagine the ecological and evolutionary consequences of releasing transgenic crops. We not only are experiencing a global experiment without controls, but also lack the tools to document it. Serious research, although extremely scarce, has already confirmed some of the theoretical fears concerning transgenesis aired by scientists a quarter century ago.

Today, we have cataclysmic world hunger paired with food surpluses. The claim that transgenesis can solve this problem is merely a diversion tailored to conceal how transgenesis manipulates the biosphere. Molecular biology might one day become part of the solution to world hunger, but it is certainly not the science most relevant to address the problem today.

What checks should be placed on the release of transgenic organisms? Every check. Through the unaccountable releases so far, we have seen enough, and possibly caused enough, environmental insult for me to say today that we should stop. We need to take stock of the consequences of transgenesis and continue researching under strictly regulated conditions.

Ignacio H. Chapela is associate professor (microbial ecology) in the Department of Environmental Science, Policy, and Management at the University of California at Berkeley. He helped found the Mycological Facility: Oaxaca, Mexico, where he serves as scientific director.

The United States Should Begin a Phased Deregulation of Biotech Crops

During the past two decades the international scientific community, biotechnology industry, and regulatory agencies in many countries have accumulated and critically evaluated a wealth of information about the production and use of biotech crops and products. Biotech crops have been planted since 1996 on more than 1 billion acres of farmland in nearly 20 countries. More than 1 billion humans and hundreds of millions of farm animals have consumed biotech foods and products. Yet there is not a single instance in which biotech crops and foods have been shown to cause illness in humans or animals or to damage the environment.

In spite of this exemplary safety record, a small but well-organized, well-financed, and vocal anti-biotechnology lobby has alleged that biotech crops and products are unsafe for humans and a danger to the environment, demanding a moratorium or outright ban on biotech crops. The rhetoric of the anti-biotechnology groups is alarming, confusing, and frightening to the public, but it is devoid of any substance, because they have never provided any credible scientific evidence to support their allegations.

Any further delay in combining the power of biotechnology with conventional breeding will seriously endanger future food security, political and economic stability, and the environment. Plant biotechnology is still the best hope for meeting the food needs of the ever-growing world population. Biotech crops are already helping to conserve valuable natural resources, reduce the use of harmful agro-chemicals, produce more nutritious foods, and promote economic development.

Twenty years ago, the United States set the precedent by developing regulations for the development and use of biotech crops. Now, as the world leader in plant biotechnology, it is imperative that it lead again by phasing out these redundant regulations in an organized and responsible manner.

Indra K. Vasil is graduate research professor emeritus at the University of Florida (Gainesville, Florida). His research focuses on the biotechnology of cereal crops, and he has been recognized as one of the world's most highly cited authors in the plant and animal sciences.

Explore this issue further by accessing **Viewpoints** at www.aw-bc.com/withgott.

decisions on environmental issues?

Issues in environmental science often lack black-and-white answers, so critical thinking skills help you navigate the gray areas. *Weighing the Issues* questions throughout each chapter encourage you to grapple with questions about science, policy, and ethics.

Weighing the Issues:
Ecosystems Where You Live

Think about the area where you live. How would you describe that area's ecosystems? How do these systems interact with one another? If one ecosystem were greatly disturbed (say, if a wetland or forest were replaced by a shopping mall), what impacts might that have on nearby natural systems?

You Decide activities on the Withgott/Brennan Companion Website allow you to play the role of decision maker as you study the data, then form your own plan for saving endangered grizzlies or stopping global warming.

Preface

We live in extraordinary times. Human impact on our environment has never been so intensive or so far-reaching. The future of Earth's systems and of our society depends more critically than ever on the way we interact with the natural systems around us. Fundamental aspects of nutrient cycling, biological diversity, atmospheric composition, and climate are changing at dizzying speeds. Yet thanks to environmental science, we now understand better than ever how our planet's systems function and how we influence these systems. Understanding environmental science helps us to characterize human-induced problems and also illuminates the tremendous opportunities we have before us for effecting positive change.

The field of environmental science captures the very essence of this unique moment in history. This interdisciplinary pursuit stands at the vanguard of the current need to synthesize academic disciplines and to incorporate their contributions into a big-picture understanding of the world and our place within it.

We wrote this book because we feel that the vital importance of environmental science in today's world makes it imperative to engage, educate, and inspire a broad audience of today's students—the citizens and leaders of tomorrow. We have therefore tried to implement the very best in modern teaching approaches and to clarify how the scientific process can inform human efforts. We also have aimed to maintain a balanced approach and to encourage critical thinking as we flesh out the social debate over many environmental issues. Finally, we have resolved to avoid gloom and doom and instead provide hope and solutions.

As environmental science has grown, so have the length and expense of the textbooks that cover it. With this volume, we aim to meet the needs of introductory environmental science courses that require a more succinct and affordable book. We have distilled the most essential content from our full-length book, *Environment: The Science behind the Stories*, now in its second edition. We have reconceptualized and streamlined the organization of chapters and sections, and we have carefully crafted our rewriting to make *Essential Environment: The Science behind the Stories* every bit as readable, informative, and engaging as its parent volume.

In this second edition of *Essential Environment*, we have updated the text with the most current information and have introduced new figures, an enhanced art style, and exercises enabling students to calculate the environmental impacts of their own choices and then see how individual impacts scale up to impacts at the societal level.

We have also retained the major features that make our books unique and that are proving so successful in classrooms across North America:

▶ **Integrated Central Case Studies.** Our teaching experiences, together with feedback from  colleagues across the continent, clearly reveal that telling compelling stories about real people and real places is the best way to capture students' interest. Providing narratives with concrete detail also helps teach abstract concepts, because it gives students a tangible framework with which to incorporate new ideas. Whereas many textbooks these days serve up case studies in isolated boxes, we have chosen to integrate each chapter's central case study into the main text, weaving information and elaboration throughout the chapter. In this way, we use the concrete realities of the central case study to help illustrate the topics we cover. We are gratified that students and instructors using our first edition have consistently applauded this approach, and we hope it can help bring about a new level of effectiveness in environmental science education.

▶ **The Science behind the Story.** Our goal is not simply to present facts, but to engage  students in the scientific process of testing and discovery. To do this, we discuss the scientific method and the social context of science in our opening chapter, and we describe hundreds of real-life studies throughout the text. We also feature in each chapter "The Science behind the Story," which elaborates on particular studies important to the chapter topic, guiding readers through the details of the research. In this way we show not merely *what* scientists discovered, but also *how* they discovered it. Instructors using our first edition have confirmed that this feature enhances student comprehension of each chapter's material and deepens understanding of the scientific process—a key component of effective citizenship in today's science-driven world.

▶ **Viewpoints.** In our text we have striven to present a balanced picture of environmental issues, informed by the best science that bears upon them. Yet we all know that sometimes intelligent people can examine the same data and come to dramatically different conclusions. To ensure that students are exposed to a diversity of interpretations on key issues, we include in each chapter the *Viewpoints* feature, which consists of paired essays authored by invited experts who present different points of view on particular questions of importance. The essays provide students a taste of informed arguments directly from individuals who are actively involved in work—and debate—on environmental issues. To encourage critical thinking, we refer students to an online resource at the book's website that presents questions they can use to critically examine and discuss the ideas in these essays and that provides links to Web sites that support the contributors' viewpoints.

▶ **Weighing the Issues.** Because the multifaceted issues in environmental science often lack black-and white answers, students need critical-thinking skills to help navigate the gray areas at the juncture of science, policy, and ethics. We have aimed to help develop these skills with our end-of-chapter questions and with our "Weighing the Issues" feature. Two to three "Weighing the Issues" questions are dispersed throughout each chapter, serving as stopping points for students to absorb and reflect on what they have read, and to wrestle with some of the complex dilemmas in environmental science.

▶ **An emphasis on solutions.** The complaint we hear most frequently from students in environmental science courses is that the deluge of environmental problems can seem overwhelming. In the face of so many problems, students often come to feel that there is no hope or that there is little they can personally do to make a difference. We have aimed to counter this impression by drawing out innovative solutions that have worked, are being implemented, or can be tried in the future. Although we do not paint an unrealistically rosy picture of the challenges that lie ahead, we portray dilemmas as opportunities and try to instill hope and encourage action. Indeed, for every problem that human carelessness has managed to create, human ingenuity can devise one—and likely multiple—solutions.

Essential Environment: The Science behind the Stories has grown directly from our professional experiences in teaching, research, and writing. Jay Withgott has synthesized and presented science to a wide readership. His experience in distilling and making accessible the fruits of scientific inquiry has shaped our book's content and the presentation of its material. Scott Brennan has taught environmental science to thousands of undergraduates and has developed an intimate feeling for what works in the classroom. His knowledge and experience have shaped the pedagogical approaches we have taken in this book.

We have also been guided in our efforts by extensive input from our professional colleagues and from hundreds of instructors from around North America who have served as reviewers for our chapters and as advisors in focus group meetings arranged by Benjamin Cummings. The participation of so many learned and thoughtful experts has improved this volume in countless ways.

We sincerely hope that our efforts will come close to being worthy of the immense importance of our subject matter. We invite you, students and instructors alike, to let us know how well we have achieved our goals and where you feel we have fallen short. We are committed to continual improvement, and we value your feedback. Please write the authors in care of Chalon Bridges (chalon.bridges @aw.com), Benjamin Cummings Publishing, 1301 Sansome Street, San Francisco, California, 94111. At this most historic time to study environmental science, we are honored to serve as your guides in the quest to better understand our world and ourselves.

Jay Withgott and Scott Brennan

Instructor Supplements

The Withgott/Brennan Media Manager
0-8053-9622-5

This powerful media package is organized chapter-by-chapter and includes all teaching resources in one convenient location. You'll find 5-minute *ABC News* Lecture Launcher videos, PowerPoint presentations, active lecture questions to facilitate class discussions (for use with or without clickers), and an image library that includes all art and tables from the text.

Instructor's Guide and Test Bank
0-8053-9433-8

This comprehensive resource provides chapter outlines, key terms, a listing of Web site and media resources, and teaching tips for lecture and classroom activities. A printed version of the Test Bank is conveniently included in the manual, offering hundreds of multiple-choice, short-answer, and essay questions to use on tests and quizzes. New to this edition are scenario-based questions to test students' critical-thinking abilities.

Computerized Test Bank
0-8053-9618-7

Hundreds of multiple-choice, short-answer, essay, and scenario-based questions are provided on a cross-platform CD-ROM, categorized by chapter objective for the instructor's ease in searching for question types.

Transparency Acetates
0-8053-9617-9

For the instructor's use, we provide 300 full-color acetates of all the art and tables from the text.

Acknowledgments

A textbook is the product of *many* more minds and hearts than one might guess from the names on the cover. The two of us are exceedingly fortunate to be supported and guided by the tremendous staff at Benjamin Cummings and by a small army of experts in environmental science who have generously shared their time and expertise. Although we alone, as authors, bear responsibility for any inaccuracies, the strengths of this book result from the collective labor and dedication of innumerable people.

We would first like to thank our acquisitions editor, Chalon Bridges, whose commitment and unremitting enthusiasm inspired us to take a successful and well-received first edition and make it still better. Chalon's extensive interaction with instructors across North America has helped us define and refine our pedagogy and our innovative features. The approach, design, and essence of this book owe a great deal to her astute guidance and vision.

Project editor Nora Lally-Graves provided a steady hand on the wheel throughout this book's development. With her keen insight, solid judgment, and level-headedness, Nora kept things on track as she deftly juggled reviews, chapter drafts, and countless questions. We thank her for her patience, encouragement, and dedication.

We thank Beth Wilbur, editor-in-chief, and Deborah Gale, director of development, for their support and for their helpful guidance at the outset of this project. Our book's newly enhanced art program owes a great deal to art developmental editor Russell Chun, while photo researcher Kristin Piljay and photo manager Travis Amos supplied us with arresting photographic images. In each chapter, our new feature, "Calculating Ecological Footprints," was imaginatively conceived and ably authored by Jonathan Frye of McPherson College, and Ned Knight of Linfield College provided perceptive reviews of these features. Significant contributions from April Lynch and Etienne Benson in the book's first edition were retained in the second. We also would like to thank the authors of our Viewpoints essays, each of whom is credited in the list that follows.

Copyeditor Sally Peyrefitte again provided thorough and meticulous examination of our text. Once the manuscript was ready, production supervisor Lori Newman and managing editor Michael Early saw it through to production. We thank production editor Holly Henjum and the rest of the staff at GGS Book Services for a tremendous job putting the book together.

We would also like to acknowledge the authors of our supplements: Kristy Manning (*Themes of the Times*); April Lynch (Powerpoint slides); and JodyLee Estrada Duek, Steven Uyeda, and Debra Socci (*Instructor's Guide and Test Bank*). All of them performed top notch work on challenging schedules. In addition, we wish to thank Ericka O'Benar and Ziki Dekel for the development and production of the innovative media that accompanies this book.

Of course, none of this has any impact on education without the marketing staff's efforts to get the book into your hands. For this we thank marketing managers Lauren Harp, and Jeff Hester, as well as Christy Lawrence, director of marketing. And last but surely not least, the many field representatives who help communicate our vision and deliver our product to instructors are absolutely vital, and we deeply appreciate their work and commitment.

We wish to express special thanks to the outside reviewers who shared their time and expertise with us to help make this book the best it could be. Nearly 400 instructors and outside experts contributed reviews for the first two editions of this book's parent volume, but in the lists below we acknowledge those who contributed particularly to this edition of *Essential Environment*. The first list includes those who reviewed multiple chapters from the first edition of *Essential Environment* and made recommendations for our second edition. The second list includes those who provided in-depth reviews of multiple chapters during their development for this second edition. If the thoughtfulness and thoroughness of these reviewers are any indication, we feel confident that the teaching of environmental science is in excellent hands!

Lastly, Jay gives loving thanks to his wife, Susan Masta, who endured this edition's preparation with tremendous patience and sacrifice and provided support and sustenance throughout. Scott would like to thank Angela, Sean, Jonathan, Korby, Karl and Jess, and Jodi and Andy.

We dedicate this book to today's students, who will shape tomorrow's world.

Jay Withgott and Scott Brennan

Reviewers and Contributors

Pre-Revision Reviewers

Christina Buffington, *University of Wisconsin, Milwaukee*
Carolann Castell, *Seattle Central Community College*
Michael L. Denniston, *Georgia Perimeter College*
Robert G. Keesee, *University at Albany, State University of New York*
Michael L. McKinney, *University of Tennessee*
Don Moll, *Southwest Missouri State University*
Kim Nelson, *New Hampshire Community Technical College*
Bruce Olszewski, *San Jose State University*
Carlton Lee Rockett, *Bowling Green State University*
Kay R.S. Williams, *Shippensburg University of Pennsylvania*
Clark Winchell, *Grossmont College*

Chapter Reviewers

Terrence G. Bensel, *Allegheny College*
William R. Epperly, Jr., *Robert Morris College*
Susan A. Habeck, *Tacoma Community College*
Thomas E. Pliske, *Florida International University*
Jamey Thompson, *Hudson Valley Community College*
Danielle M. Wirth, *Des Moines Area Community College*

Viewpoints Essayists

Frank Ackerman, *Tufts University*
James T. Carlton, *Williams College*
Ignacio Chapela, *University of California–Berkeley*
Timothy L. Cline, *Population Connection*
Thomas Flint, *AgFARMation*
Pete Geddes, *Foundation for Research on Economics and the Environment*
Eban Goodstein, *Lewis and Clark College*
Adrian Herrera, *Arctic Power*
Susan Hock, *National Renewable Energy Laboratory*
Peter Kareiva, *The Nature Conservancy*
Norman Myers, *Oxford University*
Nalini M. Nadkarni, *The Evergreen State College*
Sara Nicolas, *American Rivers*
Randal O'Toole, *American Dream Coalition*
Warren Porter, *University of Wisconsin–Madison*
Daryl Prigmore, *University of Colorado–Colorado Springs*
Terry Roberts, *Potash & Phosphate Institute*
Gavin Schmidt, *NASA/Goddard Institute for Space Studies*
Jane S. Shaw, *Property and Environment Research Center*
Daniel Simberloff, *University of Tennessee*
S. Fred Singer, *Science and Environmental Policy Project*
Jeff Speck, *National Endowment for the Arts*
Marian K. Stanley, *American Chemistry Council*
Douglas Sylva, *Catholic Family and Human Rights Institute*
Paul H. Templet, *Louisiana State University*
Indra K. Vasil, *University of Florida*
Karen Wayland, *Natural Resources Defense Council*
Nathaniel Wheelwright, *Bowdoin College*

Brief Contents

PART ONE
Foundations of Environmental Science

PART TWO
Environmental Issues and the Search for Solutions

Detailed Contents

PART TWO
Environmental Issues and the Search for Solutions

Foundations of Environmental Science

Researcher studying
eucalyptus forest,
Australia

1

An Introduction to Environmental Science

Our island, Earth

Upon successfully completing this chapter, you will be able to:

▶ Define the term *environment*

▶ Describe natural resources and explain their importance to human life

▶ Characterize the interdisciplinary nature of environmental science

▶ Explain the scientific method and describe how science operates

▶ Outline the nature, evolution, and expansion of environmental ethics in Western cultures

▶ Evaluate the concepts of sustainability and sustainable development

Our Island, Earth

Viewed from space, our home planet resembles a small blue marble suspended against a vast inky-black backdrop. Earth may seem vast to us as we go about our lives on its surface, but the astronaut's perspective shows that our planet and its natural systems are limited. From this perspective, it is clear that as our population, our technological powers, and our consumption of resources increase, so do our abilities to alter our planet and damage the very systems that keep us alive.

Our environment is the sum total of our surroundings

A photograph of Earth reveals a great deal, but it does not convey the complexity of our environment. Our **environment** includes all the living and nonliving things around us with which we interact. It includes the continents, oceans, clouds, and ice caps you can see in the photo of Earth from space, as well as the animals, plants, forests, and farms that comprise the landscapes around us. In a more inclusive sense, it encompasses our built environment—the structures, urban centers, and living spaces that people have created. In its most inclusive sense, our environment also includes the complex webs of social relationships and institutions that shape our daily lives.

People commonly use the term *environment* in the first, most narrow sense—of a nonhuman or "natural" world apart from human society. This usage is unfortunate, because it masks the very important fact that humans exist within the environment and are a part of nature. As one of many species of animals on Earth, we share with others the same dependence on a healthy functioning planet. The limitations of language make it all too easy to speak of "people and nature," or "human society and the environment," as though they were separate and did not interact. However, the fundamental insight of environmental science is that we are part of the natural world and that our interactions with other parts of it matter a great deal.

Environmental science explores interactions between humans and our environment

Understanding our interactions with the world around us is vital because we depend utterly on our environment for air, water, food, shelter, and everything else essential for living. Moreover, our actions modify our environment, whether we intend them to or not. Many of our actions have enriched our lives, bringing us longer life spans, better health, more material wealth, greater mobility, and more leisure time. However, we have often degraded the natural systems that sustain us. Impacts such as air and water pollution, soil erosion, and species extinction compromise human well-being and threaten our ability to build societies that will survive and thrive in the long term. The elements of our environment were functioning long before the human species appeared, and we would be wise to realize that we need to keep these elements in place.

Environmental science is the study of how the natural world works, how our environment affects us, and how we affect our environment. We need to understand our interactions with our environment because such knowledge is the essential first step toward devising solutions to our most pressing environmental problems.

It can be daunting to reflect on the sheer magnitude of environmental dilemmas that confront us today, but with these problems also come countless opportunities for devising creative solutions. The topics studied by environmental scientists are the most centrally important issues to our world and its future. Right now, global conditions are changing more quickly than ever. Right now, through science, we as a civilization are gaining knowledge more rapidly than ever. And right now, the window of opportunity for acting to solve problems is still open. With such bountiful challenges and opportunities, this particular moment in history is indeed an exciting time to be studying environmental science.

Natural resources are vital to our survival

An island by definition is finite and bounded, and its inhabitants must cope with limitations in the materials they need. On our island, Earth, human beings, like all living things, ultimately face environmental constraints. Specifically, there are limits to many of our **natural resources**, the various substances and energy sources we need to survive. Natural resources that are virtually unlimited or that are replenished over short periods are known as **renewable natural resources**. Some renewable resources, such as sunlight, wind, and wave energy, are perpetually available. Others, such as timber, food crops, water, and soil, renew themselves over months, years, or decades—if we are careful not to use them up too quickly. In contrast, resources such as mineral ores and crude oil are in finite supply and are formed much

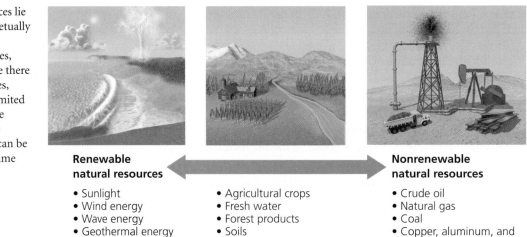

FIGURE 1.1 Natural resources lie along a continuum from perpetually renewable to nonrenewable. Perpetually renewable resources, such as sunlight, will always be there for us. Nonrenewable resources, such as oil and coal, exist in limited amounts that could one day be gone. Other resources, such as timber, soils, and food crops, can be replenished on intermediate time scales, if we are careful not to deplete them.

Renewable natural resources

- Sunlight
- Wind energy
- Wave energy
- Geothermal energy

- Agricultural crops
- Fresh water
- Forest products
- Soils

Nonrenewable natural resources

- Crude oil
- Natural gas
- Coal
- Copper, aluminum, and other metals

more slowly than we use them. These are known as **nonrenewable natural resources**. Once we deplete them, they are no longer available.

We can view the renewability of natural resources as a continuum (Figure 1.1). Some renewable resources may turn nonrenewable if we overuse them. For example, overpumping groundwater can deplete underground aquifers and turn lush landscapes into deserts. Populations of animals and plants we harvest from the wild may vanish if we overharvest them. In recent years, our consumption of natural resources has increased greatly, driven by rising affluence and the growth of the largest human population in history.

Human population growth has shaped our resource use

For nearly all of human history, only a few million people populated Earth at any one time. Figure 1.2 gives some idea of just how recently and suddenly our population has grown beyond 6 *billion* people.

Two phenomena triggered remarkable increases in population size. The first was our transition from a hunter-gatherer lifestyle to an agricultural way of life. This change began to occur around 10,000 years ago and is known as the **agricultural revolution**. As people began to grow their own crops, raise domestic animals, and live

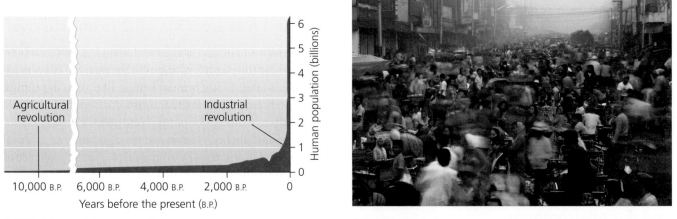

(a) World population growth

(b) Urban society

FIGURE 1.2 For almost all of human history, our population was low and relatively stable. It increased significantly as a result first of the agricultural revolution and then of the industrial revolution (**a**). Our skyrocketing population has given rise to congested urban areas, such as this city in Java, Indonesia (**b**).

sedentary lives in villages, they found it easier to meet their nutritional needs. As a result, they began to live longer and to produce more children who survived to adulthood. The second notable phenomenon, the **industrial revolution**, began in the mid-1700s. It entailed a shift from rural life, animal-powered agriculture, and manufacturing by craftsmen to an urban society powered by **fossil fuels**, which are nonrenewable energy sources such as oil, coal, and natural gas (▸ pp. 303–312). The industrial revolution introduced improvements in sanitation and medical technology, and it enhanced agricultural production with fossil-fuel-powered equipment and synthetic pesticides and fertilizers (Chapter 7).

Thomas Malthus and population growth At the outset of the industrial revolution in England, population growth was regarded as a good thing. For parents, high birth rates meant more children to support them in old age. For society, it meant a greater pool of labor for factory work.

British economist **Thomas Malthus** (1766–1834) had a different opinion. Malthus claimed that unless population growth were controlled by laws or other social strictures, the number of people would outgrow the available food supply (Figure 1.3). Malthus's most influential work, *An Essay on the Principle of Population*, published in 1798,

(a) 18th-century London, England

(b) Thomas Malthus

FIGURE 1.3 The England of Thomas Malthus's era (1766–1834), shown in this engraving (**a**), favored population growth as society industrialized. Malthus (**b**) argued that population growth could lead to disaster.

argued that if limits on births (such as abstinence and contraception) were not implemented, deaths would increase through famine, plague, and war.

Malthus's thinking was shaped by the rapid urbanization and industrialization he witnessed during the early years of the industrial revolution, but debates over his views continue today. As we will see in Chapter 6 and throughout this book, global population growth has indeed contributed to famine, disease, and conflict. However, increasing material prosperity has also helped bring down birth rates—something Malthus did not foresee.

Paul Ehrlich and the "population bomb" In our day, biologist Paul Ehrlich of Stanford University has been called a "neo-Malthusian" because he too has warned that population growth will have disastrous impacts. In his 1968 book, *The Population Bomb*, Ehrlich argued that our population was increasing faster than our ability to produce and distribute food. He predicted that our rapid growth would unleash widespread famine and conflict that would consume civilization by the end of the 20th century.

Although human population nearly quadrupled in the past 100 years—the fastest it has ever grown (see Figure 1.2a)—Ehrlich's predictions have not materialized on the scale he predicted. This is due, in part, to agricultural advances made in recent decades (▸ pp. 151–152). As a result, Ehrlich and other neo-Malthusians have revised their predictions and now warn of a postponed, but still impending, global crisis.

Resource consumption exerts social and environmental impacts

Population growth is unquestionably at the root of many environmental problems, but growth in consumption is also to blame. The industrial revolution has enhanced the material affluence of many of the world's people by greatly increasing our per-person consumption of natural resources and manufactured goods.

Garrett Hardin and the "tragedy of the commons" The late Garrett Hardin of the University of California at Santa Barbara disputed the economic theory that unfettered exercise of individual self-interest will serve the public interest (▸ p. 27). According to Hardin's best-known essay, "The Tragedy of the Commons," published in 1968, resources that are open to unregulated exploitation will eventually be depleted.

Hardin based his argument on a scenario from a pamphlet published in 1833. In a public pasture, or "common," that is open to unregulated grazing, Hardin argued, each person who grazes animals will be motivated to increase the number of his or her animals in the pasture. Ultimately, overgrazing will cause the pasture's food production to collapse. Because no single person owns the pasture, no one has incentive to expend effort taking care of it, and everyone takes what he or she can until the resource is depleted.

Some have argued that private ownership can address this problem. Others point to cases in which people sharing a common resource have voluntarily cooperated to enforce its responsible use. Still others maintain that the dilemma justifies government regulation of the use of resources held in common by the public, from forests to clean air to clean water.

Weighing the Issues:
The Tragedy of the Commons

Imagine you make your living fishing for lobster. You are free to boat anywhere and set out as many traps as you like. Your harvests have been good, and nothing is stopping you from increasing the number of your traps. However, all the other lobster fishers are thinking the same thing, and the fishing grounds are getting crowded. Catches decline year by year, until one year the fishery crashes, leaving you and all the others with catches too meager to support your families. Some of your fellow fishers call for dividing the waters and selling access to individuals plot by plot. Others urge the fishers to team up, set quotas among themselves, and prevent newcomers from entering the market. Still others are imploring the government to get involved and pass laws regulating how much fishers can catch. What do you think is the best way to combat this tragedy of the commons and restore the fishery? Why?

Wackernagel, Rees, and the ecological footprint

As global affluence has increased, human society has consumed more and more of the planet's limited resources. We can quantify resource consumption using the concept of the "ecological footprint," developed in the 1990s by environmental scientists Mathis Wackernagel and William Rees. The **ecological footprint** expresses the environmental impact of an individual or population in terms of the cumulative amount of land and water required to (1)

FIGURE 1.4 The "ecological footprint" represents the total area of land and water needed to produce the resources a given person or population uses, together with the total amount of land and water needed to dispose of their waste. The footprints of the urbanized and affluent societies of today's developed nations tend to be much larger than the geographic areas these societies take up directly.

provide the raw materials for all the goods the person or population consumes and (2) dispose of or recycle all the waste the person or population produces (Figure 1.4). An ecological footprint measures the total amount of Earth's surface "used" by a given person or population, once all direct and indirect impacts are totaled up.

For humanity as a whole, Wackernagel and Rees have calculated that our species is using 30% more resources than are available on a sustainable basis. That is, we are depleting renewable resources 30% faster than they are being replenished—like drawing the principal out of a bank account rather than living off the interest. Furthermore, people from wealthy nations have much larger ecological footprints than do people from poorer nations. If all the world's people consumed resources at the rate of North Americans, these researchers concluded, we would need the equivalent of two additional planet Earths.

Environmental science can help us avoid mistakes made by past civilizations

It remains to be seen whether the direst predictions of Malthus, Ehrlich, and others will come to pass for today's global society, but we already have historical evidence that civilizations can crumble when pressures from population and consumption overwhelm resource availability. Easter Island is a classic case (see "The Science behind the Story," ▸ pp. 8–9).

Indeed, many great civilizations have fallen after depleting resources from their environments, and each has left devastated landscapes in its wake: the Greek and Roman empires, the Angkor civilization of southeast Asia, the Norse in Greenland, and the Maya, Anasazi, and other civilizations of the New World. In Iraq and other regions of the Middle East, areas that are desert today were lush enough to support the origin of agriculture when great ancient civilizations thrived there. In his 2005 book, *Collapse*, scientist and author Jared Diamond synthesized existing research and formulated general reasons why civilizations succeed and persist, or fail and collapse. Success and persistence, it turns out, depend largely on how societies interact with their natural environments.

Today we are confronted with news and predictions of environmental catastrophes on a regular basis. Studying environmental science will outfit you with the tools that can help you evaluate information on environmental change and think critically and creatively about possible actions to take in response.

The Nature of Environmental Science

Environmental scientists aim to comprehend how Earth's natural systems function, how these systems influence us, and how we influence these systems. Many environmental scientists are motivated by a desire to develop solutions to environmental quandaries. The solutions themselves (such as new technologies, policy decisions, or resource management strategies) are *applications* of environmental science. The study of such applications and their consequences is, in turn, also part of environmental science.

Environmental science provides interdisciplinary solutions

Studying and addressing environmental problems is a complex endeavor that requires expertise from many disciplines, including ecology, earth science, chemistry, biology, economics, political science, demography, ethics, and others. Environmental science is thus an **interdisciplinary** field—one that borrows techniques from numerous disciplines and brings research results from these disciplines together into a broad synthesis (Figure 1.5). Traditional established disciplines are

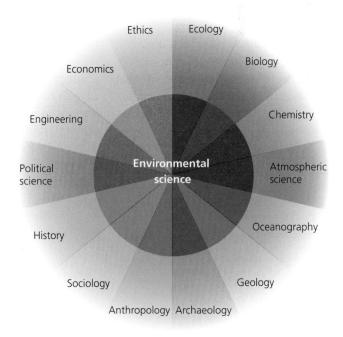

FIGURE 1.5 Environmental science is an interdisciplinary pursuit, involving input from many different established fields of study across the natural sciences and social sciences.

The Lesson of Easter Island

Easter Island is one of the most remote spots on the globe, located in the Pacific Ocean 3,750 km (2,325 mi) from South America and 2,250 km (1,395 mi) from the nearest inhabited island. When European explorers reached the island (today called Rapa Nui) in 1722, they found a barren landscape populated by fewer than 2,000 people, who lived in caves and eked out a marginal existence from a few meager crops. However, explorers also noted that the desolate island featured hundreds of gigantic statues of carved stone, evidence that a sophisticated civilization had once inhabited the island.

Historians and anthropologists wondered how people without wheels or ropes, on an island without trees, could have moved statues 10 m (33 ft) high weighing 90 metric tons (99 tons) as far as 10 km (6.2 mi) from the quarries where they were chiseled to the coastal sites where they were erected. The explanation, scientists discovered, was that the island did not always lack trees.

Indeed, scientific research tells us that the island was once lushly forested and had supported a prosperous society of 6,000 to 30,000 people. Tragically, this once-flourishing civilization overused its resources and cut down all its trees, destroying itself in a downward spi-

The haunting statues of Easter Island were erected by a sophisticated civilization that collapsed after depleting its resource base and devastating its island environment.

ral of starvation and conflict. Today Easter Island stands as a parable and a warning for what can happen when a population consumes too much of the limited resources that support it.

To solve the mystery of Easter Island's past, scientists have used various methods. British scientist John Flenley and others excavated sediments from the bottom of the island's lakes, drilling cores deep into the mud and examining ancient grains of pollen preserved there. Because pollen grains vary from one plant species to another, scientists, by identifying specific pollen grains, can reconstruct, layer by layer, the history of vegetation in a region through time. By analyzing

pollen grains under scanning electron microscopes, Flenley and other researchers found that when Polynesian people arrived (likely between A.D. 300 and 900), the island was covered with a species of palm tree related to the Chilean wine palm, a tall and thick-trunked tree. Other evidence supporting the palm's presence soon emerged: Archaeologists located ancient palm nut casings in caves and crevices, a geologist found carbon-lined channels in the soil that matched root channels typical of the Chilean wine palm, and scientists deciphering the island people's script on stone tablets discerned characters etched in the form of palm trees.

valuable because their scholars delve deeply into topics, uncovering new knowledge and developing expertise in particular areas. Interdisciplinary fields are valuable because their practitioners take specialized knowledge from different disciplines, consolidate it, and make

sense of it in a broad context to better serve the multi-faceted interests of society.

Environmental science is especially broad because it encompasses not only the **natural sciences** (disciplines that study the natural world), but also the **social sciences**

By studying pollen and the remains of wood from charcoal, scientists such as French archaeologist Catherine Orliac found that at least 21 other species of plants, many of them trees, had also been common but are now completely gone. The island had clearly supported a diverse forest. However, starting around A.D. 750, tree populations declined and ferns and grasses became more common, according to pollen analysis from one lake site. By A.D. 950, the trees were largely gone, and around A.D. 1400 overall pollen levels plummeted, indicating a dearth of vegetation. The same sequence of events occurred two centuries later at the other two lake sites, which were higher and more remote from village areas. Researchers first hypothesized that the forest loss was due to climate change, but evidence instead supported the hypothesis that the people had denuded their own island.

The trees provided fuelwood, building material for houses and canoes, fruit to eat, fiber for clothing—and presumably, logs with which to move the stone statues. By hiring groups of men to recreate the feat, several anthropologists have experimentally tested hypotheses about how the islanders moved their monoliths down from the quarries. The methods that have worked involve using numerous tree trunks as rollers or sleds, along with great quantities of rope. The only likely source of rope on the island would have been the fibrous inner bark of the hauhau tree, a species that today is near extinction.

With the trees gone, soil would have eroded away—a phenomenon confirmed by data from the bottom of Easter Island lakes, where large quantities of sediment accumulated. Faster runoff of rainwater would have meant less fresh water available for drinking. Runoff and erosion would have degraded the islanders' agricultural land, lowering yields of crops, such as bananas, sugar cane, and sweet potatoes. Reduced agricultural production would have led to starvation and population decline.

Archaeological evidence supports a scenario of environmental degradation and civilization decline. Analysis of 6,500 bones by archaeologist David Steadman has shown that at least 6 species of land birds and 25 species of seabirds nested on Easter Island and were eaten by islanders. Today, no native land birds and only one seabird species is left. Remains from charcoal fires aged by radiocarbon dating show that besides crops and birds, early islanders feasted on the bounty of the sea, including porpoises, fish, sharks, turtles, octopus, and shellfish. But analysis of islanders' diets in the later years indicate that little seafood was consumed. With the trees gone, the islanders could no longer build the great canoes their proud Polynesian ancestors had used for centuries to fish and travel among islands.

As resources declined, the islanders' main domesticated food animal, the chicken, became more valuable. Archaeologists found that later islanders kept their chickens in stone fortresses with entrances designed to prevent theft. The once prosperous and peaceful civilization fell into clan warfare, as revealed by unearthed weapons, skeletons, and skulls with head wounds.

Is the story of Easter Island as unique and isolated as the island itself, or does it hold lessons for our world today? Like the Easter Islanders, we are all stranded together on an island with limited resources. Earth may be vastly larger and richer in resources than was Easter Island, but Earth's human population is also much greater. The Easter Islanders must have seen that they were depleting their resources, but it seems that they could not stop. Whether we can learn from the history of Easter Island and act more wisely to conserve the resources on our island, Earth, is entirely up to us.

(disciplines that study human interactions and institutions). Most environmental science programs focus predominantly on the natural sciences as they pertain to environmental issues. In contrast, programs incorporating the social sciences heavily often use the term **environmental studies** to describe their academic umbrella. Whichever approach one takes, these fields reflect many diverse perspectives and sources of knowledge.

Just as an interdisciplinary approach to studying issues can help us better understand them, an integrated approach to addressing problems can produce effective and lasting solutions. One example is the dramatic improvement in one aspect of air quality in the United States over the past few decades. Lead had long been added to gasoline to make automobiles run more smoothly, even though researchers knew that lead emissions from tailpipes could cause health problems, including brain damage and premature death. In 1970 air pollution was severe, and motor vehicles accounted for 78% of U.S. lead emissions. But over the following years, engineers, medical researchers, atmospheric scientists, and politicians all merged their knowledge and skills into a process that eventually resulted in a ban on leaded gasoline. By 1996 all gasoline sold in the United States was unleaded, and the nation's largest source of atmospheric lead emissions had been completely eliminated.

Environmental science is not the same as environmentalism

Although many environmental scientists are interested in solving problems, it would be incorrect to confuse environmental science with environmentalism, or environ-

FIGURE 1.6 Environmental scientists and environmental activists play very different roles. Some scientists have become activists to promote particular solutions to environmental problems. However, most have not, and those who have generally try hard to keep their advocacy separate from their pursuit of objective scientific work.

mental activism. They are *not* the same. Environmental science is the pursuit of knowledge about the workings of the environment and our interactions with it. **Environmentalism** is a social movement dedicated to protecting the natural world—and, by extension, people—from undesirable changes brought about by human choices (Figure 1.6). Although environmental scientists study many of the same issues environmentalists care about, as scientists they attempt to maintain an objective approach in their work. Remaining free from personal or ideological bias, and open to whatever conclusions the data demand, is a hallmark of the effective scientist.

The Nature of Science

Modern scientists describe **science** as a systematic process for learning about the world and testing our understanding of it. The term *science* is also commonly used to refer to the accumulated body of knowledge that arises from this dynamic process of observation, testing, and discovery.

Knowledge gained from science can be applied to address societal problems and needs. Science is frequently used to develop technology and also to inform policy and management decisions (Figure 1.7). Many scientists are motivated by the potential for developing such useful applications, but others are motivated simply by a desire to know how the world works.

Why does science matter? The late astronomer and author Carl Sagan wrote the following in his 1995 treatise, *The Demon Haunted World: Science as a Candle in the Dark*:

> We've arranged a global civilization in which the most crucial elements—transportation, communications, and all other industries; agriculture, medicine, education, entertainment, protecting the environment; and even the key democratic institution of voting—profoundly depend on science and technology. We have also arranged things so that almost no one understands science and technology. This is a prescription for disaster. We might get away with it for a while, but sooner or later this combustible mixture of ignorance and power is going to blow up in our faces. . . . Science is an attempt, largely successful, to understand the world, to get a grip on things, to get hold of ourselves, to steer a safe course.

Sagan and many other thinkers before and since have argued that science is essential if we hope to sort fact from fiction and to develop solutions to the problems—environmental and otherwise—that we face.

(a) Prescribed burning

(b) Methanol-powered fuel-cell car

FIGURE 1.7 Scientific knowledge can be applied in policy and management decisions and in technology. Prescribed burning, shown here in the Ouachita National Forest, Arkansas (**a**), is a management practice to restore forests and is informed by scientific research into forest ecology. Energy-efficient automobiles, like this methanol-powered fuel-cell car from Daimler–Chrysler (**b**), are technological advances made possible by materials and energy research.

Scientists test ideas by critically examining evidence

The effective scientist thinks critically, and does not simply accept the conventional wisdom that he or she hears from others. The scientist becomes excited by new ideas but is a skeptic and judges ideas by the strength of evidence that supports them. In these ways, scientists are good role models for the rest of us, because every one of us can benefit from learning to think critically in our everyday lives.

Scientists examine ideas about how the world works by designing tests to determine whether these ideas are supported by evidence. Ideas can be refuted by evidence but can never be absolutely proven, so, strictly speaking, scientific testing amounts to attempting to disprove ideas. If a particular statement or explanation is testable and resists repeated attempts to disprove it, scientists are likely to accept it as a useful and true explanation. Scientific inquiry thus consists of an incremental approach to the truth.

The scientific method is the key element of science

Scientists generally follow a process called the **scientific method**. A technique for testing ideas with observations, it involves several assumptions and a series of interrelated steps. There is nothing mysterious about the scientific method; it is merely a formalized version of the procedure any of us might naturally take, using common sense, to resolve a question.

The scientific method is a theme with variations, however, and scientists pursue their work in many different ways. Because science is an active, creative, imaginative process, an innovative scientist may find good reason to stray from the traditional scientific method when a particular situation demands it. Moreover, scientists from different fields approach their work differently because they deal with dissimilar types of information. However, scientists of all persuasions broadly agree on fundamental elements of the process of scientific inquiry.

The scientific method relies on the following assumptions:

▶ The universe functions in accordance with fixed natural laws that do not change from time to time or from place to place.
▶ All events arise from some cause or causes and, in turn, cause other events.
▶ We can use our senses and reasoning abilities to detect and describe natural laws that underlie the cause-and-effect relationships we observe in nature.

As practiced by individual researchers or research teams, the scientific method (Figure 1.8) typically consists of the steps outlined below.

Make observations Advances in science typically begin with the observation of some phenomenon that the scientist wishes to explain. Observations set the scientific method in motion and also function throughout the process.

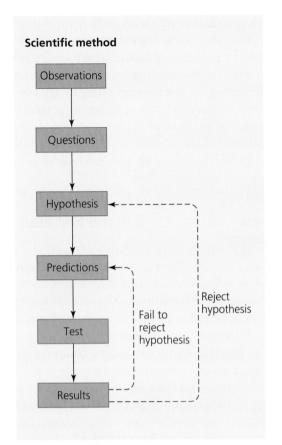

Scientific method

Observations → Questions → Hypothesis → Predictions → Test → Results

Fail to reject hypothesis

Reject hypothesis

FIGURE 1.8 The scientific method is the observation-based hypothesis-testing approach that scientists use to learn how aspects of the world work. This diagram is a simplified generalization that, although useful for instructive purposes, cannot convey the true dynamic and creative nature of science. Moreover, researchers from different disciplines may pursue their work in ways that vary from this model.

Ask questions Curiosity is a fundamental human characteristic; just observe the explorations of a young child in a new environment. Babies want to touch, taste, watch, and listen to anything that catches their attention, and as soon as they can speak, they begin asking questions. Scientists, in this respect, are kids at heart. Why are certain plants or animals less common today than they once were? Why are storms becoming more severe and flooding more frequent? What is causing excessive growth of algae in local ponds? When pesticides poison fish or frogs, does that indicate that people may also be affected? All of these are questions that environmental scientists have asked and attempted to answer.

Develop a hypothesis Scientists attempt to answer their questions by devising explanations that they can test. A **hypothesis** is an educated guess that explains a phenomenon or answers a scientific question. For example, a scientist investigating why algae are growing exces-

sively in local ponds might observe chemical fertilizers being applied on farm fields nearby. The scientist might then state a hypothesis as follows: "Agricultural fertilizers running into ponds cause algae in the ponds to increase."

Make predictions The scientist next uses the hypothesis to generate **predictions**, which are specific statements that can be directly and unequivocally tested. In our algae example, a scientist might predict, "If agricultural fertilizers are added to a pond, the quantity of algae in the pond will increase."

Test the predictions Scientists test predictions one at a time by gathering evidence that could potentially refute the prediction and thus refute the hypothesis. The strongest form of evidence comes from experimentation. An **experiment** is an activity designed to test the validity of a hypothesis. It involves manipulating **variables**, or conditions that can change. For example, a scientist could test the hypothesis linking algal growth to fertilizer by selecting two identical ponds and adding fertilizer to one while leaving the other in its natural state. In this example, fertilizer input is an **independent variable**, a variable the scientist manipulates, whereas the quantity of algae that results is the **dependent variable**, one that depends on the fertilizer input. If the two ponds are identical except for a single independent variable (fertilizer input), then any differences that arise between the ponds can be attributed to that variable. Such an experiment is known as a **controlled experiment** because the scientist controls for the effects of all variables except the one whose effect he or she is testing. In our example, the pond left unfertilized serves as a **control**, an unmanipulated point of comparison for the manipulated **treatment** pond.

Whenever possible, it is best to *replicate* one's experiment; that is, to stage multiple tests of the same comparison of control and treatment. Our scientist could perform a replicated experiment on, say, 10 pairs of ponds, adding fertilizer to one of each pair.

Analyze and interpret results Scientists record **data**, or information, from their studies. They particularly value *quantitative* data (information expressed using numbers), because numbers provide precision and are easy to compare. The scientist running the fertilization experiment might quantify the area of water surface covered by algae in each pond, or might measure the dry weight of algae in a certain volume of water taken from each. However, even with the precision that numbers provide, a scientist's results may not be clear-cut. Data from treatments and controls may vary only slightly, or different replicates may yield different results. The scientist must therefore analyze the data using statistical tests. With these mathematical methods, scientists can determine objectively and precisely the strength and reliability of patterns they find.

If experiments refute a hypothesis, the scientist will reject it and may develop a new hypothesis in its place. If experiments fail to refute a hypothesis, this outcome lends support to the hypothesis but does not *prove* it is correct. The scientist may choose to generate new predictions to test the hypothesis in a different way and further assess its likelihood of being true. Thus, the scientific method loops back on itself, often giving rise to repeated rounds of hypothesis-revision and new experimentation (see Figure 1.8).

If repeated tests fail to reject a particular hypothesis, evidence in favor of it accumulates, and the researcher may eventually conclude that the idea is well supported. Ideally, the scientist would also want to test different potential explanations. For instance, our scientist might propose an additional hypothesis that algae increase in fertilized ponds because numbers of fish or invertebrate animals that eat algae decrease. It is possible, of course, that both hypotheses could be correct and that each may explain some portion of the initial observation that local ponds were experiencing algal blooms.

There are different ways to test hypotheses

An experiment in which the researcher actively chooses and manipulates the independent variable is known as a *manipulative experiment.* A manipulative experiment provides the strongest type of evidence a scientist can obtain, because it can establish causal relationships, showing that changes in an independent variable cause changes in a dependent variable. In practice, however, we cannot run manipulative experiments for all questions, especially for processes that operate at large spatial scales or on long time scales. For example, in studying the effects of global climate change (Chapter 12), we could hardly add carbon dioxide to 10 treatment planets and compare the results to 10 control planets. Thus, in environmental science, it is common for scientists to run *natural experiments* and to search for *correlation*, or statistical association among variables.

For instance, let's suppose our scientist studying algae surveys 50 ponds, 20 of which happen to be fed by fertilizer runoff from nearby farm fields and 30 of which are not. Let's say he or she finds seven times as much algal growth in the fertilized ponds as in the unfertilized ponds. The scientist would conclude that

algal growth is correlated with fertilizer input; that is, that one tends to increase along with the other. Although this type of evidence is weaker than the causal demonstration that manipulative experiments can provide, sometimes a natural experiment is the only feasible approach.

The scientific process does not stop with the scientific method

Scientific research takes place within the context of a community of peers. To have any impact, a researcher's work must be published and made accessible to this community. Thus, the scientific method is embedded within a larger process that takes place at the level of the scientific community as a whole (Figure 1.9).

Peer review When a researcher's work is done and the results analyzed, he or she writes up the findings and submits them to a journal for publication. Several other scientists specializing in the topic of the paper examine the manuscript, provide comments and criticism (generally anonymously), and judge whether the work merits publication in the journal. This procedure, known as **peer review**, is an essential part of the scientific process. Peer review is a valuable guard against faulty science contaminating the literature on which all scientists rely. However, because scientists are human and may have their own personal biases and agendas, politics can sometimes creep into the review process. Fortunately, just as individual scientists strive their best to remain objective in conducting their research, the scientific community does its best to ensure fair review of all work. Winston Churchill once called democracy the worst form of government, except for all the others that had been tried. The same might be said about peer review; it is an imperfect system, yet no one has come up with a better one.

Conference presentations Scientists frequently present their work at professional conferences, where they interact with colleagues and receive informal comments on their research. Such feedback from colleagues can help improve the quality of a scientist's work before it is submitted for publication.

Grants and funding Research scientists spend large portions of their time writing grant applications requesting money to fund their research from private foundations or government agencies such as the National Science Foundation. Grant applications undergo peer review just as scientific papers do, and competition for

FIGURE 1.9 The scientific method (inner box) that is followed by individual researchers or research teams exists within the context of the overall process of science at the level of the scientific community (outer box). This process includes peer review and publication of research, acquisition of funding, and the development of theory through the cumulative work of many researchers.

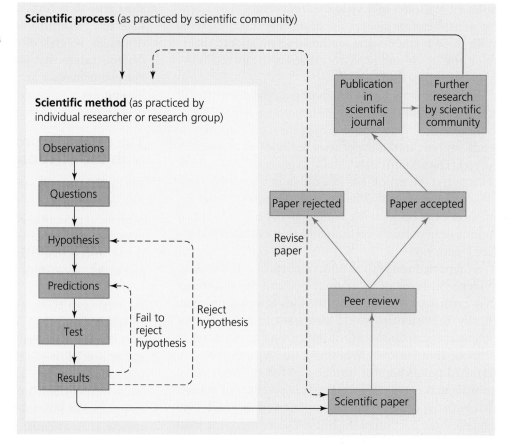

funding is often intense. Scientists' reliance on funding sources can also lead to potential conflicts of interest. A scientist who obtains data showing his or her funding source in an unfavorable light may be reluctant to publish the results for fear of losing funding—or, worse yet, may be tempted to doctor the results. This situation can arise, for instance, when an industry funds research to test its products for safety or environmental impact. When critically assessing a scientific study, therefore, you should always try to find out where the researchers obtained their funding.

Repeatability Sound science is based on repeatability rather than one-time occurrence. Even when a hypothesis appears to explain observed phenomena, scientists are inherently wary of accepting it. The careful scientist may test a hypothesis repeatedly in various ways before submitting the findings for publication. Following publication, other scientists may attempt to reproduce the results in their own experiments and analyses.

Theories If a hypothesis survives repeated testing by numerous research teams and continues to predict experimental outcomes and observations accurately, it may be incorporated into a theory. A **theory** is a widely accepted, well-tested explanation of one or more cause-and-effect relationships that has been extensively validated by a great amount of research. Whereas a hypothesis is a simple explanatory statement that may be refuted by a single experiment, a theory consolidates many related hypotheses that have been tested and have not been refuted.

Note that scientific use of the word *theory* differs from popular usage of the word. In everyday language when we say something is "just a theory," we suggest it is a speculative idea without much substance. Scientists, however, mean just the opposite when they use the term; to them, a theory is a conceptual framework that effectively explains a phenomenon and has undergone extensive and rigorous testing, such that confidence in it is extremely strong. For example, Darwin's theory of evolution by natural selection (▸ pp. 76–79) has been supported and elaborated by many thousands of studies over 150 years of intensive research. Such research has shown repeatedly and in great detail how plants and animals change over generations, or evolve, to express characteristics that best promote survival and reproduction. Because of its strong support and explanatory power, evolutionary theory is the central unifying principle of modern biology.

Science may go through "paradigm shifts"

As the scientific community accumulates data in any given area of research, interpretations may change. Thomas Kuhn's 1962 book *The Structure of Scientific Revolutions* argued that science goes through periodic revolutions: dramatic upheavals in thought in which one *paradigm*, or dominant view, is abandoned for another. For example, before the 16th century, scientists believed that Earth was at the center of the universe. Their data on the movements of planets fit that concept quite well, yet the idea eventually was disproved by Nicolaus Copernicus, who showed that placing the sun at the center of the solar system explained the planetary data even better. Another paradigm shift occurred in the 1960s, when geologists accepted plate tectonics (▶ pp. 71–72), once evidence for the movement of continents and the action of tectonic plates had accumulated and become overwhelmingly convincing. Such paradigm shifts demonstrate the strength and vitality of science, showing it to be a process that refines and improves itself through time.

Understanding how science works is vital to assessing how scientific ideas and interpretations change through time as new information accrues. This process is especially relevant in environmental science, a young field that is changing rapidly as we learn vast amounts of new information, as human impacts on the planet multiply, and as we garner lessons from the consequences of our actions.

To be able to judge good actions from bad, however, we need more than science. We also need an understanding of ethics. Science does not take place in a vacuum, but is influenced by the worldviews and cultural backgrounds of the scientists who practice it. Cultural influences also guide the ways that engineers, policymakers, and others apply scientific knowledge. Thus our brief examination of ethics here, and of economics and policy in Chapter 2, will set the stage for learning how information from the natural sciences and the social sciences are interpreted and put to use within our society.

Environmental Ethics

The field of **ethics** is a branch of philosophy that involves the study of good and bad, of right and wrong. The term *ethics* can also refer to the set of moral principles or values held by a person or a society. Ethicists help clarify how people judge right from wrong by elucidating the criteria, standards, or rules that people use in making these judgments. Such criteria are grounded in values—for instance, promoting human welfare, maximizing individual freedom, or minimizing pain and suffering.

People of different cultures or with different worldviews may differ in their values and thus may differ in the specific actions they consider to be right or wrong. Noting this phenomenon, some ethicists, called **relativists**, believe that ethics vary with social context. However, different human societies show a remarkable extent of agreement on what moral standards are appropriate. Thus many ethicists, called **universalists**, maintain that there exist objective notions of right and wrong that hold across cultures and situations. For both relativists and universalists, ethics is a *normative* or *prescriptive* pursuit; it tells us how we *ought to* behave.

Ethical standards are the criteria that help differentiate right from wrong. One classic ethical standard is *virtue*, which, as the ancient Greek philosopher Aristotle held, involves the personal achievement of moral excellence in character through reasoning and moderation. Another ethical standard is the *categorical imperative* proposed by German philosopher Immanuel Kant, which roughly approximates Christianity's "Golden Rule": to treat others as you would prefer to be treated yourself. A third standard is the principle of *utility*, elaborated by British philosophers Jeremy Bentham and John Stuart Mill. The utilitarian principle holds that something is right when it produces the greatest practical benefits for the most people. We employ such ethical standards as tools for decision making, consciously or unconsciously, innumerable times each day in our daily lives.

Environmental ethics pertains to humans and the environment

The application of ethical standards to relationships between humans and nonhuman entities is known as **environmental ethics**. Human interactions with the environment frequently give rise to ethical questions that can be difficult to resolve. Consider some examples:

▶ Does the present generation have an obligation to conserve resources for future generations? If so, how should this influence our decision making, and how much are we obligated to sacrifice?

▶ Are there situations that justify exposing some communities to a disproportionate share of pollution? If not, what actions are warranted in preventing this problem?

▶ Are humans justified in driving species to extinction? Are we justified in causing other changes in ecological systems? If destroying a forest would drive extinct an insect species few people have heard of but would create jobs for 10,000 people, would that action be ethically admissible? What if it were an owl species? What if only 100 jobs would be created?

Answers to such questions depend partly on what ethical standard(s) a person chooses to use. They also depend on the breadth and inclusiveness of the person's domain of ethical concern. A person who feels responsibility for the welfare of insects would answer the third set of questions very differently from a person whose domain of ethical concern ends with humans.

Throughout Western history, people have gradually enlarged the array of entities they feel deserve ethical consideration. For convenience, three loosely conceived categories can help us understand differences among personal domains of ethical concern. These three ethical perspectives, or *worldviews*, are anthropocentrism, biocentrism, and ecocentrism (Figure 1.10).

Anthropocentrism describes a human-centered view of our relationship with the environment. An anthropocentrist measures costs and benefits of actions solely according to their impact on people, denying or ignoring the notion that nonhuman entities can have rights. To evaluate a human action that affects the environment, an anthropocentrist might use criteria such as impacts on human health, economic costs and benefits, and aesthetic concerns. For example, if development of a mining project would provide a net economic benefit while doing no harm to human health and having little aesthetic impact, the anthropocentrist would support it, even if it might drive some native species extinct. If protecting the area from development would provide spiritual, economic, or other benefits to people, an anthropocentrist might favor its protection. In the anthropocentric perspective, anything not providing benefit to people is considered to be of negligible value.

In contrast, **biocentrism** ascribes values to actions, entities, or properties on the basis of their overall impact on living things, including—but not exclusively focusing on—human beings. In this perspective, all life has ethical standing. In the case of a proposal to develop a mining project, a biocentrist might object to a mine that posed a serious threat to the abundance and variety of living things in the area, even if it would create jobs and generate economic growth. Some biocentrists advocate equal consideration of all living things, whereas others advocate that some types of organisms should receive more than others.

Ecocentrism judges actions in terms of their benefit or harm to the integrity of whole ecological systems, which consist of living and nonliving elements and the

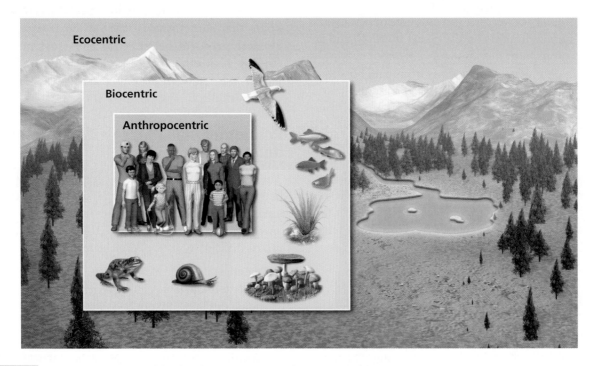

FIGURE 1.10 We can categorize people's ethical perspectives as anthropocentric, biocentric, or ecocentric. An anthropocentrist extends ethical standing only to people and judges actions in terms of their effects on people alone. A biocentrist values and considers all living things, human and otherwise. An ecocentrist extends ethical consideration to living and nonliving components of the environment and takes a holistic view of the connections among these components, valuing the larger functional systems they form.

relationships among them. For an ecocentrist, the well-being of an individual organism is less important than the well-being of a larger ecological system. An ecocentrist might approve of an action that harmed human health, caused economic loss, or took a number of lives, if such impacts were necessary to protect an entire species, community, or ecosystem (we will study species, communities, and ecosystems in Chapters 3 through 5). Ecocentrism is a more holistic perspective than biocentrism or anthropocentrism. It includes a wider variety of entities and stresses the need to preserve connections that tie the entities together into functional systems.

Conservation and preservation arose around the turn of the 20th century

Since the onset of the industrial revolution, more and more people have begun adopting biocentric and ecocentric worldviews. In the 19th and 20th centuries, worldviews of people in the United States evolved especially quickly as the nation pushed west, urbanized, and exploited the continent's resources, dramatically changing the landscape.

A key voice for restraint during this period of rapid growth was **John Muir** (1838–1914), a Scottish immigrant who eventually settled in California and made the Yosemite Valley his wilderness home. Although Muir

FIGURE 1.12 Gifford Pinchot, the first chief of what would become the U.S. Forest Service, was a leading proponent of the conservation ethic. The conservation ethic holds that humans should use natural resources but strive to ensure the greatest good for the greatest number for the longest time.

FIGURE 1.11 A pioneering advocate of the preservation ethic, John Muir is also remembered for his efforts to protect the Sierra Nevada from development and for his role in founding the Sierra Club, a leading environmental advocacy organization. Here Muir (right) stands with President Theodore Roosevelt in Yosemite National Park. After his 1903 wilderness camping trip with Muir, the president instructed his interior secretary to protect more areas in the Sierra Nevada.

chose to live in isolation in his beloved Sierra Nevada for long stretches of time, he also became politically active and won fame as a tireless advocate for the preservation of wilderness (Figure 1.11). Muir was motivated by the rapid deforestation and environmental degradation he witnessed and by his belief that we should treat the natural world with the same respect we give to beautiful cathedrals. Today he is associated with the **preservation** ethic, which holds that we should protect the natural environment in a pristine, unaltered state. Muir argued that nature deserved protection for its own inherent value (an ecocentrist argument), but he also maintained that nature played a major role in human happiness and fulfillment (an anthropocentrist argument). "Everybody needs beauty as well as bread," he wrote in 1912, "Places to play in and pray in, where nature may heal and give strength to body and soul alike."

Some of the factors that motivated Muir also inspired the forester **Gifford Pinchot** (1865–1946; Figure 1.12), who founded what would become the U.S. Forest Service and served as its chief in Theodore Roosevelt's administration. Both Muir and Pinchot opposed the deforestation and unregulated economic development of North American lands that occurred in their lifetimes. However, Pinchot took a more anthropocentric view of how and why nature should be valued. He is today the person most closely associated with the **conservation** ethic, which holds that people should put natural resources to use but that we have a responsibility to manage them wisely. Whereas preservation aims to preserve nature for its own sake and for the aesthetic and spiritual benefit of people, conservation promotes the prudent, efficient, and sustainable extraction and use of natural resources for the benefit of present and future generations. The conservation ethic employs a utilitarian standard, stating that in using resources, we should attempt to provide the greatest good to the greatest number of people for the longest time.

Pinchot and Muir came to represent different branches of the American environmental movement, and their contrasting ethical approaches often pitted them against one another on policy issues of the day. Nonetheless, they both represented reactions against a prevailing "development ethic," which holds that people are and should be masters of nature and which promotes economic development without regard to its environmental consequences. Both Pinchot and Muir left legacies that reverberate today in the different ethical approaches to environmentalism.

Weighing the Issues:
Preservation and Conservation

With which ethic do you most identify—preservation or conservation? Think of a forest or other natural resource in your region. Give an example of a situation in which you might adopt a preservation ethic and an example of one in which you might adopt a conservation ethic. Are there conditions under which you'd follow neither, but instead adopt a "development ethic"?

Aldo Leopold's land ethic arose from the conservation and preservation ethics

Aldo Leopold (1887–1949; Figure 1.13) began his career in the conservationist camp. As a young forester and wildlife manager in Arizona and New Mexico, Leopold embraced the government policy of shooting predators, such as wolves, to increase populations of deer and other game animals. At the same time, Leopold studied the development of ecological science. He eventually ceased to view certain species as "good" or "bad" and instead came to see that healthy ecological systems depend on the protection of all their interacting parts, including predators as well as prey. Drawing an analogy to mechanical maintenance, he wrote, "to keep every cog and wheel is the first precaution of intelligent tinkering."

One day Leopold shot a wolf. When he reached the animal, he was transfixed by "a fierce green fire dying in her eyes." The experience remained with him for the rest of his life and helped lead him to a more ecocentric ethi-

FIGURE 1.13 Aldo Leopold, a wildlife manager and pioneering environmental philosopher, articulated a new relationship between people and the environment. In his essay "The Land Ethic," he called on people to include the environment in their ethical framework.

cal outlook. Years later, as a University of Wisconsin professor, Leopold argued that humans should view themselves and "the land" as members of the same community and that people are obligated to treat the land in an ethical manner. In his 1949 essay "The Land Ethic," he wrote:

> All ethics so far evolved rest upon a single premise: that the individual is a member of a community of interdependent parts. . . . The land ethic simply enlarges the boundaries of the community to include soils, waters, plants, and animals, or collectively: the land. . . . A land ethic changes the role of *Homo sapiens* from conqueror of the land-community to plain member and citizen of it. . . . It implies respect for his fellow-members, and also respect for the community as such.

Leopold intended that the land ethic help guide decision making. "A thing is right," he wrote, "when it tends to preserve the integrity, stability, and beauty of the biotic community. It is wrong when it tends otherwise." Leopold died before seeing "The Land Ethic" and his best-known book, *A Sand County Almanac*, in print, but today many view him as the most eloquent and important philosopher of environmental ethics.

Environmental justice seeks equal treatment for all races and classes

Our society's domain of ethical concern has been expanding from rich to poor and from majority races and ethnic groups to minority ones. This ethical expansion involves applying a standard of fairness and equality and has given rise to the environmental justice movement. The U.S. Environmental Protection Agency (EPA) defines **environmental justice** as "the fair treatment and meaningful involvement of all people regardless of race, color, national origin, or income with respect to the development, implementation, and enforcement of environmental laws, regulations, and policies."

The environmental justice movement was fueled by the perception that poor people and minorities tend to be exposed to a greater share of pollution, hazards, and environmental degradation than are richer people and whites. A protest in the early 1980s by African Americans in Warren County, North Carolina, against a toxic waste dump in their community is widely seen as the beginning of the movement (Figure 1.14). The state had chosen to establish the dump in the county with the highest percentage of African Americans.

FIGURE 1.14 Communities of poor people and people of color have suffered more than their share of environmental problems, a situation that has given rise to the environmental justice movement. The movement first gained prominence with this protest against a toxic waste dump in Warren County, North Carolina.

In 1983, a U.S. General Accounting Office (GAO) study found that three of four toxic waste landfills in the southeastern United States were located in communities where the population of minorities exceeded that of whites. The fourth landfill was located in a community that was 38% African American. In contrast, minorities made up only 20% of the region's population. In 1987, the United Church of Christ Commission for Racial Justice found that the percentage of minorities in areas with toxic waste sites was twice that of areas without toxic waste sites. Researchers studying air pollution, lead poisoning, pesticide exposure, and workplace hazards have found similar patterns. Today the environmental justice movement has broadened to encompass equity in transportation options, redevelopment of abandoned urban sites, worker health and safety, and access to parklands.

Weighing the **issues:**
Environmental Justice

Consider the place where you grew up. Where were the factories, waste dumps, and polluting facilities located, and who lived closest to them? Who lives nearest them in the town or city that hosts your campus? Do you think the concerns of environmental justice advocates are justified? If so, what could be done to ensure that poor communities are no more polluted than wealthy ones?

Ethics and the natural sciences together argue for sustainability

Recall the question posed earlier: Does the present generation have an obligation to conserve resources for future generations? This ethical question cuts to the core of **sustainability**, a guiding principle of modern environmental science and a concept you will encounter throughout this book. Sustainability means living within our planet's means, such that Earth and its resources can sustain us, and the rest of Earth's living things, for the foreseeable future. Sustainability means not depleting Earth's natural capital, so that after we are gone our descendants will enjoy the use of resources as we have, and can inhabit a world as rich and full as ours. Sustainability means developing solutions that work in the long term. And sustainability requires maintaining fully functioning ecological systems, because we cannot sustain human civilization without sustaining the natural systems that nourish it.

We are placing a greater burden on the planet's systems each year. The ongoing rise in human population amplifies nearly all of our environmental impacts—and our consumption of resources has risen even faster than our population growth. The rise in affluence has been a positive development for humanity, and our conversion of the planet's natural capital has made life more pleasant for us so far. However, rising per capita consumption amplifies the demands we make on our environment. Moreover, affluence and consumption have not risen equally for all the world's citizens. Today the 20 wealthiest nations boast 40 times the income of the 20 poorest nations—twice the gap that existed four decades earlier. The ecological footprint of the average citizen of a developed nation such as the United States is considerably larger than that of the average resident of a developing country (Figure 1.15). Within the United States, the rich-est fifth of people claim nearly half the income, whereas the poorest fifth receive only 5%.

The most comprehensive scientific assessment of the present condition of the world's ecological systems and their ability to continue supporting our civilization was completed in 2005. In that year, over 2,000 of the world's leading environmental scientists from nearly 100 nations completed the **Millennium Ecosystem Assessment**. The four main findings of this exhaustive project are summarized in Table 1.1. The Assessment makes clear that our degradation of the world's environmental systems is having negative impacts on all of us, but that with care and diligence we can still turn many of these trends around.

Sustainability need not require great sacrifice of us. We will naturally always desire to enhance our quality of life, and there are many ways we can do so while also encouraging a more sustainable lifestyle. Indeed, this is the goal of **sustainable development**, the use of resources for economic advancement in a manner that satisfies people's current needs without compromising the future availability of resources.

Sustainability depends largely on the ability of the current human population to limit its environmental impact. Doing so will require us to make an ethical commitment and also to apply knowledge we gain from the sciences. Science can help us devise ways to limit our impact and maintain the functioning of the environmental

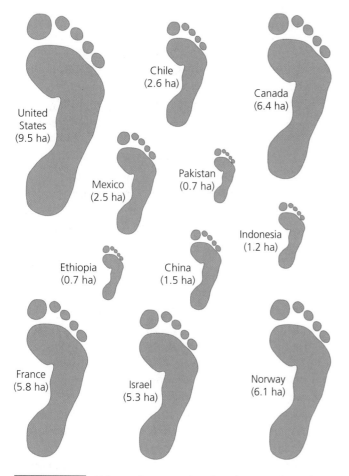

FIGURE 1.15 Citizens of some nations have larger ecological footprints than citizens of others. U.S. residents consume more resources—and thus use more land—than residents of any other nation. Shown are ecological footprints for average citizens of several developed and developing nations, as of 2001. Data from Global Footprint Network, 2005.

Table 1.1	Main Findings of the Millennium Ecosystem Assessment

▶ Over the past 50 years, humans have changed ecosystems more rapidly and extensively than in any comparable period of time in human history, largely to meet rapidly growing demands for food, freshwater, timber, fiber, and fuel. These changes have resulted in a substantial and largely irreversible loss in the diversity of life on Earth.

▶ The changes made to ecosystems have contributed to substantial net gains in human well-being and economic development, but the costs of achieving these gains have been growing. The costs include the degradation of ecosystems and the services they provide for us, and the exacerbation of poverty for some groups of people.

▶ Ecosystem degradation could grow significantly worse during the first half of this century.

▶ The challenge of reversing the degradation of ecosystems while meeting increasing demands for their services can be partially overcome. However, doing so will require that we significantly change many policies, institutions, and practices.

Adapted from *Millennium Ecosystem Assessment, Synthesis Report*, 2005.

systems on which we depend. Because so much remains unstudied and undone, and because so many issues we cannot foresee today are likely to arise in the future, environmental science will remain an exciting frontier for you to explore as a student and as an informed citizen throughout your life.

Conclusion

Finding effective ways of living peacefully, healthfully, and sustainably on our diverse and complex planet will require a sound scientific understanding of both natural and social systems. Environmental science helps us understand our intricate relationship with our environment and informs our attempts to solve and prevent environmental problems.

Identifying a problem is the first step in devising a solution to it. Many of the trends detailed in this book may cause us worry, but others give us reason to hope. One often-heard criticism of environmental science courses and textbooks is that too often they emphasize the negative. Recognizing the validity of this criticism, in this book we attempt to balance the discussion of environmental problems with a corresponding focus on solutions. Solving environmental problems can move us toward health, longevity, peace, and prosperity. Science in general, and environmental science in particular, can aid us in our efforts to develop balanced and workable solutions to the many environmental dilemmas we face today and to create a better world for ourselves and our children.

TESTING YOUR COMPREHENSION

1. How did the agricultural revolution affect human population size? How did the industrial revolution affect human population size? Explain your answers.
2. What is "the tragedy of the commons"? Explain how the concept might apply to an unregulated industry that is a source of water pollution.
3. What is environmental science? Name several disciplines involved in environmental science.
4. What are the two meanings of *science*? Name three applications of science.
5. Describe the scientific method. What is the typical sequence of steps? What generally occurs before a researcher's results are published? Why is this important?
6. What does the study of ethics encompass? Describe the three classic ethical standards. What is environmental ethics?
7. Compare and contrast the ethical perspectives, or worldviews, of anthropocentrism, biocentrism, and ecocentrism.
8. Differentiate between the preservation ethic and the conservation ethic. Explain the contributions of John Muir and Gifford Pinchot in the history of environmental ethics.
9. Describe Aldo Leopold's "land ethic." How did Leopold define the "community" to which ethical standards should be applied?
10. Describe the goals of sustainability. What is sustainable development?

SEEKING SOLUTIONS

1. Many resources are renewable if we use them in moderation but can become nonrenewable if we overexploit them. Order the following resources on a continuum of renewability (see Figure 1.1), from most renewable to least renewable: soils, timber, fresh water, food crops, and biodiversity. What factors influenced your decision? For each of these resources, what might constitute overexploitation, and what might constitute sustainable use?
2. Why do you think the Easter Islanders did not or could not stop themselves from stripping their island of all its trees? Do you see similarities between the history of the Easter Islanders and the modern history of our society? Why or why not?
3. What environmental problem do *you* feel most acutely yourself? Do you think there are people in the world who do not view your issue as an environmental problem? Who might they be, and why might they take a different view?
4. Name an environmental problem you would like to see solved or mitigated. Describe the scientific research you think would need to be completed so that workable solutions to this problem can be developed. Would more than science be needed?

5. Describe your ethical perspective, or worldview, as it pertains to your relationship with the environment. How do you think your culture has influenced your worldview? How do you think your personal experience has influenced it? Do you feel that you fit into any particular category discussed in this chapter? Why or why not?

CALCULATING ECOLOGICAL FOOTPRINTS

 Mathis Wackernagel and his colleagues have continued to refine methods of calculating ecological footprints—the amounts of land and water required to produce the energy and natural resources we consume and absorb the waste we generate. In a 1999 paper, they applied their method to 52 nations that together account for 80% of the world's population and 95% of the global gross domestic product. According to their study, there are 4.9 acres available for every person in the world.

Compare the ecological footprints of each of the countries listed in the accompanying table. Calculate their proportional relationships to the world population's average ecological footprint and to the land available globally to meet our ecological demands.

1. Why is the ecological footprint for people in Bangladesh so low?
2. Why is it so high in the United States?
3. The population of the United States is expected to grow from roughly 300 million today to about 349 million by 2025. What impact, if any, do you think this growth will have on the average global ecological footprint?
4. Based on the data in the table, what impacts do you think average family income has on ecological footprints?

Country	Ecological footprint (acres per person)	Proportion relative to world average footprint	Proportion relative to world land available
Bangladesh	1.2		
Colombia	4.9		1.0 (4.9/4.9)
Mexico	6.4		
Sweden	14.6		
Thailand	6.9		
United States	25.4		
World Average	6.9	1.0 (6.9/6.9)	1.4 (6.9/4.9)

Data from Wackernagel, M., et al. 1999. National natural capital accounting with the ecological footprint concept. *Ecological Economics* 29: 375–390.

Take It Further

Go to www.aw-bc.com/withgott, where you'll find:

▶ Suggested answers to end-of-chapter questions
▶ Quizzes, animations, and flashcards to help you study
▶ *Research Navigator*™ database of credible and reliable sources to assist you with your research projects

▶ **GRAPHit!** tutorials to help you interpret graphs
▶ **INVESTIGATEit!** current news articles that link the topics that you study to case studies from your region to around the world

Environmental Economics and Environmental Policy

Closed San Diego beach

Upon successfully completing this chapter, you will be able to:

▶ Describe precepts of classical and neoclassical economic theory and summarize their implications for the environment

▶ Compare the concepts of economic growth, economic health, and sustainability

▶ Explain the fundamentals of environmental economics and ecological economics

▶ Describe the aims of environmental policy and assess its societal context

▶ Discuss the history of U.S. environmental policy and recognize major U.S. environmental laws

▶ Characterize the institutions involved with international environmental policy

▶ Delineate the steps of the environmental policy process and evaluate its effectiveness

▶ Categorize the different approaches to environmental policy

Central Case: San Diego and Tijuana's Sewage Pollution Problems and Policy Solutions

On November 23, 1996, officials closed all the public beaches in San Diego, California. Stormwater runoff following heavy rains had washed pollutants into local rivers and coastal waters. This also occurred across the border in the Mexican city of Tijuana, whose aging sewer system became clogged with debris, causing raw sewage to overflow into streets and onto beaches. Such incidents, called "rogue flows," take place when heavy rain overwhelms the ability of sewage treatment plants to process wastewater. Rogue flows had become so common in San Diego and Tijuana that local surfers and swimmers casually referred to the initial one of each rainy season as the "first flush."

The Tijuana River winds northwestward through the arid landscape of northern Baja California, Mexico, crossing the U.S. border south of San Diego. A river's **watershed** consists of all the land from which water drains into the river, and the Tijuana River's watershed covers 4,500 km² (1,750 mi²) and is home to 2 million people of two nations. The Tijuana River watershed is a *transboundary* watershed (so named because it crosses a political boundary—in this case, a national border), with approximately 70% of its area in Mexico. On the Mexican side of the border, the river and the arroyos, or creeks, that flow into it are lined with farms, apartments, shanties, and factories, as well as leaky sewage treatment plants and toxic dump sites. Many of these sources release pollutants, which heavy rains wash through the arroyos into the Tijuana River and eventually onto U.S. and Mexican beaches.

Although pollution has flowed in the Tijuana River for at least 70 years, the problem has grown worse in recent years as the region's population has boomed, outstripping the capacity of sewage treatment facilities. Rogue flows have caused thousands of beach closures and pollution advisories in recent years. Garbage carried by the flows also litters the beaches. "Every day I find broken glass, balloons, or can pop-tops. I've even found hypodermic needles. It's really sad," one resident of Imperial Beach told her local newspaper in 2005.

The problem is worse on the Mexican side because most Mexican residents of the Tijuana River watershed live in poverty relative to their U.S. neighbors. In poor neighborhoods such as Loma Taurina, pollution of the river directly affects people's day-to-day lives. The rise of U.S.-owned factories, or *maquiladoras*, on the Mexican side of the border has contributed to the river's pollution, both through direct disposal of industrial waste and by attracting thousands of new workers to the already crowded region.

As impacts increased, people in the San Diego and Tijuana areas, from coastal residents to grassroots activists to businesspeople, pressed policymakers to do something. As we explore environmental economics and environmental policy in this chapter, we will periodically return to the Tijuana River watershed and see how citizens and policymakers together have tried to address these problems.

Economics: Approaches and Environmental Implications

Economic inequities have exacerbated the pollution problems of the Tijuana River, as they have for many other environmental troubles of the U.S.-Mexico border region. In turn, pollution has affected the economies of the region. Sewage-tainted water carries pathogens (organisms that can cause illness), posing a health threat and leading to higher medical costs. Untreated sewage also can radically alter conditions for aquatic life by lowering concentrations of dissolved oxygen, thereby increasing mortality for many species and decreasing suitability for fishing. Moreover, pollution and beach closures reduce recreation, tourism, and other economic activity—a major consideration both in Mexico and in southern California, whose beaches each year host 175 million visitors who spend over $1.5 billion.

Many environmental problems share this combination of impacts—harming human health, altering ecological systems, inflicting economic damage, and contributing to inequities among people. Thus, economic and environmental concerns are intimately connected. Although we often hear it said that environmental protection threatens economic growth, more and more economists assert that in fact, environmental protection is *good* for the economy. Which view one takes often depends on whether one thinks in the short term or the long term and whether one holds to traditional economic schools of thought or to newer ones that view human economies as coupled to the natural environment.

Economics is the study of how people decide to use scarce resources to provide goods and services in the face of demand for them. By this definition, environmental problems are also economic problems that can intensify as population and per capita resource consumption increase. For example, pollution may be viewed as depletion of the scarce resources of clean air, water, or soil. Indeed, the word *economics* and the word *ecology* come from the same Greek root, *oikos*, meaning "household." In its broadest context, the human "household" is Earth itself. Economists traditionally have studied the household of human society, and ecologists the broader household of all life.

Several types of economies exist today

An **economy** is a social system that converts resources into **goods**, material commodities manufactured for and bought by individuals and businesses; and **services**, work done for others as a form of business. The oldest type of economy is the **subsistence economy**. People in subsistence economies—who still comprise much of the human population—meet most or all of their daily needs directly from nature and do not purchase or trade for most of life's necessities.

A second type of economy is the **capitalist market economy**. In this system, buyers and sellers interact to determine which goods and services to produce, how much to produce, and how to produce and distribute them. Capitalist economies are often contrasted with state socialist economies, or **centrally planned economies**, in which government determines in a top-down manner how to allocate resources. In modern capitalist market economies, governments typically intervene for several reasons: (1) to elimi-

nate unfair advantages held by single buyers or sellers; (2) to provide social services, such as national defense, medical care, and education; (3) to provide "safety nets" (for the elderly, victims of natural disasters, and so on); (4) to manage the commons (▸ p. 5–6); and (5) to mitigate pollution.

Environment and economy are intricately linked

Economies receive inputs from the environment, process them in complex ways that enable human society to function, then discharge outputs of waste into the environment. Although these interactions between human economies and the nonhuman environment are readily apparent, traditional economic schools of thought have long overlooked the importance of these connections. Indeed, most conventional economists today still adhere to a worldview that largely ignores the environment (Figure 2.1a), and this worldview continues to drive most policy decisions. However, modern economists belonging to the fast-growing fields of environmental economics and ecological economics (▸ p. 29) explicitly accept that human economies are subsets of the environment and depend crucially on the environment (Figure 2.1b).

Economic activity uses natural resources (▸ pp. 3–4), the various substances and forces we need to survive: the sun's energy, the fresh water we drink, the trees that provide our lumber, the rocks that provide our metals, and the fossil fuels that power our machines and produce our plastics. We can think of natural resources as "goods" produced by nature.

Environmental systems also naturally function in a manner that supports economies. Earth's ecological systems purify air and water, cycle nutrients, provide for the pollination of plants by animals, and serve as receptacles and recycling systems for the waste generated by our economic activity. Such essential processes, called **ecosystem services** (Table 2.1), support the life that makes our economic activity possible. Some ecosystem services represent the very nuts-and-bolts of our survival, whereas others enhance our quality of life.

While the environment enables economic activity by providing ecosystem goods and services, economic activity can affect the environment in return. When we deplete natural resources and produce too much pollution, we can degrade the ability of ecological systems to function. In fact, the Millennium Ecosystem Assessment (▸ p. 20) concluded in 2005 that 15 of 24 ecosystem services its scientists surveyed globally were being degraded or used unsustainably. The degradation of ecosystem services can in turn disrupt economies. Currently, ecological degradation is harming poor people more than wealthy people,

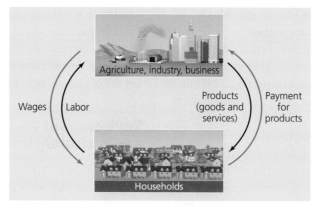

(a) Conventional view of economic activity

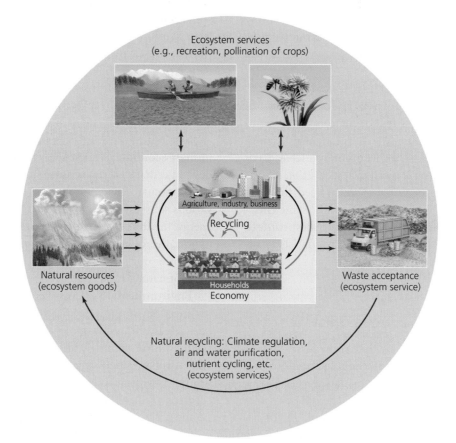

(b) Economic activity as viewed by environmental and ecological economists

FIGURE 2.1 Standard neoclassical economics focuses on processes of production and consumption between households and businesses (**a**), viewing the environment only as a "factor of production" that helps enable the production of goods. Environmental and ecological economists view the human economy as existing within the natural environment (**b**), receiving resources from it, discharging waste into it, and interacting with it through various ecosystem services.

Table 2.1 Ecosystem Services	
Type of ecosystem service*	**Example(s)**
Regulating atmospheric gases	Maintaining the ozone layer; balancing oxygen, carbon dioxide, and other gases
Regulating climate	Controlling global temperature and precipitation through oceanic and atmospheric currents, greenhouse gases, cloud formation, and so on
Damping impacts from disturbance	Providing storm protection, flood control, and drought recovery, mainly through vegetation structure
Regulating water flow	Providing water for agriculture, industry, transportation
Storing and retaining water	Providing water through watersheds, reservoirs, aquifers
Controlling erosion and promoting soil retention	Preventing soil loss from wind or runoff; storing silt in lakes and wetlands
Forming soil	Weathering rock; accumulating organic material
Cycling nutrients	Cycling carbon, nitrogen, phosphorus, sulfur, and other nutrients through ecosystems
Treating waste	Removing toxins, recovering nutrients, controlling pollution
Pollinating plants	Transporting floral gametes by wind or pollinating animals, enabling crops and wild plants to reproduce
Controlling populations biologically	Controlling prey with predators; controlling hosts with parasites; controlling herbivory on crops with predators and parasites
Providing habitat	Providing ecological settings in which creatures can breed, feed, rest, migrate, winter
Providing food	Producing fish, game, crops, nuts, and fruits that humans obtain by hunting, gathering, fishing, subsistence farming
Supplying raw materials	Producing lumber, fuel, metals, fodder
Furnishing genetic resources	Providing unique biological sources for medicine, material technologies, genes for resistance to plant pathogens and crop pests, ornamental species (pets and horticultural plant varieties)
Providing recreational opportunities	Ecotourism, sport fishing, hiking, birding, kayaking, other outdoor recreation
Providing cultural or noncommercial uses and goods	Aesthetic, artistic, educational, spiritual, and/or scientific values of ecosystems

*Ecosystem "goods" are here included in ecosystem services.
Adapted with permission from Costanza, R., et al. 1997. The value of the world's ecosystem services and natural capital. *Nature* 387: 253–260.

the Millennium Ecosystem Assessment found. As a result, restoring ecosystem services stands as a prime avenue for alleviating poverty. These interrelationships among economic and environmental conditions have only recently become widely recognized, however.

Adam Smith and other philosophers founded classical economics

When economics began to develop as a discipline in the mid-18th century, some philosophers argued that individuals acting in their own self-interest would harm society. Others believed that such behavior could benefit society, as long as the behavior was constrained by the rule of law and private property rights and operated within fairly competitive markets. The latter view was articulated by Scottish philosopher **Adam Smith** (1723–1790). Known today as the father of **classical economics**, Smith believed that when people are free to pursue their own economic

self-interest in a competitive marketplace, the marketplace will behave as if guided by "an invisible hand" ensuring that their actions will benefit society. Smith's philosophy remains a pillar of free-market thought today.

Neoclassical economics incorporates psychology and cost-benefit analysis

Economists inspired by Smith's ideas subsequently took more quantitative approaches and incorporated human psychology into their work. Modern **neoclassical economics** examines the psychological factors underlying consumer choices, explaining market prices in terms of consumer preferences for units of particular commodities. In neoclassical economic theory, buyers desire the lowest possible price, whereas sellers desire the highest possible price. This conflict between buyers and sellers results in a compromise price being reached and the "right" quantity of commodities being bought and sold.

This is often phrased in terms of *supply*, the amount of a product offered for sale at a given price, and *demand*, the amount of a product people will buy at a given price if free to do so (Figure 2.2).

To evaluate an action or decision, neoclassical economists often use **cost-benefit analysis**. In this approach, economists total up estimated costs for a proposed action and compare these to the sum of benefits estimated to result from the action. If total benefits exceed costs, the action should be pursued; if costs exceed benefits, it should not. When there are multiple alternative actions, the one with the greatest excess of benefits over costs should be chosen.

This reasoning seems eminently logical, but problems often crop up because not all costs and benefits are easily identified, defined, or quantified. For example, it may be quite feasible to quantify the costs of installing equipment to mitigate pollution, yet difficult to assess the effects of pollution on people's health or lifestyles. Because some costs and benefits cannot easily be assigned monetary values—and because it is difficult to identify and agree on all costs and benefits—cost-benefit analysis is often controversial. Moreover, because economic benefits are usually more easily quantified than environmental costs, economic benefits tend to be overrepresented in traditional cost-benefit analyses. As a result, environmental advocates often feel these analyses are biased in favor of economic development and against environmental protection.

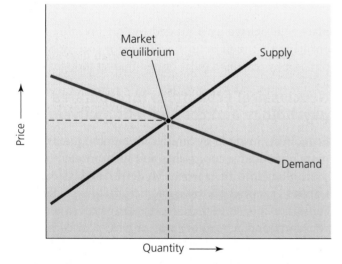

FIGURE 2.2 In a supply-and-demand graph, the demand curve indicates the quantity of a given good (or service) that consumers desire at each price, and the supply curve indicates the quantity produced at each price. The market automatically moves toward an equilibrium point at which supply equals demand.

Aspects of neoclassical economics have profound implications for the environment

Today's capitalist market systems operate largely in accord with the precepts of neoclassical economics. These systems have generated unprecedented material wealth for our societies. However, four fundamental assumptions of neoclassical economics have contributed to environmental problems.

One assumption is that workers and other resources are either infinite or largely substitutable and interchangeable. This implies that once we have depleted a resource—natural, human, or otherwise—we should be able to find a replacement for it. Certainly it is true that many resources can be replaced. However, nonrenewable resources, such as fossil fuels, can be depleted, and many renewable resources can also be used up if we exploit them faster than they are replenished.

Second, neoclassical economics gives an event far in the future far less value than one in the present; in economic terminology, future effects are "discounted." Short-term costs and benefits are granted more importance than long-term costs and benefits, causing policy to play down long-term consequences of decisions we make today.

A third assumption of neoclassical economics is that all costs and benefits associated with a particular exchange of goods or services are borne by individuals engaging directly in the transaction. In other words, it is assumed that the costs and benefits of a transaction are "internal" to the transaction, experienced by the buyer and seller alone, and do not affect other members of society. However, in many situations this is simply not the case. For example, pollution from a *maquiladora* along the Tijuana River can harm people living downstream. In such cases, someone—often taxpayers not involved in producing the pollution—ends up paying the costs of alleviating it. Market prices do not take the social, environmental, or economic costs of this pollution into account. Costs of a transaction that affect people other than the buyer or seller are known as **external costs** (Figure 2.3). External costs commonly include the following:

▶ Human health problems
▶ Property damage
▶ Declines in desirable elements of the environment, such as fewer fish in a stream
▶ Aesthetic damage, such as that resulting from air pollution or clear-cutting
▶ Stress and anxiety experienced by people downstream or downwind from a pollution source
▶ Declining real estate values resulting from these problems

FIGURE 2.3 River pollution creates external costs. This woman washing clothes in the Tijuana River may suffer upstream pollution from factories, and her use of detergents may cause further pollution for people living downstream.

By ignoring external costs, economies create a false idea of the true and complete costs of particular choices and unjustly subject people to the consequences of transactions in which they did not participate. External costs comprise one reason governments develop environmental legislation and regulations.

A fourth assumption of the neoclassical economic approach is that economic growth is required to keep employment high and maintain social order. Economic growth, it is argued, should create opportunities for the poor to become wealthier. By making the overall economic pie larger, everyone's slice becomes larger, even if some people still have much smaller slices than others. In today's economies, economic growth has become the quantitative yardstick by which progress is measured.

Economists disagree on whether economic growth is sustainable

The rate of economic growth in recent decades is unprecedented in human history, and the world economy is seven times the size it was half a century ago. All measures of economic activity—trade, rates of production, amount and value of goods manufactured—are higher than ever, and are still increasing. This growth has brought many people much greater material wealth (although not equitably, and gaps between rich and poor remain immense).

To the extent that economic growth is a means to an end—a tool with which we can achieve greater human happiness—it can be a good thing. However, many observers today worry that growth has become an end in itself and is no longer the best tool with which to pursue happiness. Critics of the growth paradigm note that runaway growth resembles the multiplication of cancer cells, which eventually overwhelm and destroy the organism in which they

grow. These critics fear that runaway economic growth will likewise destroy our economic system. Resources for growth are ultimately limited, they argue, so nonstop growth is not sustainable and will fail as a long-term strategy.

Among the proponents of unrestrained growth are those economists, businesspeople, and policymakers who believe that technology can help us overcome all our environmental limitations—a philosophy that has greatly influenced policy in market economies over the past century.

At the other end of the spectrum, **ecological economists** argue that civilizations do not, in the long run, overcome their environmental limitations. Ecological economics, which has emerged as a discipline only in the past decade or two, applies principles of ecology and systems science (Chapter 3) to the analysis of economic systems. Earth's natural systems generally operate in self-renewing cycles, not in a linear or progressive manner. Ecological economists advocate sustainability in economies and see natural systems as good models. To evaluate an economy's sustainability, ecological economists take a long-term perspective and ask, "Could we continue this activity forever and be happy with the outcome?" Most ecological economists argue that the growth paradigm will eventually fail, and many of them advocate economies that do not grow and do not shrink, but rather are stable. Such **steady-state economies** are intended to mirror natural ecological systems.

Those who oppose the idea of a steady-state economy often assume that an end to growth will mean an end to a rising quality of life. Ecological economists, however, argue that quality of life can continue to rise in a steady-state economy and, in fact, may be more likely to do so. Technological advances will not cease just because growth stabilizes, they argue, and neither will behavioral changes (such as greater use of recycling) that enhance sustainability. Instead, wealth and human happiness can continue to rise after economic growth has leveled off.

Environmental economists tend to agree with ecological economists that economies are unsustainable if population growth is not reduced and resource use is not made more efficient. Environmental economists, however, maintain that we can accomplish these changes and attain sustainability within our current economic systems. By retaining the principles of neoclassical economics but modifying them to address environmental challenges, environmental economists argue that we can keep our economies growing and that technology can continue to improve efficiency. Thus, whereas ecological economists call for revolution, environmental economists call for reform. One approach environmental economists take is to assign monetary values to ecosystem goods and services, so as to better integrate them into traditional cost-benefit analyses.

Calculating the Economic Value of Earth's Ecosystems

To Robert Costanza, the problem was like an elephant in the living room that economists had ignored for decades: Earth's ecosystems provide essential life-support services, including arable soil, waste treatment, clean water, and clean air. However, economists had failed to account for how much those services contribute economically to human welfare, as is routinely done for conventional goods and services.

So Costanza, an environmental economist at the University of Maryland, joined with 12 colleagues in 1996 at the National Center for Ecological Analysis and Synthesis in Santa Barbara, California, and combed the scientific literature. The team identified more than 100 studies that estimated the worth of such ecosystem services as water purification, greenhouse gas regulation, plant pollination, and pollution cleanup.

The studies the team reviewed had estimated values of particular ecosystem services in several ways, often by calculating the price people were willing to pay for the services. The rationale for assigning

these costs was that people clearly value such aspects of natural systems as biodiversity and aesthetics, even though no one pays actual money specifically for them. After poring over studies that examined the value of 17 services provided by oceans, forests, wetlands, and other ecosystems, Costanza and his colleagues synthesized the results to provide the first comprehensive quantitative estimate of the global value of ecosystem services.

To estimate the worth of the services more accurately, Costanza and his team reevaluated the data from the earlier studies using alternative valuation techniques. One method was to calculate the cost of replacing ecosystem services with technology. For example, marshes protect people from floods and filter out water pollutants. If a marsh were destroyed, the researchers would calculate the value of the services it had provided by measuring the cost of the levees and water-purification technology that would be needed to assume those tasks. The researchers then calculated the global monetary value of such

wetlands by multiplying those totals by the global area occupied by the ecosystem. By calculating similar totals from coral reefs, deserts, tundra, and other ecosystems, they arrived at a global value for ecosystem services.

By their calculation, the biosphere provides at least $33 trillion ($42 trillion in 2007 dollars) worth of ecosystem services each year—greater than the gross domestic product (GDP) of all nations combined. Published in the journal *Nature* in 1997, their research paper ignited a firestorm of controversy.

Some environmental advocates and ethicists argued that it was a bad idea to put a dollar figure on priceless services such as clean air and clean water. The value of these services cannot be calculated, they held, because we would all perish if they disappeared. Environmental ethicist Timothy Weiskel of Harvard Divinity School contended that to make a commodity out of biodiversity confused "sacred space with the market place."

Some economists, meanwhile, disparaged the methods by which the researchers calculated values.

We can give ecosystem goods and services monetary values

Ecosystem services are said to have **nonmarket values**, values not usually included in the price of a good or service (Table 2.2 and Figure 2.4). For example, the aesthetic and recreational pleasure we obtain from natural landscapes is something of real value. Yet this value is hard to quantify and appears in no traditional measures of economic worth. Or consider Earth's water cycle (▸ pp.

67–68, 70), by which rain fills our reservoirs with drinking water, rivers give us hydropower and flush away our waste, and water evaporates, purifying itself of contaminants and readying itself to fall again as rain. This natural cycle is absolutely vital to our existence, yet because its value is not quantified, markets impose no financial penalties when we disturb it. Because the market does not assign value to ecosystem services, debates such as those over water quality in the Tijuana River often involve comparing apples and oranges.

Replacement costs were not a legitimate way of determining value, they held, because nature's services, like other economic goods and services, are worth only what people will demonstrably pay for them. Critics also argued that combining the value of ecosystem services from various small tracts of land is meaningless because people decide whether to preserve or exploit resources based on particular local considerations, not on generalized global ones.

To address economists' criticisms, Costanza joined Andrew Balmford of Cambridge University and 17 other colleagues to conduct another analysis. They compared the benefits and costs of preserving natural systems intact with those of converting wild lands for agriculture, logging, or fish farming. Again, they synthesized studies on ecosystem services, this time focusing on just five ecosystems: west African and Malaysian tropical forests, Thai mangrove swamps, Canadian wetlands, and Philippine coral reefs.

In their paper, published in the journal *Science* in 2002, the team reported that a global network of nature reserves covering 15% of

Earth's land surface and 30% of the ocean would be worth between $4.4 and $5.2 trillion. This amount is about 100 times the value of the same areas were they to be converted for direct exploitative human use. That 100:1 benefit-cost ratio, they wrote, demonstrates clearly that "conservation in reserves represents a strikingly good bargain."

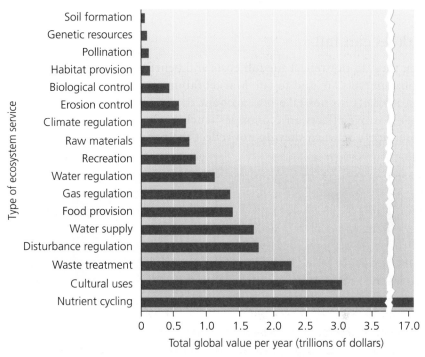

In 1997, Robert Costanza and colleagues estimated the total value of the world's ecosystem services at approximately $33 trillion. Shown are subtotals for each major class of ecosystem service. The $33 trillion figure does not include values from some ecosystems, such as deserts and tundra, for which adequate data were unavailable. Data from Costanza, R., et al. 1997. The value of the world's ecosystem services and natural capital. *Nature* 387: 253–260.

To resolve this dilemma, environmental and ecological economists have sought ways to assign values to ecosystem services. In one technique, economists use surveys to determine how much people are willing to pay to protect a resource or to restore it after damage has been done. In another approach, they measure the money, time, or effort people expend to travel to parks for recreation as a way of judging the value people place on parks. Economists also analyze housing prices, comparing homes with similar characteristics but different

environmental settings, to infer the dollar value of landscapes, views, and peace and quiet. In yet another approach, they measure the cost required to restore natural systems that have been damaged or to mitigate harm from pollution.

In 1997, one research team reviewed studies using various valuation methods, in an effort to calculate the overall economic value of all the services that ecosystems provide across the planet (see "The Science behind the Story," ▶ above).

Table 2.2 Values That Modern Market Economies Generally Do Not Address	
Nonmarket value	**Is the worth we ascribe to things . . .**
Use value	that we use directly
Option value	that we do not use now but might use later
Aesthetic value	for their beauty or emotional appeal
Cultural value	that sustain or help define our culture
Scientific value	that may be the subject of scientific research
Educational value	that may teach us about ourselves and the world
Existence value	simply because they exist, even though we may never experience them directly (e.g., an endangered species in a far-off place)

Markets can fail

When they do not reflect the full costs and benefits of actions, markets are said to fail. **Market failure** occurs when markets do not take into account the environment's positive effects on economies (such as ecosystem services) or when they do not reflect the negative effects of economic activity on the environment or on people (external costs). Traditionally, market failure has been countered by government intervention. Governments can dictate limits on corporate behavior through laws and regulations. They can institute *green taxes*, which penalize environmentally harmful activities. Or, they can design economic incentives that put

(a) Existence values

(b) Use values

(c) Option values

(d) Aesthetic values

(e) Scientific values

(f) Educational values

(g) Cultural values

FIGURE 2.4 Accounting for nonmarket values such as those shown here may help us to make better environmental and economic decisions.

Environment versus Economy?

Is environmental protection economically costly? How can we best balance the demands of economic development with the demands of environmental protection?

Economic Progress a Prerequisite for Environmental Quality

The Greek word *oikos*, meaning "household," is the root of both *economics* and *ecology*. It suggests complementarities. Protecting the environment is not free, but environmental quality and economic development are not mutually exclusive. The real enemy of the environment is poverty, not affluence.

If economic growth really destroys the environment, why do countries with large GDPs enjoy high environmental quality? The environment is cleaner in developed countries because their citizens have both the inclination and the resources to care for the environment.

Are we, as some critics assert, like the man falling from a 10-story building and concluding as he passes the second story, "so far so good"? I think not. In the long term, technological improvements and productivity gains allow us to use fewer material inputs—and to emit ever fewer pollutants—per unit of economic output. This reduces both our economic and ecological footprint.

For example, genetically modified (GM) crops and synthetic chemicals allow us greater yields on the same amount of land. As a result, the United States can afford to place 50 million acres of farmland into conservation reserves while remaining a major food exporter. The poorest countries are beneficiaries as they adopt our efficient and less environmentally damaging technologies—shortcutting the road to environmental quality.

The question we face is, in what combination and in what amounts should we seek the things we want? We value clean water and the preservation of other species. But we also value fresh produce in winter and fast and convenient transportation. Not all good things go together.

Regardless of claims, environmental quality is only one of several competing values people seek. Scarcity—the fact that virtually no resources are abundant enough to satisfy all human demands at zero cost—dictates that choices must be made among competing values. It is intellectually and ethically irresponsible to pretend away the necessity of such choices.

Pete Geddes is program director of the Foundation for Research on Economics and the Environment (FREE) and Gallatin Writers. Both are based in Bozeman, Montana.

The Trade-off Myth

Reducing pollution and protecting resources costs money. As a nation we spend over $200 billion each year, or more than 2% of GDP, on environmental protection. However, it is often contested how much environmental protection will cost in any particular case. Forecasting the costs of compliance is difficult because economists have an equally difficult time estimating future technological responses that may lower costs. For example, credible industry estimates for the costs of sulfur dioxide reduction from power plants under the 1990 Clean Air Act Amendments were eight times too high; the EPA overestimated costs by a factor of 2 to 4.

One "cost" of environmental protection that is blown out of proportion is job loss. Contrary to popular belief, there is no net "jobs-environment trade-off" in the economy, only a steady shift of jobs to cleanup work. On one hand, about 2,000 workers in the U.S. lose their jobs each year for environment-related reasons, which is less than one-tenth of 1% of all layoffs. On the other hand, as we spend more on the environment, more jobs are created. Given the industrial nature of much cleanup work, these jobs are also heavily weighted toward manufacturing and construction. Finally, and again contrary to folk wisdom, very few manufacturing plants flee the industrial countries to escape onerous environmental regulation. Plants do leave, but the overwhelming reason is the cost of labor.

What is the best way to balance costs against the benefits of environmental cleanup? Formal benefit-cost analysis is one approach, but it is of limited value when the benefits of environmental protection are highly uncertain (which is often the case), or when the costs of resource degradation are borne by a relatively small group. In these situations, the best approach is to define a health or ecological standard for cleanup and to trust democratic processes to ensure that the costs of cleanup do not rise too high.

Eban Goodstein is professor of economics at Lewis and Clark College in Portland, Oregon. He is the author of the textbook *Economics and the Environment* (John Wiley and Sons, 2004), as well as *The Trade-off Myth: Fact and Fiction about Jobs and the Environment* (Island Press, 1999).

Explore this issue further by accessing **Viewpoints** at www.aw-bc.com/withgott.

market mechanisms to work to promote fairness, resource conservation, and economic sustainability. We will now examine these approaches in our discussion of environmental policy.

Environmental Policy: An Overview

When a society reaches broad agreement that a problem exists, it may persuade its leaders to try to resolve the problem through the making of policy. **Policy** consists of a formal set of general plans and principles intended to address problems and guide decision making in specific instances. **Public policy** is policy made by governments, including those at the local, state, federal, and international levels. Public policy consists of laws, regulations, orders, incentives, and practices intended to advance societal welfare. **Environmental policy** is policy that pertains to human interactions with the environment. It generally aims to regulate resource use or reduce pollution to promote human welfare and/or protect natural systems.

Forging effective policy requires input from science, ethics, and economics. Science provides the information and analysis needed to identify and understand environmental problems and devise potential solutions to them. Ethics and economics offer criteria to assess the extent and nature of problems, and they help clarify how society might like to address the problems. Government interacts with individual citizens, organizations, and the private sector in a variety of ways to formulate policy (Figure 2.5).

Environmental policy addresses issues of equity and resource use

The capitalist market economic systems of modern constitutional democracies are largely driven by incentives for short-term economic gain rather than long-term social and environmental stability. Market capitalism provides little incentive for businesses or individuals to behave in ways that minimize environmental impact or equalize costs and benefits among parties. As we noted, such *market failure* has traditionally been viewed as justification for government intervention. Environmental policy

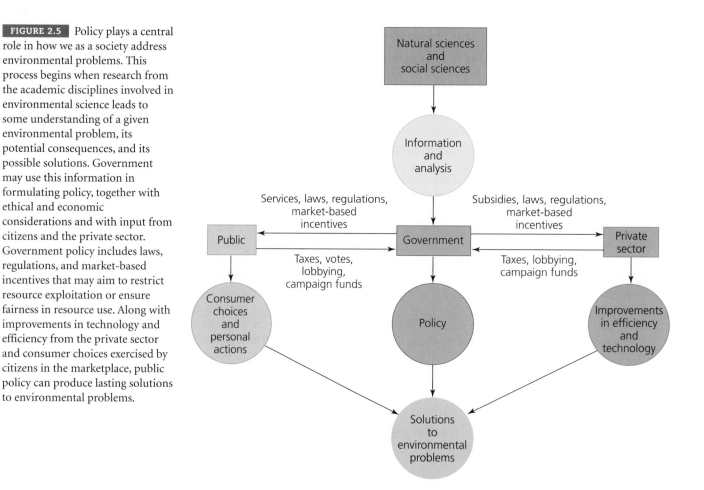

FIGURE 2.5 Policy plays a central role in how we as a society address environmental problems. This process begins when research from the academic disciplines involved in environmental science leads to some understanding of a given environmental problem, its potential consequences, and its possible solutions. Government may use this information in formulating policy, together with ethical and economic considerations and with input from citizens and the private sector. Government policy includes laws, regulations, and market-based incentives that may aim to restrict resource exploitation or ensure fairness in resource use. Along with improvements in technology and efficiency from the private sector and consumer choices exercised by citizens in the marketplace, public policy can produce lasting solutions to environmental problems.

aims to protect environmental quality and the natural resources people use, and also to promote equity in people's use of resources.

The tragedy of the commons Policy to protect resources held and used in common by the public is intended to safeguard them from depletion or degradation. As Garrett Hardin explained in his essay, "The Tragedy of the Commons" (▸ p. 5), a resource held in common that is unregulated will eventually become overused and degraded. Therefore, he argued, it is in our best interest to develop guidelines for the use of common resources. In Hardin's illustrative example of a common pasture, such guidelines might limit the number of animals each individual is allowed to graze or might require pasture users to pay to restore and manage the shared resource. These two concepts, restriction of use and active management, are central to environmental policy today.

Free riders Another reason to develop policy for publicly held resources is the **free rider** predicament. Let's say a community on a river suffers from water pollution that emanates from 10 different factories. The problem could in theory be solved if every factory voluntarily agreed to reduce its own pollution. However, once they all begin reducing their pollution, it becomes tempting for any one of them to stop doing so. Such a factory, by avoiding the sacrifices others are making, would in essence get a "free ride" on the efforts of others. If enough factories take a free ride, the whole effort will collapse. Because of the free rider problem, private voluntary efforts are often less effective than efforts mandated by public policy, which can ensure that all parties sacrifice equitably.

External costs Environmental policy is also developed to ensure that some parties do not use resources in ways that harm others. One way to promote fairness is to deal with external costs. For example, a factory that discharges waste freely into a river may reap greater profits by avoiding paying for waste disposal or for recycling. Its actions, however, impose external costs (water pollution, decreased fish populations, aesthetic degradation, or other problems) on downstream users of the river. U.S.-owned *maquiladoras* in the Tijuana River watershed dump waste that affects Mexican families downstream (see Figure 2.3). Likewise, sewage from the growing number of people living in the watershed further pollutes the river, creating additional external costs for families farther downstream and for beachgoers in Mexico and California.

Weighing the **Issues:**
Do We Really Need Environmental Policy?

Many free-market advocates maintain that environmental laws and regulations are an unnecessary government intrusion into private affairs. Adam Smith (▸ p. 27) argued that individuals will benefit both themselves and society by pursuing their own self-interest. Do you agree? Can you describe a situation in which an individual acting in his or her self-interest could harm society by causing an environmental problem? Can you describe how environmental policy might rectify the situation? What are some advantages and disadvantages of instituting environmental laws and regulations, versus allowing unfettered exchange of materials and services?

U.S. Environmental Policy

The United States provides a good focus for understanding environmental policy in constitutional democracies worldwide, for several reasons. First, the United States historically pioneered innovative environmental policy. Second, U.S. policies have served as models—both of success and of failure—for many other nations and international government bodies. Third, the United States exerts a great deal of influence on the affairs of other nations. In addition, understanding U.S. environmental policy on the federal level enables us to better understand environmental policy at local, state, and international levels.

The three branches of the U.S. federal government—legislative, executive, and judicial—are each involved in aspects of environmental policy. Once **legislation**, or statutory law, is passed by Congress and signed into law by the president, implementation and enforcement of the legislation is assigned to the appropriate *administrative agency* within the executive branch. These agencies, which may be established by Congress or by presidential order, are sometimes nicknamed the "fourth branch" of government because they are the source of a great deal of policy, in the form of regulations. **Regulations** are specific rules based on the more broadly written statutory law. Besides issuing regulations, administrative agencies monitor compliance with laws and regulations and enforce them when individuals or corporations violate them.

The structure of the federal government is mirrored at the state level with governors, legislatures, judiciaries, and agencies. State laws cannot violate principles of the U.S. Constitution, and if state and federal laws are found to be in direct conflict, federal laws take precedence. Many states with dense urban populations, such as California, New York, and Massachusetts, have strong environmental laws and well-funded environmental agencies, whereas many less-populous states, such as those of the interior West, put less priority on environmental protection.

Early U.S. environmental policy addressed public land management

The laws that comprise U.S. environmental policy were created largely in three periods. Laws enacted during the first period, from the 1780s to the late 1800s, dealt primarily with the management of public lands and accompanied the westward expansion of the nation. Environmental laws of this period were intended mainly to promote settlement and the extraction and use of the West's abundant natural resources.

Among these early laws were the *General Land Ordinances of 1785* and *1787*, which gave the federal government the right to manage Western lands and created a grid system for surveying them and readying them for private ownership. Between 1785 and the 1870s, the federal government promoted settlement in the West on lands it had expropriated from Native Americans, by doling out these lands to its citizens. Western settlement provided these citizens with means to achieve prosperity, while relieving crowding in Eastern cities. It expanded the nation's geographical reach at a time when the young United States was still jostling with European powers for control of the continent. It also wholly displaced the millions of Native Americans who had long inhabited these lands. U.S. environmental policy of this era reflected the public perception that Western lands were practically infinite and inexhaustible in natural resources. The following are a few laws typical of this era:

▶ The Homestead Act of 1862 allowed any citizen to claim 65 ha (160 acres) of public land by living there for 5 years and cultivating the land or building a home, for a $16 fee (Figure 2.6a). A waiting period of only 14 months was available to those who could pay $176 for the land.

▶ The Mineral Lands Act of 1866 provided land for $5 per acre to promote mining and settlement. It allowed mining to occur subject to local customs, with no government oversight (Figure 2.6b).

▶ The Timber Culture Act of 1873 granted 65 ha (160 acres) to any citizen promising to cultivate trees on one-quarter of that area (Figure 2.6c).

Such laws encouraged settlers, entrepreneurs, and land speculators to move west, hastening the closing of the frontier.

The second wave of U.S. environmental policy addressed impacts of the first

In the late 1800s, as the West became more populated and its resources were increasingly exploited, public perception and government policy toward natural resources began to shift. Laws of this period aimed to mitigate some of the environmental problems associated with westward expansion. In 1872 Congress designated Yellowstone as the world's first national park. In 1891 Congress, to prevent overharvesting and protect forested watersheds, passed a law authorizing the president to create "forest reserves" off-limits to logging. In 1903, President Theodore Roosevelt created the first national wildlife refuge. These acts enabled the creation, over the next few decades, of a national park system, national forest system, and national wildlife refuge system that still stand as global models (▶ pp. 208, 212–213). These developments reflected a new understanding that the West's resources were exhaustible, and required legal protection.

Land management policies continued through the 20th century, targeting soil conservation in the Dust Bowl years (▶ p. 146) and extending through the Wilderness Act of 1964 (▶ p. 213), which sought to preserve still-pristine lands "untrammeled by man, where man himself is a visitor who does not remain."

The third wave of U.S. environmental policy responded largely to pollution

Further social changes in the mid- to late 20th century gave rise to the third major period of U.S. environmental policy. In a more densely populated country driven by technology, heavy industry, and intensive resource consumption, Americans found themselves better off economically but living amid dirtier air, dirtier water, and more waste and toxic chemicals. During the 1960s and 1970s, several events triggered increased awareness of environmental problems and brought about a shift in public priorities and important changes in public policy.

A landmark event was the 1962 publication of *Silent Spring*, a book by American scientist and writer Rachel Carson

FIGURE 2.6 Settlers (**a**) took advantage of the federal government's early environmental policies, including the Homestead Act of 1862. Early mining activities on public lands were largely unregulated (**b**). The Mineral Lands Act of 1866 required that mining could occur "subject . . . to the local customs or rules of miners," rather than in a way that would protect the environment. Although the Timber Culture Act of 1873 promoted tree-planting on settled agricultural lands, elsewhere the timber industry was allowed to clear-cut the nation's ancient trees (**c**) with little government policy that limited logging or required replanting or conservation.

(a) Settlers in Custer County, Nebraska, circa 1860

(b) Nineteenth-century mining operation, Lynx Creek, Alaska

(c) Loggers felling an old-growth tree, Washington

(Figure 2.7). *Silent Spring* awakened the public to the negative ecological and health effects of pesticides and industrial chemicals (▶ pp. 222–223). (The book's title refers to Carson's warning that pesticides might kill so many birds that few would be left to sing in springtime.)

The Cuyahoga River (Figure 2.8) also did its part to raise attention to pollution hazards. The Cuyahoga was so polluted with oil and industrial waste that the river actually caught fire near Cleveland, Ohio, more than half a dozen times during the 1950s and 1960s. This spectacle, coupled with an oil spill off the Pacific coast near Santa Barbara, California, in 1969, moved the public to prompt Congress and the president to do more to protect the environment.

Today, largely because of environmental policies enacted since the 1960s, pesticides are more strictly regulated, and the nation's air and water are considerably

FIGURE 2.7 Scientist, writer, and citizen activist Rachel Carson illuminated the problem of pollution from DDT and other pesticides in her 1962 book, *Silent Spring*.

cleaner. The public enthusiasm for environmental protection that spurred such advances remains strong today. Polls repeatedly show that an overwhelming majority of Americans favor environmental protection. Such support is evident each year in April, when millions of people worldwide celebrate Earth Day in thousands of locally based events featuring speeches, lectures, demonstrations, hikes, bird-walks, and more. In the decades since the first Earth Day, in 1970, participation in this event has grown and spread to nearly every country in the world (Figure 2.9).

NEPA gives citizens input into environmental policy decisions

Besides Earth Day, two federal actions marked 1970 as the dawn of the modern era of environmental policy in the United States. On January 1, 1970, President Richard Nixon signed the **National Environmental Policy Act (NEPA)** into law. NEPA created an agency called the Council on Environmental Quality and required that an **environmental impact statement (EIS)** be prepared for any major federal action that might significantly affect environmental quality. An EIS is a report of results from

FIGURE 2.8 In a spectacular display of the need for better control over water pollution, Ohio's Cuyahoga River caught fire several times in the 1950s and 1960s. The Cuyahoga was so polluted with oil and industrial waste that the river would burn for days at a time.

studies that assess the potential impacts on the environment that would likely result from development projects undertaken or funded by the federal government.

NEPA's effects have been far-reaching. The EIS process forces government agencies and any businesses that contract with them to evaluate environmental impacts before proceeding with a new dam, highway, or construction project. Although the EIS process generally does not halt such projects, it can serve as an incentive to minimize environmental damage. NEPA also grants ordinary citizens input in the policy process by requiring that environmental impact statements be made publicly available and that public comment on them be solicited and considered.

FIGURE 2.9 The first Earth Day celebration in 1970 demonstrated public support for environmental protection in the United States. Today, Earth Day is celebrated each year by millions of people across the globe, as shown here in Nepal.

The creation of the EPA marked a shift in federal environmental policy

Six months after signing NEPA into law, Nixon issued an executive order calling for a new integrated approach to environmental policy. "The Government's environmentally-related activities have grown up piecemeal over the years," the order stated. "The time has come to organize them rationally and systematically." Nixon's order moved elements of agencies regulating water quality, air pollution, solid waste, and other issues into the newly created **Environmental Protection Agency (EPA).** The order charged the EPA with conducting and evaluating research, monitoring environmental quality, setting and enforcing standards for pollution levels, assisting the states in meeting standards and goals, and educating the public.

Other prominent laws followed

Ongoing public demand for a cleaner environment during this period resulted in a number of key laws that remain linchpins of U.S. environmental policy (Figure 2.10). For problems like the Tijuana River's pollution, a crucial law has been the Clean Water Act of 1977. Prior to passage of federal legislation such as the Clean Water Act, pollution problems were left largely to local and state governments or were addressed through lawsuits. The flaming waters of the Cuyahoga, however, indicated to many people that tough legislation was needed. Thanks to restrictions on pollutants by the Federal Water Pollution Control Acts of 1965 and 1972, and then the Clean Water Act, U.S. waterways finally began to recover. These laws regulated the discharge of wastes, especially from industry, into rivers and streams. The Clean Water Act also aimed to protect wildlife and establish a system for granting permits for the discharge of pollutants. Today there are thousands of federal, state, and local environmental laws in the United States, and thousands more abroad.

The social context for environmental policy changes over time

Historians have suggested that significant environmental policy was implemented in the 1960s and 1970s because three factors converged. First, evidence of environmental problems became widely and readily apparent. Second, people could visualize policies to deal with the problems. Third, the political climate was ripe, with a supportive public and leaders who were willing to act.

By the 1980s, the political climate in the United States had changed. Although public support for the goals of environmental protection remained high, many citizens and policy experts began to feel that the legislative and regulatory means used to achieve environmental policy goals too often imposed economic burdens on businesses or personal burdens on individuals. Since 1980, numerous efforts have been made at the federal level to roll back or reform environmental laws, culminating in an array of efforts by the George W. Bush administration and the Republican-controlled Congress in power during the Bush years.

As the United States retreated from its leadership in environmental policy during the past two decades, other nations were increasing their political attention to environmental issues. The 1992 Earth Summit at Rio de Janeiro,

Key Environmental Protection Laws, 1963–1985

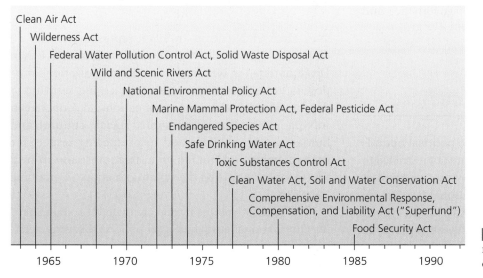

Clean Air Act
Wilderness Act
Federal Water Pollution Control Act, Solid Waste Disposal Act
Wild and Scenic Rivers Act
National Environmental Policy Act
Marine Mammal Protection Act, Federal Pesticide Act
Endangered Species Act
Safe Drinking Water Act
Toxic Substances Control Act
Clean Water Act, Soil and Water Conservation Act
Comprehensive Environmental Response, Compensation, and Liability Act ("Superfund")
Food Security Act

1965   1970   1975   1980   1985   1990

FIGURE 2.10 Most major laws in modern U.S. environmental policy were enacted in the 1960s and 1970s.

FIGURE 2.11 Many nations are shifting policies to support sustainable development efforts to increase standards of living while safeguarding the environment. Here, a woman stirs rice on a solar-powered oven at a restaurant built of recycled drink cans, showcased at the Ubunto Village at the U.N. World Summit on Sustainable Development in Johannesburg, South Africa, in 2002.

Brazil, was the largest international diplomatic conference ever held, drawing representatives from 179 nations and unifying these leaders around the idea of sustainable development (▶ p. 20, 372). Indeed, we may now be embarking on a fourth wave of environmental policy, one focused on sustainable development (Figure 2.11). This new policy approach tries to find ways to safeguard the functionality of natural systems while raising living standards for the world's poorer people. As the world's nations continue to feel the social, economic, and ecological effects of environmental degradation, environmental policy will without doubt become a more central part of governance and everyday life in all nations in the years ahead.

International Environmental Policy

Environmental systems pay no heed to political boundaries, so environmental problems often are not restricted to the confines of particular countries. For instance, most of the world's major rivers cross international borders, so problems like those along the Tijuana River are frequently international in scope. Because U.S. law has no authority in Mexico or any other nation outside the United States, international law is vital to solving transboundary problems.

International law includes customary law and conventional law

International law known as **customary law** arises from long-standing practices, or customs, held in common by most cultures. International law known as **conventional law** arises from *conventions*, or *treaties*, into which nations enter. One example of conventional law is the Montreal Protocol, a 1987 accord among more than 160 nations to reduce the emission of airborne chemicals that thin the ozone layer (▶ p. 281). Another example is the Kyoto Protocol to reduce emissions that contribute to global climate change (▶ pp. 297–298). This agreement took effect in February 2005 without United States participation, after a quorum of nations ratified the accord.

In 1990, Mexico and the United States signed a treaty and agreed to build an international wastewater treatment plant to handle excess sewage from Tijuana. The plant began operating just north of the border in 1997 and collects and treats up to 95 million L (25 million gal) of wastewater each day. In this case the treaty process worked well, but the results fell short of expectations. The facility reached its capacity within 3 years and then began discharging material to the ocean that did not meet safety standards under U.S. law.

Several organizations shape international environmental policy

Although there is no real mechanism for enforcing international environmental law, a number of international organizations regularly act to influence the behavior of nations by providing funding, applying political or economic pressure, and/or directing media attention.

The United Nations In 1945, representatives of 50 countries founded the *United Nations (U.N.)*. Headquartered in New York City, this organization's purpose is "to maintain international peace and security; to develop friendly relations among nations; to cooperate in solving international economic, social, cultural and humanitarian problems and in promoting respect for human rights and fundamental freedoms; and to be a centre for harmonizing the actions of nations in attaining these ends."

The United Nations has taken an active role in shaping international environmental policy. Of several agencies within it that influence environmental policy, most notable is the *United Nations Environment Programme*

(*UNEP*), created in 1972, which helps nations understand and solve environmental problems. Based in Nairobi, Kenya, its mission is sustainability, enabling countries and their citizens "to improve their quality of life without compromising that of future generations." UNEP's research and outreach activities provide a wealth of information useful to policymakers and scientists throughout the world.

The World Bank Established in 1944 and based in Washington, D.C., the *World Bank* is one of the globe's largest sources of funding for economic development. This institution can shape environmental policy through its funding of dams, irrigation infrastructure, and other major development projects. In fiscal year 2005 the World Bank provided over $22.3 billion in loans for projects designed to benefit the poorest people in the poorest countries.

Despite its admirable mission, the World Bank has frequently been criticized for funding unsustainable projects that cause more environmental problems than they solve. Providing for the needs of growing populations in poor nations while minimizing damage to the environmental systems on which people depend can be a tough balancing act. Environmental scientists today agree that sustainable development must be the guiding principle for such efforts.

The European Union The *European Union* (*EU*) was not established primarily with environmental problem solving in mind, but the treaty that created it held as one of its goals the promotion of solutions to environmental problems. Formed after World War II, the EU as of early 2006 contained 25 member nations. It seeks to promote Europe's unity and its economic and social progress and to "assert Europe's role in the world." The EU can sign binding treaties on behalf of its members and can enact regulations that have the same authority as national laws in each member nation. It can also issue *directives*, which are more advisory in nature. The EU's European Environment Agency works to address waste management, noise pollution, water pollution, air pollution, habitat degradation, and natural hazards. The EU also seeks to remove trade barriers among member nations. It has classified some nations' environmental regulations as barriers to trade; for example, some northern European nations have traditionally had more stringent environmental laws that prevent the import and sale of environmentally harmful products from other member nations.

The World Trade Organization Based in Geneva, Switzerland, the *World Trade Organization* (*WTO*) was established in 1995, having grown from a 50-year-old international trade agreement. The WTO represents multinational corporations and promotes free trade by reducing obstacles to international commerce and enforcing fairness among nations in trading practices. Whereas the United Nations and the European Union have limited influence over nations' internal affairs, the WTO has real authority to impose financial penalties on nations that do not comply with its directives. These penalties can sometimes play major roles in shaping environmental policy.

Like the EU, the WTO has interpreted some national environmental laws as unfair barriers to trade. For instance, in 1995 the U.S. EPA issued regulations requiring cleaner-burning gasoline in U.S. cities, following Congress's amendments of the Clean Air Act. Brazil and Venezuela filed a complaint with the WTO, saying the new rules unfairly discriminated against the petroleum they exported to the United States, which did not burn as cleanly. The WTO agreed, ruling that even though the South American gasoline posed a threat to human health in the United States, the EPA rules were an illegal trade barrier. The ruling forced the United States to alter its approach to regulating gasoline. Not surprisingly, critics have frequently charged that the WTO aggravates environmental problems.

Weighing the **Issues:**
Trade Barriers and Environmental Protection

If Nation A has stricter laws for environmental protection than Nation B, and if these laws restrict the ability of Nation B to export its goods to Nation A, then by the policy of the WTO and the EU, Nation A's environmental protection laws could be overruled in the name of free trade. Do you think this is right? What if Nation A is a wealthy industrialized country and Nation B is a poor developing country that needs every economic boost it can get?

Nongovernmental organizations A number of *nongovernmental organizations* (*NGOs*) have grown to become international in scope and exert influence over international environmental policy. Some, such as The Nature Conservancy, focus on accomplishing conservation objectives on the ground (in its case, purchasing and managing land and habitat for rare species) without becoming politically involved. Other groups, including

Conservation International, the World Wide Fund for Nature, Greenpeace, Population Connection, and many others, attempt to shape policy directly or indirectly through research, education, lobbying, or protest. NGOs apply more funding and expertise to environmental problems, and conduct more research intended to solve them, than do many national governments.

The Environmental Policy Process

Anyone can become involved in helping ideas become public policy. In constitutional democracies such as the United States, it is true that each and every person has a political voice and can make a difference. Unfortunately, it is also true that money wields influence and that some people and organizations are far more politically connected and influential than others. We will explore some of the ways people make themselves influential as we examine the main steps of the policymaking process. Our discussion pertains both to citizens at the grassroots level and to large organizations and corporations.

The environmental policy process begins when a problem is identified

The first step in the environmental policy process (Figure 2.12) is to identify an environmental problem. Identifying a problem requires curiosity, observation, record keeping, and an awareness of our relationship with the environment. For example, assessing the contamination of San Diego- and Tijuana-area beaches required understanding the ecological and health impacts of untreated wastewater. It also required being able to detect contamination on beaches and understanding water flow dynamics among the beaches, the Pacific Ocean, and the Tijuana River watershed.

Identifying causes of the problem is the second step in the policy process

Once an individual or group has defined a particular environmental problem, discovering specific causes of the problem is next. Identifying causes often requires scientific research, giving science a role in the policy process. A person seeking causes for pollution in the Tijuana

1 Identify problem

2 Identify specific causes of the problem

3 Envision solution and set goals

4 Get organized

5 Cultivate access and influence

6 Manage development of policy

FIGURE 2.12 Understanding the steps of the policy process is an essential element of solving environmental problems.

River watershed might notice that transboundary sewage spills became more toxic and industrial once U.S.-based companies began opening *maquiladoras* on the Mexican side of the border. Advocates of the *maquiladora* system argue that these factories provide much-needed jobs south of the border while keeping companies' costs low by paying Mexican workers far less than U.S. workers. Critics argue that the factories are waste-generating, water-guzzling polluters whose transboundary nature makes them particularly difficult to regulate.

The third step is envisioning a solution

The better one can identify specific causes of a problem, the more effectively one will be able to envision solutions to it and argue for implementing those solutions. Science plays a vital role here too, although solutions frequently involve primarily social or political action. In San Diego, citizen activists wanted Tijuana to enforce its own pollution laws more effectively—something that, once visualized, started to happen when San Diego city employees began training and working with their Mexican counterparts to keep hazardous wastes out of the sewage treatment system.

Getting organized is the fourth step

When it comes to gaining the ear of elected officials and influencing policy, organizations are generally more effective than lone individuals. The sole critic or crusader is easily dismissed as a crackpot or troublemaker, but a group of hundreds or thousands of individuals is not as easily dismissed. Furthermore, organizations are more effective at raising funds, which by U.S. law they may contribute to political campaigns.

As effective as large organizations can be, however, small coalitions and even individual citizens who are motivated, informed, and organized can solve environmental problems. As the anthropologist Margaret Meade once remarked, "Never doubt that a small group of thoughtful, committed citizens can change the world—indeed it is the only thing that ever has." San Diego-area resident Lori Saldaña provides an example. Concerned about the Tijuana River's pollution, Saldaña reviewed plans for the international wastewater treatment plant that the U.S. government proposed to build. She concluded that it would merely shift pollution from the river to the ocean, where sewage would be released 5.6 km (3.5 mi) offshore. Working with her local Sierra Club chapter, Saldaña protested the plant's design and participated in a

lawsuit that forced the government to conduct further studies. The EPA finally agreed to a design change, although funding for it has stalled in Congress. For her efforts, Saldaña received awards and was appointed to a commission on border environmental issues by President Bill Clinton. After a decade of activism, Saldaña ran for the California State Assembly in 2004 and won. She is now the representative from California's 76th district.

Gaining access to political powerbrokers is the fifth step

The next step in the policy process entails gaining access to policymakers who have the clout to help enact the desired changes. People gain access and influence through lobbying and campaign contributions.

Anyone who spends time or money trying to change an elected official's mind is engaged in *lobbying*. Although anyone can lobby, it is much more difficult for an ordinary citizen than for the thousands of full-time professional lobbyists employed by the many businesses and organizations seeking a voice in politics. Supporting a candidate's reelection efforts with money is another way to make one's voice heard, and any individual can donate money to political campaigns.

Shepherding a solution into law is the sixth step in the policy process

Having gained access to elected officials, the next step is to prepare a bill, or draft law, that embodies the desired solutions. Anyone can draft a bill. The hard part is finding members of the House and Senate willing to introduce the bill and shepherd it from subcommittee through full committee and on to passage by the full Congress. Lobbying and media attention intensify as the bill progresses through this process (Figure 2.13). If it passes through all of these steps, the bill may become law, but it can die in countless fashions along the way.

The policy process does not end with the enactment of legislation. Following a law's enactment, administrative agencies implement regulations. Policymakers also evaluate the policy's successes and failures and may revise the policy as necessary. Moreover, the judicial branch interprets law in response to suits in the courts, and a great deal of environmental policy has lived or died by judicial interpretation. The full policy process is long and often cumbersome, but it has yielded effective results in constitutional democracies in the United States and many other nations.

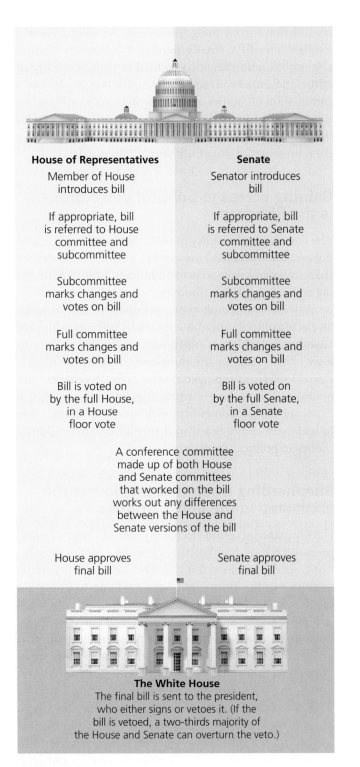

House of Representatives

Member of House introduces bill

If appropriate, bill is referred to House committee and subcommittee

Subcommittee marks changes and votes on bill

Full committee marks changes and votes on bill

Bill is voted on by the full House, in a House floor vote

Senate

Senator introduces bill

If appropriate, bill is referred to Senate committee and subcommittee

Subcommittee marks changes and votes on bill

Full committee marks changes and votes on bill

Bill is voted on by the full Senate, in a Senate floor vote

A conference committee made up of both House and Senate committees that worked on the bill works out any differences between the House and Senate versions of the bill

House approves final bill

Senate approves final bill

The White House
The final bill is sent to the president, who either signs or vetoes it. (If the bill is vetoed, a two-thirds majority of the House and Senate can overturn the veto.)

FIGURE 2.13 Before a bill becomes a law, it must clear a number of hurdles in both legislative bodies. If the bill passes the House and Senate, a conference committee works out any differences between the House and Senate versions before the bill is sent to the president. The president may then sign or veto the bill.

Approaches to Environmental Policy

Many environmental laws and regulations—for instance, those aiming to reduce pollution—have simply set strict legal limits and threatened punishment for violating these limits. This approach is sometimes called **command-and-control**. The command-and-control approach has resulted in some major successes, as evidenced by the cleaner air and water U.S. residents enjoy today. Without doubt, our environment would be in far worse shape were it not for government regulatory intervention. However, many people have grown disenchanted with the top-down, sometimes heavy-handed nature of the command-and-control approach.

Sometimes government actions are well intentioned but not well enough informed, so they can lead to unforeseen consequences. Policy can also fail if a government does not live up to its responsibilities to protect its citizens or treat them equitably. This may occur when leaders allow themselves to be unduly influenced by *interest groups*, small groups of people seeking private gain that work against the larger public interest. Finally, the command-and-control approach can fail if it generates opposition to government policy by causing citizens to view laws and regulations primarily as restrictions on their freedom.

Criticism of the command-and-control approach has spurred policymakers and the experts who advise them to devise alternative pathways for attaining environmental policy goals. The most widely developed alternatives involve the creative use of economic incentives.

Subsidies are a widespread economic policy tool

One set of economic policy tools aims to promote industries or activities that are deemed desirable. Governments may give *tax breaks* to certain types of businesses or individuals, for instance. Relieving the tax burden lowers costs for the business or individual, thus assisting the desirable industry or activity.

A similar economic policy tool is the **subsidy**, a government giveaway of cash or publicly owned resources. Subsidies can be used to promote environmentally sustainable activities, but all too often they have been used to prop up unsustainable ones. Subsidies judged to be harmful to the environment and to the economy total roughly $1.45 *trillion* yearly across the globe, according to British environmental scientist Norman Myers—an amount larger than the economies of all but five nations. The average U.S. taxpayer pays $2,000 per year in environmentally harmful subsidies, plus $2,000 more through increased prices for goods and through degradation of ecosystem services, Myers estimates. Among the more prominent examples are subsidies for nonrenewable fossil fuels, road building in national forests, and mining on public lands.

Green taxes discourage undesirable activities

Another economic policy tool—taxation—can be used to discourage undesirable activities. Taxing undesirable activities helps to "internalize" external costs by making them part of the overall cost of doing business. Taxes on environmentally harmful activities and products are called **green taxes**. By taxing activities and products that cause undesirable environmental impacts, a tax becomes a tool for policy as well as simply a way to fund government.

Green taxes have yet to gain widespread support in the United States, although similar "sin taxes" on cigarettes and alcohol are tools of U.S. social policy. Taxes on pollution have been widely instituted in Europe, where many nations have adopted the *"polluter pays" principle.*

Under green taxation, a factory that pollutes a waterway would pay taxes based on the amount of pollution it discharges. The idea is to give companies a financial incentive to reduce pollution, while allowing the polluter the freedom to decide how best to minimize its expenses. One polluter might choose to invest in technologies to reduce its pollution if doing so is less costly than paying the taxes. Another polluter might find that abating its pollution would be more costly and could choose to pay the taxes instead—funds the government might then apply toward mitigating pollution in some other way. Green taxation provides incentive for industry to lower emissions not merely to a level specified in a regulation, but to still-lower levels. However, green taxes do have disadvantages. One is that businesses will most likely pass on their tax expenses to consumers.

Markets in permits can save money and produce results

A creative market-based approach is for government to sell or give to companies the right to pollute, by establishing markets in tradable pollution permits. After determining the overall amount of pollution it will allow an entire industry to produce, the government issues permits to individual polluters that allow them each to emit a certain fraction of that amount. Polluters are then allowed to buy, sell, and trade these permits with other polluters. With such **marketable emissions permits**, governments create incentives for firms to reduce their pollution in a way that is compatible with market capitalism.

Suppose, for example, you are a plant owner with permits to release 10 units of pollution, but you find that you can become more efficient and release only 5 units of pollution instead. You then have a surplus of permits, which might be very valuable to some other plant that is having trouble reducing its pollution or to one that wants to expand production. In such a case, you can sell your extra permits. Doing so meets the needs of the other plant and generates income for you, while preventing any increase in total pollution.

Such a system of marketable emissions permits has been in place in the United States since 1990, under amendments to the Clean Air Act that mandated reduced emissions of sulfur dioxide, a major contributor to acidic deposition (▶ pp. 282–285). Estimated savings from the Clean Air Act permits add up to several billion dollars per year. Although such "cap-and-trade" programs can succeed in reducing the overall amount of pollution, they have a drawback: They allow hotspots of pollution to occur around plants that buy permits to pollute more. Currently, a global market in carbon emissions is developing as a result of the Kyoto Protocol to address climate change (▶ pp. 297–298).

Ecolabeling empowers consumers

Another strategy that uses the marketplace to counteract market failure allows consumers to play the key role. Governments may allow or require manufacturers of certain products to designate on their labels how the products were grown, harvested, or manufactured. This approach, called **ecolabeling**, tells consumers which brands use environmentally benign processes (Figure 2.14). By preferentially buying

FIGURE 2.14 Ecolabeling allows businesses to promote products that have low environmental impact. Organic juices and produce are examples of ecolabeled products that are becoming widely available in the marketplace.

ecolabeled products, consumers can provide businesses a powerful incentive to switch to more environmentally friendly processes. In one early example, cans of tuna were labeled "dolphin-safe," to indicate that methods used to catch the tuna avoid the accidental capture of dolphins. Other examples include labeling recycled paper, organically grown foods, and lumber harvested through sustainable forestry.

Market incentives are being tried widely on the local level

You may have already taken part in transactions involving financial incentives as policy tools. Many municipalities charge residents for waste disposal according to the amount of waste they generate. Other cities place taxes or disposal fees on items whose safe disposal is costly, such as tires and motor oil. Still others give rebates to residents who buy water-efficient toilets and appliances, because the rebates cost the city less than upgrading its wastewater treatment system. Likewise, power companies sometimes offer discounts to customers who buy high-efficiency lightbulbs and appliances, because the discounts cost the utilities less than expanding the generating capacity of their plants. The widespread and growing application of such approaches at the local level is an encouraging sign.

Conclusion

Environmental policymaking is a problem-solving pursuit that makes use of science, ethics, and economics and that requires an astute understanding of the political process. Conventional command-and-control approaches of legislation and regulation are the most common approaches to policymaking, but various innovative economic policy tools have also been developed. These tools, as well as the valuation of ecosystem services by environmental and ecological economists, are helping bring economic approaches to bear on environmental protection and resource conservation. Equating economic well-being with economic growth, as most economists and policymakers traditionally have, suggests that economic welfare entails a trade-off with environmental quality. However, if economic welfare can be enhanced in the absence of growth, we can envision economies and environmental quality benefiting mutually.

TESTING YOUR COMPREHENSION

1. Name two key ways in which human economies are linked to the natural environment.
2. Describe four ways in which neoclassical economic approaches negatively affect the environment, according to critics.
3. Compare and contrast the views and approaches of neoclassical economists, environmental economists, and ecological economists.
4. What are ecosystem services? Describe at least two ways in which some economists have tried to assign market values to ecosystem services.
5. Describe and critique three common justifications for environmental policy. Explain the concept of external costs, and state why it is relevant to environmental policy.
6. Summarize the differences between the first, second, and third waves of environmental policy in U.S. history.
7. What did the National Environmental Policy Act accomplish? Briefly describe the origin and mission of the U.S. Environmental Protection Agency.
8. What is the difference between customary law and conventional law? What special challenges do transboundary environmental problems present to policymakers?
9. List the steps of the environmental policy process, from identification of a problem through enactment of a federal law.
10. Differentiate between a green tax, a subsidy, a tax break, and a marketable emissions permit.

SEEKING SOLUTIONS

1. What is a steady-state economy? Do you think this model is a practical alternative to the growth paradigm? Why or why not?
2. Do you think we should attempt to quantify and assign market values to ecosystem services and other entities that have only nonmarket values? Why or why not?
3. Reflect on the causes for the transitions in U.S. history from one type of environmental policy to another. Now peer into the future, and think about how life

might be different in 25, 50, or 100 years. What would you speculate about the environmental policy of the future? What issues might it address? Do you predict we will have more or less environmental policy?

4. Compare the roles of the United Nations, the European Union, the World Bank, the World Trade Organization, and nongovernmental organizations. If you could gain the support of just one of these institu-

tions for a policy you favored, which would you choose? Why?

5. Think of one environmental problem that you would like to see solved. From what you've learned in this chapter, what policy approach do you think would be most effective in addressing it? How do you think you could best shepherd your ideas through the policymaking process?

CALCULATING ECOLOGICAL FOOTPRINTS

 How many of us think about the destination of our waste when we flush the toilet? Some nutrient pollution from sewage generally ends up in a local waterway, but the U.S. Environmental Protection Agency establishes strict discharge standards following amendments to the U.S. Clean Water Act. One measure used is total suspended solids (TSS). To comply with federal regulations, wastewater treatment facilities in the United States must discharge water with a monthly average TSS no greater than 30 mg per liter.

Assuming that the average toilet flush is 3 gallons and occurs 4 times per person per day, calculate the total discharge of TSS in wastewater with the maximum permissible content. Note that 1 gal = 3.7 L, 1,000 mg = 1 g, and 1,000 g = 2.2 lb.

1. Assuming that water in an average toilet flush contains 100 g TSS, calculate the amount of TSS removed from the nation's wastewater each day, month, and year, if the EPA standard is met.

2. Do you think the standards should be stricter? What would be at least one advantage and one disadvantage of stricter standards?

	TSS discharged per day	TSS discharged per month	TSS discharged per year
You	0.440 g	13.4 g	160.6 g
Your class			
Your hometown			
United States			

Take It Further

Go to www.aw-bc.com/withgott, where you'll find:

▶ Suggested answers to end-of-chapter questions
▶ Quizzes, animations, and flashcards to help you study

▶ *Research Navigator*™ database of credible and reliable sources to assist you with your research projects

▶ **GRAPHit!** tutorials to help you interpret graphs

▶ **INVESTIGATEit!** current news articles that link the topics that you study to case studies from your region to around the world

3

Environmental Systems: Chemistry, Energy, and Ecosystems

Fisherman in the Gulf of Mexico's "dead zone"

Upon successfully completing this chapter, you will be able to:

▶ Describe the nature of environmental systems

▶ Explain and apply the fundamentals of environmental chemistry

▶ Describe the molecular building blocks of organisms

▶ Differentiate among the types of energy and recite the basics of energy flow

▶ Distinguish photosynthesis from respiration and summarize their importance to living things

▶ Define ecosystems and evaluate how living and nonliving entities interact in ecosystem-level ecology

▶ Compare and contrast how carbon, phosphorus, nitrogen, and water cycle through the environment

▶ Explain how plate tectonics and the rock cycle shape Earth's physical characteristics

Central Case: The Gulf of Mexico's "Dead Zone"

> "In nature there is no 'above' or 'below,' and there are no hierarchies. There are only networks nesting within other networks."
> —FRITJOF CAPRA, THEORETICAL PHYSICIST

> "Let's say you put Saran Wrap over south Louisiana and suck the oxygen out. Where would all the people go?"
> —NANCY RABALAIS, BIOLOGIST FOR THE LOUISIANA UNIVERSITIES MARINE CONSORTIUM

Louisiana fishermen have long hauled in more seafood than those of any other U.S. state except Alaska. Each year they have plied the rich waters of the northern Gulf of Mexico and have sent nearly 600 million kg (1.3 billion lb) of shrimp, fish, and shellfish to our dinner tables. Then in 2005, Hurricane Katrina and Hurricane Rita pummeled the Gulf Coast and left Louisiana's fisheries in ruin. Boats, docks, marinas, and fueling stations were destroyed, and thousands of people were suddenly out of work. Today, fishermen are struggling to reestablish their livelihoods as the industry continues to rebuild itself.

But for years before the hurricanes hit, fishing had become increasingly difficult. In the words of longtime Louisiana fisherman Johnny Glover, it was "getting harder and harder to make a living." The reason? Each year billions of organisms were suffocating in the Gulf's "dead zone," a region of water so depleted of oxygen that marine organisms are killed or driven away.

The low concentrations of dissolved oxygen in the bottom waters of this region represent a condition called **hypoxia** (see "The Science behind the Story," ▶ pp. 54–55 and Figure 3.3, ▶ p. 52). Aquatic animals obtain oxygen by respiring through their gills, and, like us, these animals will asphyxiate if deprived of oxygen. Fully oxygenated water contains up to 10 parts per million (ppm) of oxygen, but when concentrations drop below 2 ppm, creatures that can leave an affected area will do so. Below 1.5 ppm, most marine organisms die. In the Gulf's hypoxic zone, oxygen concentrations frequently drop well below these levels.

The dead zone appears each spring and grows through the summer and fall, starting near the mouths of the Mississippi and Atchafalaya rivers off the Louisiana coast. In 2002 the dead zone reached a record 22,000 km^2 (8,500 mi^2)—an area larger than New Jersey. Shrimp boats came up with nets nearly empty. One shrimper derided his meager catch as "cat food." Others,

ironically, said they hoped a hurricane would strike and stir some oxygen into the Gulf's stagnant waters.

What's starving these waters of oxygen? Scientists studying the dead zone have identified modern Midwestern farm practices and other human impacts hundreds of kilometers away. The Gulf, they say, is being over-enriched by nitrogen and phosphorus flushed down the Mississippi River. This nutrient pollution comes from fertilizers used on farms far upstream in the Mississippi River basin, as well as from urban runoff, industrial discharges, atmospheric deposition from fossil fuel combustion, and municipal sewage outflow.

The U.S. government has acted on these findings, proposing that farmers in states such as Ohio, Iowa, and Illinois cut down on fertilizer use. Farmers' advocates protest that farmers are being singled out while urban pollution sources are being ignored. Meanwhile, coastal dead zones have appeared in 150 other areas throughout the world, from Chesapeake Bay to Oregon to Denmark to the Black Sea. The story of how scientists have determined the causes of these dead zones involves understanding environmental systems and the often complex behavior they exhibit.

Earth's Environmental Systems

Our planet's environment consists of complex networks of interlinked systems. These systems include the ecological webs of relationships among species and the interaction of living organisms with the nonliving entities around them. Earth's systems also include cycles that guide the flow of key chemical elements and compounds that support life and regulate climate. We depend on these systems for our very survival. Scientists are increasingly taking a "systems approach" in environmental science research, and such a holistic approach is also ideal for designing solutions to complex environmental problems.

Systems involve feedback loops

A **system** is a network of relationships among parts, elements, or components that interact with and influence one another through the exchange of energy, matter, or information. Systems receive inputs of energy, matter, or information; process these inputs; and produce outputs. For example, the Gulf of Mexico receives inputs of freshwater, sediments, nutrients, and pollutants from the Mississippi and other rivers. Shrimpers harvest some of the Gulf system's output: matter and energy in the form of shrimp. This output subsequently becomes input to the human economic system and to the digestive systems of the many people who consume the shrimp.

Sometimes a system's output can serve as input to that same system, a circular process known as a **feedback loop**. Feedback loops are of two types, negative and positive. In a **negative feedback loop** (Figure 3.1a), output that results from a system moving in one direction acts as input that moves the system in the other direction. Input

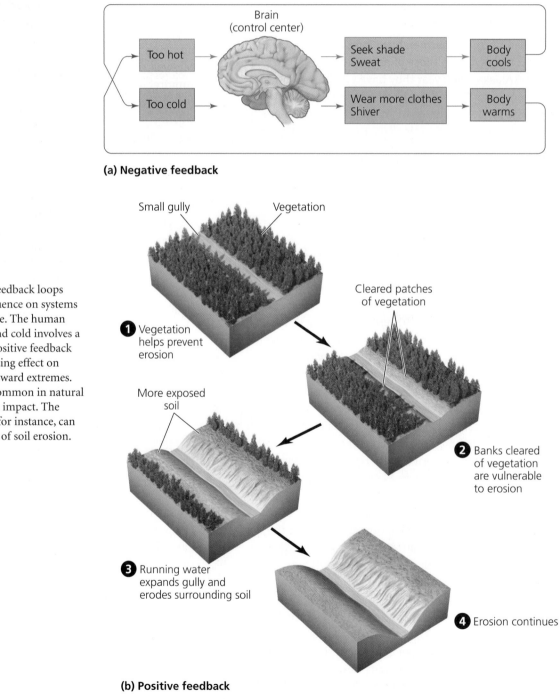

FIGURE 3.1 Negative feedback loops **(a)** exert a stabilizing influence on systems and are common in nature. The human body's response to heat and cold involves a negative feedback loop. Positive feedback loops **(b)** have a destabilizing effect on systems and push them toward extremes. Rare in nature, they are common in natural systems altered by human impact. The clearing of forested land, for instance, can lead to a runaway process of soil erosion.

Brain (control center)

Too hot → Seek shade / Sweat → Body cools

Too cold → Wear more clothes / Shiver → Body warms

(a) Negative feedback

Small gully   Vegetation

❶ Vegetation helps prevent erosion

Cleared patches of vegetation

❷ Banks cleared of vegetation are vulnerable to erosion

More exposed soil

❸ Running water expands gully and erodes surrounding soil

❹ Erosion continues

(b) Positive feedback

and output essentially neutralize one another's effects, stabilizing the system. A thermostat, for instance, stabilizes a room's temperature by turning the furnace on when the room gets cold and shutting it off when the room gets hot. Similarly, negative feedback regulates our body temperature. If we get too hot, our sweat glands pump out moisture that evaporates to cool us down, or we may move from sun to shade. If we get too cold, we shiver, creating heat, or we move into the sun or put on more clothing. Most systems in nature involve negative feedback loops. Negative feedback loops enhance stability, and in the long run, only those systems that are stable will persist.

Positive feedback loops have the opposite effect. Rather than stabilizing a system, they drive it further toward one extreme or another. This can occur with the process of erosion, the removal of soil by water or wind (▶ p. 145). Once vegetation is cleared and exposes soil, erosion may become progressively more severe if the forces of water or wind surpass the rate of vegetative regrowth (Figure 3.1b). Positive feedback can alter a system substantially. Positive feedback loops are rare in nature, but common in systems altered by human impact.

Understanding the dead zone requires considering Mississippi River and Gulf of Mexico systems together

The Gulf of Mexico and the Mississippi River are systems that interact with one another. On a map, the Mississippi River appears as a branched and braided network of water channels lined by farms, cities, and forests (Figure 3.2). But where are this system's boundaries? For a scientist interested in runoff and the flow of water, sediment, or pollutants, it may make most sense to view the Mississippi River's *watershed* (▶ p. 24) as a system. However, for a scientist interested in the Gulf of Mexico's dead zone, it may be best to view the watershed together with the Gulf as the system of interest, because their interaction is central to the problem. In environmental science, one's delineation of a system can and should depend on the questions one is addressing.

The reason for the Gulf of Mexico's dangerously low levels of oxygen, scientists have concluded, is abnormally high levels of nutrients such as **nitrogen** and **phosphorus**. The excess nutrients originate with sources in the Mississippi River watershed, particularly nitrogen- and phosphorus-rich fertilizers applied to crops. Inorganic nitrogen fertilizers used on Midwestern farms account for roughly 30% of total nitrogen contributions to the river, an amount equal to that from natural soil decomposition. The remainder comes from various sources, including animal manure, nitrogen-fixing crops, sewage treatment facilities, street runoff, and industrial and automobile emissions. All together, agricultural sources are thought to contribute 74% of the nitrate and 65% of the total nitrogen carried in the river. Much of the nitrate originates from farms growing corn and soybeans in Iowa, Illinois, Indiana, Minnesota, and Ohio. Nitrogen fertilizer input to farmland in the Mississippi River watershed has increased dramatically since 1950. Since 1980, the Mississippi River and the Atchafalaya River (which drains a third of the Mississippi's water through a

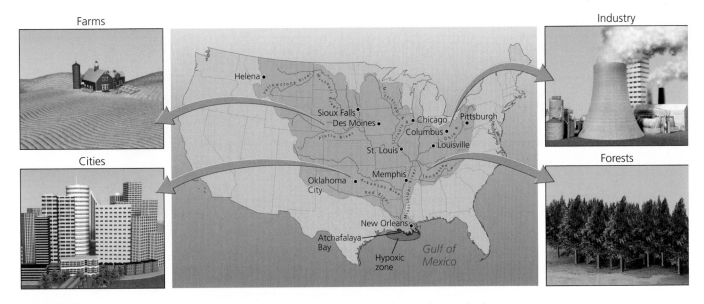

FIGURE 3.2 The Mississippi River system is the largest in North America. The river's watershed encompasses 3.2 million km² (1.2 million mi²), or 41% of the area of the lower 48 U.S. states. The river carries water, sediment, and pollutants from a variety of sources downriver to the Gulf of Mexico, where nutrient pollution has given rise to a hypoxic zone.

second delta) have pumped about 1 million metric tons of nitrates into the Gulf of Mexico each year, three times as much as during the 1960s.

The enhanced nitrogen input to the Gulf boosts the growth of *phytoplankton*, microscopic photosynthetic algae, protists, and cyanobacteria that drift near the surface. Phytoplankton ordinarily are limited in their growth by scarcity of nutrients such as nitrogen and phosphorus. As phytoplankton flourish at the surface and as the tiny animals called *zooplankton* consume them, more dead phytoplankton and waste products of phytoplankton and zooplankton drift to the bottom, providing food for bacteria that decompose them. The result is a population explosion of bacteria. These decomposers consume enough oxygen to cause oxygen concentrations in bottom waters to plummet, suffocating shrimp and fish that live at the bottom and creating the dead zone. The fresh water from the river remains naturally stratified in a layer at the surface that mixes only very slowly with the denser salty ocean water, so that oxygenated surface water does not make its way down to the bottom-dwelling life that needs it. This process of nutrient overenrichment, blooms of algae, increased production of organic matter, and subsequent ecosystem degradation is known as **eutrophication** (Figure 3.3).

Moderate amounts of additional nutrients may increase the productivity of fisheries, but at higher concentrations, this fertilizing effect is offset by hypoxia, and fishery yields decline.

Chemistry and the Environment

Chemistry plays a central role in the environmental challenges facing the Gulf of Mexico. Understanding how too much nitrogen or phosphorus in one part of a system can lead to too little oxygen in another requires a good working knowledge of chemistry.

Indeed, examine any environmental issue, and you will likely discover chemistry playing a key role. Chemistry is crucial to understanding how gases such as carbon dioxide and methane contribute to global climate change, how pollutants such as sulfur dioxide and nitric oxide cause acid precipitation, and how pesticides and other artificial compounds we release into the environment affect the health of wildlife and people. Chemistry is central, too, in understanding water pollution and wastewater treatment,

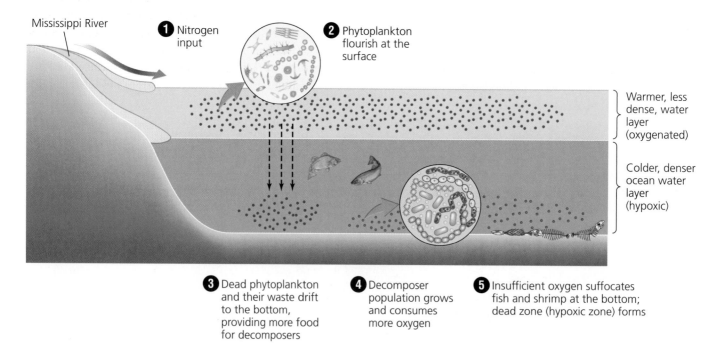

FIGURE 3.3 Excess nitrogen causes eutrophication in coastal marine systems such as the Gulf of Mexico. Coupled with stratification (layering) of water, eutrophication can severely deplete dissolved oxygen. Nitrogen from river water (1) boosts growth of phytoplankton (2), which die and are decomposed at the bottom by bacteria (3). Stability of the surface layer prevents deeper water from absorbing oxygen to replace that consumed by decomposers (4), and the oxygen depletion suffocates or drives away bottom-dwelling marine life (5). This process gives rise to hypoxic zones like that of the Gulf of Mexico.

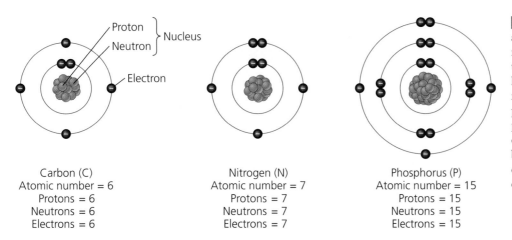

Carbon (C)
Atomic number = 6
Protons = 6
Neutrons = 6
Electrons = 6

Nitrogen (N)
Atomic number = 7
Protons = 7
Neutrons = 7
Electrons = 7

Phosphorus (P)
Atomic number = 15
Protons = 15
Neutrons = 15
Electrons = 15

FIGURE 3.4 In an atom, protons and neutrons are held in the nucleus, and electrons move around the nucleus. Each chemical element has a different number of protons, neutrons, and electrons. Carbon possesses 6 of each, nitrogen 7, and phosphorus 15. Electrons do *not* orbit the nucleus in rings as shown, but these diagrams are meant to clearly show and compare numbers of electrons.

hazardous waste and its cleanup and disposal, atmospheric ozone depletion, and most energy issues. Environmental chemists today are excited about the countless current and future applications of chemistry that can help us address environmental problems.

Atoms and elements are chemical building blocks

To appreciate the complex chemistry involved in environmental science, we must begin with a grasp of the fundamentals. The nitrogen, phosphorus, and oxygen that play key roles in the Gulf of Mexico's situation are each elements. An **element** is a fundamental type of matter, a chemical substance with a given set of properties, which cannot be broken down into substances with other properties. Chemists currently recognize 92 elements occurring in nature, as well as more than 20 others that have been artificially created. Besides **carbon** and nitrogen, elements especially abundant in living organisms include **hydrogen** and **oxygen**. Each element is assigned an abbreviation, or chemical symbol. The *periodic table of the elements* (see Appendix C) summarizes information on the elements in a comprehensive and elegant way.

Elements are composed of **atoms**, the smallest components that maintain the chemical properties of the element. Every atom has a nucleus of **protons** (positively charged particles) and **neutrons** (particles lacking electric charge). The atoms of each element have a defined number of protons, called the *atomic number*. (Elemental carbon, for instance, has six protons; thus, its atomic number is 6.) An atom's nucleus is surrounded by negatively charged particles known as **electrons**, which balance the positive charge of the protons (Figure 3.4).

Although all atoms of a given element contain the same number of protons, they do not necessarily contain the same number of neutrons. Atoms with differing numbers of

neutrons are referred to as **isotopes** (Figure 3.5a). Isotopes are denoted by their elemental symbol preceded by the *mass number*, or combined number of protons and neutrons in the atom. For example, ^{14}C (carbon-14) is an isotope of carbon with 8 neutrons (and 6 protons) in the nucleus rather than the normal 6 neutrons of ^{12}C (carbon-12). Because they differ slightly in mass, isotopes of an element differ slightly in their behavior.

Some isotopes are *radioactive* and "decay," changing their chemical identity as they shed subatomic particles and emit high-energy radiation. *Radioisotopes* decay into lighter and lighter radioisotopes, until they become *stable isotopes*, isotopes that are not radioactive. Each radioisotope decays at a rate determined by that isotope's *half-life*, the amount of time it takes for one-half the atoms to give

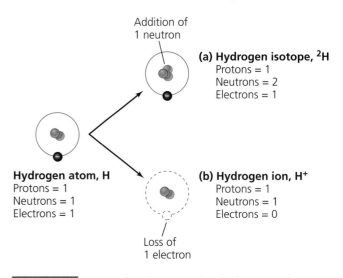

Addition of 1 neutron

(a) Hydrogen isotope, ^2H
Protons = 1
Neutrons = 2
Electrons = 1

Hydrogen atom, H
Protons = 1
Neutrons = 1
Electrons = 1

(b) Hydrogen ion, H^+
Protons = 1
Neutrons = 1
Electrons = 0

Loss of 1 electron

FIGURE 3.5 Atoms of an element such as hydrogen can become chemically altered to form isotopes and ions. Shown in **(a)** is an isotope of hydrogen, hydrogen-2 (^2H), or deuterium. This isotope contains two neutrons rather than one, and thus it contains greater mass than a typical hydrogen atom. Shown in **(b)** is the hydrogen ion, H^+. By losing its electron, it gains a positive charge.

Hypoxia and the Gulf of Mexico's "Dead Zone"

She was prone to sea-sickness, but Nancy Rabalais cared too much about the Gulf of Mexico to let that stop her. Leaning over the side of an open boat idling miles from shore, she hauled a water sample aboard—and helped launch the long effort to breathe life back into the Gulf's "dead zone."

Since that expedition in 1985, Rabalais, her colleague and husband Eugene Turner, and fellow scientists at the Louisiana Universities Marine Consortium (LUMCON) and Louisiana State University have made great progress in unraveling the mysteries of the region's hypoxia—and in getting it on the political radar screen.

Rabalais and other researchers started tracking oxygen levels at nine sites in the Gulf every month, and continued those measurements for five years. At dozens of other spots near the shore and in deep water, they took less frequent oxygen readings. For some of this work, the researchers have relied on mobile oxygen probes. Sensors, as they are lowered into the water, measure oxygen levels and send continuous readings back to a shipboard computer. Further data have come from fixed, submerged oxygen meters that continuously measure dissolved oxygen and store the data.

The team also collected hundreds of coastal and Gulf water samples, using lab tests to measure levels of nitrogen, salt, bacteria, and phytoplankton. LUMCON scientists logged hundreds of miles in their boats, regularly monitoring more than 70 sites in the Gulf. They also donned scuba gear to view firsthand the condition of shrimp, fish, and other sea life. Such a range of long-term data allowed the scientists to build a "map" of the dead zone, tracking its location and effects.

In 1991, Rabalais made that map public, earning immediate headlines. That year, her group mapped the size of the zone at more than 10,000 km^2 (about 4,000 mi^2). Bottom-dwelling shrimp were stretching out of their burrows, straining for oxygen. Many fish had fled. The bottom waters, infused with sulfur from bacterial decomposition, smelled of rotten eggs.

The group's years of continuous tracking also explained the dead zone's predictable emergence. As rivers rose each spring (and as fertilizers were applied in the Midwestern farm states), oxygen would start to disappear in the northern Gulf. The hypoxia would last through the summer or fall, until seasonal storms mixed oxygen into hypoxic areas. Over time, monitoring linked the dead zone's size to the volume of river flow and its nutrient load; the 1993 flooding of the Mississippi created a zone much larger than the year before. Conversely, a drought in 2000 brought lower river flows, lower nutrient loads, and a smaller dead zone.

The source of the problem, Rabalais said, lay back on land. The Mississippi and Atchafalaya Rivers were carrying high concentrations of nitrates and other chemicals from agricultural fertilizers and other sources. That pollution spurred algal blooms whose decomposition snuffed out oxygen in wide stretches of ocean water.

Many Midwestern farming advocates and some scientists, such as Derek Winstanley, chief of the Illinois State Water Survey, argued that the Mississippi naturally carries high loads of nitrogen from the rich prairie soil and that the Rabalais team had not ruled out upwelling in the Gulf as a source of nutrients. But sediment analyses showed that Mississippi River mud was much lower in nitrates early in the century, and Rabalais and Turner found that silica residue from phytoplankton blooms had increased in Gulf sediments between 1970 and 1989, paralleling rising nitrogen levels. In 2000, the federal integrative assessment involving dozens of scientists laid the blame for the dead zone on nutrients from fertilizers and other sources.

Then in 2004, while representatives of farmers and fishermen bickered over political fixes, Environmental Protection Agency (EPA) water quality scientist Howard Marshall suggested that

(a) Frequency of hypoxia, 1985–1999

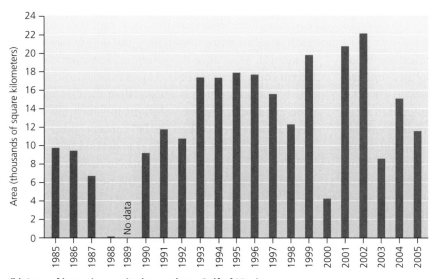

(b) Area of hypoxic zone in the northern Gulf of Mexico

Some parts of the Gulf of Mexico suffer from hypoxia more frequently than others (**a**). Areas in red have experienced oxygen concentrations below 2 ppm in more than 75% of surveys; those in orange experienced hypoxia in 50–75% of surveys; those in yellow were affected in 25–50% of surveys; and those in green were affected in less than 25% of surveys. The size of the Gulf's hypoxic zone varies (**b**) as a result of several factors. These include floods, which increase its size by bringing additional runoff (as with the Mississippi River floods of 1993), and tropical storms, which decrease its size by mixing oxygen-rich water into the dead zone (as in 2003). Between 1985 and 2005, the hypoxic zone averaged 12,700 km^2 in size and ranged up to 22,000 km^2. Data from Rabalais, N., et al. 2002. Beyond science into policy: Gulf of Mexico hypoxia and the Mississippi. *BioScience* 52: 129–142 (a); and N. Rabalais, LUMCON (b).

phosphorus pollution from industry and sewage treatment might instead be at fault. His reasoning: the ratio of nitrogen to phosphorus in the Gulf was so biased toward nitrogen that phosphorus had become the limiting factor. Phytoplankton now had more nitrogen than they could use but not enough phosphorus. Thus, Marshall suggested, we'd be best off

reducing phosphorus if we want to alleviate the dead zone.

Other scientists are giving this idea a mixed reception, and many have proposed that nitrogen and phosphorus should be managed jointly. Further suggestions have come from still more research. One is that the federally mandated 30% reduction in nitrogen in the river will not be enough. Another is that

large-scale restoration of wetlands along the river and at the river's delta would best filter pollutants before they reached the Gulf. All this research is guiding a federal plan to reduce farm runoff, clean up the Mississippi, restore wetlands, and shrink the Gulf's dead zone—and it has also led to a better understanding of hypoxic zones around the world.

off radiation and decay. Different radioisotopes have very different half-lives, ranging from fractions of a second to billions of years. The radioisotope uranium-235 (^{235}U) is our society's source of energy for commercial nuclear power ($\blacktriangleright$ pp. 319–321). It decays into a series of daughter isotopes, eventually forming lead-207 (^{207}Pb), and has a half-life of about 700 million years.

Atoms may also gain or lose electrons to become **ions**, electrically charged atoms or combinations of atoms (Figure 3.5b). Ions are denoted by their elemental symbol followed by their ionic charge. For instance, a common ion used by mussels and clams to form shells is Ca^{2+}, a calcium atom that has lost two electrons and so has a charge of positive 2.

Atoms bond to form molecules and compounds

Atoms can bond together and form **molecules**, combinations of two or more atoms. Common molecules containing only a single element include those of oxygen gas (O_2) and nitrogen gas (N_2), both of which are abundant in air. A molecule composed of atoms of two or more different elements is called a **compound**. Water is a compound; composed of two hydrogen atoms bonded to one oxygen atom, it is denoted by the chemical formula H_2O. Another compound is **carbon dioxide**, consisting of one carbon atom bonded to two oxygen atoms; its chemical formula is CO_2.

Atoms bond together because of an attraction for one another's electrons. Because the strength of this attraction varies among elements, atoms may be held together in different ways, according to whether and how they share or transfer electrons. When atoms in a molecule share electrons, they generate a *covalent bond*. For instance, two atoms of hydrogen bond to form hydrogen gas, H_2, by sharing electrons equally. Atoms in a covalent bond can also share electrons unequally, with one atom exerting a greater pull. Such is the case with water, in which oxygen attracts electrons more strongly than hydrogen, forming what are termed *polar* covalent bonds. In contrast, if the strength of attraction is unequal enough, an electron may be transferred from one atom to another. Such a transfer creates oppositely charged ions that are said to form *ionic bonds*. These associations are called *ionic compounds*, or *salts*. Table salt (NaCl) contains ionic bonds between positively charged sodium ions (Na^+), each of which donated an electron, and negatively charged chloride ions (Cl^-), each of which received an electron.

Elements, molecules, and compounds can also come together in mixtures without chemically bonding.

Homogenous mixtures of substances are called *solutions*, a term most often applied to liquids but also applicable to some gases and solids. Air in the atmosphere is a solution formed of constituents such as nitrogen, oxygen, water, carbon dioxide, **methane** (CH_4), and **ozone** (O_3). Human blood, ocean water, plant sap, and metal alloys such as brass are all solutions.

Hydrogen ions determine acidity

In any aqueous solution, a small number of water molecules dissociate, each forming a hydrogen ion (H^+) and a hydroxide ion (OH^-). The product of hydrogen and hydroxide ion concentrations is always 10^{-14}. As the concentration of one increases, the concentration of the other decreases, and the product of their concentrations remains constant. Pure water contains equal numbers of these ions, each at a concentration of 10^{-7}, and we say that this water is neutral. Most aqueous solutions, however, contain different concentrations of these two ions. Solutions in which the H^+ concentration is greater than the OH^- concentration are **acidic**, whereas solutions in which the OH^- concentration is greater than the H^+ concentration are **basic**.

The **pH** scale (Figure 3.6) was devised to quantify the acidity or basicity of solutions. It runs from 0 to 14, because these numbers reflect the negative logarithm of

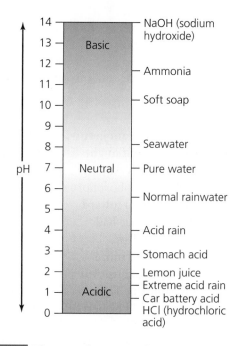

FIGURE 3.6 The pH scale measures how acidic or basic a solution is. The pH of pure water is 7, the midpoint of the scale. Acidic solutions have higher hydrogen ion concentrations and lower pH, whereas basic solutions have lower hydrogen ion concentrations and higher pH.

the hydrogen ion concentration. Thus pure water has a pH of 7, because its hydrogen ion concentration is 10^{-7}. Seawater has a greater concentration of hydroxide ions, close to 10^{-6}. This means that its hydrogen ion concentration is about 10^{-8}, and thus its pH is close to 8. Solutions with pH less than 7 are acidic, those with pH greater than 7 are basic, and those with pH of 7 are neutral. Because the pH scale is logarithmic, each step on the scale represents a tenfold difference in hydrogen ion concentration. Thus, a substance with pH of 6 contains 10 times as many hydrogen ions as a substance with pH of 7, and a substance with pH of 5 contains 100 times as many hydrogen ions as one with pH of 7. Figure 3.6 shows pH for a number of common substances. Industrial air pollution has intensified the acidity of precipitation (▶ pp. 282–285), and rain in parts of the northeastern and midwestern United States now frequently dips to pH of 4 or lower.

Matter is composed of organic and inorganic compounds

Beyond their need for water, living things also depend on organic compounds, which they create and of which they are created. **Organic compounds** consist of carbon atoms (and generally hydrogen atoms) joined by covalent bonds, and they may include other elements, such as nitrogen, oxygen, sulfur, and phosphorus. Carbon's unusual ability to build elaborate molecules has resulted in millions of different organic compounds that show various degrees of complexity. Because of the diversity of organic compounds and their importance in living organisms, chemists differentiate organic compounds from inorganic compounds, which lack carbon–carbon bonds.

One class of organic compound that is important in environmental science is hydrocarbons. **Hydrocarbons**, which consist solely of atoms of carbon and hydrogen, make up the fossil fuels we combust for so many of our energy needs. The simplest hydrocarbon is methane (CH_4), the key component of natural gas; it has one carbon atom bonded to four hydrogen atoms. Crude oil is a complex mixture of hundreds of types of hydrocarbons.

Macromolecules are building blocks of life

Just as the carbon atoms in hydrocarbons may be strung together in chains, other organic compounds sometimes combine to form long chains of repeated molecules. Some of these chains, called *polymers*, play key roles as building blocks of life. Three types of polymers are essential to life: proteins, nucleic acids, and carbohydrates. Lipids are not considered polymers but are also essential.

These four types of molecules are referred to as **macromolecules** because of their large size.

Proteins are made up of long chains of organic molecules called *amino acids*. Organisms combine up to 20 types of amino acids into long chains to build proteins. The various kinds of proteins comprise the majority of each organism's matter and serve many functions. Some help produce tissues and provide structural support for the organism. For example, animals use proteins to generate skin, hair, muscles, and tendons. Some proteins help store energy, and others transport substances. Some function as components of the immune system, defending the organism against foreign attackers. Still others act as hormones, molecules that serve as chemical messengers within an organism. Finally, proteins can serve as enzymes, molecules that catalyze, or promote, certain chemical reactions.

Nucleic acids direct the production of proteins. The two nucleic acids—**deoxyribonucleic acid (DNA)** and **ribonucleic acid (RNA)**—carry the hereditary information for organisms and are responsible for passing traits from parents to offspring. Nucleic acids are composed of series of nucleotides, each of which contains a sugar molecule, a phosphate group, and a nitrogenous base. DNA includes four types of nucleotides and can be pictured as a ladder twisted into a spiral, giving the molecule a shape called a double helix (Figure 3.7).

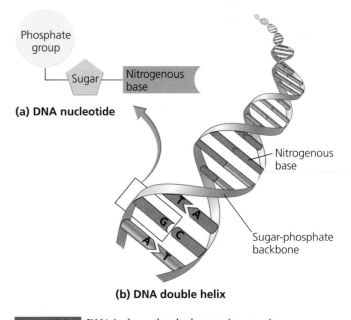

(a) DNA nucleotide

(b) DNA double helix

FIGURE 3.7 DNA is the molecule that carries genetic information from parent to offspring across the generations. The information is coded in the sequence of nucleotides (a), small molecules that pair together like rungs of a ladder that twist into the shape of a double helix (b).

Carbohydrates include simple sugars that are three to seven carbon atoms long. Glucose ($C_6H_{12}O_6$) fuels living cells and serves as a building block for complex carbohydrates, such as starch. Plants use starch to store energy, and animals eat plants to acquire starch. Plants and animals also use complex carbohydrates to build structure. Insects and crustaceans form hard shells from the carbohydrate chitin. Cellulose, the most abundant organic compound on Earth, is a complex carbohydrate found in the cell walls of leaves, bark, stems, and roots.

Lipids are a chemically diverse group of compounds, classified together because they do not dissolve in water. Lipids include fats and oils (for energy storage), phospholipids (for membranes), waxes (for structure), and steroids (for hormone production).

Organisms use cells to compartmentalize macromolecules

All living things are composed of **cells**, the most basic unit of organismal organization. Organisms range in complexity from single-celled bacteria to plants and animals that contain millions of cells. Cells vary greatly in size, shape, and function. Biologists classify organisms into two groups based on the structure of their cells. The cells of *eukaryotes* (plants, animals, fungi, and protists) contain a membrane-enclosed nucleus and various organelles that perform specific functions. *Prokaryotes* (bacteria and archaea) are generally single-celled, and their cells lack organelles and a nucleus.

Energy Fundamentals

Creating and maintaining organized complexity, whether of a cell, an organism, or an ecological system, requires energy. Energy is needed to organize matter into complex forms such as polymers, to build and maintain cellular structure, to power interactions among species, and to drive the geological forces that shape our planet. Energy is somehow involved in nearly every biological, chemical, and physical event.

But what, exactly, is energy? An intangible phenomenon, **energy** is that which can change the position, physical composition, or temperature of matter. Scientists differentiate between two types of energy: **potential energy**, energy of position; and **kinetic energy**, energy of motion. Consider river water held behind a dam. By preventing water from moving downstream, the dam causes the water to accumulate potential energy. When the dam gates are opened, the potential energy is converted to kinetic energy, in the form of water's motion as it rushes downstream.

Such energy transfers take place at the atomic level every time a chemical bond is broken or formed. **Chemical energy** is potential energy held in the bonds between atoms. Bonds differ in their amounts of chemical energy, depending on the atoms they hold together. Converting a molecule with high-energy bonds (such as the carbon–carbon bonds of petroleum products) into molecules with lower-energy bonds (such as the bonds in water or carbon dioxide) releases energy by changing potential energy into kinetic energy and produces motion, action, or heat. Just as our automobile engines split the hydrocarbons of gasoline to release chemical energy and generate movement, our bodies split glucose molecules from our food for the same purpose.

Energy is always conserved, but it changes in quality

Although energy can change from one form to another, it cannot be created or lost. The total energy in the universe remains constant and thus is said to be conserved.

FIGURE 3.8 The burning of firewood demonstrates energy transfer leading from a more-ordered to a less-ordered state. This increase in entropy reflects the second law of thermodynamics.

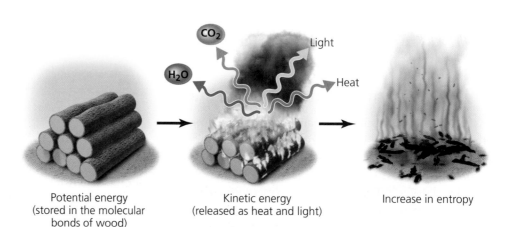

Potential energy
(stored in the molecular bonds of wood)

Kinetic energy
(released as heat and light)

Increase in entropy

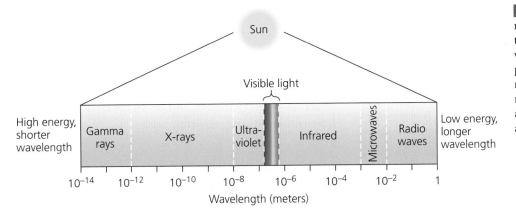

FIGURE 3.9 The sun emits radiation from many portions of the electromagnetic spectrum, and visible light makes up only a small proportion of this energy. Some radiation that reaches our planet is reflected back; some is absorbed by air, land, and water; and a small amount powers photosynthesis.

Scientists have dubbed this principle the **first law of thermodynamics**. The potential energy of the water behind a dam will equal the kinetic energy of its eventual movement down the riverbed. Similarly, burning converts the potential energy in a log of firewood to an equal amount of energy produced as heat and light.

Although the overall amount of energy is conserved in any process of energy transfer, the **second law of thermodynamics** states that the nature of energy will change from a more-ordered state to a less-ordered state, if no force counteracts this tendency. That is, systems tend to move toward increasing disorder, or *entropy*. For instance, after death every organism undergoes decomposition and loses its structure. A log of firewood—the highly organized and structurally complex product of many years of slow tree growth—transforms in the campfire to a residue of carbon ash, smoke, and gases such as carbon dioxide and water vapor, as well as the light and heat of the flame (Figure 3.8). With the help of oxygen, the complex biological polymers making up the wood are converted into a disorganized assortment of rudimentary molecules and heat and light energy.

Light energy from the sun powers most living systems

The energy that powers Earth's ecological systems comes primarily from the sun. The sun releases radiation from large portions of the electromagnetic spectrum, although our atmosphere filters much of this out and we see only some of this radiation as visible light (Figure 3.9).

The sun's radiation is used directly by some organisms to produce their own food. Such organisms, called **autotrophs** or **producers**, include green plants, algae, and bacteria called cyanobacteria. Autotrophs turn light energy from the sun into chemical energy through the process called **photosynthesis** (Figure 3.10). In photosynthesis, sunlight powers a series of chemical reactions that convert carbon dioxide and

water into sugars, transforming low-quality energy from the sun into high-quality energy the organism can use. It is an example of moving toward a state of lower entropy, and as such it requires a substantial input of outside energy.

Photosynthesis occurs within cell organelles called *chloroplasts*, where the light-absorbing pigment *chlorophyll*

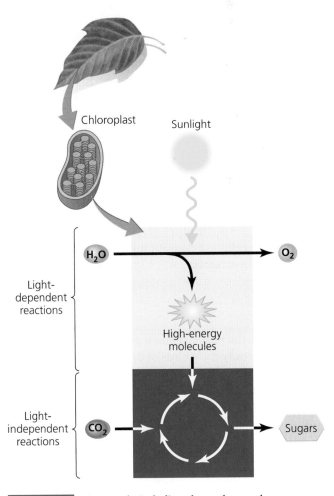

FIGURE 3.10 Autotrophs including plants, algae, and cyanobacteria use sunlight to convert carbon dioxide and water into sugars and oxygen in photosynthesis. Autotrophs provide themselves and the many heterotrophs that eat them with energy for life.

(which is what makes plants green) uses solar energy to initiate a series of chemical reactions called *light-dependent reactions*. During these reactions, water molecules are split, and they react to form hydrogen ions (H^+) and molecular oxygen (O_2), thus creating the oxygen that we breathe. The light-dependent reactions also produce small, high-energy molecules that are used to fuel a set of *light-independent reactions*. In these reactions, carbon atoms from carbon dioxide are linked together to manufacture sugars. Photosynthesis is a complex process, but the overall reaction can be summarized in the following equation:

$$6CO_2 + 6H_2O \xrightarrow{\text{+ the sun's energy}} \underset{\text{(sugar)}}{C_6H_{12}O_6} + 6O_2$$

Thus in photosynthesis, green plants draw up water from the ground through their roots, absorb carbon dioxide from the air through their leaves, and harness sunlight. With these ingredients, they create sugars for their growth and maintenance and release as a by-product the oxygen that we, and all other animals, breathe.

Cellular respiration releases chemical energy

The chemical energy created by photosynthesis can later be used by organisms in the process known as **cellular respiration**. To release the chemical energy of glucose, cells use oxygen to convert glucose back into its original starting materials, water and carbon dioxide. The energy released during this process is used to form chemical bonds or to perform other tasks within cells. The net equation for cellular respiration is the exact opposite of that for photosynthesis:

$$\underset{\text{(sugar)}}{C_6H_{12}O_6} + 6O_2 \longrightarrow 6CO_2 + 6H_2O + \text{energy}$$

However, the energy gained per glucose molecule in respiration is only two-thirds of the energy input per glucose molecule in photosynthesis—a prime example of the second law of thermodynamics in action. The extraction of energy from glucose through respiration occurs in the autotrophs that created the glucose and also in the animals that obtain glucose by consuming autotrophs. Organisms that consume autotrophs are called **heterotrophs**, or **consumers**, and include most animals, as well as the fungi and microbes that decompose organic matter. In most ecological systems, plants, algae, or cyanobacteria form the base of a food chain through which energy passes to heterotrophs (▶ pp. 101–102).

How Environmental Systems Work

Let's now apply our knowledge of chemistry and energy to see how energy, matter, and nutrients cycle through the living and nonliving environment.

Ecosystems are systems of interacting living and nonliving entities

An **ecosystem** consists of all organisms and nonliving entities that occur and interact in a particular area at the same time. Animals, plants, air, water, soil, nutrients—all these and more help comprise ecosystems. An ecosystem can be as small as an ephemeral puddle of water where brine shrimp and tadpoles feed on algae and detritus with mad abandon as the pool dries up. Or an ecosystem might be as large as a lake or a forest. For some purposes, scientists even view the entire planet as a single all-encompassing ecosystem. The term is most often used, however, to refer to systems of moderate geographic extent with somewhat discrete boundaries. For example, the salt marshes that line the outer delta of the Mississippi River where its waters mix with those of the Gulf of Mexico may be classified as an ecosystem.

Ecosystems that physically abut one another may interact extensively. For instance, coastal dunes, the ocean, and the lagoon or salt marsh between them all interact, as do forests and prairie where they converge. Areas where ecosystems meet may consist of transitional zones called *ecotones*, in which elements of each ecosystem mix. Because of this mixing, ecologists sometimes find it useful to view these systems at a larger landscape scale that focuses on geographic areas that include multiple ecosystems. Such a broad-scale approach, often called *landscape ecology*, is important in studying birds that migrate between continents, for example, or fish, such as salmon, that move between marine and freshwater ecosystems.

Weighing the **Issues:**
Ecosystems Where You Live

Think about the area where you live. How would you describe this area's ecosystems? How do these systems interact with one another? If one ecosystem were greatly disturbed (say, if a wetland or forest were replaced by a shopping mall), what impacts might that have on nearby natural systems?

Energy is converted to biomass

Energy flows in one direction through ecosystems; most arrives as radiation from the sun, powers the system, and exits in the form of heat. Matter, in contrast, is generally recycled within ecosystems. We will see in Chapter 5 (► pp. 101–103) how energy and matter are passed among organisms through food web relationships, from producers to primary consumers to secondary consumers. Matter is recycled because when organisms die and decay their nutrients remain in the system. In contrast, most energy that organisms gain is lost during their lifetimes through respiration.

As autotrophs convert solar energy to the energy of chemical bonds in sugars during photosynthesis, they perform *primary production.* Specifically, the assimilation of energy by autotrophs is termed *gross primary production.* Autotrophs use a portion of this production to power their own metabolism by respiration. The energy that remains after respiration is used to generate biomass, a process that ecologists call **net primary production.** Thus, net primary production equals gross primary production minus respiration. Net primary production can be measured by the energy or the organic matter stored by plants or other autotrophs after they have metabolized enough for their own maintenance.

Ecosystems vary in the rate at which autotrophs convert energy to biomass. The rate at which production occurs is termed *productivity*, and ecosystems whose producers convert solar energy to biomass rapidly are said to have high **net primary productivity.** Freshwater wetlands, tropical forests, and coral reefs and algal beds tend to have the highest net primary productivities, whereas deserts, tundra, and open ocean tend to have the lowest (Figure 3.11a). Variation among ecosystem types in net primary productivity results in geographic patterns of variation across the globe (Figure 3.11b). In terrestrial ecosystems, net primary productivity tends to increase with temperature and precipitation. In aquatic ecosystems, net primary productivity tends to rise with light and the availability of nutrients.

Nutrients can limit ecosystem productivity

Nutrients are elements and compounds that organisms consume and require for survival. Organisms need several dozen naturally occurring chemical elements to survive; among these are nitrogen, carbon, and phosphorus. Nutrients stimulate production by autotrophs, and a lack of nutrients can limit production. The availability of nitrogen or phosphorus frequently is a limiting factor (► pp. 88–89) for plant or algal growth. When these nutrients are added to a system, producers show the greatest response to whichever nutrient has been in shortest supply. Research has shown that phosphorus tends to be limiting in freshwater systems, and nitrogen in marine systems. Thus the Gulf of Mexico's hypoxic zone is thought to result primarily from excess nitrogen, whereas ponds and lakes in the Mississippi River Valley tend to suffer eutrophication when they receive too much phosphorus.

Because nutrients run off from land worldwide, primary productivity in the oceans tends to be greatest in nearshore waters and lowest in open ocean areas far from land (see Figure 3.11b). As a result of increased nutrient pollution from farms, cities, and industry, the number of known dead zones is increasing globally, with about 150 documented so far. Most are located off the coasts of Europe and the eastern United States. Increasingly, scientists are proposing innovative and economically acceptable ways to reduce nutrient runoff.

Nutrients circulate through ecosystems in biogeochemical cycles

Nutrients move through ecosystems in **nutrient cycles** or **biogeochemical cycles.** A carbon atom in your fingernail today might have helped comprise the muscle of a cow a year earlier, may have resided in a blade of grass a month before that, and may have been part of a dinosaur's tooth 100 million years ago. After we die, the nutrients in our bodies will spread widely through the environment, eventually being incorporated by an untold number of organisms far into the future.

Nutrients move from one *pool,* or *reservoir,* to another, remaining for varying amounts of time *(residence time)* in each. The dinosaur, the grass, the cow, and you are each reservoirs for carbon atoms. The movement of nutrients among pools is termed *flux,* and the rates of flux between any given pair of pools can change over time. Human activity has influenced certain flux rates.

The carbon cycle circulates a vital organic nutrient

The **carbon cycle** describes the routes that carbon atoms take through the environment (Figure 3.12). Producers, including terrestrial and aquatic plants, algae, and cyanobacteria, pull carbon dioxide out of the atmosphere and out of surface water to use in photosynthesis. Autotrophs and the heterotrophs that consume them use carbohydrates produced in photosynthesis for structural growth and also to fuel their respiration, releasing carbon

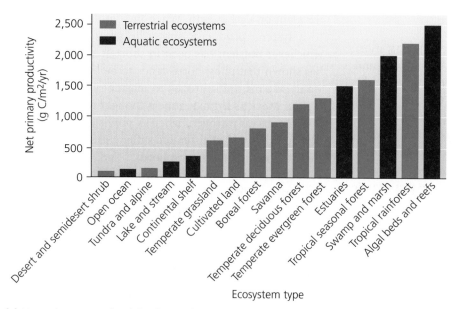

(a) Net primary productivity for major ecosystem types

FIGURE 3.11 Freshwater wetlands, tropical forests, coral reefs, and algal beds tend to show high net primary productivities (**a**), whereas deserts, tundra, and the open ocean show low values. A world map of net primary production created from satellite data (**b**) shows that on land, net primary production varies geographically with temperature and precipitation. In the world's oceans, net primary production is highest around the margins of continents, where nutrients (of both natural and human origin) run off from land. Data from Whittaker, R. H. 1975. *Communities and ecosystems*, 2nd ed. New York: MacMillan (a); and Field, C. B., et al. 1998. Primary production of the biosphere: Integrating terrestrial and oceanic components. *Science* 281: 237–240 (b).

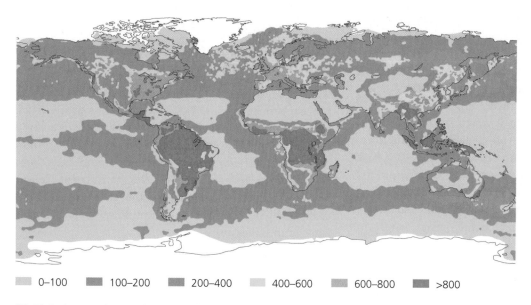

(b) Global map of net primary productivity

back into the atmosphere and oceans as CO_2. Because CO_2 is a greenhouse gas (▶ pp. 288–289), much research on global climate change has tried to measure the amount of CO_2 that plants take in.

As organisms die, their remains may settle in sediments in ocean basins or freshwater wetlands. As layers of sediment accumulate, older layers are buried more deeply, experiencing high pressure over long periods of time. These conditions can convert soft tissues into fossil fuels—coal, oil, and natural gas—and shells and skeletons into sedimentary rock (▶ pp. 70–71), such as limestone. Sedimentary rock comprises the largest single reservoir in the carbon cycle. Although any given carbon

atom spends a relatively short time in the atmosphere, carbon trapped in sedimentary rock may reside there for hundreds of millions of years. Carbon trapped in sediments and fossil fuel deposits may eventually be released into the oceans or atmosphere by geological processes such as uplift, erosion, and volcanic eruptions. It also reenters the atmosphere when we extract and burn fossil fuels.

The world's oceans are the second-largest reservoir in the carbon cycle. They absorb carbon-containing compounds from the atmosphere, terrestrial runoff, undersea volcanoes, and the detritus of marine organisms. Some carbon absorbed by ocean water combines with calcium

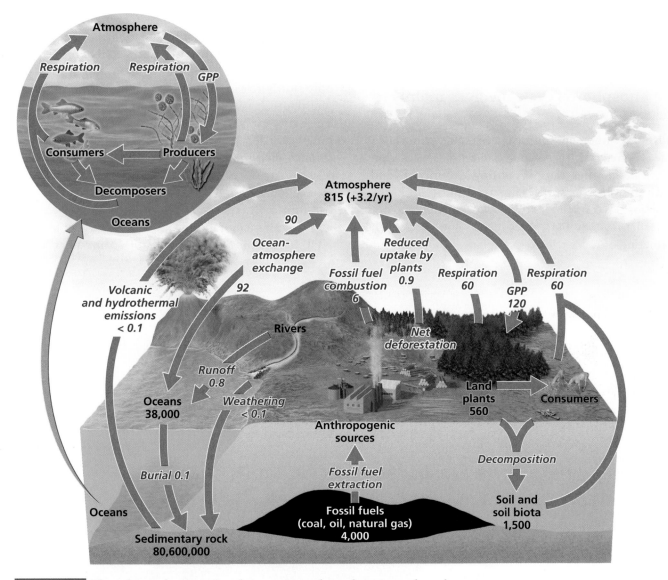

FIGURE 3.12 The carbon cycle summarizes the many routes that carbon atoms take as they move through the environment. Gray arrows represent fluxes among reservoirs, or pools, for carbon. In the carbon cycle, plants use carbon dioxide from the atmosphere for photosynthesis (gross primary production, or "GPP" in the figure). Carbon dioxide is returned to the atmosphere through respiration by plants, their consumers, and decomposers. The oceans sequester carbon in their water and in deep sediments. The vast majority of the planet's carbon is stored in sedimentary rock. In the figure, pool names are printed in black type, and numbers in black type represent pool sizes, expressed in units of 10^{15} g C. Processes are printed in italic red type; they give rise to fluxes, also printed in italic red type and expressed in units of 10^{15} g C per year. Data from Schlesinger, W. H. 1997. *Biogeochemistry: An analysis of global change.* 2nd ed. London: Academic Press.

ions to form calcium carbonate ($CaCO_3$), an ingredient in the skeletons and shells of microscopic marine organisms. As these organisms die, their shells sink to the ocean floor and begin to form sedimentary rock.

By mining fossil fuel deposits, humans are essentially removing carbon from an underground reservoir with a residence time of millions of years. By combusting fossil fuels in our automobiles, homes, and industries, we release

carbon dioxide and greatly increase the flux of carbon from the ground to the air. Moreover, cutting down forests and burning fields removes carbon from the pool of vegetation and releases it to the air. And if less vegetation is left on the surface, there are fewer plants to draw CO_2 back out of the atmosphere. As a result, scientists estimate that today's atmospheric carbon dioxide reservoir is the largest that Earth has experienced in the past 420,000 years, and perhaps in the past 20 million years. The ongoing flux of carbon out of the fossil-fuel reservoir and into the atmosphere is a driving force behind global climate change (Chapter 12).

The phosphorus cycle involves mainly rocks and ocean

Unlike the carbon and nitrogen cycles, the **phosphorus cycle** (Figure 3.13) has no appreciable atmospheric component besides the transport of tiny amounts of wind-blown dust and sea spray. The vast majority of Earth's phosphorus is contained within rocks and is released only by weathering (▸ p. 144), an extremely slow process. The weathering of rocks releases phosphate (PO_4^{3-}) ions into

water. Phosphates dissolved in lakes or in the oceans precipitate into solid form and settle to the bottom in sediments, which are eventually compressed into rock.

Because most phosphorus is bound up in rock and only slowly released, environmental concentrations of phosphorus available to organisms tend to be very low. This rarity explains why phosphorus is frequently a limiting factor for growth. Plants can take up phosphorus through their roots only when phosphate is dissolved in water. Primary consumers acquire phosphorus from water and plants and pass it on to secondary and tertiary consumers. In organisms, phosphorus is a key component of cell membranes, DNA, RNA, and a number of vital biochemical compounds. Decomposers break

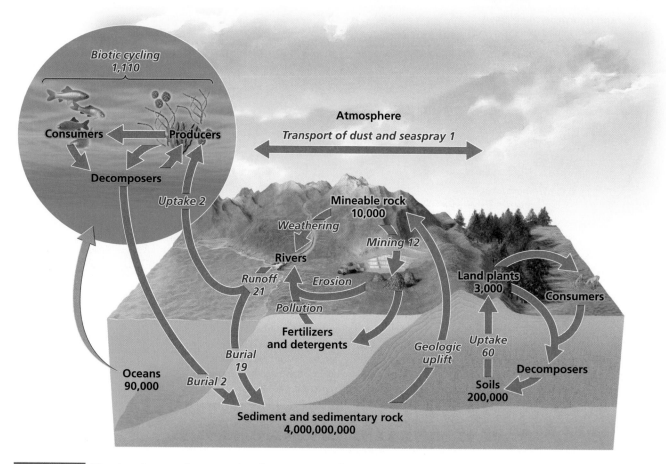

FIGURE 3.13 The phosphorus cycle summarizes the many routes that phosphorus atoms take as they move through the environment. Gray arrows represent fluxes among reservoirs, or pools, for phosphorus. Most phosphorus resides underground in rock and sediment, but the phosphorus cycle moves this element through the soil, the oceans, and freshwater and terrestrial ecosystems. Rocks containing phosphorus are uplifted geologically and weathered away in this slow process. Small amounts of phosphorus cycle through food webs, where this nutrient is often a limiting factor for plant growth. In the figure, pool names are printed in black type, and numbers in black type represent pool sizes, expressed in units of 10^{12} g P. Processes, which are printed in italic red type, give rise to fluxes, also printed in italic red type and expressed in units of 10^{12} g P per year. Data from Schlesinger, W. H. 1997. *Biogeochemistry: An analysis of global change.* 2nd ed. London: Academic Press.

down phosphorus-rich organisms and their wastes and, in so doing, return phosphorus to the soil.

Humans influence the phosphorus cycle in several ways. We mine rocks to extract phosphorus for the inorganic fertilizers we use on crops and lawns. Our wastewater discharge tends to be rich in phosphates, and phosphates that run off into waterways can boost algal growth and cause eutrophication. Phosphates are also present in detergents, so one way each of us can reduce phosphorus input into the environment is to purchase phosphate-free detergents.

The nitrogen cycle involves specialized bacteria

Nitrogen makes up 78% of our atmosphere by mass and is the sixth most abundant element on Earth. It is an essential ingredient in the proteins, DNA, and RNA that build our bodies and, like phosphorus, is an essential nutrient for plant growth. Thus the **nitrogen cycle** (Figure 3.14) is of vital importance to us and to all other organisms. Despite its abundance in the air, nitrogen gas (N_2) is

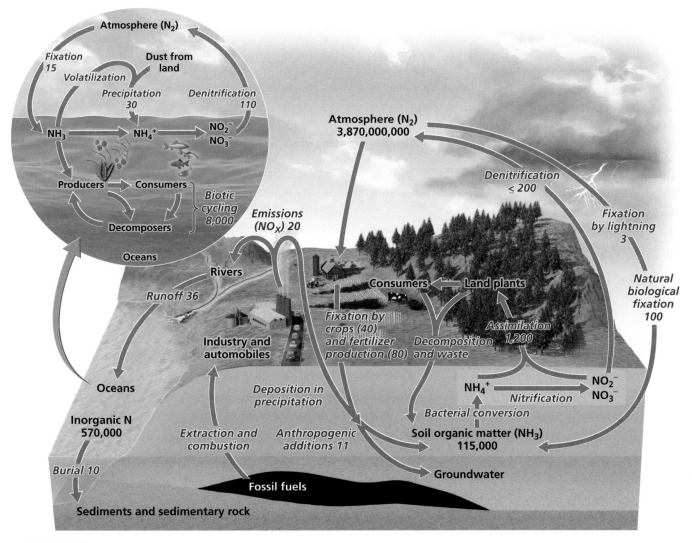

FIGURE 3.14 The nitrogen cycle summarizes the many routes that nitrogen atoms take as they move through the environment. Gray arrows represent fluxes among reservoirs, or pools, for nitrogen. In the nitrogen cycle, specialized bacteria play key roles in "fixing" atmospheric nitrogen and converting it to chemical forms that plants can use, while other types of bacteria convert nitrogen compounds back to the atmospheric gas N_2. In the oceans, inorganic nitrogen is buried in sediments while nitrogen compounds are cycled through food webs as they are on land. In the figure, pool names are printed in black type, and numbers in black type represent pool sizes, expressed in units of 10^{12} g N. Processes, which are printed in italic red type, give rise to fluxes, also printed in italic red type and expressed in units of 10^{12} g N per year. Data from Schlesinger, W. H. 1997. *Biogeochemistry: An analysis of global change.* 2nd ed. London: Academic Press.

chemically inert and cannot cycle out of the atmosphere and into living organisms without assistance from lightning, highly specialized bacteria, or human intervention. However, once nitrogen undergoes the right kind of chemical change, it becomes biologically active and available to the organisms that need it, and it can act as a potent fertilizer.

To become biologically available, inert nitrogen gas (N_2) must be "fixed," or combined with hydrogen in nature to form ammonia (NH_3), whose water-soluble ions of ammonium (NH_4^+) can be taken up by plants. **Nitrogen fixation** can be accomplished in two ways: by the intense energy of lightning strikes, or when air in the top layer of soil comes in contact with particular types of **nitrogen-fixing bacteria**. These bacteria live in a mutualistic relationship (▶ p. 101) with many types of plants, including soybeans and other legumes, providing them nutrients by converting nitrogen to a usable form. Other types of specialized bacteria then perform a process known as **nitrification**, converting ammonium ions first into nitrite ions (NO_2^-), then into nitrate ions (NO_3^-). Plants can take up these ions, which also become available after atmospheric deposition on soils or in water or after application of nitrate-based fertilizer.

Animals obtain the nitrogen they need by consuming plants or other animals. Decomposers obtain nitrogen from dead and decaying plant and animal matter and from the urine and feces of animals. Once decomposers process the nitrogen-rich compounds they take in, they release ammonium ions, making these available to nitrifying bacteria to convert again to nitrates and nitrites. The next step in the nitrogen cycle occurs when **denitrifying bacteria** convert nitrates in soil or water to gaseous nitrogen. Denitrification thereby completes the cycle by releasing nitrogen back into the atmosphere as a gas.

Humans have greatly influenced the nitrogen cycle

The impacts of excess nitrogen from agriculture and other human activities in the Mississippi River watershed have become painfully evident to shrimpers and scientists with an interest in the Gulf of Mexico (see "The Science behind the Story" on ▶ pp. 54–55). But hypoxia in the Gulf and other coastal locations around the world is hardly the only problem resulting from human manipulation of the nitrogen cycle.

Historically, nitrogen fixation was a *bottleneck*, a step that limited the flux of nitrogen out of the atmosphere. But the development of synthetic nitrogen-based fertilizers has allowed us to fix nitrogen on a massive scale and accelerate its flux into other reservoirs. Today, our species is fixing at least as much nitrogen artificially as is being fixed naturally. We have effectively doubled the natural rate of nitrogen fixation on Earth (Figure 3.15).

Moreover, when we burn forests and fields, we force nitrogen out of soils and vegetation and into the atmosphere. When we burn fossil fuels, we increase the rate at

FIGURE 3.15 In the past half century, human inputs of nitrogen into the environment have greatly increased, such that today fully half of the nitrogen entering the environment is of human origin. Agricultural fertilizer has for the past 35 years been the leading source of all nitrogen inputs, natural and artificial. Data from National Science and Technology Council: Committee on Environment and Natural Resources. May 2000. Hypoxia: An integrated assessment in the northern Gulf of Mexico.

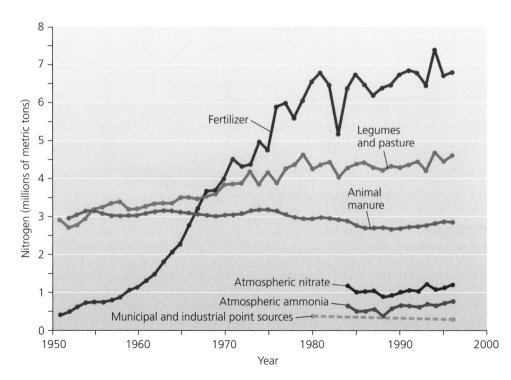

which nitric oxide (NO) enters the atmosphere and reacts to form nitrogen dioxide (NO_2). This compound is a precursor to nitric acid (HNO_3), a key component of acid precipitation (▶ pp. 282–285). We introduce another nitrogen-containing gas, nitrous oxide (N_2O), by allowing anaerobic bacteria to break down the animal waste produced in agricultural feedlots (▶ p. 162). We have also accelerated the introduction of nitrogen-rich compounds into terrestrial and aquatic systems by destroying wetlands and cultivating legume crops that host nitrogen-fixing bacteria in their roots.

In 1997, a team of scientists led by Peter Vitousek of Stanford University summarized the changes humans have caused in the global nitrogen cycle:

▶ Doubled the rate that fixed nitrogen enters terrestrial ecosystems (and the rate is still increasing)
▶ Increased atmospheric concentrations of the greenhouse gas N_2O and of other oxides of nitrogen that produce smog
▶ Depleted essential nutrients, such as calcium and potassium, from soils, because fertilizer helps flush them out
▶ Acidified surface water and soils
▶ Greatly increased transfer of nitrogen from rivers to oceans
▶ Encouraged plant growth, causing more carbon to be stored within terrestrial ecosystems
▶ Reduced biological diversity, especially plants adapted to low nitrogen concentrations
▶ Changed the composition and function of estuaries and coastal ecosystems
▶ Harmed many coastal marine fisheries

Weighing the Issues:
Nitrogen Pollution and Its Financial Impacts

Most nitrate that enters the Gulf of Mexico originates from farms and other sources in the upper Midwest, yet many of its negative impacts are borne by downstream users, such as Gulf Coast fishermen. Who do you believe should be responsible for addressing this problem? Should environmental policies on this issue be developed and enforced by state governments, the federal government, both, or neither? Explain the reasons for your answer.

In 1998, the U.S. Congress passed the Harmful Algal Bloom and Hypoxia Research and Control Act. This law called for an "integrated assessment" of hypoxia in the northern Gulf to address the extent, nature, and causes of the dead zone, as well as its ecological and economic impacts. The assessment report in 2000 proposed that the federal government work with Gulf Coast and Midwestern communities to:

▶ Reduce nitrogen fertilizer use on Midwestern farms.
▶ Change the timing of fertilizer application to minimize rainy-season runoff.
▶ Use alternative crops.
▶ Manage nitrogen-rich livestock manure.
▶ Restore nitrogen-absorbing wetlands along the Mississippi River.
▶ Use artificial wetlands to filter farm runoff.
▶ Improve technologies in sewage treatment plants.
▶ Restore frequently flooded lands to reduce runoff.
▶ Evaluate how these approaches work.

As scientists, farmers, and policymakers searched for innovative solutions to alleviate pollution while not hurting agriculture, Congress in 2003 reauthorized the Harmful Algal Bloom and Hypoxia Research and Control Act for five more years, promising to fund further research and the development of solutions. One proposal offers farmers insurance and economic incentives for not using excess fertilizer. Another program tests new farming strategies that decrease fertilizer use while maintaining crop yields. A third approach involves planting cover crops in the off-season to reduce runoff from bare fields. Yet another encourages farmers to maintain artificial wetlands on their lands to serve as natural buffers against pollution. Many Midwestern farmers are taking part in these strategies.

The hydrologic cycle influences all other cycles

Water is so integral to life that we frequently take it for granted. The essential medium for all manner of biochemical reactions, water plays key roles in nearly every environmental system, including each nutrient cycle. Water carries nutrients and sediments from the continents to the oceans via rivers, streams, and surface runoff, and it distributes sediments onward in ocean currents. Increasingly, water also distributes artificial pollutants. The water cycle, or **hydrologic cycle** (Figure 3.16), summarizes how water—in liquid, gaseous, and solid forms—flows through our environment.

The oceans are the main reservoir in the hydrologic cycle, holding 97% of all water on Earth. The freshwater we depend on for our survival accounts for less than 3%, and two-thirds of this small amount is tied up in glaciers, snowfields, and ice caps (▶ pp. 240–241). Thus, considerably less than 1% of the planet's water is in a form that we

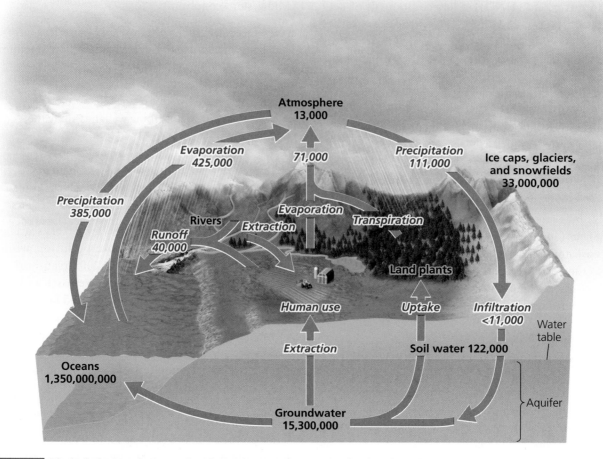

FIGURE 3.16 The hydrologic cycle summarizes the many routes that water molecules take as they move through the environment. Gray arrows represent fluxes among reservoirs, or pools, for water. The hydrologic cycle is a system unto itself but also plays key roles in other biogeochemical cycles. Oceans hold 97% of our planet's water, but most freshwater resides in groundwater and icecaps. Water vapor in the atmosphere condenses and falls to the surface as precipitation, then evaporates from land and transpires from plants to return to the atmosphere. Water flows downhill into rivers, eventually reaching the oceans. In the figure, pool names are printed in black type, and numbers in black type represent pool sizes, expressed in units of km³. Processes, which are printed in italic red type, give rise to fluxes, also printed in italic red type and expressed in units of km³ per year. Data from Schlesinger, W. H. 1997. *Biogeochemistry: An analysis of global change*, 2nd ed. London: Academic Press.

can readily use—groundwater, surface freshwater, and rain from atmospheric water vapor.

Water moves from oceans, lakes, ponds, rivers, and moist soil into the atmosphere by **evaporation**, the conversion of a liquid to gaseous form. Water also enters the atmosphere by **transpiration**, the release of water vapor by plants through their leaves. Transpiration and evaporation act as natural processes of distillation, effectively creating pure water by filtering out minerals carried in solution. Water returns from the atmosphere to Earth's surface as **precipitation** when water vapor condenses and falls as rain or snow. Precipitation may be taken up by

plants and used by animals, but much of it flows as **runoff** into streams, rivers, lakes, ponds, and oceans.

Some water soaks down through soil and rock to recharge underground reservoirs known as **aquifers**. Aquifers are spongelike regions of rock and soil that hold **groundwater**, water found underground beneath layers of soil. The upper limit of groundwater held in an aquifer is referred to as the **water table**. Aquifers may hold groundwater for long periods of time, so the water may be quite ancient.

Human activity has affected every aspect of the water cycle. By damming rivers to create reservoirs, we have

The Dead Zone

Scientific research has indicated that nitrogen fertilizers from the Mississippi River watershed are contributing to hypoxia in the Gulf of Mexico. **Do you agree this is occurring? Why or why not? If so, what steps should be taken to solve the problem?**

Evidence Not Conclusive

Scientific evidence that nitrogen (N) fertilizer is polluting the northern Gulf of Mexico is not conclusive. Hypoxia in the Gulf has been recognized since 1935, long before fertilizer use became widespread in the 1960s. Nitrogen fertilizer use in the Mississippi River basin has remained relatively stable in the last two decades, but the size of the hypoxic zone has fluctuated markedly, especially since 1988.

According to the U.S. Geological Survey, the annual discharge of N from the Mississippi River has tripled in the last 30 years, with most of the increase occurring from 1970 to 1983. However, since 1980, river N discharge has changed very little, whereas N fertilizer use has grown by almost 10%. From 1980 to 1999, N fertilizer sales in the Mississippi River basin explained 8% of the variation in river nitrate-N discharge, whereas the Mississippi River's average annual flow explained nearly 80% of its variation.

Recent data suggest that the molar ratio of inorganic N to inorganic phosphorus (P) is the principal indicator of phytoplankton blooms that cause hypoxia, and that the N:P ratio in freshwater entering the Gulf is high enough that any increase in inorganic P loading will increase the hypoxic zone. The blame is shifting from N to P fertilizer. However, P fertilizer sales in the Mississippi River Basin have declined by 17% since 1980.

Can hypoxia be blamed on sales of N or P fertilizer? Numerous nutrient sources contribute to Gulf loading. Atmospheric deposition, decomposition of crop residue and soil organic matter, legumes, animal manure, municipal sewage sludge and effluent, and composted household wastes all contribute nutrients to the Gulf.

Hypoxia results from a complex interaction of chemical, biological, and physical factors. Fertilizer is a potential pollutant if used improperly, but used correctly, it increases food production and helps protect the environment.

Terry L. Roberts is vice president of the Potash & Phosphate Institute (PPI) and vice president of the Foundation for Agronomic Research (FAR), located in Norcross, Georgia. Dr. Roberts is a Certified Crop Adviser and a Fellow of the American Society of Agronomy.

Act Now to Save These Resources

The springtime area of low-oxygen (anoxic) water in the Gulf of Mexico, known as the dead zone, is driven by a massive influx of nutrients into a system no longer able to process them. Eutrophication begins when nutrients from farmlands in the floodplain states wash to the sea. These nutrients (nitrogen fertilizers) now present in the water lead to plankton blooms, which in turn reduce dissolved oxygen in the water and eventually kill fish.

Taking a system view is slightly more complicated, but understanding the system is important for the most effective long-term management. Before people built levees all along the delta, the Mississippi River flooded each spring, and the waters of the river covered the extensive wetlands. This important renewal process deposited sediment on the wetlands to build up the soil base while the plants of the wetlands made use of the nutrient pollutants in the water. The result was cleaner water, richer wetlands, and a sustained environment. Levees now prevent the flooding, the dead zone emerges, and the wetlands are lost as they sink below sea level. Rises in sea level speed the loss.

Loss of wetlands is serious. The wetlands are the base of the fisheries of the Gulf of Mexico, and their loss is irreversible. Saving the wetlands and reversing the dead zone requires a twofold approach. First, reduce the amount of fertilizer so that plants use it more efficiently and so that less enters streams. This has the added benefit of saving money and reducing energy consumption (making fertilizers is energy intensive). Second, reinstate the flooding of the wetlands.

Should we wait to act? No. We know enough now to design strategies that can sustain these resources, and new information is unlikely to change what we know. The precautionary principle, which environmental managers use, says that even if information is imperfect, it is important to act before the resource is lost entirely and while any possible cost of error is small and manageable. We need to act now to save these resources.

Paul Templet is a professor at the Institute for Environmental Studies at Louisiana State University. He organized the first Earth Day at LSU in 1970 and served as the secretary of the Louisiana Department of Environmental Quality from 1988 to 1992.

Explore this issue further by accessing **Viewpoints** at www.aw-bc.com/withgott.

increased evaporation and, in some cases, infiltration of surface water into aquifers. By altering Earth's surface and vegetation, we have increased surface runoff and erosion. By spreading water on agricultural fields, we have depleted rivers, lakes, and streams, and have increased evaporation. By exploiting groundwater for drinking, irrigation, and industrial use, we have begun to deplete groundwater resources. By removing forests and other vegetation, we have reduced transpiration and have lowered water tables. And by emitting into the atmosphere pollutants that dissolve in water droplets, we have changed the chemical nature of precipitation, in effect sabotaging the natural distillation process that evaporation and transpiration provide. We will revisit the water cycle, water resources, and human impacts in more detail in Chapter 11.

Even rocks go through a cycle

We tend to think of rock as pretty solid stuff. Yet in the long run, over geological time, rocks do change. Rocks and the minerals that comprise them are heated, melted, cooled, broken down, and reassembled in a very slow process called the **rock cycle** (Figure 3.17). The type of

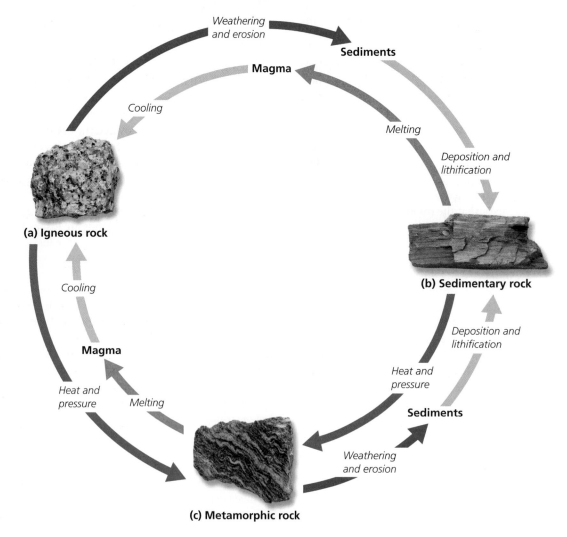

FIGURE 3.17 The rock cycle is a slow cycle that shapes Earth's crust and affects Earth's nutrient cycles. Igneous rock (**a**) is formed when rock melts to form magma, and the magma then cools. Sedimentary rock (**b**) is formed when rock is weathered and eroded, and the resulting sediments are compressed to form new rock. Metamorphic rock (**c**) is formed when rock is subjected to intense heat and pressure underground. Through these several processes (shown by differently colored arrows in the figure), each type of rock can be converted into either of the other two types.

rock in a given region helps determine soil chemistry and thereby influences the biotic components of the region's ecosystems.

All rocks can melt. At high enough temperatures, rock enters a molten, liquid state called **magma**. If magma is released in a volcanic eruption, it may flow or spatter across Earth's surface as **lava**. Rock that forms when magma cools is called **igneous rock** (Figure 3.17a). Igneous rock comes in several types, because magma can solidify in different ways.

All rock weathers away with time. The relentless forces of wind, water, freezing, and thawing eat away at rocks, stripping off one tiny grain (or large chunk) at a time. Particles of rock blown by wind or washed away by water come to rest downhill, downstream, or downwind from their sources, forming **sediments**. Sediment layers accumulate over time, increasing the weight and pressure on underlying layers. **Sedimentary rock** (Figure 3.17b) is formed when dissolved minerals seep through sediment layers and act as a kind of glue, crystallizing and binding sediment particles together.

Geological forces may bend, uplift, compress, or stretch rock. When great heat or pressure is exerted, rock may change its form, becoming **metamorphic rock** (Figure 3.17c). The forces that metamorphose rock occur at temperatures lower than the rock's melting point but high enough to reshape crystals within the rock and change its appearance and physical properties.

Plate tectonics shapes Earth's geography

The rock cycle takes place within the broader context of **plate tectonics**, a process that underlies earthquakes and volcanoes and that determines the geography of Earth's surface. Earth's surface consists of a thin, lightweight *crust* of rock floating atop a malleable *mantle*, which in turn surrounds a molten heavy *core* made mostly of iron. Earth's internal heat drives convection currents that flow in loops in the mantle, pushing the mantle's soft rock cyclically upward (as it warms) and downward (as it cools), like a gigantic conveyor belt. As the mantle material moves, it drags large plates of crust along its surface edge.

Earth's surface consists of about 15 major tectonic plates, most including some combination of ocean and continent (Figure 3.18). Imagine peeling an orange and putting the pieces of peel back onto the fruit. The ragged pieces of peel are like the plates of crust riding atop Earth's surface. These plates move at rates of roughly 2–15 cm

(1–6 in.) per year. This movement has influenced Earth's climate and life's evolution throughout our planet's history as the continents combined, separated, and recombined in various configurations. By studying ancient rock formations throughout the world, geologists have determined that at least twice, all landmasses were joined together in a supercontinent scientists have dubbed *Pangaea* (see Figure 3.18).

At **divergent plate boundaries**, magma surging upward to the surface divides plates and pushes them apart, creating new crust as it cools and spreads (Figure 3.19a). A prime example is the mid-Atlantic Ridge, part of a 74,000-km (46,000-mi) system of magmatic extrusion cutting across the seafloor. Plates expanding outward from divergent plate boundaries at mid-ocean ridges bump against other plates, creating different types of plate boundaries.

When two plates meet, they may slip and grind alongside one another, forming a **transform plate boundary** (Figure 3.19b), and creating friction that spawns earthquakes. The Pacific Plate and the North American Plate rub against each other along California's San Andreas Fault. Southern California is slowly inching its way toward northern California along this fault, and Los Angeles will eventually reach San Francisco.

When plates collide at **convergent plate boundaries** (Figure 3.19c), one plate of crust may slide beneath another in a process called **subduction**. The subducted crust is heated as it dives into the mantle, and it may send up magma that erupts through the surface in volcanoes. When denser ocean crust slides beneath lighter continental crust, volcanic mountain ranges are formed that parallel coastlines. Examples are the Cascades (which include Mount Saint Helens) and South America's Andes Mountains. Alternatively, two colliding plates of continental crust may slowly lift material from each plate. The Himalayas, the world's highest mountains, result from the Indian-Australian Plate's collision with the Eurasian Plate 40–50 million years ago—and these mountains are still being uplifted today.

Amazingly, this environmental system of such fundamental importance was completely unknown to us just half a century ago. Our civilization was sending people to the moon by the time our geologists were explaining the movement of land under our very feet. It is humbling to reflect on this; what other fundamental systems might we not yet appreciate or understand while our technology—and our ability to affect Earth's processes—continues racing ahead?

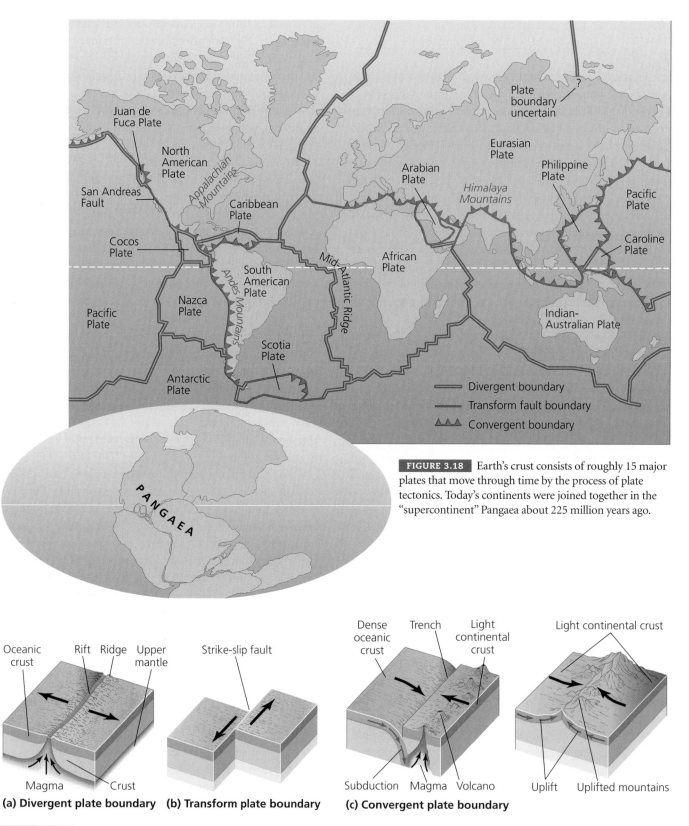

FIGURE 3.18 Earth's crust consists of roughly 15 major plates that move through time by the process of plate tectonics. Today's continents were joined together in the "supercontinent" Pangaea about 225 million years ago.

(a) Divergent plate boundary   **(b) Transform plate boundary**   **(c) Convergent plate boundary**

FIGURE 3.19 At a divergent plate boundary (**a**), magma extrudes from beneath the crust, and the two plates move away in the manner of conveyor belts. At a transform plate boundary (**b**), two plates slide alongside one another, creating friction that leads to earthquakes. Where plates collide at a convergent plate boundary (**c**), one plate may be subducted beneath another, leading to volcanism, or both plates may be uplifted, leading to the formation of mountain ranges.

Conclusion

Earth hosts many interacting systems, and the way one perceives them depends on the questions in which one is interested. Physical systems and processes such as the hydrologic cycle, the rock cycle, and plate tectonics lay the groundwork for the ways in which life spreads itself across the planet. Life interacts with its abiotic environment in ecosystems, systems through which energy flows and materials are recycled. Understanding the biogeochemical cycles that describe the movement of nutrients within and among ecosystems is crucial, because human activities are causing significant changes in the ways these cycles function.

Understanding energy, energy flow, and chemistry enhances our comprehension of how organisms interact with one another, how they relate to their nonliving environment, and how environmental systems function. Energy and chemistry are in some way tied to nearly every significant process involved in environmental science. Moreover, applications of chemistry can provide solutions to environmental problems involving agricultural practices, water resources, air quality, energy policy, and environmental health.

Thinking in terms of systems is important in understanding how Earth works, so that we may learn how to avoid disrupting its processes and how to mitigate any disruptions we cause. By studying the environment from a systems perspective and by integrating scientific findings with the policy process, people who care about the Mississippi River and the Gulf of Mexico are working today to solve their pressing problems. Their model is one that we can adapt to many other issues in environmental science.

TESTING YOUR COMPREHENSION

1. Describe how hypoxic conditions can develop in coastal marine ecosystems such as the northern Gulf of Mexico.
2. Differentiate between an ion and an isotope.
3. Describe the two major forms of energy, and give examples of each. Compare and contrast the first law of thermodynamics and the second law of thermodynamics.
4. What substances are produced by photosynthesis? By cellular respiration?
5. Describe the typical movement of energy through an ecosystem. Describe the typical movement of matter through an ecosystem.
6. What role does each of the following play in the carbon cycle?

   ▸ Cars
   ▸ Photosynthesis

   ▸ The oceans
   ▸ Earth's crust

7. Distinguish between the function performed by nitrogen-fixing bacteria and that performed by denitrifying bacteria.
8. How has human activity altered the carbon cycle? The phosphorus cycle? The nitrogen cycle? The hydrologic cycle? To what environmental problems have these changes given rise?
9. Name the three main types of rocks, and describe how each type may be converted to the others via the rock cycle.
10. How does plate tectonics account for mountains? For volcanoes? For earthquakes? Why do you think it took so long for scientists to discover such a fundamental environmental system as plate tectonics?

SEEKING SOLUTIONS

1. Can you think of an example of an environmental problem not mentioned in this chapter that a good knowledge of chemistry could help us solve? Explain your answer.
2. Referring to the chemical reactions for photosynthesis and respiration, provide an argument for why increases in the amount of carbon dioxide in the atmosphere due to global climate change might potentially increase amounts of oxygen in the atmosphere. Now give an argument for why increases in carbon dioxide might potentially decrease amounts of atmospheric oxygen.

What would you need to know to determine which of these two outcomes might occur?
3. Consider the ecosystem(s) that surround(s) your campus. How, specifically, do some of the principles from our discussion on ecosystems apply to them?
4. Imagine that you are a shrimper on the Louisiana coast and that your income is decreasing year by year because the dead zone is making it harder and harder to catch shrimp. One day your senator comes to town, and you have a one-minute audience with her. What

steps would you urge her to take in Washington, D.C., to try to help alleviate the dead zone and bring back the shrimp fishery?

5. Now imagine that you are an Iowa farmer and that you have learned that the federal government is insisting that you use 30% less fertilizer on your crops each year. You know that in good growing years you could do without that fertilizer, and you'd be glad not to have to pay for it. But in bad growing years, you need the fertilizer to ensure a harvest so that you can continue making a living. And you must apply the fertilizer each spring before you know whether it will be a good or bad year. What would you tell your senator when she comes to town?

CALCULATING ECOLOGICAL FOOTPRINTS

 In ecological systems, a rough rule of thumb is that when energy is transferred from plants to plant-eaters or from prey to predator, the efficiency is only about 10% (▸ p. 101). Much of this inefficiency is a consequence of the second law of thermodynamics. Another way to think of this is that eating 10 calories of plant material is the ecological equivalent of eating 1 calorie of material from an animal.

Humans are considered omnivores because we can eat both plants and animals. The choices we make about what to eat have significant ecological impacts. With this in mind, calculate the ecological energy requirements for four different diets, each of which provides a total of 2,000 dietary calories per day.

Diet	Source of calories	Number of calories consumed, by source	Ecologically equivalent calories, by source	Total ecologically equivalent calories per day
100% plant	Plant			
0% animal	Animal			
90% plant	Plant	1,800	1,800	3,800
10% animal	Animal	200	2,000	
50% plant	Plant			
50% animal	Animal			
0% plant	Plant			
100% animal	Anima			

1. How many ecologically equivalent calories would it take to support you for a year on each of the four diets listed?
2. What is the relative ecological impact of including as little as 10% of your calories from animal sources (e.g., milk, dairy products, eggs, and meat)? What is the ecological impact of a strictly carnivorous diet compared with a strict vegetarian diet?

3. What percentages of the calories in your own diet do you think come from plant versus animal sources? Estimate the ecological impact of your diet, relative to a strictly vegetarian one.
4. Describe some challenges of providing food for the growing human population, especially as people in many poorer nations develop a taste for an American-style diet rich in animal protein and fat.

Take It Further

Go to www.aw-bc.com/withgott, where you'll find:

▸ Suggested answers to end-of-chapter questions
▸ Quizzes, animations, and flashcards to help you study

▸ *Research Navigator*™ database of credible and reliable sources to assist you with your research projects
▸ **GRAPHit!** tutorials to help you interpret graphs
▸ **INVESTIGATEit!** current news articles that link the topics that you study to case studies from your region to around the world

Evolution, Biodiversity, and Population Ecology

4

Golden toads at
Monteverde

Upon successfully completing this chapter, you will be able to:

▶ Explain the process of natural selection, and cite evidence for this process

▶ Describe the ways in which evolution results in biodiversity

▶ Discuss reasons for species extinction and mass extinction events

▶ List the levels of ecological organization

▶ Outline the characteristics of populations that help predict population growth

▶ Assess logistic growth, carrying capacity, limiting factors, and other fundamental concepts of population ecology

Central Case: Striking Gold in a Costa Rican Cloud Forest

> "I must confess that my initial response when I saw them was one of disbelief and suspicion that someone had dipped the examples in enamel paint."
> —Dr. Jay M. Savage, describing the golden toad in 1966

> "What a terrible feeling to realize that within my own lifetime, a species of such unusual beauty, one that I had discovered, should disappear from our planet."
> —Dr. Jay M. Savage, describing the golden toad in 1998

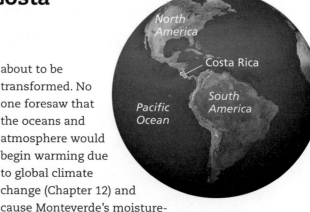

During a 1963 visit to Central America, biologist Jay M. Savage heard rumors of a previously undocumented toad living in Costa Rica's mountainous Monteverde region. The elusive amphibian, according to local residents, was best known for its color: a brilliant golden yellow-orange. Savage was told the toad was hard to find because it appeared only during the early part of the region's rainy season.

Monteverde means "green mountain" in Spanish, and the name couldn't be more appropriate. The village of Monteverde sits beneath the verdant slopes of the Cordillera de Tilarán, mountains that receive over 400 cm (157 in.) of annual rainfall. Some of the lush forests above Monteverde, which begin around 1,600 m (5,249 ft, just under a mile high), are known as *cloud forests* because much of the moisture they receive arrives from low-moving clouds that blow inland from the Caribbean Sea. Monteverde's cloud forest was not fully explored at the time of Savage's first visit, and researchers who had been there described the area as pristine, with a rich bounty of ferns, liverworts, mosses, clinging vines, orchids, and other organisms that thrive in cool, misty environments. Savage knew that such conditions create ideal habitat for toads and other amphibians.

In May of 1964, Savage organized an expedition into the muddy mountains above Monteverde to try to document the existence of the previously unknown toad species in its natural habitat. Late on the afternoon of May 14, he and his colleagues found what they were looking for. Approaching the mountain's crest, they spotted bright orange patches on the forest's black floor. In one area that was only 5 m (16.4 ft) in diameter, they counted 200 golden toads.

The discovery received international attention, making a celebrity of the tiny amphibian—which Savage named *Bufo periglenes* (meaning "brilliant toad")—and making a travel destination of its mountain home. At the time, no one knew that the Monteverde ecosystem was about to be transformed. No one foresaw that the oceans and atmosphere would begin warming due to global climate change (Chapter 12) and cause Monteverde's moisture-bearing clouds to rise, drying the forest. No one could guess that this newly discovered species of toad would become extinct in less than 25 years.

Evolution as the Wellspring of Earth's Biodiversity

The golden toad was new to science, and countless species still await discovery, but scientists understand quite well how the world became populated with the remarkable diversity of organisms we see today. We know that the process of biological evolution has brought us from a stark planet inhabited solely by microbes to a lush world of 1.5 million (and likely millions more) species (Figure 4.1).

The term *evolution* in the broad sense means change over time, and biological **evolution** consists of genetic change in organisms across generations. This genetic change often leads to change in the appearance, functioning, or behavior of organisms through time. Biological evolution results from random genetic modifications, and it may proceed randomly or may be directed by natural selection. **Natural selection** is the process by which traits that enhance survival and reproduction are passed on more frequently to future generations than those that do not, thus altering the genetic makeup of populations through time.

Evolution by natural selection is one of the best-supported and most illuminating concepts in all of science, yet it has remained socially controversial among some nonscientists who fear it may threaten their religious beliefs. Although scientists sometimes disagree about the specific mechanisms thought to drive evolution in particular cases, or about the time scales on which it takes place, this routine scientific debate should not be equated with the socially driven opposition of some nonscientists. From a scientific standpoint, evolutionary

(c) Harlequin frog

(a) Resplendent quetzal

(b) Puffball mushroom

(d) Scutellerid bug

FIGURE 4.1 Much of our planet's biological diversity resides in tropical rainforests. Monteverde's cloud-forest community includes organisms such as this **(a)** resplendent quetzal (*Pharomachrus mocinno*), **(b)** puffball mushroom (*Calostoma cinnabarina*), **(c)** harlequin frog (*Atelopus varius*), and **(d)** scutellerid bug (*Pachycoris torridus*).

theory is indispensable, because it is the foundation of modern biology.

Understanding evolution is also vital for a full appreciation of environmental science. Perceiving how organisms adapt to their environments and change over time is crucial for understanding the history of life. It is also needed for ecology, a central component of environmental science. Evolutionary processes are relevant to many aspects of environmental science, including pesticide resistance, agriculture, medicine, and environmental health.

Natural selection shapes organisms and diversity

Charles Darwin and **Alfred Russell Wallace** each proposed the concept of natural selection as a mechanism for evolution and as a way to explain the great variety of living things. Darwin and Wallace were exceptionally keen naturalists from England who had studied plants and animals in such exotic locales as the Galapagos Islands and the Malay Archipelago. Both men recognized that organisms face a constant struggle to gain sufficient resources to survive and reproduce. They each observed that organisms produce more offspring than can possibly survive. They also saw that because individuals vary in their characteristics, some offspring would happen to be better suited to their environment than others, and thus they would likely survive longer or reproduce more. Finally, they recognized that whichever characteristics give certain individuals advantages in surviv-

ing and reproducing might be inherited by their offspring. As a result, such characteristics would tend to become more prevalent in the population in future generations.

Natural selection is a simple concept that offers an astonishingly powerful explanation for patterns apparent in nature. Simply put, a parent that produces many offspring will pass on more genes than an individual that produces only a few offspring. From one generation to another through time, the genes of more-successful individuals will become more and more prevalent over those of less-successful individuals. As a result, species will evolve to possess characteristics that lead to better and better success in a given environment. A trait that promotes success is called an **adaptive trait**, or an **adaptation**.

Natural selection acts on genetic variation

For an organism to pass a trait along to future generations, genes in the organism's DNA (▸ p. 57) must code for the trait. In an organism's lifetime, its DNA will be copied millions of times by millions of cells. In all this copying and recopying, sometimes a mistake is made. Accidental alterations that arise during DNA replication, called **mutations**, give rise to genetic variation among individuals. If a mutation occurs in a sperm or egg cell, it may be passed on to the next generation. Although most mutations have little effect, some can be deadly, whereas others can be beneficial. Those that are not lethal provide the genetic variation on which natural selection acts.

Sexual reproduction also generates variation. When organisms reproduce through sex, they mix, or recombine, their genetic material, so that some of each parent's genes are included in the genes of the offspring. This process of *recombination* produces novel combinations of genes, generating variation among individuals.

An organism's environment determines what pressures natural selection will exert on the organism. Genes and environments interact as organisms engage in a perpetual process of adapting to the changing conditions around them. During this process, natural selection does not simply weed out unfit individuals. It also helps to elaborate and diversify traits that may lead to the formation of new species and whole new types of organisms.

Evidence of selection is all around us

The results of natural selection are all around us, visible in every adaptation of every organism (Figure 4.2). In addition, countless lab experiments (mostly with fast-reproducing organisms, such as bacteria and fruit flies) have demonstrated rapid evolution of traits. The evidence for selection that may be most familiar to us is that which Darwin himself cited prominently in his work 150 years ago: our breeding of domesticated animals. In our dogs, our cats, and our livestock, we have conducted selection under our own direction. We have chosen animals with traits we like and bred them together, while not breeding those with variants we do not like. Through such *selective breeding*, we have been

FIGURE 4.2 Natural selection has produced tremendous diversity among organisms in the wild. In the group of birds known as Hawaiian honeycreepers, closely related species have adapted to different food resources, habitats, or ways of life, as indicated by the diversity in their plumage colors and the shapes of their bills. Such a burst of species formation due to natural selection is known as an *adaptive radiation*.

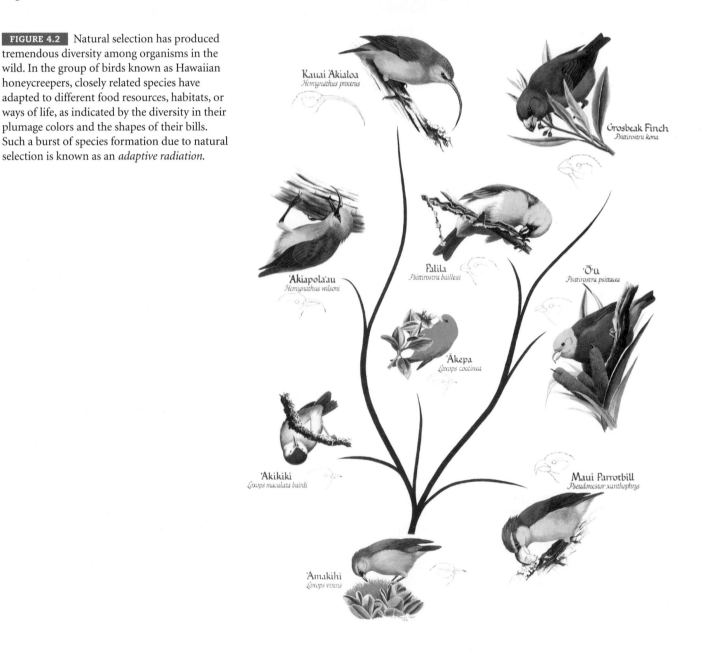

able to exaggerate particular traits we prefer. Consider the great diversity of dog breeds (Figure 4.3a), all of which comprise variations on a single species. From Great Dane to Chihuahua, they can interbreed freely and produce viable offspring, yet breeders maintain striking differences among them by allowing only like individuals to breed with like. This process of selection conducted under human direction is termed **artificial selection**.

Artificial selection has also given us the many crop plants we depend on for food, all of which were domesticated from wild ancestors and carefully bred over years, centuries, or millennia (Figure 4.3b). Through selective breeding, we have created corn with larger sweeter kernels; wheat and rice with larger and more numerous grains; and apples, pears, and oranges with better taste. We have diversified single types into many, for instance, breeding variants of the plant *Brassica oleracea* to create broccoli, cauliflower, cabbage, and Brussels sprouts. Our entire agricultural system is based on artificial selection.

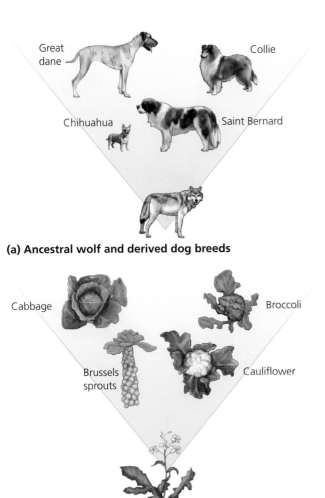

(a) Ancestral wolf and derived dog breeds

(b) Ancestral *Brassica oleracea* and derived crops

FIGURE 4.3 Selection imposed by humans (selective breeding, or artificial selection) has resulted in the numerous breeds of dogs **(a)**. By starting with the gray wolf (*Canis lupus*) as the ancestral wild species, and by breeding like with like and selecting for the traits we prefer, we have evolved breeds as different as Great Danes and Chihuahuas. By this same process we have created our immense variety of crop plants. Cabbage, Brussels sprouts, broccoli, and cauliflower were all evolved from a single ancestral species, *Brassica oleracea* **(b)**.

Evolution generates biological diversity

Evolution has generated a world full of diverse organisms. **Biological diversity**, or **biodiversity** for short, refers to the sum total of all organisms in an area, taking into account the diversity of species, their genes, their populations, and their communities. A **species** is a particular type of organism or, more precisely, a population or group of populations whose members share certain characteristics and can freely breed with one another and produce fertile offspring. A **population** is a group of individuals of a particular species that live in the same area. Genes are functional stretches of DNA (▸ p. 57), and we will introduce communities shortly (▸ pp. 83–84; and Chapter 5).

Scientists have described between 1.5 million and 1.8 million species, but many more remain undiscovered or unnamed. Estimates for the total number of species in the world range up to 100 million, with many of them thought to occur in tropical forests. In this light, the discovery of a new toad species in Costa Rica in 1964 seems far less surprising. Although Costa Rica covers a tiny fraction (0.01%) of Earth's surface area, it is home to 5–6% of all species known to scientists. And of the 500,000 species scientists estimate exist in the country, only 87,000 (17%) have been inventoried and described.

Tropical rainforests such as Costa Rica's, however, are by no means the only places rich in biodiversity. Step outside anywhere on Earth, even in a major city, and you will find numerous species within easy reach. They may not always be large and conspicuous like Yellowstone's bears or Africa's elephants, but they will be there. Plants poke up from cracks in asphalt in every city in the world, and even Antarctic ice harbors microbes. In a handful of backyard soil there may exist an entire miniature world of life, including several insect species, several types of mites, a millipede or two, many nematode worms, a few plant seeds, countless fungi, and millions upon millions of bacteria. We will examine Earth's biodiversity in detail in Chapter 8.

Speciation produces new types of organisms

How did Earth come to have so many species? Whether there are 1.5 million or 100 million, such large numbers require scientific explanation. The process by which new species are generated is termed **speciation**. Speciation can occur in a number of ways, but most biologists consider the main mode of species formation to be *allopatric speciation*, species formation due to the physical separation of populations over some geographic distance. To understand allopatric speciation, begin by picturing a population of organisms. Individuals within the population possess many similarities that unify them as a species, because they reproduce with one another and share genetic information. However, if the population is broken up into two or more populations that become isolated from one another, individuals from one population cannot reproduce with individuals from the others.

When a mutation arises in the DNA of an organism in one of these isolated populations, it cannot spread to the other populations. Over time, each population will independently accumulate its own set of mutations. Eventually, the populations may diverge, or grow different enough, that their members can no longer mate with one another. Once this has happened, there is no going back; the two populations cannot interbreed, and they have embarked on their own independent evolutionary trajectories as separate species (Figure 4.4). The populations will continue diverging as chance mutations accumulate that confer traits causing the populations to become different in random ways. If environmental conditions happen to be different for the two populations, then natural selection may accelerate this divergence.

The long-term geographic isolation of populations that can lead to allopatric speciation can occur in a variety of ways (Table 4.1). Through the speciation process, single species can generate multiple species, each of which may in turn generate more.

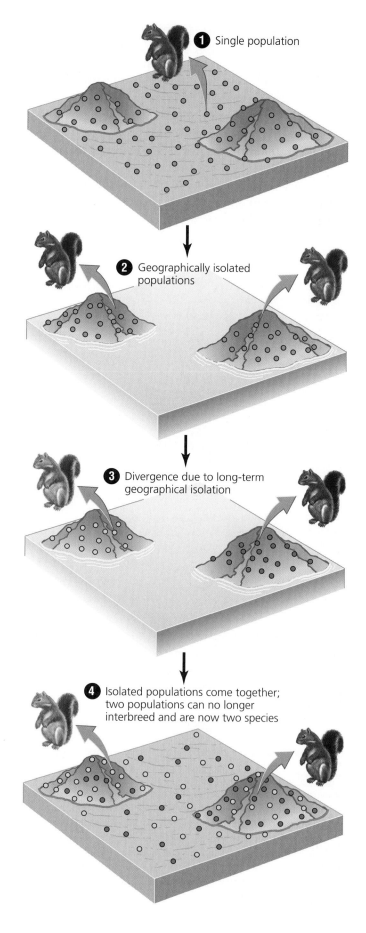

❶ Single population

❷ Geographically isolated populations

❸ Divergence due to long-term geographical isolation

❹ Isolated populations come together; two populations can no longer interbreed and are now two species

FIGURE 4.4 Allopatric speciation has generated much of Earth's diversity. In this process, some geographical barrier splits a population. In this diagram, two mountaintops (1) are turned into islands by rising sea level (2), isolating populations of squirrels. Each isolated population accumulates its own independent set of genetic changes over time, until individuals become genetically distinct and unable to breed with individuals from the other population (3). The two populations now represent separate species and will remain so even if the geographical barrier is removed and the new species intermix (4).

Table 4.1 Mechanisms of Population Isolation That Can Give Rise to Allopatric Speciation
▶ Glacial ice sheets advance
▶ Mountain chains are uplifted
▶ Major rivers change course
▶ Sea level rises, creating islands (see Figure 4.4)
▶ Climate warms, pushing vegetation up mountain slopes and fragmenting it
▶ Climate dries, dividing lakes into smaller ones
▶ Ocean current patterns shift
▶ Volcanism forms islands in the ocean

Life's diversification results from numerous speciation events

Evolutionary biologists represent life's history by using branching, treelike diagrams called **phylogenetic trees** (Figure 4.5). Similar to family genealogies, phylogenetic trees can show relationships among species, major groups of species, populations, or genes. In such trees of species, each branching point represents a speciation event. Scientists construct these trees by analyzing patterns of similarity among present-day organisms and inferring which are most closely related.

Scientists also learn about life's history by studying fossils. As organisms die, some are buried by sediment. Under the right conditions, the hard parts of their bodies—such as bones, shells, and teeth—may be preserved as sediments are compressed into rock (▶ pp. 70–71). Minerals replace the organic material, leaving behind a **fossil**, an imprint in stone of the dead organism. In thousands of locations across the world, geological processes over millions of years have buried sedimentary rock layers and later brought them to the surface, revealing assemblages of fossilized plants and animals from different time periods. The cumulative body of fossils worldwide is known as the **fossil record**.

Life's history, as revealed by phylogenetic trees and by the fossil record, is complex indeed, but a few big-picture trends are apparent. Life in its roughly 3.5 billion years has evolved complex structures from simple ones, and large sizes from small ones. However, these are only generalizations. Many organisms have evolved to become simpler or smaller when natural selection favored it. Many very complex life forms have disappeared, and it is easy to argue that Earth still belongs to the bacteria and other microbes, some of them little changed over eons.

Even fans of microbes, however, must marvel at some of the exquisite adaptations that animals, plants, and fungi have evolved: The heart that beats so reliably for an animal's entire lifetime that we take it for granted. The complex organ system to which the heart belongs. The stunning plumage of a peacock in display. The ability of each and every plant on the planet to lift water and nutrients from the soil, gather light from the sun, and turn it into food. The staggering diversity of beetles and other insects. The human brain and its ability to reason. All these and more have resulted from the process of evolution as it has generated new species and whole new branches on the tree of life.

Although speciation generates Earth's biodiversity, it is only one part of the equation—because the vast majority of species that once lived are now gone. The disappearance of a species from Earth is called **extinction**. From studying the fossil record, paleontologists calculate that the average time a species spends on Earth is 1–10 million years. The number of species in existence at any one time is equal to the number added by speciation minus the number removed by extinction.

Some species are more vulnerable to extinction than others

In general, extinction occurs when environmental conditions change rapidly or severely enough that a species cannot adapt genetically to the change; natural selection simply does not have enough time to work. All manner of environmental events can cause extinction, from climate change to the rise and fall of sea level, to the arrival of new harmful species, to severe weather events such as extended droughts. In general, small populations and species narrowly specialized on some particular resource or way of life are most vulnerable to extinction from environmental change.

The golden toad was a prime example of a vulnerable species. It was *endemic* to the Monteverde cloud forest, meaning that it occurred nowhere else on the planet. Endemic species face relatively high risks of extinction because all their members belong to a single, sometimes small, population. At the time of its discovery, the golden toad was known from only a 4-km² (988 acre) area of Monteverde. It also required very specific conditions to breed successfully. During the spring at Monteverde, water collects in shallow pools within the network of roots that span the cloud forest's floor. The golden toad gathered to breed in these root-bound reservoirs, and it was here that Jay Savage and his companions collected their specimens in 1964. Monteverde provided ideal

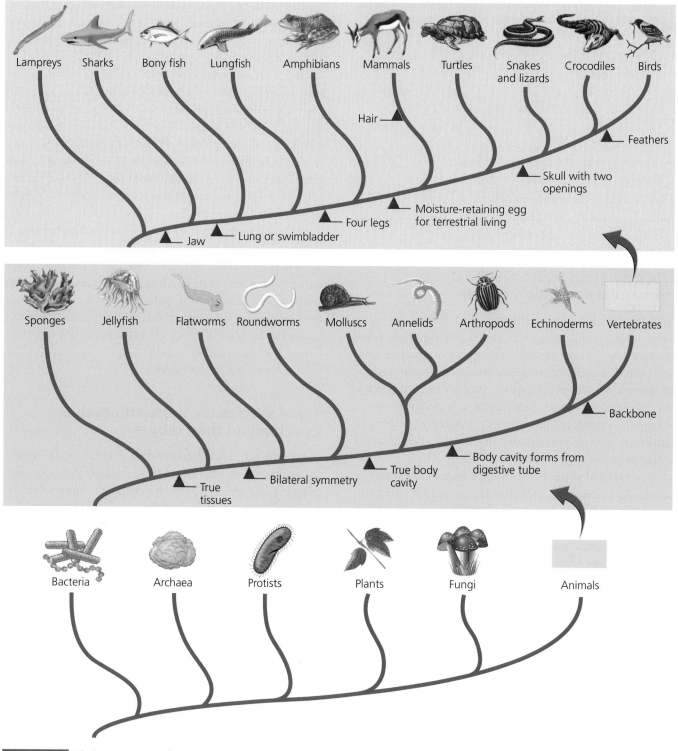

FIGURE 4.5 Phylogenetic trees show the history of life's divergence. Similar to family genealogies, these trees illustrate relationships among groups of organisms, as inferred from the study of similarities and differences among present-day creatures. The diagram here is a greatly simplified representation of relationships among a few major groups—one small portion of the huge and complex "tree of life." Each branch results from a speciation event, and time proceeds upward from bottom to top. By mapping traits onto phylogenetic trees, biologists can study how traits have evolved over time. In this diagram, several major traits are mapped, using triangular arrows indicating the point at which they originated. For instance, all vertebrates "above" the point at which jaws are indicated have jaws, whereas lampreys diverged before jaws originated and thus lack them.

habitat for the golden toad, but the extent of that habitat was minuscule—so any environmental stresses that deprived the toad of the resources it needed to survive might doom the entire world population of the species.

Earth has seen several episodes of mass extinction

Most extinction occurs gradually, one species at a time. The rate at which this type of extinction occurs is referred to as the *background extinction rate*. However, Earth has seen five events of staggering proportions that killed off massive numbers of species at once. These episodes, called **mass extinction events**, have occurred at widely spaced intervals in Earth history and have wiped out half to 95% of our planet's species each time.

The best-known mass extinction occurred 65 million years ago and brought an end to the dinosaurs (although birds are modern representatives of dinosaurs). Evidence suggests that the impact of a gigantic asteroid caused this event. As massive as this extinction event was, however, it was moderate compared to the mass extinction at the end of the Permian period 250 million years ago. Paleontologists estimate that 75–95% of all species on Earth may have perished during this event. Precisely what caused this event scientists do not know. The hypothesis with the most support so far is that massive volcanism threw into the atmosphere a global blanket of soot and sulfur, smothering the planet, reducing sunlight, and inducing severe climate change.

The sixth mass extinction is upon us

Many biologists have concluded that Earth is currently entering its sixth mass extinction event—and that we are the cause. The changes to Earth's natural systems set in motion by human population growth, development, and resource depletion have driven many species extinct and are threatening countless more. The alteration and outright destruction of natural habitats, the hunting and harvesting of species, and the introduction of invasive species from one place to another where they can harm native species—these processes and more have combined to threaten Earth's biodiversity.

As we will see in Chapter 8, the loss of our planet's biological diversity affects people directly, because other organisms provide us with life's necessities—food, fiber, medicine, and ecosystem services (▸ pp. 25–27, 30–31). Species extinction brought about by human impact may well be the single biggest environmental problem we face, because the loss of a species is irreversible.

Levels of Ecological Organization

Life occurs in a hierarchy of levels, from the atoms, molecules, and cells we reviewed in Chapter 3 (▸ pp. 53, 56–58) up through the **biosphere**, which is the cumulative total of living things on Earth and the areas they inhabit. *Ecologists* study relationships on the higher levels of this hierarchy (Figure 4.6), namely on the organismal, population, community, and ecosystem levels. **Ecology** is the study of interactions among organisms and between organisms and their environments.

At the organismal level, the science of ecology describes relationships between organisms and their physical environments. **Population ecology** investigates the quantitative dynamics of how individuals within a species interact with one another. **Community ecology** (Chapter 5) focuses on interactions among species, from one-to-one interactions to complex interrelationships involving entire communities. And as we saw in Chapter 3 (▸ pp. 60–62), ecology at the ecosystem level reveals patterns, such as energy and nutrient flow, by studying living and nonliving components of systems in conjunction. As improving technologies allow scientists to learn more about the complex operations of natural systems on a global scale, ecologists are increasingly expanding their horizons beyond ecosystems to the biosphere as a whole.

Habitat, niche, and specialization are important in organismal ecology

On the organismal level, each organism relates to its environment in ways that tend to maximize its survival and reproduction. One key relationship involves the specific environment in which an organism lives—its **habitat**. A species' habitat consists of both living and

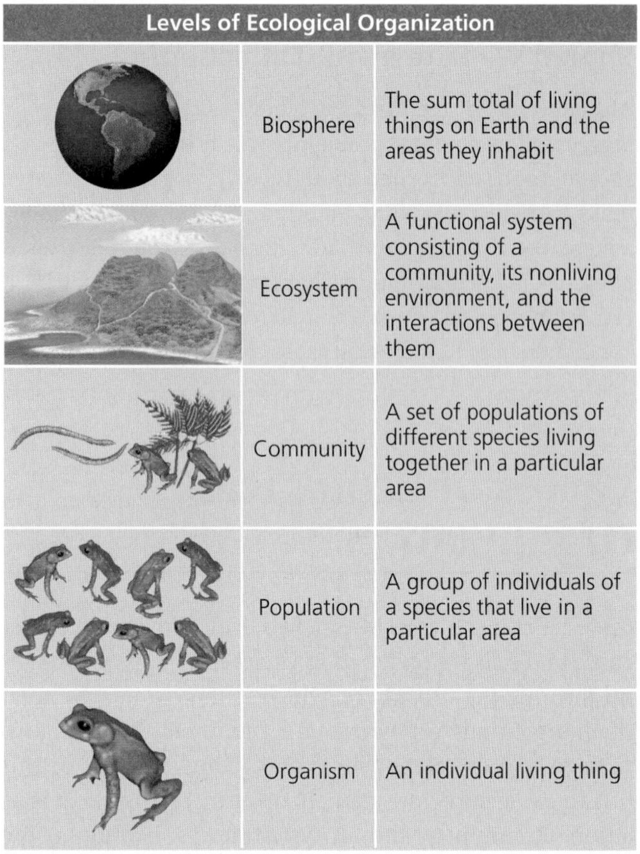

Levels of Ecological Organization		
	Biosphere	The sum total of living things on Earth and the areas they inhabit
	Ecosystem	A functional system consisting of a community, its nonliving environment, and the interactions between them
	Community	A set of populations of different species living together in a particular area
	Population	A group of individuals of a species that live in a particular area
	Organism	An individual living thing

FIGURE 4.6 Life exists in a hierarchy of levels. Ecology includes the study of the organismal, population, community, and ecosystem levels and, increasingly, the level of the biosphere.

nonliving elements—of rock, soil, leaf litter, and humidity, as well as the other organisms around it. The golden toad lived in a habitat of cloud forest, and specifically, on the moist forest floor, using seasonal pools for breeding and burrows for shelter. Habitats are scale-dependent. A tiny soil mite may perceive its habitat as a mere square meter of soil. A vulture, in contrast, may view its habitat in terms of miles upon miles of hills and valleys that it easily traverses by air.

Each organism thrives in certain habitats and not in others, leading to nonrandom patterns of *habitat use.* Mobile organisms actively select habitats in which to live from among the range of options they encounter, a process called *habitat selection.* In the case of plants and sessile animals, whose progeny disperse passively, patterns of habitat use result from success in some habitats and failure in others. The criteria by which organisms favor some habitats over others can vary greatly. The soil mite may judge available habitats in terms of the chemistry, moisture, and compactness of the soil and the percentage and type of organic matter. The vulture may

ignore not only soil but also topography and vegetation, focusing largely on the abundance of dead animals that it scavenges for food. Every species judges habitats differently because every species has different needs.

Another way in which an organism relates to its environment is through its niche. A species' **niche** reflects its use of resources and its functional role in a community. This includes its habitat use, its consumption of certain foods, its role in the flow of energy and matter, and its interactions with other organisms. The niche is a multidimensional concept, a kind of summary of everything an organism does.

Organisms vary in the breadth of their niche. Species with narrow breadth, and thus very specific requirements, are said to be **specialists**. Those with broad tolerances, able to use a wide array of habitats or resources, are **generalists**. An organism's habitat, niche, and degree of specialization each reflect the adaptations of the species and are products of natural selection.

Population Ecology

Individuals of a given species that inhabit a particular area make up a population. Species may consist of multiple isolated populations. This is the case with a species characteristic of Monteverde—the resplendent quetzal (*Pharomachrus mocinno*), considered one of the world's most spectacular birds (see Figure 4.1a). Although it ranges from southernmost Mexico to Panama, the resplendent quetzal lives only in high-elevation tropical forest and is absent from low-elevation areas. Moreover, human development has destroyed much of its forest habitat. Thus, the species today exists in many separate populations scattered across Central America.

In contrast, humans have spread into nearly every corner of the planet. As a result, it is difficult to define distinct human populations. Some would maintain that in the ecological sense of the word, all 6.5 billion of us comprise one population.

Populations exhibit characteristics that help predict their dynamics

All populations—from humans to quetzals to golden toads—show characteristics that help population ecologists predict the future dynamics of the population.

Population size Expressed as the number of individuals present at a given time, **population size** may increase, decrease, undergo cyclical change, or remain the same over time. Extinctions are generally preceded by population declines. As late as 1987, scientists documented more than 1,500 individuals in Monteverde's golden toad population, but in 1988 and 1989 scientists sighted only a single toad. By 1990, the species had disappeared.

The passenger pigeon *(Ectopistes migratorius)*, also now extinct, illustrates the extremes of population size (Figure 4.7). Once the most abundant bird in North America, flocks of passenger pigeons literally darkened the skies. In the early 1800s, ornithologist Alexander Wilson wrote of watching a flock of 2 billion birds 390 km (240 mi) long that took 5 hours to fly over and sounded like a tornado. Passenger pigeons nested in gigantic colonies in the forests of the upper Midwest and southern Canada. Once people began cutting the forests, the birds' great concentrations made them easy targets for market hunters, who gunned down thousands at a time and shipped them to market by the wagonload. By the end of the 19th century, the passenger pigeon population had declined to such a low number that they could not form the large colonies they apparently needed to breed successfully. In 1914, the last known passenger pigeon on Earth died in the Cincinnati Zoo, bringing the continent's most numerous bird to extinction within just a few decades.

Population density The flocks and breeding colonies of passenger pigeons showed high population density, another attribute that ecologists assess to understand populations. **Population density** describes the number of individuals within a population per unit area. For instance, the 1,500 golden toads counted in 1987 were found within 4 km^2 (988 acres), indicating a density of 375 toads/km^2. In general, larger organisms have lower population densities because they require more resources—and thus more area—to survive.

High population density can make it easier for organisms to group together and find mates, but it can also lead to conflict in the form of competition if space, food, or mates are in limited supply. Overcrowded organisms may also become more vulnerable to the predators that feed on them, and close contact among individuals can increase the transmission of infectious disease. For these reasons, organisms sometimes leave an area when densities become too high. In contrast, at low population densities, organisms benefit from more space and resources but may find it harder to locate mates and companions.

Overcrowding at high population densities is thought to have doomed Monteverde's harlequin frog (see Figure 4.1c), an amphibian that disappeared from the cloud

(a) Passenger pigeon

(b) 19th-century lithograph of pigeon hunting in Iowa

FIGURE 4.7 The passenger pigeon was once North America's most numerous bird, and its flocks literally darkened the skies as millions of birds passed overhead. However, human cutting of forests and hunting drove the species to extinction within a few decades.

forest at the same time as the golden toad. The harlequin frog was a habitat specialist, favoring "splash zones," areas alongside rivers and streams that receive spray from waterfalls and rapids. As Monteverde's climate grew warmer and drier in the 1980s and 1990s, water flow decreased, and harlequin frogs were forced to cluster together in what remained of the splash-zone habitat. Researchers J. Alan Pounds and Martha Crump recorded frog population densities up to 4.4 times higher than normal, with more than two frogs per meter (3.3 ft) of stream. Such crowding likely made the frogs vulnerable to disease transmission, predator attack, and assault from parasitic flies.

Population distribution It was not simply the harlequin frog's density, but also its distribution in space that led to its demise at Monteverde. **Population distribution**, or **population dispersion**, describes the spatial arrangement of organisms within an area. Ecologists define three distribution types: random, uniform, and clumped (Figure 4.8). In a *random distribution*, individuals are located haphazardly in space in no particular pattern. This type of distribution can occur when the resources an organism needs are found throughout an area and other organisms do not strongly influence where members of a population settle.

A *uniform distribution* is one in which individuals are evenly spaced. This can occur when individuals hold territories or otherwise compete for space. In a desert where water is scarce, each plant may need a certain amount of space for its roots to gather adequate moisture. As a result, each individual plant may be equidistant from others.

In a *clumped distribution*, organisms arrange themselves according to the availability of the resources they need. Many desert plants grow in patches around isolated springs or along creeks that flow with water after rainstorms. During their mating season, golden toads would cluster at seasonal breeding pools. Humans, too, exhibit clumped distribution; people aggregate together in villages and urban centers. Clumped distributions often indicate habitat selection.

Sex ratios For organisms that reproduce sexually and have distinct male and female individuals, the sex ratio of a population can help determine whether it will increase or decrease in size over time. A population's **sex ratio** is its proportion of males to females. In monogamous species (in which each sex takes a single mate), a 1:1 sex ratio maximizes population growth, whereas an unbalanced ratio leaves many individuals of one sex without mates.

(a) Random

(b) Uniform

(c) Clumped

FIGURE 4.8 Individuals in a population can be spatially distributed over a landscape in three fundamental ways. In a random distribution (**a**), organisms are dispersed at random through the environment. In a uniform distribution (**b**), individuals are spaced evenly, at equal distances from one another. Territoriality can result in such a pattern. In a clumped distribution (**c**), individuals occur in patches, concentrated more heavily in some areas than in others. Habitat selection or flocking to avoid predators can result in such a pattern.

Age structure Many populations consist of individuals of different ages. **Age distribution**, or **age structure**, describes the relative numbers of organisms of each age within a population. A population made up mostly of individuals past reproductive age will tend to decline over time. In contrast, a population with many individuals of reproductive age or soon to be of reproductive age is likely to increase. A population with an even age distribution will likely remain stable as births keep pace with deaths.

Age structure diagrams, often called *age pyramids*, are visual tools scientists use to show the age structure of populations (Figure 4.9). The width of each horizontal bar represents the relative size of each age class. A pyramid with a wide base has many pre-reproductive individuals, indicating a population capable of rapid growth. In this respect, a wide base of an age pyramid is like an oversized engine on a rocket—the bigger the booster, the faster the increase. We will examine age pyramids further in Chapter 6 (▸ pp. 128–130) in reference to human populations.

Populations may grow, shrink, or remain stable

Now that we have outlined some key attributes of populations, we are ready to take a quantitative view of population change by examining some simple mathematical concepts used by population ecologists and *demographers* (those who study human populations). Population growth, or decline, is determined by four factors:

1. Births within the population (*natality*)
2. Deaths within the population (*mortality*)
3. *Immigration* (arrival of individuals from outside the population)
4. *Emigration* (departure of individuals from the population)

To understand how a population changes, we measure its **growth rate**, which can be calculated as the birth rate plus the immigration rate, minus the death rate plus the emigration rate, each expressed as the number per 1,000 individuals per year:

$$\text{(birth rate + immigration rate)} - \text{(death rate + emigration rate)} = \text{growth rate}$$

The resulting number tells us the net change in a population's size per 1,000 individuals. For example, a population with a birth rate of 18 per 1,000, a death rate of 10 per 1,000, an immigration rate of 5 per 1,000, and an emigration rate of 7 per 1,000 would have a growth rate of 6 per 1,000:

$$(18/1,000 + 5/1,000) - (10/1,000 + 7/1,000) = 6/1,000$$

Thus, a population of 1,000 in one year will reach 1,006 in the next. If the population is 1,000,000, it will reach 1,006,000 the next year. These population increases

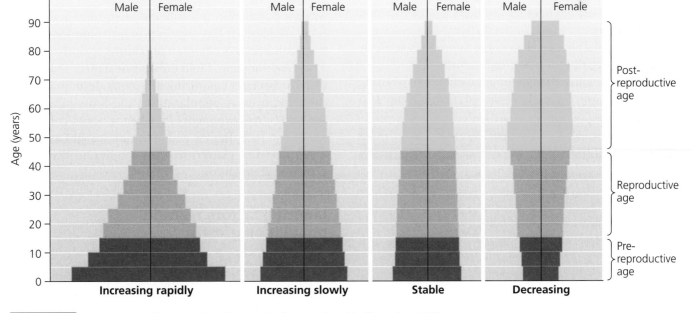

FIGURE 4.9 Age structure diagrams show the relative frequencies of individuals of different age classes in a population. In this example for humans, populations heavily weighted toward young age classes (at left) grow most quickly, whereas those weighted heavily toward old age classes (at right) decline.

are often expressed as percentages, which we can calculate using the formula:

$$\text{growth rate} \times 100\%$$

Thus, a growth rate of 6/1,000 would be expressed as:

$$6/1,000 \times 100\% = 0.6\%$$

By measuring population growth in terms of percentages, scientists can compare increases and decreases in populations that have far different sizes. They can also project changes that will occur in the population over longer periods, much like you might calculate the amount of interest your savings account will earn over time.

Unregulated populations increase by exponential growth

When a population, or anything else, increases by a fixed percentage each year, it is said to undergo **exponential growth**. A savings account is a familiar frame of reference for describing exponential growth. If at the time of your birth your parents had invested $1,000 in a savings account earning 5% interest compounded each year, you would have only $1,629 by age 10, and $2,653 by age 20, but you would have over $30,000 when you turn 70. If you could wait just 10 years more, that figure would rise to nearly $50,000 (Table 4.2). Only $629 was added during your first decade, but approximately $19,000 was added during the decade between ages 70 and 80. The reason is that a fixed percentage of a small number makes for a small increase, but that same percentage of a large number produces a large increase. Thus, as savings accounts (or populations) become larger, each incremental increase likewise gets larger. Such acceleration is a characteristic of exponential growth.

We can visualize changes in population size by using population growth curves. The J-shaped curve in Figure 4.10

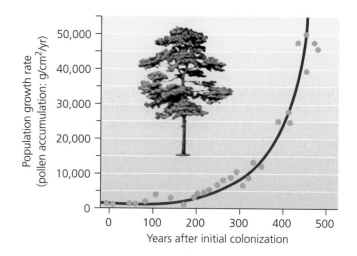

FIGURE 4.10 No species can maintain exponential growth indefinitely, but some may grow exponentially for a time when colonizing an unoccupied environment or exploiting an unused resource. Scientists have used pollen records to determine that the Scots pine (*Pinus sylvestris*) increased exponentially after the retreat of glaciers following the last ice age around 9,500 years ago. Go to **GRAPHIT!** at www.aw-bc.com/withgott. Data from Bennett, K. D. 1983. Postglacial population expansion of forest trees in Norfolk, U.K. *Nature* 303: 164–167.

shows exponential increase. As Thomas Malthus (▶ p. 5) realized, populations of all organisms increase exponentially unless they meet constraints. Each organism reproduces by a certain number, and as populations grow, more individuals reproduce by that number. If there are no external limits, ecologists theoretically expect exponential growth.

Normally, exponential growth occurs in nature only when a population is small and environmental conditions are ideal for the organism in question. Most often, these conditions occur when organisms are introduced to a new environment. Mold growing on a piece of bread or fruit, or bacteria colonizing a recently dead animal, are cases in point. A population of the Scots pine, *Pinus sylvestris*, grew exponentially when it began colonizing the British Isles after the end of the last ice age (see Figure 4.10). Receding glaciers had left conditions ideal for its exponential expansion.

Limiting factors restrain population growth

Exponential growth rarely lasts long. If even a single species in Earth's history had increased exponentially for very many generations, it would have blanketed the planet's surface, and nothing else could have survived. Instead, every population eventually is constrained by **limiting factors**, which are the physical, chemical, and biological characteristics of the environment that restrain

Table 4.2 Exponential Growth in a Savings Account with 5% Annual Compound Interest	
Age (in years)	**Principal**
0 (birth)	$1,000
10	$1,629
20	$2,653
30	$4,322
40	$7,040
50	$11,467
60	$18,679
70	$30,426
80	$49,561

population growth. The interaction of these factors determines the **carrying capacity**, the maximum population size of a species that a given environment can sustain.

Ecologists use the curve in Figure 4.11 to show how an initial exponential increase is slowed and finally brought to a standstill by limiting factors. Called the **logistic growth curve**, it rises sharply at first but then begins to level off as the effects of limiting factors become stronger. Eventually the force of these factors—which taken together are termed *environmental resistance*—stabilizes the population size at its carrying capacity.

For animals in terrestrial environments, limiting factors include the availability of food, water, mates, shelter, and suitable breeding sites; temperature extremes; prevalence of disease; and abundance of predators. Plants are often limited by amounts of sunlight and moisture and the type of soil chemistry, in addition to disease and attack from plant-eating animals. In aquatic systems, limiting factors include salinity, sunlight, temperature, dissolved oxygen, fertilizers, and pollutants. Sometimes one limiting factor may outweigh all others and restrict population growth. For example, scientists hypothesize that Monteverde's population of golden toads had plenty of space, food, and shelter but came to lack adequate moisture (see "The Science behind the Story," ▸ pp. 90–91).

A population's density can increase or decrease the impact of certain limiting factors on that population. Recall that high population density can help organisms find mates but can also increase competition and the risk of predation and disease. Such factors are said to be **density-dependent** factors, because their influence waxes and wanes according to population density. The logistic growth curve in Figure 4.11 represents the effects of density dependence. The more population size rises, the more environmental resistance kicks in. **Density-independent** factors are limiting factors whose influence is not affected by population density. Temperature extremes and catastrophic events such as floods, fires, and landslides are examples of density-independent factors, because they can eliminate large numbers of individuals without regard to their density.

The logistic curve is a simplified model, and real populations can behave differently. Some may oscillate indefinitely above and below the carrying capacity. Some may show cycles that become less extreme and approach the carrying capacity. Others may overshoot the carrying capacity and then crash, fated either for extinction or recovery (Figure 4.12, ▸ p. 92).

Because environments are complex and ever-changing, carrying capacities can vary. Moreover, organisms may be capable of altering their environment to reduce environmental resistance. Our own species has proved particularly effective at this. When our ancestors began to build shelters and use fire for heating and cooking, they reduced the environmental resistance of areas with cold climates and were able to expand into new territory. As limiting factors are overcome (through development of new technologies or through natural environmental change), the carrying capacity for a species may increase.

Weighing the ISSUES:
Carrying Capacity and Human Population Growth

As we have seen (▸ p. 4), the global human population has risen from fewer than 1 billion 200 years ago to 6.5 billion today, and we have far exceeded our historic carrying capacity. What factors increased Earth's carrying capacity for people? Do you think there are limiting factors for the human population? What might they be? Do you think we can keep raising our carrying capacity in the future? Might Earth's carrying capacity for us decrease?

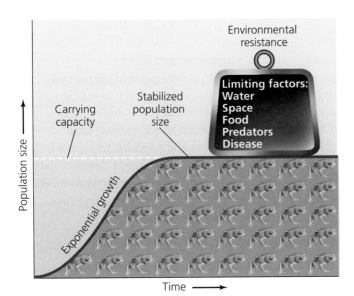
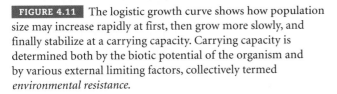

FIGURE 4.11 The logistic growth curve shows how population size may increase rapidly at first, then grow more slowly, and finally stabilize at a carrying capacity. Carrying capacity is determined both by the biotic potential of the organism and by various external limiting factors, collectively termed *environmental resistance.*

Biotic potential and reproductive strategies vary from species to species

Organisms differ in their *biotic potential*, or ability to produce offspring. A fish with a short gestation period that lays thousands of eggs at a time has high biotic

Climate Change and Its Effects on Monteverde

Soon after the golden toad's disappearance, scientists began to examine the potential role of climate change in driving cloud-forest species toward extinction. They noted that the period from July 1986 to June 1987 was the driest on record in Monteverde, with unusually high temperatures and record-low stream flows. These conditions had caused the golden toad's breeding pools to dry up in the spring of 1987, likely killing nearly all the eggs and tadpoles in the pools.

Scientists began reviewing reams of weather data and found that the number of dry periods each winter in the Monteverde region had increased between 1973 and 1998. Because amphibians breathe and absorb moisture through their skin, they are susceptible to dry conditions, high temperatures, acid rain, and pollutants concentrated by reduced water levels. Based on these facts, herpetologists J. Alan Pounds and Martha Crump in 1994 hypothesized that hot, dry conditions were to blame for increased adult mortality and breeding problems among golden toads and other amphibians.

Throughout this period, scientists worldwide were realizing that the oceans and atmosphere were warming because of human release of carbon dioxide and other gases

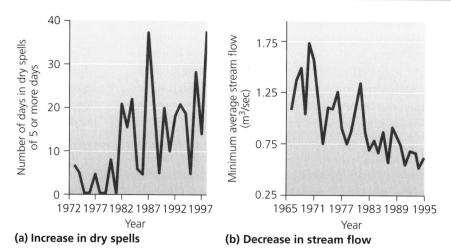

(a) Increase in dry spells　　**(b) Decrease in stream flow**

Warming and drying trends in Monteverde's climate may have contributed to the region's amphibian declines. Evidence gathered over 25 years shows (**a**) an increase in the annual number of dry days and (**b**) a decrease in the amount of annual stream flow. Data from Pounds, J. A., et al. 1999. Biological response to climate change on a tropical mountain. *Nature* 398: 611–615.

into the atmosphere (Chapter 12). With this in mind, Pounds and others reviewed the scientific literature on ocean and atmospheric science to analyze the effects on Monteverde's local climate of warming patterns in the ocean regions around Costa Rica.

These researchers determined that Monteverde's cloud forest was becoming drier because the clouds that had given the forest its name and much of its moisture now passed by at higher elevations, where they were no longer in contact with the trees. The primary factor determining the clouds' altitude,

the researchers found, is nearby ocean temperatures; as ocean temperatures increase, clouds pass over Monteverde at higher elevations. Once the cloud forest's water supply was pushed upward, out of reach of the mountaintops, the cloud forest began to dry out.

In a 1999 paper in the journal *Nature*, Pounds and two colleagues reported that broad-scale climate modification was causing local changes at the species, population, and community levels. Rising cloud levels and decreasing moisture could explain not only the disappearance of the golden toad and harlequin

potential, whereas a whale with a long gestation period that gives birth to a single calf at a time has low biotic potential.

Giraffes, elephants, humans, and other large animals with low biotic potential produce a relatively small number of offspring and take a long time to gestate and raise

each of their young. Species that take this approach to reproduction compensate by devoting large amounts of energy and resources to caring for the relatively few offspring they produce. Such species are said to be **K-selected** (so named because their populations tend to stabilize near carrying capacity, commonly symbolized as K). Because

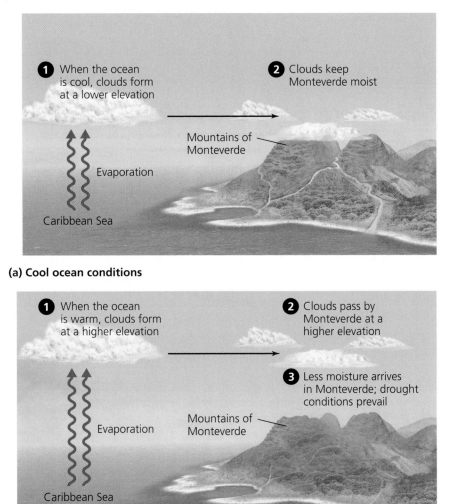

(a) Cool ocean conditions

1 When the ocean is cool, clouds form at a lower elevation

2 Clouds keep Monteverde moist

Mountains of Monteverde

Evaporation

Caribbean Sea

(b) Warm ocean conditions

1 When the ocean is warm, clouds form at a higher elevation

2 Clouds pass by Monteverde at a higher elevation

3 Less moisture arrives in Monteverde; drought conditions prevail

Mountains of Monteverde

Evaporation

Caribbean Sea

Monteverde's cloud forest gets its name and life-giving moisture from clouds that sweep inland from the oceans. When ocean temperatures are cool (**a**), the clouds keep Monteverde moist. Warmer ocean conditions resulting from global climate change (**b**) cause clouds to form at higher elevations and pass over the mountains, drying the cloud forest.

frog, but also the concurrent population crashes in 1987 and subsequent disappearance of 20 species of frogs and toads from the Monteverde region. Amphibians that survived underwent population crashes in each of the region's three driest years.

Pounds and colleagues followed this up in 2006 with another *Nature* paper, showing that at Monteverde and other tropical locations, daytime and nighttime temperatures were becoming more similar. Such conditions are optimal for chytrid fungi, which lethally infect amphibians and may now have led to the extinction of two-thirds of the world's harlequin frog species.

In their 1999 paper, Pounds and his co-workers further described "a constellation of demographic changes that have altered communities of birds, reptiles and amphibians" in the area as likely additional consequences of the shift in moisture availability. As these mountaintop forests dried out, dry-tolerant species crept in, and moisture-dependent species were stranded at the mountaintops by a rising tide of dryness. Although organisms may in general be driven from one area to another by changing environmental conditions, if a species has nowhere to go, then extinction may result.

their populations stay close to carrying capacity, these organisms must be good competitors, able to hold their own in a crowded world. Thus in these species, natural selection favors individuals that invest in producing offspring of high quality that can be good competitors.

In contrast, species that are **r-selected** focus on quantity, not quality. They have high biotic potential and devote their energy and resources to producing as many offspring as possible in a relatively short time. Their offspring do not require parental care after birth, so r-selected species simply

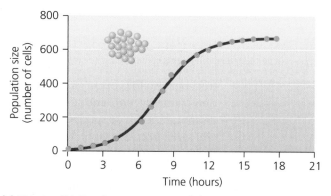

(a) Yeast cells, *Saccharomyces cerevisiae*

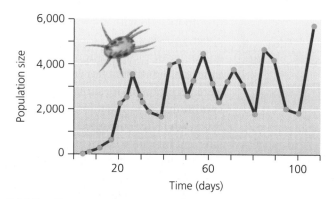

(b) Mite, *Eotetranychus sexmaculatus*

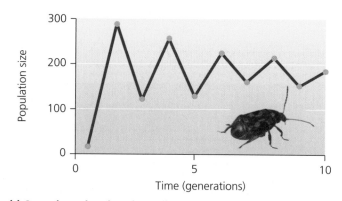

(c) Stored-product beetle, *Callosobruchus maculatus*

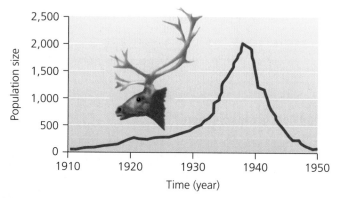

(d) St. Paul reindeer, *Rangifer tarandus*

FIGURE 4.12 Population growth in nature often departs from the stereotypical logistic growth curve, and it can do so in several fundamental ways. Yeast cells from an early lab experiment show logistic growth (**a**) that closely matches the theoretical model. Some organisms, like the mite shown here, show cycles in which population fluctuates indefinitely above and below the carrying capacity (**b**). Population oscillations can also dampen, lessening in intensity and eventually stabilizing at carrying capacity (**c**), as in a lab experiment with the stored-product beetle. Populations that rise too fast and deplete resources may crash just as suddenly (**d**), like the population of reindeer introduced to the Bering Sea island of St. Paul. Data from Pearl, R. 1927. The growth of populations. *Quarterly Review of Biology*, 2: 532–548 (a); Huffaker, C. B. 1958. Experimental studies on predation: dispersion factors and predator-prey oscillations, *Hilgardia*, 27: 343–383 (b); Utida, S. 1967. Damped oscillation of population density at equilibrium, *Researches on Population Ecology*, 9: 1–9 (c); Scheffer, V. C. 1951. Rise and fall of a reindeer herd, *Scientific Monthly*, 73: 356–362 (d).

leave their survival to chance. The abbreviation *r* denotes the rate at which a population increases in the absence of limiting factors. Populations of r-selected species fluctuate greatly, so they are often well below carrying capacity. This is why natural selection in these species favors traits that lead to rapid population growth. Many fish, plants, frogs, insects, and others are r-selected. The golden toad is one example; each adult female laid 200–400 eggs, and its tadpoles spent 5 weeks unsupervised in the breeding pools metamorphosing into adults. Table 4.3 summarizes stereotypical traits of r-selected and K-selected species. However, these are two extremes on a very messy continuum. Most species fall somewhere between these endpoints or show combinations of these traits.

Changes in populations influence the composition of communities

In the late 1980s, the golden toad and the harlequin frog were the most diligently studied species affected by changing environmental conditions in the Costa Rican cloud forest. Once scientists began looking at populations of other species at Monteverde, they began to notice more troubling changes. By the early 1990s, not only had golden toads, harlequin frogs, and other organisms been pushed from their cloud-forest habitat into apparent extinction, but also many species from lower, drier habitats had begun to appear at Monteverde. By the year 2000, 15 dry-forest species had moved into the cloud forest and

VIEWPOINTS

Conservation at Monteverde

What lessons, if any, can we learn from ecological changes that have occurred in the Monteverde region since the golden toad's discovery in 1964?

Lessons from the Green Mountain: Changes in Ecological and Human Communities

Monteverde—the "Green Mountain"—is one of the best-studied cloud forests in the world. Over the years, researchers have documented negative effects of human activities on the diverse biota. These include forest fragmentation and its isolation of plant and animal populations; lengthened dry seasons due to regional deforestation; and the upward shift of lower-elevation animal and plant populations due to global climate change.

One complex aspect of increasing human presence in Monteverde is ecotourism. The annual influx of nearly 80,000 visitors has had negative effects on the human community. Consumerism has increased. Television has replaced square-dancing. Cell telephones have replaced the single community party line used in previous decades. Guides rely on having cars rather than feet to show visitors around.

However, the presence of outside visitors who are deeply interested in nature has also affected Monteverde in positive ways. Some Monteverde families have abandoned dairy farming as their source of income and have been able to turn to natural history guiding or managing small hotels. This has led to the reforestation of pastures and the coalescence of formerly fragmented forest patches.

In addition, each visitor to Monteverde becomes a potential conservation activist. By seeing the unique wildlife and habitats—and learning about the causes that threaten it—visitors can take political and economic action to promote conservation when they return home. Grass roots conservation and educational organizations such as the Monteverde Conservation League and the Cloud Forest School have done much to promote conservation. These organizations rely on the contributions of ecotourists who have been touched by their experiences of seeing a quetzal or flowering orchid and become moved to preserve the habitat of the region.

Nalini M. Nadkarni is a member of the faculty at The Evergreen State College in Olympia, Washington. Her research focuses on the ecological interactions that occur in tropical and temperate rainforest canopies. The co-founder and president of the International Canopy Network, she is co-editor of the book *Monteverde: Ecology and Conservation of a Tropical Cloud Forest.*

Conservation Successes in the Shadow of the Golden Toad

In the late 1970s, the world began to discover the pristine montane forests and peaceful farming community of Monteverde, Costa Rica. Overnight, it seemed, ecotourism boomed, new houses, hotels, and restaurants sprung up, and tens of thousands of visitors arrived annually, eager to see resplendent quetzals and epiphyte-laden cloud forests. But a decade later, equally suddenly, the golden toad, a stunning amphibian found nowhere else on Earth, vanished. Shortly afterward, 19 other species of frogs and salamanders—40% of all the amphibian species that were present when I began my dissertation research there in 1979—disappeared from the area. Fortunately, none but the golden toad was endemic to Monteverde.

Yet, surprisingly, there have also been impressive achievements in conservation at Monteverde over this same time period. In 1979, poaching of large mammals and birds was commonplace. Now the very people who hunted scarce animals with rifles use binoculars instead as they lead natural history tours. Tapirs and guans are more common today than they have been for more than half a century. As the Monteverde Cloud Forest Preserve has expanded tenfold, clearings on the Atlantic slope have reverted to lush forest. The Guacimal River, formerly rancid because of waste dumped by the community dairy plant, is much cleaner now.

Thus, although ecotourism can have a negative effect on local populations of some plants and animals, it can also have enduring positive impacts on conservation—not just locally by increasing economic opportunities and incentives for land preservation, but also globally by educating the public about environmental values. Monteverde has inspired visitors to appreciate tropical forests and, through that experience, their own natural heritage—and that may help protect threatened habitats worldwide.

Nathaniel Wheelwright is professor of biology at Bowdoin College in Brunswick, Maine, director of the Bowdoin Scientific Station on Kent Island, New Brunswick, and co-editor of *Monteverde: Ecology and Conservation of a Tropical Cloud Forest.*

Explore this issue further by accessing **Viewpoints** at www.aw-bc.com/withgott.

Table 4.3 Traits of r-selected and K-selected Species

r-selected species	K-selected species
Small size	Large size
Fast development	Slow development
Short-lived	Long-lived
Reproduction early in life	Reproduction later in life
Many small offspring	Few large offspring
Fast population growth rate	Slow population growth rate
No parental care	Parental care
Weak competitive ability	Strong competitive ability
Variable population size, often well below carrying capacity	Constant population size, close to carrying capacity
Variable and unpredictable mortality	More constant and predictable mortality

begun to breed. Meanwhile, population sizes of several cloud-forest bird species had declined. After 1987, 20 of 50 frog species vanished from one part of Monteverde, and ecologists later reported more disappearances, including those of two lizards native to the cloud forest. Scientists hypothesized that the warming, drying trends that researchers were documenting (see "The Science behind the Story," ▶ pp. 90–91) were causing population fluctuations and unleashing changes in the composition of the community.

We can address population and community change with biodiversity conservation

Changes in populations and communities have taken place naturally as long as life has existed, but today human development, resource extraction, and popula-

tion pressure are speeding the rate of change and altering the types of change. The phenomena that threaten biodiversity today have complex social, economic, and political roots, and we must understand these aspects if we are to develop solutions.

Fortunately, people can act to forestall population declines of species threatened with extinction. Millions of people around the world are already taking action to safeguard the Earth's biodiversity and ecological and evolutionary processes (Chapter 8). Costa Ricans have been confronting the challenges to their nation's biodiversity, and their actions so far show what even a small country of modest means can do. Fully 12% of this nation's area is contained in national parks, and a further 16% is devoted to other types of wildlife and conservation reserves. By working to protect endangered species and recover their populations, Costa Rica and its citizens are now reaping ecological and economic benefits from their conservation efforts.

Conclusion

The golden toad and other organisms of the Monteverde cloud forest help illuminate the fundamentals of evolutionary biology and population ecology that are integral to environmental science. The evolutionary processes of natural selection, speciation, and extinction help determine Earth's biodiversity. Understanding how ecological processes function at the population level is crucial to protecting biodiversity threatened by the mass extinction event that many biologists maintain is already under way. Understanding the basics of population ecology also informs the study of human populations, another key endeavor in environmental science, which we will take up in Chapter 6.

TESTING YOUR COMPREHENSION

1. Explain the premises and logic that supports the concept of natural selection.
2. Describe two examples of evidence for selection.
3. How does allopatric speciation occur?
4. Name two organisms that have gone extinct, and give a probable reason for each extinction.
5. What is the difference between a species and a population? Between a population and a community?
6. Contrast the concepts of habitat and niche.
7. List and describe each of the five major population characteristics discussed in this chapter. Explain how each shapes population dynamics.
8. Could any species undergo exponential growth forever? Explain your answer.
9. Describe how limiting factors relate to carrying capacity.
10. Explain the difference between K-selected species and r-selected species. Can you think of examples of each that were not mentioned in the chapter?

SEEKING SOLUTIONS

1. In what ways has artificial selection changed people's quality of life? Give examples. Can you imagine a way in which artificial selection could be used to improve our quality of life further? Can you imagine a way it could be used to reduce our environmental impact?

2. What types of species are most vulnerable to extinction, and what kinds of factors threaten them? Can you think of any species in your region that are threatened with extinction? What reasons lie behind their endangerment? What could be done to restore their populations?

3. Do you think the human species can continue raising its global carrying capacity? How so, or why not? Do you think we *should* try to keep raising our carrying capacity? Why or why not?

4. Describe the evidence suggesting that changes in temperature and precipitation led to the extinction of the golden toad and to population crashes for other amphibians at Monteverde. What do you think could be done to help make future such declines less likely?

5. As Monteverde changed and some species disappeared, scientists reported that others moved in from lower, drier areas. If this is true, should we be concerned about the extinction of the golden toad and disappearance of other species from Monteverde? Explain your answer.

CALCULATING ECOLOGICAL FOOTPRINTS

In 2004, coffee consumption in the United States topped 2.7 billion pounds (out of 14.8 billion pounds produced globally). Most coffee is produced on large tropical plantations, where coffee is the only tree species and is grown in full sun. However, approximately 2% of coffee is produced in small groves where coffee trees and other species are intermingled. These *shade-grown* coffee forests maintain greater habitat diversity for tropical rainforest wildlife. Given the information above, estimate the coffee consumption rates in the table.

	Population	Pounds of coffee per day	Pounds of coffee per year
You (or the average American)	1	0.025	9
Your class			
Your hometown			
Your state			
United States			

Data from O'Brien, T. G. and M. F. Kinnaird. 2003. Caffeine and conservation. *Science* 300: 587; and International Coffee Organization.

1. What percentage of global coffee production is consumed in the United States? If only shade-grown coffee were consumed in the United States, how much would shade-grown production need to increase to meet that demand?

2. How much extra would you be willing to pay for a pound of shade-grown coffee, if you knew that your money would help to prevent habitat loss or extinction for animals such as Sumatran tigers, rhinoceroses, and the many songbirds that migrate between Latin America and North America each year?

3. If everyone in the United States were willing to pay as much extra per pound for shade-grown coffee as you are, how much additional money would that provide for conservation of biodiversity in the tropics each year?

Take It Further

Go to www.aw-bc.com/withgott, where you'll find:

▶ Suggested answers to end-of-chapter questions
▶ Quizzes, animations, and flashcards to help you study

▶ *Research Navigator*™ database of credible and reliable sources to assist you with your research projects

▶  **GRAPHIT!** tutorials to help you interpret graphs

▶ **INVESTIGATEIT!** current news articles that link the topics that you study to case studies from your region to around the world

5

Species Interactions and Community Ecology

Aggregation of zebra mussels

Upon successfully completing this chapter, you will be able to:

▶ Compare and contrast the major types of species interactions

▶ Characterize feeding relationships and energy flow, using them to construct trophic levels and food webs

▶ Distinguish characteristics of a keystone species

▶ Characterize the process of succession

▶ Perceive and predict the potential impacts of invasive species in communities

▶ Explain the goals and methods of ecological restoration

▶ Describe and illustrate the terrestrial biomes of the world

Central Case: Black and White, and Spread All Over: Zebra Mussels Invade the Great Lakes

As if the Great Lakes hadn't been through enough already, the last thing they needed was the zebra mussel. The pollution-fouled waters of Lake Erie and the other Great Lakes shared by Canada and the United States had become gradually cleaner in the years following the U.S. Clean Water Act of 1970. As government regulation brought industrial discharges under control, people once again began to use the lakes for recreation, and populations of fish rebounded.

Then the zebra mussel arrived. Black-and-white-striped shellfish the size of a dime, zebra mussels attach to hard surfaces and open their paired shells, feeding on algae by filtering water through their gills. This mollusc, given the scientific name *Dreissena polymorpha*, is native to the Caspian Sea, Black Sea, and Azov Sea in western Asia and eastern Europe. It made its North American debut in 1988 when it was discovered in Canadian waters at Lake St. Clair, which connects Lake Erie with Lake Huron. Evidently, ships arriving from Europe had discharged ballast water containing the mussels or their larvae into the Great Lakes.

Within just two years of their discovery, zebra mussels had reached all five of the Great Lakes. The next year, these invaders entered New York's Hudson River to the east, and the Illinois River at Chicago to the west. From the Illinois River and its canals, they soon reached the Mississippi River, giving them access to a vast watershed covering 40% of the United States. By 1994 zebra mussels had colonized waters in 19 U.S. states and two Canadian provinces.

How could a mussel spread so quickly? The zebra mussel's larval stage is well adapted for long-distance dispersal. Its tiny larvae drift freely for several weeks, traveling as far as the currents take them. Adults that attach themselves to boats and ships may be transported from one place to another, even to small isolated lakes and ponds well away from major rivers. Moreover, in North America the mussels encountered none of the particular species of predators, competitors, and parasites that had evolved to limit their population growth in the Old World.

But why all the fuss? Zebra mussels clog up water intake pipes at factories, power plants, municipal water supplies, and wastewater treatment facilities. At one Michigan power plant, workers counted 700,000 mussels per square meter of pipe surface. Great densities of these organisms can damage boat engines, degrade docks, foul fishing gear, and sink buoys that ships use for navigation. Through such impacts, it is estimated that zebra mussels cost the U.S. economy hundreds of millions of dollars each year.

Zebra mussels also have severe impacts on the ecological systems they invade. Because each mussel filters a liter or more of water every day, they consume so much phytoplankton (▸ p. 52) that they can deplete populations. Phytoplankton is the foundation of the Great Lakes food web, so its depletion is bad news for zooplankton (▸ p. 52), the tiny aquatic animals that eat phytoplankton—and for the fish that eat both. Water bodies with zebra mussels have fewer zooplankton and open-water fish than water bodies without them, researchers are finding. Zebra mussels also suffocate native molluscs by attaching to their shells.

However, zebra mussels also benefit some bottom-feeding invertebrates and fish. By filtering algae and organic matter from open water and depositing nutrients in feces that sink, they shift the community's nutrient balance to the bottom and benefit the species that feed there. Once they have cleared the water, sunlight penetrates more deeply, spurring the growth of large-leafed underwater plants and algae.

In the past few years, scientists have noted a surprising new twist: The invader is being displaced by another invader. The quagga mussel, a close relative of the zebra mussel, is spreading through the Great Lakes and beyond. This species, named after an extinct zebra-like animal, appears to be replacing the zebra mussel in many locations. Scientists are only beginning to understand what consequences this shift may have for ecological communities.

Species Interactions

By interacting with many species in a variety of ways, zebra mussels have set in motion an array of changes in the ecological communities they have invaded. Interactions among species are the threads in the fabric of communities. Ecologists have organized species interactions into several fundamental categories (Table 5.1).

Competition can occur when resources are limited

When multiple organisms seek the same limited resource, their relationship is said to be one of **competition**. Competing organisms do not usually fight with one another directly and physically. Competition is generally more subtle, involving the consequences of one organism's ability to match or outdo others in procuring resources. The resources for which organisms compete can include just about anything an organism might need to survive, including food, water, space, shelter, mates, sunlight, and more. Competitive interactions can take place among members of the same species (*intraspecific competition*) or among members of two or more different species (*interspecific competition*).

Coexisting species can adapt to competition over evolutionary time by evolving to use slightly different resources or to use shared resources in different ways. If two bird species eat the same type of seeds, one might come to specialize on larger seeds and the other to specialize on smaller seeds. Or one bird might become more active in the morning and the other more active in the evening, thus avoiding direct interference. This process is called **resource partitioning**, because the species divide, or partition, the resource they use in common by specializing in different ways (Figure 5.1).

In competitive interactions, each participant has a negative effect on other participants, because each takes resources the others could have used. This is reflected in the two minus signs shown for competition in Table 5.1. In other types of interactions, some participants benefit while others are harmed; that is, one species exploits the other (note the +/− interactions in Table 5.1). Such *exploitative* interactions include predation, parasitism, and herbivory.

Predators kill and consume prey

Predation is the process by which individuals of one species, a *predator*, hunt, capture, kill, and consume individuals of another species, the *prey*. Interactions among predators and prey structure the food webs we will examine shortly, and also influence community composition by helping determine the relative numbers of predators and prey.

Zebra mussels consume the smaller types of zooplankton, and this predation has decreased zooplankton population sizes and biomass by up to 70% in Lake Erie and the Hudson River. (Also contributing to this decline is the fact that zebra mussels, by eating phytoplankton, compete with zooplankton for food.) Most predators are also prey, however, and zebra mussels have become a food source for a variety of North American fish, ducks, muskrats, and crayfish.

Predation sometimes drives population dynamics. For instance, an increase in the population size of prey creates more food for predators, which may survive and reproduce more effectively as a result. As the predator population rises, additional predation drives down the population of prey. Fewer prey in turn causes some predators to starve, so that the predator population declines. This allows the prey population to begin rising again, starting the cycle anew (Figure 5.2).

Predation also has evolutionary ramifications. Individual predators that are more adept at capturing prey will likely live longer, healthier lives and be better able to provide for their offspring than will less adept individuals. Thus, natural selection (▸ pp. 76–78) on individuals

Table 5.1 Effects of Species Interactions on Their Participants		
Type of interaction	Effect on species 1	Effect on species 2
Mutualism	+	+
Commensalism	+	0
Predation, parasitism, herbivory	+	−
Neutralism	0	0
Amensalism	−	0
Competition	−	−

"+" denotes a positive effect; "−" denotes a negative effect; "0" denotes no effect.

FIGURE 5.1 When species compete, they tend to partition resources, each specializing on a slightly different resource or way of attaining a shared resource. Various types of birds—including the woodpeckers, creeper, and nuthatch shown here—feed on insects from tree trunks, but they use different portions of the trunk, seeking different foods in different ways.

White-breasted nuthatch climbs downward along trunk looking for insects

Yellow-bellied sapsucker drills rows of holes in trunk and consumes sap and insects that get stuck in sap

Pileated woodpecker digs deeply into wood to find large insects

Brown creeper climbs upward along trunk looking for tiny insects

within a predator species leads to the evolution of adaptations that make them better hunters. Prey face an even stronger selective pressure—the risk of immediate death. For this reason, predation pressure has caused organisms to evolve an elaborate array of defenses against being eaten (Figure 5.3).

Parasites exploit living hosts

Organisms can exploit other organisms without killing them. **Parasitism** is a relationship in which one organism, the *parasite*, depends on another, the *host*, for nourishment or some other benefit while simultaneously doing

FIGURE 5.2 Predator-prey systems sometimes show paired cycles, in which increases and decreases in one organism apparently drive increases and decreases in the other. Although such cycles are predicted by theory and are seen in lab experiments, they are very difficult to document conclusively in natural systems. Data from MacLulich, D. A. 1937. Fluctuation in the numbers of varying hare (*Lepus americanus*). *Univ. Toronto Stud. Biol. Ser. 43*, Toronto: University of Toronto Press.

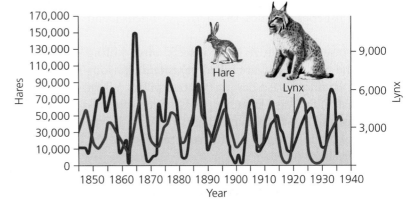

(a) Cryptic coloration

(b) Warning coloration

(c) Mimicry

FIGURE 5.3 Natural selection to avoid predation has resulted in many fabulous adaptations. Some prey hide from predators by *crypsis*, or camouflage, such as this gecko on tree bark (**a**). Other prey are brightly colored to warn predators that they are toxic or distasteful, such as this poison dart frog (**b**). Still others fool predators with mimicry. Some, like walking sticks imitating twigs, mimic for crypsis. Others mimic toxic, distasteful, or dangerous organisms, like this caterpillar (**c**); when it is disturbed, the caterpillar swells and curves its tail end and shows eyespots, to look like a snake's head.

the host harm. Unlike predation, parasitism usually does not result in an organism's immediate death.

Some types of parasites are free-living and come into contact with their hosts infrequently. For example, the cuckoos of Eurasia and the cowbirds of the Americas parasitize other birds by laying eggs in their nests and letting the host bird raise the parasite's young. However, many parasites live inside their hosts. These parasites include disease pathogens, such as the protists that cause malaria and dysentery, as well as animals, such as tapeworms, that live in their hosts' digestive tracts. Other parasites live on the exterior of their hosts. For example, the sea lamprey (*Petromyzon marinus*) is a tube-shaped vertebrate that grasps the bodies of fish with a suction-cup mouth and a rasping tongue, sucking their blood for days or weeks. Sea lampreys invaded the Great Lakes from the Atlantic Ocean after people dug canals to connect the lakes for shipping, and the lampreys soon devastated economically important fisheries of chubs, lake herring, whitefish, and lake trout.

Many insects parasitize other insects, often killing them in the process, and are called *parasitoids*. Various species of parasitoid wasps lay eggs on caterpillars. When the eggs hatch, the wasp larvae burrow into the caterpillar's tissues and slowly consume them. The wasp larvae metamorphose into adults and fly from the body of the dying caterpillar.

Just as predators and prey evolve in response to one another, so do parasites and hosts, in a process termed

coevolution. Hosts and parasites can become locked in a duel of escalating adaptations, known as an *evolutionary arms race.* Like rival nations racing to stay ahead of one another in military technology, host and parasite may repeatedly evolve new responses to the other's latest advance. In the long run, though, it may not be in a parasite's best interest to become too harmful to its host. Instead, a parasite might leave more offspring in the next generation—and thus be favored by natural selection—if it allows its host to live a longer time, or even to thrive.

Herbivores exploit plants

One of the most common exploitative interactions is **herbivory**, which occurs when animals feed on the tissues of plants. Insects that feed on plants are the most widespread herbivores; just about every plant in the world is attacked by some type of insect. In most cases, herbivory does not kill a plant outright but may affect its growth and reproduction.

Like animal prey, plants have evolved a wide array of defenses against the animals that feed on them. Many plants produce chemicals that are toxic or distasteful to herbivores. Others arm themselves with thorns, spines, or irritating hairs. In response, herbivores may evolve ways to overcome these defenses, and the plant and the animal may embark on an evolutionary arms race.

Mutualists help one another

Unlike exploitative interactions, **mutualism** is a relationship in which interacting species benefit from one another. Generally each partner provides some resource or service that the other needs. Many mutualistic relationships—like many parasitic relationships—occur between organisms that live in close physical contact. Such physically close association is called **symbiosis**. Thousands of terrestrial plant species depend on mutualisms with fungi; plant roots and some fungi together form symbiotic associations called mycorrhizae. In these symbioses, the plant provides energy and protection to the fungus, while the fungus helps the plant absorb nutrients from the soil. In the ocean, coral polyps, the tiny animals that build coral reefs (▸ p. 260), share beneficial arrangements with algae known as zooxanthellae. The coral provide housing and nutrients for the algae in exchange for a steady supply of food. You, too, are part of a symbiotic association. Your digestive tract is filled with microbes that help you digest food—microbes you are providing a place to live.

Not all mutualists live in close proximity. *Pollination*, an interaction vital to agriculture and our food supply (▸ pp. 155–156), involves free-living organisms that may encounter each other only once. Insects, birds, bats, and other creatures transfer pollen (male sex cells) from one flower to another, fertilizing eggs that become embryos within seeds. Most pollinating animals visit flowers for their nectar, a reward the plant uses to entice them. The pollinators receive food, and the plants are pollinated and reproduce. Various types of bees alone pollinate 73% of our crops, one expert has estimated—from soybeans to potatoes to tomatoes to beans to cabbage to oranges.

Ecological Communities

As we saw in Chapter 4 (Figure 4.6, ▸ p. 84), a **community** is an assemblage of species living in the same place at the same time. Members of a community interact with one another in the ways discussed above, and these species interactions help determine the composition, structure, and function of communities. *Community ecologists* are interested in which species coexist, how they relate to one another, how communities change through time, and why these patterns exist.

Energy passes among trophic levels

The interactions among community members are many and varied, but some of the most important involve who eats whom. As organisms feed on one another, matter and energy moves through the community, from one rank in the feeding hierarchy, or **trophic level**, to another (Figure 5.4).

Producers Producers, or autotrophs ("self-feeders"), comprise the first trophic level. As we saw in Chapter 3 (▸ pp. 59–60), terrestrial green plants, cyanobacteria, and algae capture solar energy and use photosynthesis to produce sugars.

Consumers Organisms that consume producers are known as *primary consumers* and comprise the second trophic level. Herbivorous grazing animals, such as deer and grasshoppers, are primary consumers. The third trophic level consists of *secondary consumers*, which prey on primary consumers. Wolves that prey on deer are considered secondary consumers, as are rodents and birds that prey on grasshoppers. Predators that feed at still higher trophic levels are known as *tertiary consumers*. Examples of tertiary consumers include hawks and owls that eat rodents that have eaten grasshoppers.

Detritivores and decomposers *Detritivores* and *decomposers* consume nonliving organic matter. Detritivores, such as millipedes and soil insects, scavenge the waste products or dead bodies of other community members. Decomposers, such as fungi and bacteria, break down leaf litter and other nonliving matter into simple constituents that can be taken up and used by plants. These organisms enhance the topmost soil layers (▸ p. 144) and play essential roles as the community's recyclers, making nutrients from organic matter available for reuse by living members of the community.

Energy decreases at higher trophic levels

At each trophic level, most of the energy that organisms obtain and use is lost as waste heat through respiration (▸ p. 60). Only a small portion of the energy is transferred to the next trophic level through predation, herbivory, or parasitism. A rough rule of thumb is that each trophic level contains just 10% of the energy of the trophic level below it.

This pattern can be visualized as a pyramid, which illustrates why eating lower on the food chain—being a vegetarian instead of a meat-eater, for instance—decreases a person's ecological footprint. Each amount of meat or other animal product we eat requires the input of a considerably greater amount of plant material (see Figure 7.19, ▸ p. 163). Thus, when we eat animal products

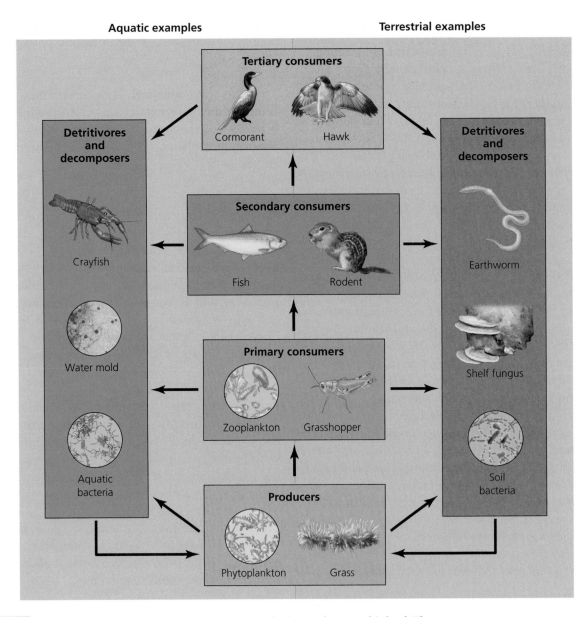

Aquatic examples   **Terrestrial examples**

FIGURE 5.4 Ecologists organize species hierarchically by their feeding rank, or trophic level. The diagram shows aquatic (left) and terrestrial (right) examples at each level. Arrows indicate the direction of energy flow. Producers produce food by photosynthesis, primary consumers (herbivores) feed on producers, secondary consumers eat primary consumers, and tertiary consumers eat secondary consumers. Communities can have more or fewer trophic levels than in this example. Detritivores and decomposers feed on nonliving organic matter and the remains of dead organisms from all trophic levels, and they "close the loop" by returning nutrients to the soil or the water column for use by producers.

we consume far more energy per calorie gained than when we eat plant products.

The pyramid pattern of energy loss from lower trophic levels to higher ones also generally holds for biomass and for numbers of organisms. This is why predators such as wolves and hawks are less numerous than the deer and the mice they prey on, and why deer and mice in turn are outnumbered by the plants they feed on.

Food webs show feeding relationships and energy flow

As energy is transferred from lower to higher trophic levels, it is said to pass up a *food chain*, a linear series of feeding relationships. Thinking in terms of food chains is conceptually useful, but in reality ecological systems are far more complex than simple linear chains. A more accurate representation of the feeding relationships in a community is a **food web**, a visual map of feeding relationships and energy flow, showing the many paths by which energy passes among organisms as they consume one another.

Figure 5.5 shows a food web from a temperate deciduous forest of eastern North America. Like virtually all diagrams of ecological systems, it is greatly simplified, leaving out the vast majority of species and interactions that occur. Note, however, that even within this simplified diagram, we can pick out a number of different food chains involving different sets of species.

A Great Lakes food web would involve phytoplankton and cyanobacteria, producers that photosynthesize near the water's surface; zooplankton, primary consumers that

FIGURE 5.5 Food webs are conceptual representations of feeding relationships in a community. This food web pertains to eastern North America's temperate deciduous forest and includes organisms on several trophic levels. In a food web diagram, arrows are drawn from one organism to another to indicate the direction of energy flow as a result of predation, parasitism, or herbivory. For example, an arrow leads from the grass to the cottontail rabbit to indicate that cottontails consume grasses. The arrow from the cottontail to the tick indicates that parasitic ticks derive nourishment from cottontails. Communities include so many species and are complex enough, however, that most food web diagrams are bound to be gross simplifications.

Inferring Zebra Mussels' Impacts on Fish Communities

When zebra mussels appeared in the Great Lakes, people feared for sport fisheries and estimated that fish population declines could cost billions of dollars. The mussels would deplete the phytoplankton and zooplankton that fish depended on, people reasoned, and many fewer fish would survive. However, food webs are complicated systems, and disentangling them to infer the effects of any one species is fraught with difficulty. Thus, even after 15 years, scientists had little solid evidence of widespread harm to fish populations.

So, aquatic biologist David Strayer of the Institute of Ecosystem Studies in Millbrook, New York, joined Kathryn Hattala and Andrew Kahnle of New York State's Department of Environmental Conservation (DEC). They mined datasets on fish populations in the Hudson River, which zebra mussels had invaded in 1991.

Strayer and others had already been studying effects of zebra mussels on aspects of the community for years. Their data showed that since the species' introduction to the Hudson:

▶ Biomass of phytoplankton fell 80%.
▶ Biomass of small zooplankton fell 76%.
▶ Biomass of large zooplankton fell 52%.

Zebra mussels increased filter-feeding in the community 30-fold, thereby depleting the phytoplankton and small zooplankton, and leaving all sizes of zooplankton with less phytoplankton to eat. Overall, the zooplankton and invertebrate animals of the open water that are eaten by open-water fish declined by 70%.

However, Strayer's work had also found that *benthic*, or bottom-dwelling, invertebrates in shallow water (especially in the nearshore, or *littoral*, zone) had increased by 10%, and likely much more, because the mussels' shells provide habitat structure and their feces provide nutrients.

These contrasting trends in the benthic shallows and the open deep water led Strayer's team to hypothesize that zebra mussels would harm open-water fish that ate plankton but would help littoral-feeding fish. They predicted that after zebra mussel introduction, larvae and juveniles of six common open-water fish species would decline in number, decline in growth rate, and shift downriver toward saltier water, where mussels are absent. Conversely, they predicted that larvae and juveniles of 10 littoral fish species would increase in number, increase in growth rate, and shift upriver to regions of greatest zebra mussel density.

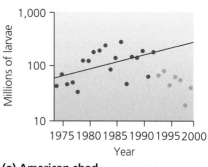

(a) American shad

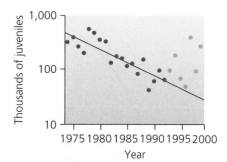

(b) Tessellated darter

Larvae of American shad (**a**), an open-water fish, had been increasing in abundance before zebra mussels were introduced (red points and trend line). After zebra mussel introduction, shad larvae decreased in abundance (orange points). Juveniles of the tessellated darter (**b**), a littoral zone fish, had been decreasing in abundance before zebra mussels were introduced (red points and trend line). After zebra mussel introduction, they increased in abundance (orange points). *Source:* Strayer, D., et al. 2004. Effects of an invasive bivalve (*Dreissena polymorpha*) on fish in the Hudson River estuary. *Can. J. Fish. Aquat. Sci.* 61: 924–941.

To test their predictions, the researchers analyzed data from

eat them; fish that eat phytoplankton and zooplankton; larger fish that eat the smaller fish; and lampreys that parasitize the fish. The food web would include a number of native mussels and clams and, since 1988, the zebra mussel that is outcompeting them. It would include diving ducks that used to feed on native bivalves and now are preying on zebra mussels. This food web would also show that crayfish and other bottom-dwelling invertebrates feed from the refuse of zebra mussels. Finally, the food web would include underwater plants and macroscopic algae, whose

three types of fish surveys carried out over 26 years spanning periods before and after the zebra mussel's arrival. One data set came from surveys conducted from 1985 to 1999 by the DEC. The other two came from surveys conducted from 1974 to 1999 by biologists hired by electric utilities, which in New York are required to monitor fish populations in return for using the Hudson's water for cooling at their power plants.

The researchers compared values for abundance, growth, and distribution for young fish before 1991 with values after 1991. The results supported their predictions. Larvae and juveniles of open-water fish, such as American shad, blueback herring, and alewife, tended to decline in abundance in the years after zebra mussel introduction (first figure, part (a)). Those of littoral fish, such as tessellated darter, bluegill, and largemouth bass, tended to increase (first figure, part (b)). Growth rates showed the same trend: Open-water fish grew more slowly after zebra mussel introduction, whereas littoral fish grew more quickly. In terms of distribution in the 248-km (154-mi) stretch of river studied, open-water fish shifted downstream toward areas with fewer zebra mussels, whereas littoral fish shifted upstream

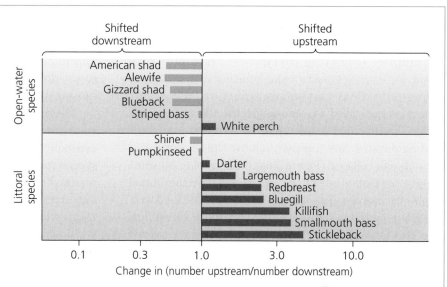

Young of open-water fish, such as American shad, blueback herring, and alewife, tended to shift downstream toward areas with fewer zebra mussels in the years following zebra mussel arrival. Young of littoral fish, such as killifish, bluegill, and largemouth bass, tended to shift upstream toward areas with more zebra mussels. *Source:* Strayer, D., et al. 2004. Effects of an invasive bivalve (*Dreissena polymorpha*) on fish in the Hudson River estuary. *Can. J. Fish. Aquat. Sci.* 61: 924–941.

toward areas with more zebra mussels (second figure). The results were published in 2004 in the *Canadian Journal of Fisheries and Aquatic Sciences.*

Overall, the results supported the hypothesis that the fish community would respond to changes in its food resources caused by zebra mussels. The results are correlative, and correlation does not prove causation. However, previous attempts to address these questions experimentally were so limited in time and scale that they could not reflect true effects in natural systems.

Research such as this helps illuminate the often obscure connections and impacts that particular species interactions have on communities as a whole. In this case, the research may also help fisheries managers predict changes in commercially and recreationally important fish populations. With this knowledge, biologists may be able to manage fisheries more effectively in the Hudson and other areas invaded by zebra mussels.

growth is promoted when zebra mussels clarify the water by filtering out phytoplankton, allowing sunlight to penetrate more deeply, spurring photosynthesis. (See Figure 5.9a, ▶ p. 109, for an illustration of some of these effects.)

Overall, zebra mussels alter the Great Lakes food web

essentially by shifting productivity from the open-water regions to the benthic and *littoral* (nearshore) regions. As such, zebra mussels help benthic and littoral fishes and make life harder for open-water fishes (see "The Science behind the Story," above).

Some organisms play bigger roles in communities than others

"Some animals are more equal than others," George Orwell wrote in his 1945 book *Animal Farm*. Although Orwell was making wry sociopolitical commentary, his remark hints at a truth in ecology. In communities, ecologists have found, some species exert greater influence than do others. A species that has particularly strong or far-reaching impact is often called a **keystone species**. A keystone is the wedge-shaped stone at the top of an arch that is vital for holding the structure together; remove the keystone, and the arch will collapse (Figure 5.6a). In an ecological community, removal of a keystone species will have substantial ripple effects and will alter a large portion of the food web.

Often, large-bodied secondary or tertiary consumers near the tops of food chains are considered keystone species. Top predators control populations of herbivores,

which would otherwise multiply and could, through increased herbivory, greatly modify the plant community (Figure 5.6b). In the United States, for example, government bounties promoted the hunting of wolves and mountain lions, which were largely exterminated by the middle of the 20th century. In the absence of these predators, unnaturally dense deer populations have overgrazed forest-floor vegetation and eliminated tree seedlings, causing major changes in forest structure.

The removal of top predators in the United States was an uncontrolled large-scale experiment with unintended consequences. But ecologists have verified the keystone species concept in controlled experiments. Classic work by marine biologist Robert Paine established that the predatory starfish *Pisaster ochraceus* influences the community composition of intertidal organisms (▶ pp. 260–261) on the Pacific coast of North America. When *Pisaster* is present in this community, species diversity is high, with several

FIGURE 5.6 A keystone is the wedge-shaped stone at the top of an arch that holds its structure together (**a**). A keystone species, such as the sea otter, is one that exerts great influence on a community's composition and structure (**b**). Sea otters consume sea urchins that eat kelp in marine nearshore environments of the Pacific. When otters are present, they keep urchin numbers down, which allows lush underwater forests of kelp to grow and provide habitat for many other species. When otters are absent, urchin populations increase and the kelp is devoured, destroying habitat and depressing species diversity.

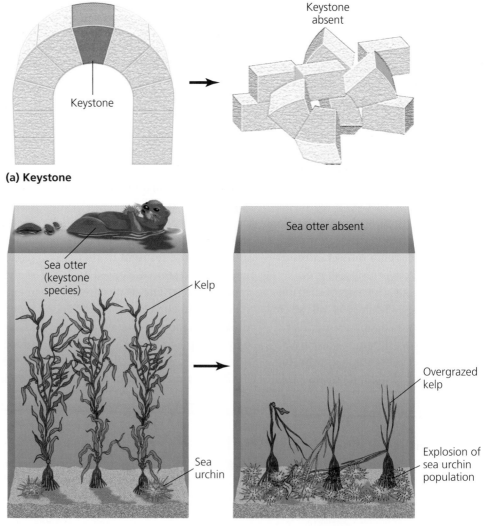

(a) Keystone

(b) A keystone species

types of barnacles, mussels, and algae. When *Pisaster* is removed, the mussels it preys on become numerous and displace other species, suppressing species diversity.

Animals at high trophic levels, such as wolves, starfish, and sea otters, can be keystone species, but other types of organisms also can exert strong community-wide effects. "Ecosystem engineers" physically modify the environment shared by community members. Beavers build dams and turn streams into ponds, flooding acres of dry land and turning them to swamp. Prairie dogs dig burrows that aerate the soil and serve as homes for other animals. Less conspicuous organisms and those toward the bottoms of food chains may have even greater impact. Remove the fungi that decompose dead matter, or the insects that control plant growth, or the phytoplankton that are the base of the marine food chain, and a community may change very rapidly indeed. Because there are usually more species at lower trophic levels, however, it is less likely that any one of them alone might have wide influence; if one species is removed, other species that remain may be able to perform many of its functions.

Communities respond to disturbance in different ways

The removal of a keystone species and the spread of an invasive species are just two of many types of disturbance that can modify the composition, structure, or function of an ecological community. Over time, any given community may experience natural disturbances ranging from gradual phenomena such as climate change to sudden events such as hurricanes, floods, or avalanches.

Communities are dynamic systems and may respond to disturbance in several ways. A community that resists change and remains stable despite disturbance is said to show **resistance** to the disturbance. Alternatively, a community may show **resilience**, meaning that it changes in response to disturbance but later returns to its original state. Or, a community may be modified by disturbance permanently and may never return to its original state.

Succession follows severe disturbance

If a disturbance is severe enough to eliminate all or most of the species in a community, the affected site will undergo a somewhat predictable series of changes that ecologists call **succession**. In the traditional view of this process, ecologists described two types of succession. **Primary succession** follows a disturbance so severe that no vegetation or soil life remains from the community that occupied the site. In primary succession, a biotic community is built essentially from scratch. In contrast, **secondary succession** begins when a disturbance dramatically alters an existing community but does not destroy all life or all organic matter in the soil. In secondary succession, vestiges of the previous community remain, and these building blocks help shape the process.

At terrestrial sites, primary succession takes place after a bare expanse of rock, sand, or sediment becomes newly exposed to the atmosphere. This can occur when glaciers retreat, lakes dry up, or volcanic lava flows spread across the landscape. Species that arrive first and colonize the new substrate are referred to as **pioneer species**. Pioneer species are well adapted for colonization, having traits such as spores or seeds that can travel long distances. The pioneers best suited to colonizing bare rock are the mutualistic aggregates of fungi and algae known as *lichens*. Lichens succeed because their algal component provides

Oaks, hardwoods

Pines

Shrubs, poplar trees

Shrubs, seedlings

Grasses, herbs, forbs

Time ⟶

FIGURE 5.7 Secondary succession in a terrestrial setting occurs after a disturbance, such as fire, landslides, or farming, removes most vegetation from an area. Here is shown a typical series of changes in a plant community of eastern North America following the abandonment of a farmed field.

food and energy via photosynthesis while the fungal component takes a firm hold on rock and captures moisture. As lichens grow, they secrete acids that break down the rock surface. This begins the formation of soil, and soon small plants, insects, and worms find the rocky outcrops more hospitable. As new organisms arrive, they provide more nutrients and habitat for future arrivals. As time passes, larger plants establish themselves, the amount of vegetation increases, and species diversity rises.

Secondary succession on land begins when a fire, a hurricane, logging, or farming removes much of the biotic community. Consider a farmed field in eastern North America that has been abandoned (Figure 5.7). After farming ends, the site will be colonized by pioneer species of grasses and herbs that disperse well or were already in the vicinity. As time passes, shrubs and fast-growing trees such as aspens rise from the field. Pine trees subsequently rise above the aspens and shrubs, forming a pine-dominated forest. This pine forest develops an understory of hardwood trees, because pine seedlings do not grow well under mature pines, whereas some hardwood seedlings do. Eventually the hardwoods outgrow the pines, creating a hardwood forest.

Succession also occurs in aquatic systems. A lake or pond begins to undergo succession as it is colonized by algae, microbes, plants, and zooplankton. As these organisms grow, reproduce, and die, the water body slowly fills with organic matter. The lake or pond acquires further organic matter and sediments from the water it receives from rivers, streams, and surface runoff. Eventually, the water body fills in and undergoes a gradual transition to a terrestrial system (Figure 5.8).

In the traditional view of succession described here, the transitions between stages eventually lead to a *climax community*, which remains in place, with little modification, until some disturbance restarts succession. Early ecologists felt that each region had its own characteristic climax community, determined by the region's climate. Today, ecologists recognize that succession is far more variable and less predictable than originally thought. Assemblages of species may form complex mosaics in space and time. Moreover, once a community is disturbed and succession is set in motion, there is no guarantee that the community will ever return to its original state.

Invasive species pose new threats to community stability

Traditional concepts of communities involve sets of organisms understood to be native to an area. But what if a new organism arrives from elsewhere? And what if this non-native organism turns *invasive*, spreading widely

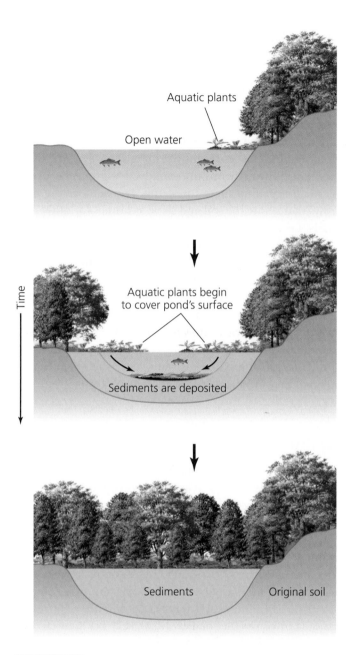

FIGURE 5.8 Primary aquatic succession occurs when plant growth gradually fills in a pond or lake and converts an aquatic system to a wet meadow and ultimately to a terrestrial system. Increased nutrient input can accelerate this process.

and becoming dominant in a community? Such **invasive species** can potentially alter a community substantially, and they are one of the central ecological forces in today's world.

Most often, invasive species are non-native species that people have introduced, intentionally or by accident, from elsewhere in the world. Species become invasive when limiting factors (▸ p. 88–89) that regulate their population growth are removed. Plants and animals brought to a new area may leave behind the predators, parasites,

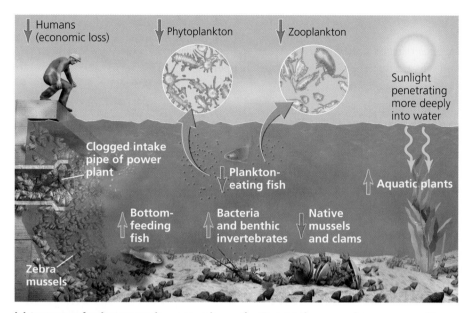

Humans (economic loss)

Phytoplankton

Zooplankton

Sunlight penetrating more deeply into water

Clogged intake pipe of power plant

Plankton-eating fish

Aquatic plants

Bottom-feeding fish

Bacteria and benthic invertebrates

Native mussels and clams

Zebra mussels

FIGURE 5.9 The zebra mussel is a biological invader that has modified an ecological community. By filtering phytoplankton and small zooplankton from open water, it generates a number of impacts on other species, both negative (red downward arrows) and positive (green upward arrows) **(a)**. This map **(b)** shows the range of the zebra mussel in the United States as of 2005. In less than two decades, it has spread from the Great Lakes east to Vermont and Connecticut; west to Nebraska, Kansas, and Oklahoma; and south to Louisiana and Mississippi (shaded states and red dots). It has been transported by people to other states as well (blue dots). *Source:* (b) U.S. Geological Survey.

(a) Impacts of zebra mussels on members of a Great Lakes nearshore community

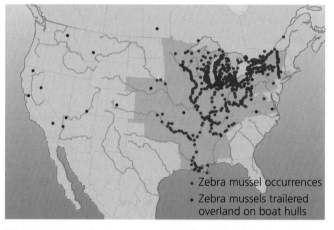

• Zebra mussel occurrences
• Zebra mussels trailered overland on boat hulls

(b) Occurrence of zebra mussels in the United States, 2005

and competitors that had evolved to exploit them in their native land. If there happen to be few organisms in the new environment that can act as predators, parasites, or competitors, the introduced species can do very well. As it proliferates, it may exert diverse influences on its fellow community members (Figure 5.9).

For example, various grasses introduced in the American West by ranchers have overrun entire regions, pushing out native vegetation. Fish introduced into streams for sport compete with and exclude native fish. Hundreds of island-dwelling animals and plants worldwide have been driven extinct by the goats, pigs, and rats introduced by human colonists. We will examine more examples in our discussion of biodiversity in Chapter 8 (▸ pp. 177–178).

Ecologists generally view the impacts of invasive species as overwhelmingly negative. Yet many people enjoy the exotic beauty of introduced ornamental plants,

even those that turn invasive—and some invasive species have provided benefits to our society, such as the European honeybee (▸ p. 156). Whatever view one takes, the impacts of invasive species on native species and ecological communities are significant, and they are growing year by year with the increasing mobility of people and the globalization of our society.

Weighing the issues:
Are Invasive Species All Bad?

Some ethicists have questioned the notion that all invasive species should automatically be considered bad. If we introduce a non-native species to a community and it greatly modifies the community, do you think that is a bad thing? What if it drives another species extinct? What if the invasive species arrived on its own, rather than through human intervention? What ethical standard(s) (▸ p. 15) would you apply to determine whether we should battle or accept a given invasive species?

We can try to restore altered communities to their former condition

Invasive species are adding to the tremendous transformations that humans have already forced on natural landscapes and communities through habitat alteration, deforestation, hunting of keystone species, pollution, and other activities. With so much of Earth's landscape altered by human impact, it is impossible to find areas that are

Invasive Species

Invasive species can substantially modify the ecological communities they invade. **How should we as a society respond to invasive species?**

Prevention Is the Best Strategy

Non-native species have been introduced globally for centuries for food and other human uses, but many human-mediated invasions have caused widespread impacts in freshwater, marine, and terrestrial ecosystems. Environmental impacts include displacement of native fauna and flora, modification of nutrient and chemical cycles and food webs, reduction of groundwater levels, and erosion. Economic problems include extensive impacts on industry, recreation, and fishing. Numerous diseases and pathogenic organisms have been introduced as well.

All of these considerations have led to efforts to *prevent* further invasions and *control* those exotic species that have become established. *Prevention* is one of the most important tools of invasion management. Public education, starting in elementary school, is the long-term foundation of prevention, such that individuals are aware of the concern about introduced species and take responsibility for their personal roles in accidentally or intentionally moving non-native species. For terrestrial species, preventing the arrival of non-native species includes thorough examination of imported goods, shipping containers, or luggage at airports, as well as chemical treatment of imported wood and other products. Non-native aquatic species may arrive by the movement of fishing bait, live seafood, aquaculture species, and particularly ships' ballast water. Prevention measures for ballast water include filtering, heating, ultraviolet treatment, oxygen deprivation, and other techniques.

Controlling invaders that have become established involves a broad range of tools (such as chemical treatments, barriers, and physical removal) focused on limiting the spread and abundance of the target species. Eradication is sometimes possible for terrestrial plants and animals, but it is generally impossible in aquatic environments, where populations often become too widespread to locate and remove all individuals.

Stopping the arrival of unwanted species in the first place is the best long-term management strategy: If it's raining, close the windows.

James T. Carlton is professor of marine sciences at Williams College in Williamstown, Massachusetts, and director of Williams-Mystic, the maritime studies program of Williams College and Mystic Seaport (Mystic, Connecticut). His research is on global marine bioinvasions and marine extinctions in modern times. He is the founding editor-in-chief of the international journal *Biological Invasions*.

Deal with Invasions on a Case-By-Case Basis

Certain introduced species invade natural ecosystems and damage human enterprises such as agriculture and fisheries. However, these harmful invaders are a minority of introduced species. Other introduced species are useful as food crops, as pets, or in other ways. Controversy therefore arises over how to deal with introduced species: Which ones should we target for management, and what sort of management?

It is more efficient to keep introduced species out in the first place than to try to eradicate or manage established populations. For deliberately introduced species, most nations now use risk assessment methods to try to quantify the likelihood that a species will become a pest and then to decide whether to permit introduction. For species that arrive on their own, biologists and policymakers now focus on restricting frequently used pathways, such as ballast water and untreated wood.

Once a species establishes itself, it is often possible to eradicate it, especially if it has not spread far. Because of early failed eradication attempts that wasted much money and harmed nontarget species, many managers hesitate to attempt eradication. However, technologies have improved to the point where there are now many successes.

If eradication fails or is not attempted, there are three general approaches to managing introduced species. Mechanical or physical control entails pulling out weeds by hand or machine, shooting vertebrates, hand-collecting snails, and the like. Chemical control involves using pesticides, but this approach is controversial because the chemicals can have nontarget impacts. Biological control uses natural enemies—predators, parasites, and pathogens— imported from the native range of the introduced pest, but this approach is also controversial because it entails introducing new species, some of which have become invasive pests themselves.

Pronouncements that chemicals are dangerous or biological control is dangerous are far too sweeping, and what is needed is case-by-case examination of all available technologies to deal with established invaders.

Daniel Simberloff is the Nancy Gore Hunger Professor of Environmental Studies at the University of Tennessee. His research is on ecology, biogeography, and evolution and often relates to the causes and consequences of species associations in communities. Much of his research has focused on conservation issues, especially the impacts of introduced species.

Explore this issue further by accessing **Viewpoints** at www.aw-bc.com/withgott

truly pristine. This realization has given rise to the conservation effort known as **ecological restoration**. The practice of ecological restoration is informed by the science of **restoration ecology**. Restoration ecologists research the historical conditions of ecological communities as they existed before our industrialized civilization altered them. They then try to devise ways to restore some of these areas to an earlier condition, often to a natural "presettlement" condition.

For instance, in the United States nearly every last scrap of tallgrass prairie that once covered the eastern Great Plains and parts of the Midwest was converted to agriculture in the 19th century. Now a number of efforts are underway to restore small patches of prairie by planting native prairie plants, weeding out invaders and competitors, and introducing controlled fire to mimic the fires that historically maintained this community.

The world's largest restoration project is the ongoing effort to restore the Florida Everglades. The Everglades, a 7,500-km^2 (4,700-mi^2) system of marshes and seasonally flooded grasslands, has been drying out for decades because the water that feeds it has been managed for flood control and overdrawn for irrigation and development. Economically important fisheries have suffered greatly as a result, and populations of wading birds have dropped by 90–95%. The 30-year, $7.8-billion restoration project intends to restore water by undoing damming and diversions of 1,600 km (1,000 mi) of canals, 1,150 km (720 mi) of levees, and 200 water control structures. Because the Everglades provides drinking water for millions of Florida citizens, as well as considerable tourism revenue, restoring the ecosystem services (▶ pp. 25, 27) of this system should prove economically beneficial as well as ecologically valuable.

As our population grows and development spreads, ecological restoration is becoming an increasingly vital conservation strategy. However, restoration is difficult and expensive, and it rarely succeeds in creating systems that function as effectively as natural systems. It is therefore best, whenever possible, to protect natural systems from degradation so that restoration does not become necessary.

Weighing the Issues:
Restoring "Natural" Communities

Practitioners of ecological restoration in North America aim to restore communities to their natural state. But what is meant by "natural"? Does it mean the state of the community before industrialization? Before Europeans came to the New World? Before any people laid eyes on

the community? Let's say Native Americans altered a forest community 8,000 years ago by burning the underbrush regularly to improve hunting, and continued doing so until Europeans arrived 400 years ago and cut down the forest for farming. Today the area's inhabitants want to restore the land to its "natural" forested state. Should restorationists try to recreate the forest of the Native Americans, or the forest that existed before Native Americans arrived? What values do you think underlie the desire for restoration?

Earth's Biomes

Across the world, each portion of each continent has different sets of species, leading to endless variety in community composition. However, communities in far-flung places often share strong similarities in their structure and function. This allows us to classify communities into broad types. A **biome** is a major regional complex of similar communities—a large ecological unit recognized primarily by its dominant plant type and vegetation structure. The world contains a number of biomes, each covering large contiguous geographic areas (Figure 5.10).

Which biome covers any particular portion of the planet depends on a variety of abiotic factors, including temperature, precipitation, atmospheric circulation, and soil characteristics. Among these factors, temperature and precipitation exert the greatest influence (Figure 5.11). Because biome type is largely a function of climate, and because average monthly temperature and precipitation are among the best indicators of an area's climate, scientists often use climate diagrams, or *climatographs*, to depict such information.

We can divide the world into roughly ten terrestrial biomes

Temperate deciduous forest The **temperate deciduous forest** (Figure 5.12, ▶ p. 114) that dominates the landscape around the southern Great Lakes is characterized by broad-leafed trees that are *deciduous*, meaning that they lose their leaves each fall and remain dormant during winter, when hard freezes would endanger leaves. These midlatitude forests occur in much of Europe and eastern China as well as in eastern North America—all areas where precipitation is spread relatively evenly throughout the year. Although soils of the temperate deciduous forest are relatively fertile, the biome generally consists of far

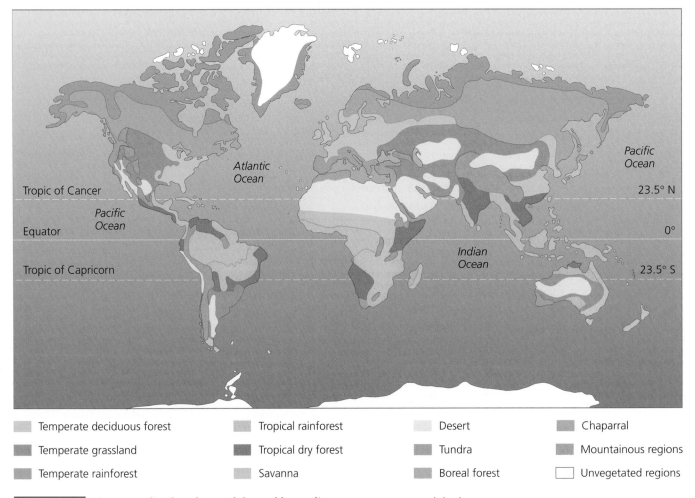

Temperate deciduous forest   Tropical rainforest   Desert   Chaparral

Temperate grassland   Tropical dry forest   Tundra   Mountainous regions

Temperate rainforest   Savanna   Boreal forest   Unvegetated regions

FIGURE 5.10   Biomes are distributed around the world according to temperature, precipitation, atmospheric and oceanic circulation patterns, and other factors.

fewer tree species than are found in tropical rainforests. Oaks, beeches, and maples are a few of the most common trees in these forests. Some typical animals of the temperate deciduous forest of eastern North America are shown in Figure 5.5 (▸ p. 103).

Temperate grassland Traveling westward from the Great Lakes, we find **temperate grasslands** (Figure 5.13, ▸ p. 114). Temperate grasslands arise where temperature differences between winter and summer become more extreme and rainfall diminishes. The limited precipitation in the Great Plains region can support grasses more easily than trees. Also known as steppe or prairie, temperate grasslands were once widespread throughout parts of North and South America and much of central Asia. Today people have converted most of the world's grasslands for agriculture. Characteristic vertebrates of the North American grasslands include American bison,

prairie dogs, pronghorn antelope, and ground-nesting birds such as meadowlarks.

Temperate rainforest Further west in North America, the topography becomes varied, and biome types intermix. The coastal Pacific Northwest region, with its heavy rainfall, features **temperate rainforest** (Figure 5.14, ▸ p. 115). Coniferous trees, such as cedars, spruces, hemlocks, and Douglas fir, grow very tall in the temperate rainforest, and the forest interior is shaded and damp. Moisture-loving animals such as the bright yellow banana slug are common, and old-growth stands hold the endangered spotted owl. The soils of temperate rainforests are usually quite fertile but are susceptible to landslides and erosion if forests are cleared. Temperate rainforests have been the focus of controversy in the Pacific Northwest, where overharvesting of old-growth timber has driven some species toward extinction and

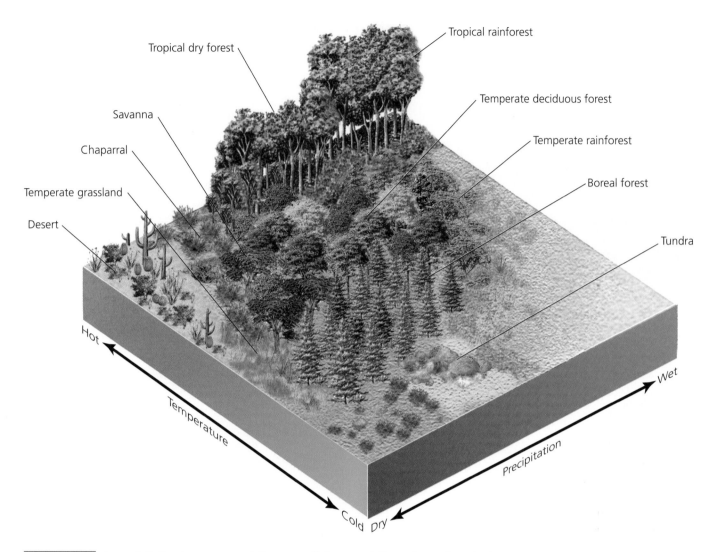

FIGURE 5.11 As precipitation increases, vegetation generally becomes taller and more luxuriant. As temperature increases, types of plant communities change. Together, temperature and precipitation are the main factors determining which biome occurs in a given area.

pushed many forest-dependent human communities toward economic stagnation.

Tropical rainforest In tropical regions we see the same pattern found in temperate regions: Areas of high rainfall grow rainforests, areas of intermediate rainfall host dry or deciduous forests, and areas of lower rainfall become dominated by grasses. However, tropical biomes differ from their temperate counterparts in other ways because they are closer to the equator and therefore warmer on average year-round. For one thing, they hold far greater biodiversity.

Tropical rainforest (Figure 5.15, ▶ p. 115) is found in Central America, South America, southeast Asia, west Africa, and other tropical regions and is characterized by year-round rain and uniformly warm temperatures. Tropical rainforests have dark, damp interiors, lush vegetation, and highly diverse communities, with greater numbers of species of insects, birds, amphibians, and various other animals than any other biome. These forests are not dominated by single species of trees, as are forests closer to the poles, but instead consist of very high numbers of tree species intermixed, each at a low density. Any given tree may be draped with vines, enveloped by strangler figs, and loaded with epiphytes (orchids and other plants that grow in trees), such that trees occasionally collapse under the weight of all the life they support. Despite this profusion of life, tropical rainforests often have poor, acidic soils that are low in organic matter. Nearly all nutrients in this biome are contained in the plants, not in the soil. An unfortunate

(a) Temperate deciduous forest

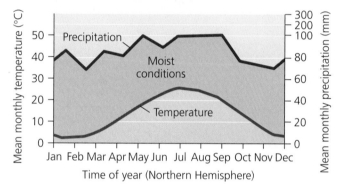

(b) Washington, D.C., USA

FIGURE 5.12 Temperate deciduous forests experience relatively stable seasonal precipitation but variation in seasonal temperatures. Scientists use climate diagrams to illustrate an area's average monthly precipitation and temperature. Typically in these diagrams, the *x* axis marks months of the year, and paired *y* axes denote average monthly temperature and precipitation. The twin curves plotted indicate trends in precipitation (blue) and temperature (red) from month to month. When the precipitation curve lies above the temperature curve, as is the case year-round in the temperate deciduous forest around Washington, D.C., the region experiences relatively "moist" conditions, indicated with green coloration. Climatograph adapted from Breckle, S. W. 1999. *Walter's vegetation of the Earth: The ecological systems of the geo-biosphere*, 4th ed. Berlin: Springer-Verlag.

consequence is that once tropical rainforests are cleared, the nutrient-poor soil can support agriculture for only a short time. As a result, farmed areas are abandoned quickly, and the soil and forest vegetation recover slowly.

Tropical dry forest Tropical areas that are warm year-round but where rainfall is lower overall and highly seasonal give rise to **tropical dry forest**, or tropical deciduous

forest (Figure 5.16, ▶ p. 116), a biome widespread in India, Africa, South America, and northern Australia. Wet and dry seasons each span about half a year in tropical dry forest. Rains during the wet season can be extremely heavy and, coupled with erosion-prone soils, can lead to severe soil loss where people have cleared forest. Across the globe, much tropical dry forest has been converted to agriculture. Clearing for farming or ranching is made easier by the fact that vegetation heights are much lower and canopies less dense than in tropical rainforest. Organisms that inhabit tropical dry forest have adapted to seasonal fluctuations in precipitation and temperature. For instance, plants are deciduous and often leaf out and grow profusely with the rains, then drop their leaves during the driest times of year.

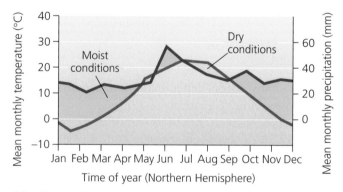

(a) Temperate grassland

(b) Odessa, Ukraine

FIGURE 5.13 Temperate grasslands experience seasonal temperature variation and too little precipitation for many trees to grow. This climatograph indicates "moist" (green) climate conditions, as well as "dry" climate conditions (in yellow, when the temperature curve is above the precipitation curve in May and mid-June through September). Climatograph adapted from Breckle, S. W., 1999.

Savanna Drier tropical regions give rise to **savanna** (Figure 5.17 ▸ p. 116), tropical grassland interspersed with clusters of acacias or other trees. The savanna biome is found today across stretches of Africa, South America, Australia, India, and other dry tropical regions. Precipitation in savannas usually arrives during distinct rainy seasons and concentrates grazing animals near widely spaced water holes. Common herbivores on the African savanna include zebras, gazelles, and giraffes, and the predators of these grazers include lions, hyenas, and other highly mobile carnivores.

Desert Where rainfall is very sparse, **desert** (Figure 5.18, ▸ p. 117) forms. The driest biome on Earth, most deserts receive well under 25 cm (9.8 in.) of precipitation per year, much of it during isolated storms months or years apart. Some deserts, like Africa's Sahara and Namib deserts, are

(a) Tropical rainforest

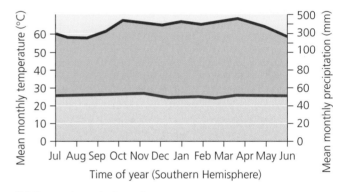

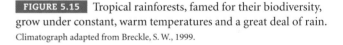

(b) Bogor, Java, Indonesia

FIGURE 5.15 Tropical rainforests, famed for their biodiversity, grow under constant, warm temperatures and a great deal of rain. Climatograph adapted from Breckle, S. W., 1999.

(a) Temperate rainforest

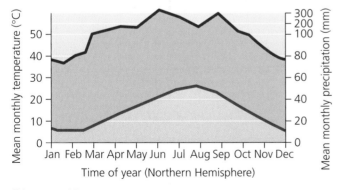

(b) Nagasaki, Japan

FIGURE 5.14 Temperate rainforests receive a great deal of precipitation and feature moist, mossy interiors. Climatograph adapted from Breckle, S. W., 1999.

mostly bare sand dunes; others, like the Sonoran Desert of Arizona and northwest Mexico, receive more rain and are more heavily vegetated. Because deserts have low humidity and relatively little vegetation to insulate them from temperature extremes, sunlight readily heats them in the daytime, but daytime heat is quickly lost at night. As a result, temperatures vary widely from day to night and across seasons of the year.

Desert soils can often be quite saline and are sometimes known as lithosols, or stone soils, for their high mineral and low organic-matter content. Desert animals and plants show many adaptations to deal with a harsh climate. Most reptiles and mammals, such as rattlesnakes and kangaroo mice, are active in the cool of night, and many Australian desert birds are nomadic, wandering long distances to find areas of recent rainfall and plant growth. Many desert plants have thick leathery leaves to

(a) Tropical dry forest

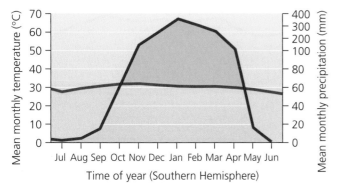

(b) Darwin, Australia

FIGURE 5.16 Tropical dry forests experience significant seasonal variations in precipitation and relatively stable warm temperatures. Climatograph adapted from Breckle, S. W., 1999.

reduce water loss, or green trunks that enable the plant to photosynthesize without leaves, which lose water. The spines of cacti and many other desert plants guard those plants from being eaten by herbivores desperate for the precious water they hold.

Tundra Nearly as dry as desert, **tundra** (Figure 5.19) is located at very high latitudes along the northern edges of Russia, Canada, and Scandinavia. Extremely cold winters with little daylight and moderately cool summers with lengthy days characterize this landscape of lichens and low, scrubby vegetation without trees. The great seasonal variation in temperature and day length results from this biome's position close to the poles, which are angled toward the sun in the summer and away from the sun in the winter. Because of the cold climate, underground soil remains more or less permanently frozen, and is called *permafrost*. During the long, cold winters, the surface

soils freeze as well; then, when the weather warms, they melt and produce seasonal accumulations of surface water that make ideal habitat for mosquitoes and other insects. The swarms of insects benefit bird species that migrate long distances to breed during the brief but productive summer. Caribou also migrate to the tundra to breed, and they then leave for the winter. Only a few animals, such as polar bears and musk oxen, can survive year-round in this extreme climate. Tundra also occurs as *alpine tundra* at the tops of high mountains in temperate and tropical regions.

Boreal forest The **boreal forest**, often called *taiga* (Figure 5.20, ▶ p. 118), stretches in a broad band across much of Canada, Alaska, Russia, and Scandinavia. It consists of a few species of evergreen trees, such as black spruce, that dominate large stretches of forest interspersed with occasional bogs and lakes. These forests

(a) Savanna

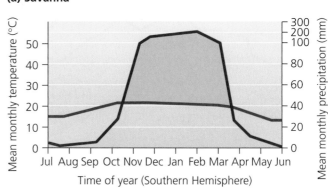

(b) Harare, Zimbabwe

FIGURE 5.17 Savannas are grasslands with clusters of trees. They experience slight seasonal variation in temperature but significant variation in rainfall. Climatograph adapted from Breckle, S. W., 1999.

develop in cooler, drier regions than do temperate forests, and they experience long, cold winters and short, cool summers. Soils are typically nutrient-poor and somewhat acidic. As a result of the strong seasonal variation in day length, temperature, and precipitation, many organisms compress a year's worth of feeding and breeding into a few warm, wet months. Year-round residents of boreal forest include mammals such as moose, wolves, bears, lynx, and many rodents. This biome also hosts many insect-eating birds that migrate from the tropics to breed during the brief, intensely productive, summer season.

Chaparral In contrast to the boreal forest's broad, continuous distribution, **chaparral** (Figure 5.21, ▸ p. 118) is limited to fairly small patches widely flung around the

(a) Tundra

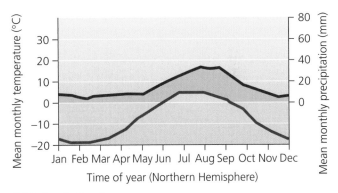

(b) Vaigach, Russia

FIGURE 5.19 Tundra is a cold, dry biome found near the poles and atop high mountains at lower latitudes. Climatograph adapted from Breckle, S. W., 1999.

(a) Desert

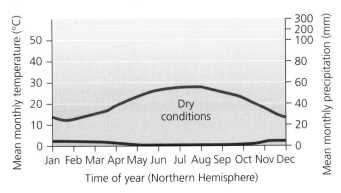

(b) Cairo, Egypt

FIGURE 5.18 Deserts are dry year-round, but they are not always hot. The temperature curve is consistently above the precipitation curve in this climatograph of Cairo, Egypt, indicating that the region experiences "dry" conditions all year. The photograph, from the Sonoran Desert in Arizona, shows the maximum amount of vegetation a desert can support. Climatograph adapted from Breckle, S. W., 1999.

globe. Chaparral consists mostly of evergreen shrubs and is densely thicketed. This biome is also highly seasonal, with mild, wet winters and warm, dry summers. This type of climate is induced by oceanic influences and is often termed "Mediterranean." In addition to ringing the Mediterranean Sea, chaparral occurs along the coasts of California, Chile, and southern Australia. Chaparral communities experience frequent fire, and their plant species are adapted to resist fire or even to depend on it for germination of their seeds.

Aquatic systems also show biome-like patterns

We have focused exclusively on terrestrial systems because the biome concept, as traditionally developed and applied, has been limited to terrestrial systems. Areas equivalent to biomes also exist in the oceans, but their geographic shapes would look very different from those of terrestrial biomes if plotted on a world map. One might consider the thin strips along the world's coastlines to represent one aquatic system, the continental shelves another, and the open ocean, the deep sea, coral reefs, and kelp forests as still others. There are also many coastal systems that straddle the line between terrestrial and aquatic, such as salt marshes, rocky intertidal communities, mangrove forests, and estuaries. And of course there are freshwater systems such as those of the Great Lakes.

Unlike terrestrial biomes, aquatic systems are shaped not by air temperature and precipitation, but by factors such as water temperature, salinity, dissolved nutrients, wave action, currents, depth, and type of substrate (e.g., sandy, muddy, or rocky bottom). Marine communities are also more clearly delineated by their animal life than by their plant life. We will examine freshwater and marine systems in the greater detail they deserve in Chapter 11.

(a) Boreal forest

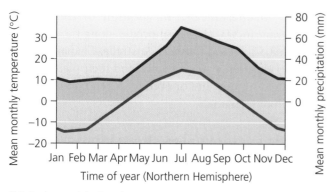

(b) Archangelsk, Russia

FIGURE 5.20 Boreal forest is defined by long, cold winters, relatively cool summers, and moderate precipitation. Climatograph adapted from Breckle, S. W., 1999.

(a) Chaparral

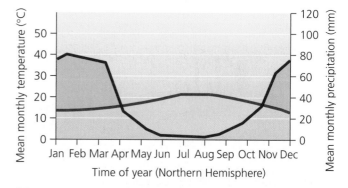

(b) Los Angeles, California, USA

FIGURE 5.21 Chaparral is a highly seasonal biome dominated by shrubs, influenced by marine weather, and dependent on fire. Climatograph adapted from Breckle, S. W., 1999.

Conclusion

The natural world is so complex that we can visualize it in many ways and at various scales. Dividing the world's communities into major types, or biomes, is informative at the broadest geographic scales. Understanding how communities function at more local scales requires understanding how species interact with one another. Species interactions such as competition, predation, parasitism, and mutualism give rise to effects both weak and strong, direct and indirect. Feeding relationships can be represented by the concepts of trophic levels and food webs, and particularly influential species are sometimes called keystone species. Increasingly humans are altering communities, partly by introducing non-native species that may turn invasive. But increasingly, through ecological restoration, we are also attempting to undo the changes we have caused.

TESTING YOUR COMPREHENSION

1. How does competition promote resource partitioning?
2. Contrast the several types of exploitative species interactions. How do predation, parasitism, and herbivory differ?
3. Give examples of symbiotic and nonsymbiotic mutualisms. Describe at least one way in which mutualisms affect your daily life.
4. What are trophic levels? Use the concepts of trophic levels and energy flow to state why the ecological footprint of a vegetarian is smaller than that of a meat eater.
5. Differentiate between a food chain and a food web. Which best represents the reality of communities, and why?
6. What is meant by a keystone species, and what types of organisms are most often considered keystone species?

7. Explain primary succession. How does it differ from secondary succession?
8. What factors most strongly influence the type of biome that forms in a particular place on land? What factors determine the type of aquatic system that may form in a given location?
9. Draw a typical climate diagram for a tropical rainforest. Label all parts of the diagram, and describe all of the types of information an ecologist could glean from such a diagram.
10. Now draw a climate diagram for a desert. How does it differ from the rainforest climatograph, and what does this tell you about how the two biomes differ?

SEEKING SOLUTIONS

1. Imagine that you spot two species of birds feeding side by side, eating seeds from the same plant. You begin to wonder whether competition is at work. Describe how you might design scientific research to address this question. What observations would you try to make at the outset? Would you try to manipulate the system to test your hypothesis that the two birds are competing? If so, how?
2. Spend some time outside on your campus or in your yard or in the nearest park or natural area. Find at least 10 species of organisms (plants, animals, or others), and observe them long enough to watch them feed or to make an educated guess about how they derive their nutrition. Now, using Figure 5.5 as a model, draw a simple food web involving all the organisms you observed.

3. Can you think of one organism not mentioned in this chapter as a keystone species that you believe may be a keystone species? For what reasons do you suspect this? How could you experimentally test whether an organism is a keystone species?
4. Why do scientists consider invasive species to be a problem? What makes a species "invasive," and what ecological effects can invasive species have?
5. Can you devise possible responses to the zebra mussel invasion? What strategies would you consider if you were put in charge of the effort to control this species' spread and reduce its impacts? Name one advantage of each of your ideas, and identify one obstacle that each might face in being implemented.

CALCULATING ECOLOGICAL FOOTPRINTS

Species appearing in a new area are generally called "invasive" if they increase markedly in population and also increase in their impacts on the biotic communities and landscapes around them. By these measures, are human beings an invasive species? The table shows human population and per capita energy consumption for the United States across the time period of two generations, from 1950 to 1975 and from 1975 to 2000. Total energy consumption gives us a rough measure of total environmental impact. Calculate total energy consumption by multiplying the population and per capita consumption values.

Year	Human population (U.S., in millions)	Per capita energy consumption (U.S., in million BTU)	Total energy consumption (U.S., in quadrillion BTU)
1950	151	229	
1975	216	334	
2000	281	352	

Data source: U.S. Census Bureau and U.S. Energy Information Administration.

1. In percentage terms, how much did total energy consumption, as calculated here, increase between 1950 and 1975? Between 1975 and 2000?
2. What effects on biotic communities do you think this increase in total energy consumption has had? Speculate on overall effects, and give several specific known or likely examples.
3. After considering these data, do you consider yourself to be a member of an invasive species? Why or why not?

Take It Further

Go to www.aw-bc.com/withgott, where you'll find:

▶ Suggested answers to end-of-chapter questions
▶ Quizzes, animations, and flashcards to help you study
▶ *Research Navigator*™ database of credible and reliable sources to assist you with your research projects

▶ **GRAPHit!** tutorials to help you interpret graphs
▶ **INVESTIGATEit!** current news articles that link the topics that your study to case studies from your region to around the world

Environmental Issues
and the Search for Solutions

Canal Street,
New Orleans, after
Hurricane Katrina

6 Human Population

Upon successfully completing this chapter, you will be able to:

▶ Assess the scope of human population growth

▶ Evaluate how human population, affluence, and technology affect the environment

▶ Explain and apply the fundamentals of demography

▶ Outline and assess the concept of demographic transition

▶ Describe how wealth and poverty, the status of women, and family-planning programs affect population growth

Central Case: China's One-Child Policy

"Population growth is analogous to a plague of locusts. What we have on this earth today is a plague of people."
—TED TURNER, MEDIA
MAGNATE AND SUPPORTER OF
THE UNITED NATIONS
POPULATION FUND

"There is no population problem."
—SHELDON RICHMAN, SENIOR
EDITOR, CATO INSTITUTE

The People's Republic of China is the world's most populous nation, home to one-fifth of the 6.5 billion people living on Earth at the start of 2006. When Mao Zedong founded the country's current regime 57 years earlier, roughly 540 million people lived in a mostly rural, war-torn, impoverished nation. Mao believed population growth was desirable, and under his leadership China grew and changed. By 1970, improvements in food production, food distribution, and public health allowed China's population to swell to approximately 790 million people. At that time, the average Chinese woman gave birth to 5.8 children in her lifetime.

Unfortunately, the country's burgeoning population and its industrial and agricultural development were eroding the nation's soils, depleting its water, leveling its forests, and polluting its air. Chinese leaders realized that the nation might not be able to feed its people if their numbers grew much larger. They saw that continued population growth could exhaust resources and threaten the stability and economic progress of Chinese society. The government decided to institute a population-control program that precluded large numbers of Chinese couples from having more than one child.

The program began with education and outreach efforts encouraging people to marry later and have fewer children. Along with these efforts, the Chinese government increased the accessibility of contraceptives and abortion. By 1975, China's annual population growth rate had dropped from 2.8% to 1.8%. To further decrease birth rates, in 1979 the government took the more drastic step of instituting a system of rewards and punishments to enforce a one-child limit. One-child families received better access to schools, medical care, housing, and government jobs, and mothers with only one child were given longer maternity leaves. Families with more than one child, meanwhile, were subjected to social scorn and ridicule, employment discrimination, and monetary fines. In some cases, the fines exceeded half the offending couple's annual income.

In enforcing these policies, China has been conducting one of the largest and most controversial social experiments in history. In purely quantitative terms, the experiment has been a major success; the nation's growth rate is now down to 0.6%, making it easier for the country to deal with its many social, economic, and environmental challenges. However, China's population control policies have also produced unintended consequences, such as widespread killing of female infants, an unbalanced sex ratio, and a black-market trade in teenaged girls. Moreover, the policies have elicited intense criticism from people who oppose government intrusion into personal reproductive choices.

China embarked on its policy because its leaders felt the steps were necessary. As other nations become more and more crowded, might their governments also feel forced to turn to drastic policies that restrict individual freedoms? In this chapter, we examine human population dynamics worldwide, consider their causes, and assess their consequences for the environment and our society.

Human Population Growth: Baby 6 Billion and Beyond

While China was working to slow its population growth and speed its economic growth, on the other side of the Eurasian continent a milestone was reached in 1999. On the morning of October 12 of that year, the first cries of a newborn baby in Sarajevo, Bosnia-Herzegovina, marked the arrival of the six-billionth human being on our planet (Figure 6.1). At least that was how the milestone was symbolically marked by the United Nations, which monitors human population growth, among other global trends.

Just how much is 6 billion? We often have trouble conceptualizing the scale of such huge numbers. A billion is 1,000 times greater than a million. If you were to count once each second without ever sleeping, it would take over 30 years to reach a billion. To put a billion miles on your car, you would need to drive from New York to Los Angeles more than 350,000 times.

FIGURE 6.1 U.N. Secretary-General Kofi Annan recognized the newborn son of Fatima Nevic and her husband, Jasminko, as our six-billionth neighbor. Although it is impossible to know the precise moment the world's population reached 6 billion, U.N. population experts pinpointed October 12, 1999, as the best approximation to make the symbolic declaration. Many observers interpreted the selection of a child born in war-ravaged Sarajevo as a harbinger of the hard times that could face future generations as population grows and competition for scarce resources increases.

The human population is growing nearly as fast as ever

As we saw in Chapter 1 (▸ p. 4), the human population has been growing at a tremendous rate. Our population has doubled just since 1964 and is growing by roughly 78 million people annually (nearly 2.5 people every *second*). This is the equivalent of adding all the people of California, Texas, and New York to the world each year. It took until after 1800, virtually all of human history, for our population to reach 1 billion. Yet we expanded from 5 billion to 6 billion in just 12 years (Figure 6.2). Think about when you were born and how many people have been added to the planet just since that time. This unprecedented growth means that today's generations are in circumstances that previous generations never experienced. Our grandparents never had to deal with the number of people that crowd our planet today.

We saw in Chapter 4 (▸ p. 88) how exponential growth—the increase in a quantity by a fixed percentage per unit time—accelerates the absolute increase of population size over time, just as compound interest accrues in a savings account. The reason, you will recall, is that a given percentage of a large number is a greater quantity than the same percentage of a small number. Thus, even if the growth rate remains steady, popula-

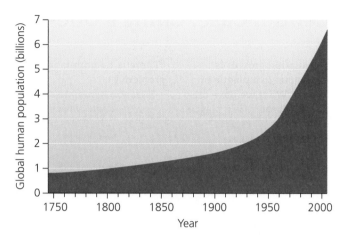

FIGURE 6.2 The global human population has grown exponentially, rising from less than 1 billion in 1800 to over 6.5 billion today. Data from U.S. Bureau of the Census.

tion size will increase by greater increments with each successive generation.

In fact, our growth rate has not remained steady. Instead, for much of the 20th century, the growth rate of the human population actually rose from year to year. It peaked at 2.1% during the 1960s and has declined to 1.2% since then. Although 1.2% may sound small, exponential growth endows small numbers with large consequences. For instance, a hypothetical population starting with 1 man and 1 woman that grows at 1.2% gives rise to a population of 2,939 after 40 generations and 112,695 after 60 generations. In today's world, rates of annual growth vary greatly from region to region (Figure 6.3).

At a 2.1% annual growth rate, a population doubles in size in only 33 years. For low rates of increase, we can estimate doubling times with a handy rule of thumb. Just take the number 70, and divide it by the annual percentage growth rate: 70 ÷ 2.1 = 33.3. Had China not instituted its one-child policy—and had its growth rate remained unchecked at 2.8%—it would have taken only 25 years to double in size.

Is population growth really a "problem"?

Our ongoing population growth has resulted largely from technological innovations, improved sanitation, better medical care, increased agricultural output, and other factors that have led to a decline in death rates, particularly a drop in rates of infant mortality. Birth rates have not declined as much, so births have outpaced deaths for many years now. Thus, the so-called population problem actually arises from a very good thing—our ability to keep more people alive longer.

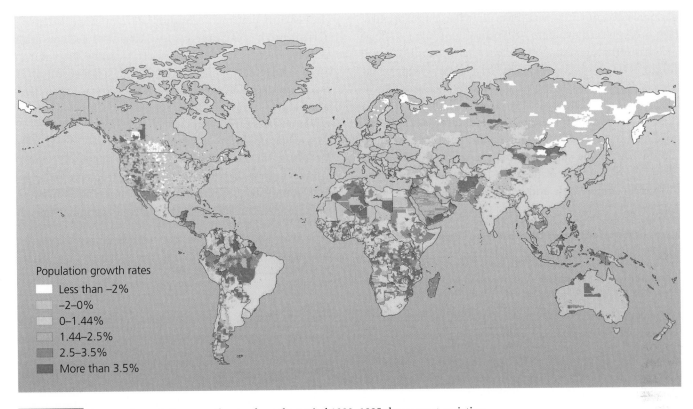

FIGURE 6.3 A map of population growth rates from the period 1990–1995 shows great variation from place to place. Population is growing fastest in tropical regions and in some desert and rainforest areas that have historically been sparsely populated. Data from Center for International Earth Science Information Network (CIESIN), Columbia University; and Harrison, P. and F. Pearce. 2000. *AAAS atlas of population and environment.* Berkeley, CA: University of California Press.

Indeed, just as the mainstream view in the day of Thomas Malthus (▸ p. 5) held that population increase was a good thing, many people today argue that population growth poses no problems. Under the Cornucopian view that many economists hold, resource depletion due to population increase is not a problem if new resources can be found to replace depleted resources (▸ p. 28). Libertarian writer Sheldon Richman expressed this view at the time the six-billionth baby was born:

> The idea of carrying capacity doesn't apply to the human world because humans aren't passive with respect to their environment. Human beings create resources. We find potential stuff and human intelligence turns it into resources. The computer revolution is based on sand; human intelligence turned that common stuff into the main component [silicon] of an amazing technology.

In contrast to Richman's point of view, environmental scientists recognize that few resources are actually created by people and that not all resources can be replaced once they are depleted. For example, once a species has gone extinct, we cannot replicate its exact function in an ecosystem, or

know what medicines or other practical applications we might have obtained from it, or regain the educational and aesthetic value of observing it. Another irreplaceable resource is land, that is, space in which to live; we cannot expand Earth like a balloon to increase its surface area.

Even if resource substitution could hypothetically enable population growth to continue indefinitely, could we maintain the *quality* of life that we would desire for ourselves and our descendants? Surely some of today's resources are bound to be easier or cheaper to use, or less environmentally destructive to harvest or mine, than any resources that can replace them. Replacing such resources might make our lives more difficult or less pleasant. In any case, unless resource availability keeps pace with population growth, the average person in the future will have less space in which to live, less food to eat, and less material wealth than the average person does today. Thus population increases are indeed a problem if they create stress on resources, social systems, or the natural environment, such that our quality of life declines.

Nonetheless, many governments have found it difficult to let go of the notion that population growth increases a nation's economic, political, or military strength. Many national governments, even those that view global population increase as a problem, still offer financial and social incentives that encourage their citizens to produce more children. Governments of countries currently experiencing population declines (such as many in Europe) feel especially uneasy. According to the Population Reference Bureau, more than three of every five European national governments now take the view that their birth rates are too low, and none state that theirs is too high. However, outside Europe, 56% of national governments feel their birth rates are too high, and only 8% feel they are too low.

Population is one of several factors that affect the environment

The extent to which population increase can be considered a problem involves more than just numbers of people. One widely used formula gives us a handy way to think about factors that affect the environment. Nicknamed the **IPAT model**, it is a variation of a formula proposed in 1974 by Paul Ehrlich (▸ p. 5) and John Holdren, a professor of environmental policy at Harvard University. The IPAT model represents how our total impact (I) on the environment results from the interaction among population (P), affluence (A), and technology (T):

$$I = P \times A \times T$$

Increased population intensifies impact on the environment as more individuals take up space, use natural resources, and generate waste. Increased affluence magnifies environmental impact through the greater per capita resource consumption that generally has accompanied enhanced wealth. Changes in technology may either decrease or increase human impact on the environment. Technology that enhances our abilities to exploit minerals, fossil fuels, old-growth forests, or ocean fisheries generally increases impact, but technology to reduce smokestack emissions, harness renewable energy, or improve manufacturing efficiency can decrease impact.

We can refine the IPAT equation by adding terms for the sensitivity of given environments to human impact, and for the effects of social institutions such as education, laws and their enforcement, stable and cohesive societies, and ethical standards that promote environmental well-being. Factors like these all affect how population, affluence, and technology translate into environmental impact.

Modern-day China shows how elements of the IPAT formula can combine to cause rapid environmental impact. While millions of Chinese are increasing their material wealth and resource consumption, the country is battling unprecedented environmental challenges brought about by its pell-mell economic development. Intensive agriculture has expanded westward out of the country's historic moist rice-growing areas, causing farmland to erode and literally blow away. China has overpumped many of its aquifers and has drawn so much water for irrigation from the Yellow River that the once-mighty waterway now dries up in many stretches. The nation now faces new urban pollution and congestion threats from rapidly increasing numbers of automobiles. As the world's developing countries try to attain the level of material prosperity that industrialized nations enjoy, China is a window on what much of the rest of the world could soon become.

Demography

Humans exist within their environment as one species out of many. As such, all the principles of population ecology we outlined in Chapter 4 that apply to toads, frogs, and passenger pigeons apply to humans as well. Environmental factors set limits on our population growth, and the environment has a carrying capacity (▸ p. 89) for our species, just as it does for every other. Environmental scientists who have tried to pin a number to the human carrying capacity have come up with wildly differing estimates. The most rigorous estimates range from 1–2 billion people living prosperously in a healthy environment to 33 billion living in extreme poverty in a degraded world of intensive cultivation without natural areas.

We happen to be a particularly successful organism, however—one that has repeatedly raised its carrying capacity by developing technology to overcome the natural limits on its growth. We did so with the agricultural and the industrial revolutions (▸ pp. 4–5) and likely before that with our invention of tools (Figure 6.4). As our population climbs toward 7 billion and beyond, we may yet continue to find ways to raise our carrying capacity. Given our knowledge of population ecology, however, we know for certain that human numbers cannot go on growing indefinitely.

Demography is the study of human population

The application of population ecology principles to the study of statistical change in human populations is the focus of the social science of **demography**. Data gathered by demographers help us understand how differences in population characteristics and related phenomena (for instance, decisions about reproduction) affect human communities and their environments. Demographers study population size, density, distribution, age structure, sex ratio, and rates

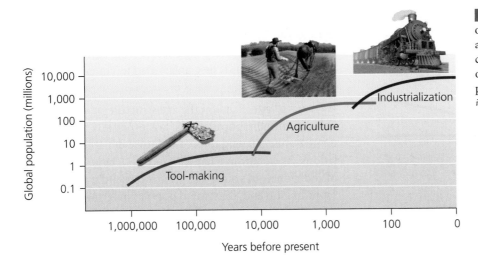

FIGURE 6.4 Tool making, the advent of agriculture, and industrialization each allowed our species to raise its global carrying capacity. The logarithmic scale of the axes makes it easier to visualize this pattern. Data from Goudie, A. 2000. *The human impact.* Cambridge, MA: MIT Press.

of birth, death, immigration, and emigration of people, just as population ecologists study these characteristics in other organisms. Each of these characteristics helps predict population dynamics and environmental impacts.

Population size The global human population of more than 6.5 billion consists of well over 200 nations with populations ranging from China's 1.3 billion, India's 1.1 billion, and the 300 million of the United States down to a number of island nations with populations below 100,000 (Figure 6.5). The size that our global population will eventually reach remains to be seen (Figure 6.6). However, population size alone—the absolute number of individuals—doesn't tell the whole story. Rather, a population's environmental impact depends on its density, distribution, and composition (as well as on affluence, technology, and other factors outlined earlier).

Population density and distribution People are distributed very unevenly over the globe. In ecological terms, our distribution is clumped (▶ p. 86) at all spatial scales. At the largest scales (Figure 6.7), population density is high in regions with temperate, subtropical, and tropical climates, such as China, Europe, Mexico, southern Africa, and India. Population density is low in regions with extreme-climate biomes, such as desert, deep rainforest, and tundra. Dense along seacoasts and rivers, human population is less dense at locations far from water. At intermediate scales, we cluster together in cities and suburbs and are spread more sparsely across rural areas. At small scales, we cluster in neighborhoods and households.

This uneven distribution means that some areas bear far more environmental impact than others. Just as the Yellow River has experienced intense pressure from millions of Chinese farmers, the world's other major rivers, from the

Nile to the Danube to the Ganges to the Mississippi, have all received more than their share of human impact. The urban way of life entails the packaging and transport of goods, intensive fossil fuel consumption, and hotspots of pollution. However, people's concentration in cities relieves pressure on ecosystems in less-populated areas by releasing some of them from direct human development (▶ p. 204).

At the same time, areas with low population density are often vulnerable to environmental impacts, because the reason they have low populations in the first place is that they

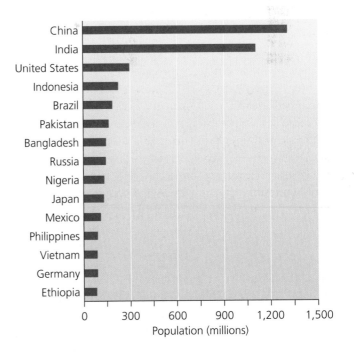

FIGURE 6.5 The world's nations range in human population from several thousand (on some South Pacific islands) up to China's 1.3 billion. Shown here are the 2005 populations for the world's most populous 15 nations. Data from Population Reference Bureau. 2005. *2005 world population data sheet.*

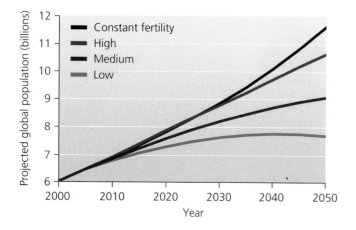

FIGURE 6.6 The United Nations predicts trajectories of world population growth, presenting its estimates in several scenarios based on different assumptions of fertility rates. In this 2004 projection, population is estimated to reach 11.7 billion in the year 2050 if fertility rates remain constant at 2004 levels (top line in graph). However, U.N. demographers expect fertility rates to continue falling, so they arrived at a best guess (*medium* scenario) of 9.1 billion for the human population in 2050. In the *high* scenario, if women on average have 0.5 child more than in the medium scenario, population will reach 10.6 billion in 2050. In the *low* scenario, if women have 0.5 child less than in the medium scenario, the world will contain 7.7 billion people in 2050. Data from United Nations Population Division. 2004. *World population prospects: The 2004 revision.*

are sensitive and cannot support many people. Deserts, for instance, are easily affected by development that commandeers a substantial share of available water. Grasslands can be turned to deserts if they are farmed too intensively.

Age structure Data on the age structure or age distribution of human populations are especially valuable to demographers trying to predict future dynamics of populations. As we saw in Chapter 4 (▶ p. 87), large proportions of individuals in young age groups portend a great deal of reproduction and, thus, rapid population growth. Examine age pyramids for the nations of Canada and Madagascar (Figure 6.8). Not surprisingly, it is Madagascar that has the greater population growth rate. In fact, its annual growth rate, 2.7%, is 9 times that of Canada's 0.3%.

In most industrialized nations today, populations are aging. Older populations will present challenges for many societies because increasing numbers of elderly will require care and financial assistance from relatively fewer working-age citizens. In China, the one-child policy has accentuated this trend. By causing dramatic reductions in the number of children born since 1970, China virtually guaranteed that its population age structure would change. Indeed, in 1995 the median age in China was 27; by 2030 it will be 39. The number of people older than 65 will rise from 100 million in 2005 to 236 million in 2030 (Figure 6.9). This dramatic shift in age structure will challenge

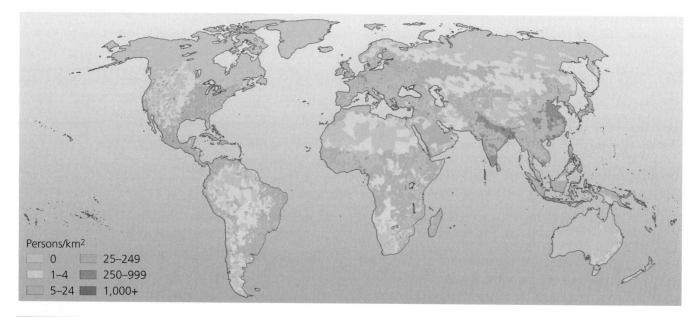

Persons/km²

0	25–249
1–4	250–999
5–24	1,000+

FIGURE 6.7 Human population density varies tremendously from one region to another. Arctic and desert regions have the lowest population densities, whereas areas of India, Bangladesh, and eastern China have the densest populations. Data are for 2000, from Center for International Earth Science Information Network (CIESIN), Columbia University; and Centro Internacional de Agricultura Tropical (CIAT), 2004.

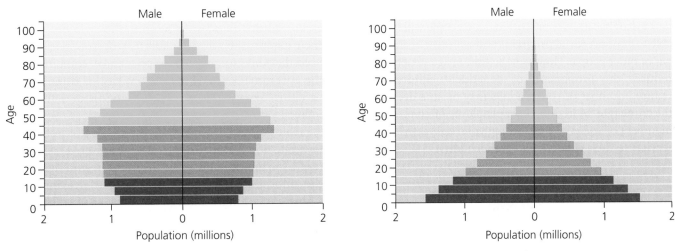

(a) **Age pyramid of Canada in 2005**

(b) **Age pyramid of Madagascar in 2005**

FIGURE 6.8 Canada (**a**) shows a balanced age structure, with relatively even numbers of individuals in various age classes. Madagascar (**b**) shows an age distribution heavily weighted toward young people. Madagascar's population growth rate is 9 times that of Canada's. Go to **GRAPHit!** at www.aw-bc.com/withgott. Data from U.N. Population Division.

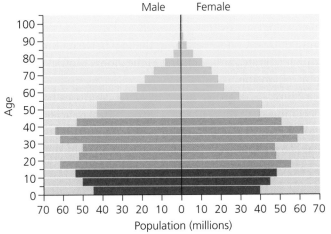

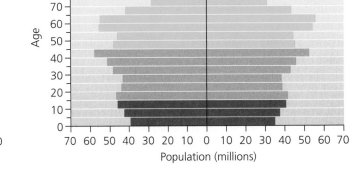

(a) **Age pyramid of China in 2005**

(b) **Projected age pyramid of China in 2030**

FIGURE 6.9 As China's population ages, older people will outnumber the young. Age pyramids show the predicted graying of the Chinese population between 2005 (**a**) and 2030 (**b**). Today's children may, as working-age adults (**c**), face pressures to support greater numbers of older citizens than has any previous generation. Data from U.N. Population Division.

(c) **Young female factory workers in Hong Kong**

China's economy, health care systems, families, and military forces because fewer working-age people will be available to support social programs that assist the rising number of older people. However, the shift in age structure also reduces the proportion of dependent children. The reduced number of young adults may mean a decrease in the crime rate. Moreover, older people are often productive members of society, contributing volunteer activities and services to their children and grandchildren.

Weighing the **Issues:**
China's Reproductive Policy

Consider the benefits as well as the problems associated with a reproductive policy such as China's. Do you think a government should be able to enforce strict penalties for citizens who fail to abide by such a policy? If you disagree with China's policy, what alternatives can you suggest for dealing with the resource demands of a quickly growing population?

Sex ratios The ratio of males to females also can affect population dynamics. Imagine two islands, one populated by 99 men and 1 woman and the other by 50 men and 50 women. Where would we be likely to see the greatest population increase over time? Of course, the island with an equal number of men and women would have a greater number of potential mothers and thus a greater potential for population growth.

The naturally occurring sex ratio in human populations at birth features a slight preponderance of males; for every 100 female infants born, 105 to 106 male infants are born. This may be an evolutionary adaptation to the fact that males are slightly more prone to dying during any given year of life. It usually ensures that the ratio of men to women is approximately equal at the time they reach reproductive age. Thus, a slightly uneven sex ratio at birth may be beneficial. However, a greatly distorted ratio can lead to problems.

In recent years, demographers have witnessed an unsettling trend in China: The ratio of newborn boys to girls has become strongly skewed. In the 2000 census, 120 boys were reported born for every 100 girls. Some provinces reported sex ratios as high as 138 boys for every 100 girls. A leading hypothesis for these unusual sex ratios is that some parents, having learned the gender of their fetuses by ultrasound, are selectively aborting female fetuses. Traditionally, Chinese culture has valued sons because they carry on the family name, assist with farm labor in rural areas, and care for aging parents. Daughters,

in contrast, will most likely marry and leave their parents, as the culture dictates. As a result, they will not provide their parents the same benefits as will sons. Sociologists hold that this cultural gender preference, combined with the government's one-child policy, has led some couples to abort female fetuses or to abandon or kill female infants.

China's skewed sex ratio may have the effect of further lowering population growth rates. However, it has proved tragic for the "missing girls." It is also beginning to have the undesirable social consequence of leaving many Chinese men single. This, in turn, has resulted in a grim new phenomenon. In parts of rural China, teenaged girls are being kidnapped and sold to families in other parts of the country as brides for single men.

Population growth depends on rates of birth, death, immigration, and emigration

The formula for measuring population growth that we used in Chapter 4 (▸ p. 88) also pertains to humans: birth and immigration add individuals to a population, whereas death and emigration remove individuals. Technological advances have led to a dramatic decline in human death rates, widening the gap between birth rates and death rates and resulting in the global human population expansion.

Since 1970, growth rates in many countries have been declining, even without population-control policies, and the global growth rate has declined (Figure 6.10). This decline has come about, in part, from a steep drop in birth rates.

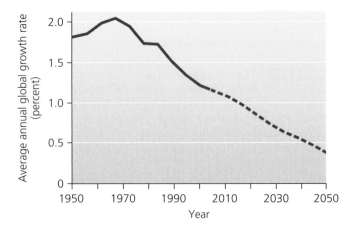

FIGURE 6.10 The annual growth rate of the global human population peaked in the 1960s and has been declining since then. The dashed line indicates projected future trends. Data from United Nations Population Division. 2004. *World population prospects: The 2004 revision.*

A population's total fertility rate influences population growth

One key statistic demographers calculate to examine a population's potential for growth is the **total fertility rate (TFR)**, or the average number of children born per female member of a population during her lifetime. **Replacement fertility** is the TFR that keeps the size of a population stable. For humans, replacement fertility is equal to a TFR of 2.1. When the TFR drops below 2.1, population size, in the absence of immigration, will shrink.

Various factors influence TFR and have acted to drive it downward in many countries in recent years. Historically, people tended to conceive many children, which helped ensure that at least some would survive, but lower infant mortality rates have made this less necessary. Increasing urbanization has also driven TFR down; whereas rural families need children to contribute to farm labor, in urban areas children are usually excluded from the labor market, are required to go to school, and impose economic costs on their families. If a government provides some form of social security, as many do these days, parents need fewer children to support them in their old age when they can no longer work. Finally, with greater education and changing roles in society, women tend to shift into the labor force, putting less emphasis on child rearing.

All these factors have come together in Europe, where TFR has dropped from 2.6 to 1.4 in the past half century. Every European nation now has a fertility rate below the replacement level, and populations are declining in 18 of 43 European nations. In 2005, Europe's overall annual *natural rate of population change* (change due to birth and death rates alone, excluding migration) was −0.1%. Worldwide by 2005, a total of 71 countries had fallen below the replacement fertility of 2.1. These countries made up roughly 45% of the world's population and included China (with a TFR of 1.6). Table 6.1 shows the TFRs of major continental regions.

Table 6.1 Total Fertility Rates for Major Regions	
Region	**Total fertility rate (TFR)**
Africa	5.1
Latin America and the Caribbean	2.6
Asia	2.5
Oceania	2.1
North America	2.0
Europe	1.4

Data from Population Reference Bureau. 2005. *2005 World population data sheet.*

Some nations have experienced the demographic transition

Many nations that have lowered their birth rates and TFRs have been going through a similar set of interrelated changes. In countries with good sanitation, good health care, and reliable food supplies, more people than ever before are living long lives. As a result, over the past half century the *life expectancy* (the time a person can expect to live) for the average person has increased from 46 years to 67. Much of the increase in life expectancy is due to reduced rates of infant mortality. Societies going through these changes are mostly the ones that have undergone urbanization and industrialization and have been able to generate personal wealth for their citizens.

To make sense of these trends, demographers developed a concept called the **demographic transition**. This is a model of economic and cultural change proposed in the 1940s and 1950s by demographer Frank Notestein and elaborated on by others to explain the declining death rates and birth rates that have occurred in Western nations as they became industrialized. Notestein believed nations moved from a stable pre-industrial state of high birth and death rates to a stable post-industrial state of low birth and death rates. Industrialization, he proposed, caused these rates to fall naturally by first decreasing mortality and then lessening the need for large families. Parents would thereafter choose to invest in quality of life rather than quantity of children. Because death rates fall before birth rates fall, a period of net population growth results. Thus, under the demographic transition model, population growth is viewed as a temporary phenomenon that occurs as societies move from one condition to another.

Notestein's model proceeds in several stages (Figure 6.11). The first is the **pre-industrial stage**, characterized by conditions that have defined most of human history. In pre-industrial societies, both death rates and birth rates are high. Death rates are high because disease is widespread, medical care rudimentary, and food supplies unreliable and difficult to obtain. Birth rates are high because people must compensate for high mortality rates in infants and young children by having several children. In this stage, children are valuable as additional workers who can help meet a family's basic needs. Populations within the pre-industrial stage are not likely to experience much growth, which is why the human population was relatively stable from Neolithic times until the industrial revolution.

Industrialization initiates the second stage of the demographic transition, known as the **transitional stage**. This stage is generally characterized by declining death rates due to increased food production and

FIGURE 6.11 The demographic transition is an idealized process that has taken some populations from a pre-industrial state of high birth rates and high death rates to a post-industrial state of low birth rates and low death rates. In this diagram, the wide green area between the two curves illustrates the gap between birth and death rates that causes rapid population growth during the middle portion of this process. Adapted from Kent, M. M. and K. A. Crews. 1990. *World population: Fundamentals of growth*. Population Reference Bureau.

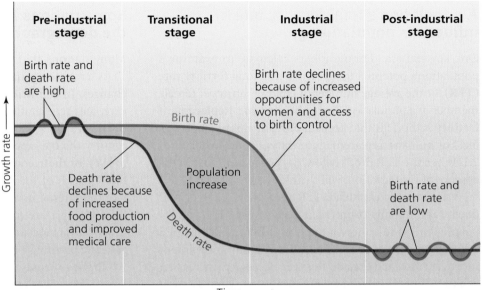

improved medical care. Birth rates remain high, however, because people have not yet grown used to the new economic and social conditions. As a result, population growth surges.

The third stage in the demographic transition is the **industrial stage**. Industrialization increases opportunities for employment outside the home, particularly for women. Children become less valuable, in economic terms, because they do not help meet family food needs as they did in the pre-industrial stage. If couples are aware of this, and if they have access to birth control, they may choose to have fewer children. Birth rates fall, closing the gap with death rates and reducing the rate of population growth.

In the final stage, the **post-industrial stage**, both birth and death rates have fallen to low and stable levels. Population sizes stabilize or decline slightly. The society enjoys the fruits of industrialization without the threat of runaway population growth.

The demographic transition has been occurring in nearly all developed nations over the past 200 to 300 years, but some social scientists think it may not apply to all developing nations as they industrialize in the future. These scientists point out that population dynamics may be different for developing nations that adopt the Western world's industrial model rather than devising their own. Other demographers assert that the transition may fail in cultures that place greater value on childbirth or that grant women fewer freedoms. Moreover, natural scientists warn that the world has too few resources to enable all countries to attain the standard of living that developed countries now enjoy. It has been estimated that for

all nations to enjoy the quality of life that United States citizens enjoy, we would need the natural resources of two more planet Earths. Whether developing nations pass through the demographic transition, as developed nations have, is one of the most important and far-reaching questions for the future of our civilization and Earth's environment.

Population and Society

Demographic transition theory links the statistical study of human populations with societal factors that influence, and are influenced by, population dynamics. Let's now examine a few of these societal factors more closely.

Women's empowerment affects population growth rates

Many demographers had long believed that fertility rates were influenced largely by degrees of wealth or poverty. However, affluence alone cannot determine TFR, because a number of developing countries now have fertility rates lower than that of the United States. Recent research is highlighting factors pertaining to the social empowerment of women. Drops in TFR have been most noticeable in countries where women have gained better access to contraceptives and education, particularly family-planning education (Figure 6.12; Figure 6.13; and "The Science behind the Story," ▶ pp. 134–135).

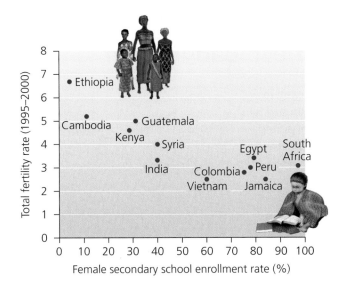

FIGURE 6.12 Increasing female literacy is strongly associated with reduced birth rates in many nations. Data from McDonald, M., and D. Nierenberg. 2003. Linking population, women, and biodiversity. *State of the World 2003*. Washington, D.C.: Worldwatch Institute.

In 2005, 53% of married women worldwide (ages 15–49) reported using some modern method of contraception to plan or prevent pregnancy. China, at 86%, had the highest rate of contraceptive use of any nation. Six western European nations showed rates of contraceptive use above 70%, as did Costa Rica, Cuba, New Zealand, Canada, Brazil, and Thailand (the U.S. rate was 68%). At the other end of the spectrum, 23 African nations had rates below 10%. These low rates of contraceptive use contribute to high fertility rates in sub-Saharan Africa, where the region's TFR is 5.6 children per woman. By comparison, in Asia, where the TFR in 1950 was 5.9, today it is 2.5—in part a legacy of the population control policies of China and some other Asian countries.

These data clearly indicate that in societies where women have little power, many pregnancies are unintended. Unfortunately, many women still lack the information or personal freedom of choice to allow them to make decisions about when to have children and how many to have. Studies show that in societies in which women are freer to decide whether and when to have children, fertility rates have fallen, and the resulting children are better cared for, healthier, and better educated.

Population policies and family-planning programs are working around the globe

Data show that funding and policies that encourage family planning have been effective in lowering population growth rates in all types of nations, even those that are least industrialized. The government of Thailand has relied on an education-based approach to family planning that has reduced birth rates and slowed population growth. In the 1960s, Thailand's growth rate was 2.3%, but by 2005 it had declined to 0.7%. This decline was

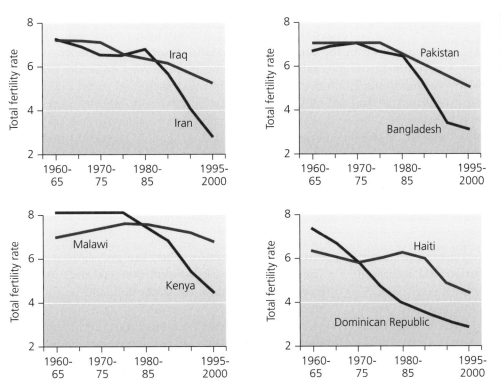

FIGURE 6.13 Data from four pairs of neighboring countries demonstrate the effectiveness of family planning in reducing fertility rates. In each case, the nation that invested in family planning and (in some cases) made other reproductive rights, education, and health care more available to women (blue lines) reduced its total fertility rate (TFR) far more dramatically than its neighbor (red lines). Data from U.N. Population Division; and Harrison, P., and F. Pearce. 2000. *AAAS atlas of population and environment* Berkeley, CA: University of California Press.

Causes of Fertility Decline in Bangladesh

Research in developing countries indicates that poverty and overpopulation can create a vicious cycle, in which poverty encourages high fertility and high fertility obstructs economic development. Are there policy steps that such countries can take to bring down fertility rates? Scientific analysis of family-planning programs in the South Asian nation of Bangladesh suggests that the answer is yes.

Bangladesh is one of the poorest, most densely populated countries on the planet. Its 145 million people live in an area about the size of Wisconsin, and 45% of them live below the poverty line. With few natural resources and a population density twice that of New Jersey, limiting population growth is critically important. As Bangladeshi president Ziaur Rahman declared in 1976, "If we cannot do something about population, nothing else that we accomplish will matter much."

Fortunately, Bangladesh has made striking progress in controlling population growth in the past three decades. Despite stagnant economic development, low literacy rates, poor health care, and limited rights for women, the nation's total fertility rate (TFR) has dropped markedly (see the first figure).

Researchers seeking to explain Bangladesh's rapid reduction in TFR hypothesized that family-planning programs were responsible. Because conducting an experiment to test such a hypothesis is difficult, some researchers took advantage of a natural experiment (▶ p. 13). By comparing Bangladesh to countries that are socioeconomically similar but have had less success in lowering TFR, such as Pakistan, researchers concluded that Bangladesh succeeded because of aggressive, well-funded outreach efforts that were sensitive to the values of its traditional society.

However, no two countries are identical, so it is difficult to draw firm conclusions from broad-scale studies. This is why the Matlab Family Planning and Health Services Project has become one of the best-known experiments in family planning in developing countries. The Matlab Project was an intensive outreach program run collaboratively by the Bangladeshi government and international aid organizations. Households in the isolated rural area of Matlab, Bangladesh, received biweekly visits from local women offering counseling, education, and free contraceptives (see the second figure). Compared to a similar government-

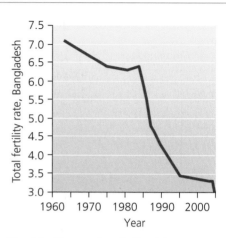

Total fertility rate has declined from 7.1 to 3.0 in Bangladesh in the past 40 years, in part because of the enhanced availability of contraceptives. However, TFR has leveled off in recent years, suggesting that other societal changes are needed to lower it further. Data from Bangladesh Bureau of Statistics; Bangladesh Fertility Survey; The Global Reproductive Health Forum at Harvard; International Centre for Diarrhoeal Disease Research, Bangladesh; National Family Planning and Fertility Survey; United States Agency for International Development.

run program in a nearby area, the Matlab Project featured more training, more services, and more frequent visits. In both areas, a highly organized health surveillance system gave researchers detailed information about births, deaths, and health-related behaviors such as contraceptive use. The result was an experiment comparing the Matlab

achieved largely via government-sponsored programs devoted to family-planning education and increased availability of contraceptives.

India has had long-standing policies, but many observers think they are too weak. Unless it strengthens its efforts to slow population growth, India seems set to overtake China in population soon. Brazil, Mexico, Iran, Cuba, and many other developing countries have insti-

tuted active programs to reduce population growth. These programs entail setting targets and providing incentives, education, contraception, and reproductive health care.

In 1994, the United Nations hosted a milestone conference on population and development in Cairo, Egypt, at which 179 nations endorsed a platform calling on all governments to offer universal access to reproductive

Project with the government-run area.

When Matlab Project director James Phillips and his colleagues reviewed a decade's worth of data in 1988, they found that fertility rates had declined in both areas. The decline appeared to be due almost entirely to a rise in contraceptive use, because other factors—such as the average age of marriage—remained the same. Phillips and his team also found that the declines had been greater in the Matlab area than in the government-run area, suggesting that high-intensity outreach efforts can affect fertility rates even in the absence of improvements in women's status, education, or economic development.

Why exactly was the outreach program successful? One hypothesis was that visits from health care workers helped convince local women that small families are desirable. However, in 1999, a team led by University of Michigan graduate student Mary Arends-Kuenning reported that there was no relationship between women's perception of the ideal family size and the number of visits made by outreach workers, either in Matlab or nearby comparison areas. Ideal family size declined equally in all areas. Instead of creating new demand for birth

In the Matlab Project, Bangladeshi households received visits from local women offering counseling, education, and free contraceptives.

control, the Matlab Project appears to have helped women convert an already-existing desire for fewer children into behaviors, such as contraceptive use, that reduce fertility.

Bangladesh's ability to rein in fertility rates despite unfavorable social and economic conditions bodes well for impoverished nations facing explosive population growth. However, significant challenges remain. Since the 1990s, Bangladesh's TFR has leveled off at slightly more than 3 children per woman. If rates fail to decline further, the country's population could double to 290 million—nearly the size of today's U.S. population—within 30 years. Scientific research has helped illuminate the impact of family-planning programs on fertility, but further reductions may require fundamental social, political, and economic changes that are difficult to implement in traditional, resource-strapped countries such as Bangladesh.

health care within 20 years. The conference marked a turn away from older notions of command-and-control (▸ p. 44) population policy geared toward pushing contraception and lowering population to preset targets. Instead, it urged governments to offer better education and health care and to address social needs that bear indirectly on population (such as alleviating poverty, disease, and sexism).

Weighing the Issues:
U.S. Involvement in International Family Planning

From 1998 to 2001, the U.S. government provided $46.5 million to the United Nations Population Fund (UNFPA), whose programs provide education in family planning, HIV/AIDS prevention, and teen pregnancy prevention in many nations, including China. Since then, the Bush administration has withheld funds, pointing out that U.S. law prohibits funding any organization that "supports or participates in the management of a program of coercive abortion or involuntary sterilization," and claiming that the Chinese government has been implicated in both these activities. Many nations and organizations criticized the U.S. decision, and the European Union offered additional funding to UNFPA to offset the loss of U.S. contributions. What do you think of the U.S. decision? Should the United States fund family-planning efforts in other nations? What conditions, if any, should it place on the use of such funds?

Poverty is strongly correlated with population growth

Poorer societies tend to show higher population growth rates than do wealthier societies—a pattern consistent with demographic transition theory. Poorer nations tend to have higher fertility rates, birth rates, and infant mortality rates, along with lower rates of contraceptive use.

Trends such as these affect the distribution of people on the planet. In 1960, 70% of all people lived in developing nations. By 2005, 81% of the world's population was living in these countries. Moreover, fully 98% of the next billion people to be added to the global population will be born in these poor, less developed regions (Figure 6.14). This is unfortunate from a social standpoint, because these people will be added to the countries that are least able to provide for them. It is also unfortunate from an environmental standpoint, because poverty often results in environmental degradation. For instance, people dependent on agriculture in a region of poor farmland may need to try to farm even if doing so degrades the soil and is not sustainable. This is largely why Africa's once-productive Sahel region, like many regions of western China, is turning to desert.

Consumption from affluence creates environmental impact

Poverty can lead people into environmentally destructive behavior, but wealth can produce even more severe and far-reaching environmental impacts. The affluence that characterizes a society such as the United States, Japan, or the Netherlands is built on massive and unprecedented levels of resource consumption. Much of this chapter has dealt with numbers of people rather than on the amount

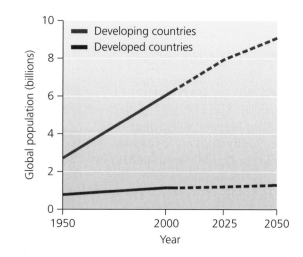

FIGURE 6.14 Nearly 98% of the next 1 billion people added to Earth's human population will reside in the less developed, poorer parts of the world. The dashed line indicates projected future trends. *Data from U.N. Population Division; and Harrison, P., and F. Pearce. 2000. AAAS atlas of population and environment. 2000. Berkeley, CA: University of California Press.*

of resources each member of the population consumes or the amount of waste each member produces. However, recall the A for affluence in the IPAT equation. The environmental impact of human activities depends not only on the number of people involved but also on the way those people live.

In Chapter 1 (▸ pp. 6–7), we introduced the concept of the *ecological footprint*, the cumulative amount of Earth's surface area required to provide the raw materials a person or population consumes and to dispose of or recycle the waste that they produce. Individuals from affluent societies leave a considerably larger per capita ecological footprint (see Figure 1.15, ▸ p. 20). This fact should remind us that the "population problem" does not lie entirely with the developing world. Just as population is rising, so is consumption, and environmental scientists have calculated that we are already living beyond the planet's means to support us sustainably. One recent analysis concluded that humanity's global ecological footprint surpassed Earth's capacity to support us in 1987 and that our species is now living more than 20% beyond its means (Figure 6.15).

The wealth gap and population growth contribute to conflict

The stark contrast between affluent and poor societies in today's world is the cause of social as well as environmental stress. Over half the world's people live below the internationally defined poverty line of U.S. $2 per day. The richest

Population Control

Debate over human population growth and environmental problems is often contentious. **Do you believe that national governments should implement policies, subsidies, or other programs to reduce birth rates?**

Implement ICPD Program of Action

Access to reproductive health care, including family planning, is a basic human right. To exercise this right, men and women need to be informed about family planning. They also need to have access to safe, effective, affordable, and acceptable methods of family planning of their choice.

All national governments should adopt policies, subsidies, and other programs to help implement the Program of Action, which the United Nations agreed to at the International Conference on Population and Development (ICPD), held in Cairo in 1994.

The ICPD is based on principles virtually every country in the world agreed to, one of which is as follows:

. . . States should take all appropriate measures to ensure, on a basis of equality of men and women, universal access to health-care services, including those related to reproductive health care, which includes family planning and sexual health. Reproductive health-care programs should provide the widest range of services without any form of coercion. All couples and individuals have the basic right to decide freely and responsibly the number and spacing of their children and to have the information, education and means to do so.

In the same way that democratic nations are obligated to assist emerging democracies in holding fair and free elections, developed nations ought to assist developing nations in implementing the ICPD Program of Action. In every society where the principles of the ICPD have been implemented, birth rates have gone down, infant survival rates have gone up, maternal mortality has declined, and the quality of life has improved.

Timothy Cline is director of communication at Population Connection, where he has held a variety of positions, including publications manager. He has also served as chief of advocacy and policy research at the Johns Hopkins University Center for Communication Programs. Before joining Population Connection, Cline served as operations manager for the Washington Regional Alliance and was a senior editor at Ecomedia.

Population Control: A Bad Idea for Governments

To address this issue, we must ask three questions. First, is population growing at an unsustainable rate? Second, are the world's problems—poverty, hunger, ecological degradation, and so on—caused by population growth? And third, do governments possess rightful authority to engage in population control?

1. The world is experiencing historically unprecedented fertility decline. In about half the countries on Earth— including rich and poor, developed and developing—fertility is at or below replacement rate, meaning that in many countries women are not having enough babies to maintain current population levels. Some demographers now worry that entire economies and social welfare systems may face profound difficulties because of this decline.

2. The population growth that has occurred over the past century (from roughly 2 billion to 6 billion) is not the direct cause of severe development problems. For instance, according to the U.N. Population Division, "Even for those environmental problems that are concentrated in countries with rapid population growth, it is not necessarily the case that population increase is the main cause, nor that slowing population growth would make an important contribution to resolving the problem." Persistent development problems are more likely linked to types of economic and political organization (nondemocratic countries tend to have worse environmental records, for instance) and should be addressed through political reform.

3. The quest for government control over such an intimate matter as family size seems prone to abuse, which has occurred in such diverse nations as India, Peru, and China. In China, where the "one-child policy" has often been touted as history's most successful population control program, over 100 million women have been forced to abort unborn babies or to be sterilized.

For these practical and ethical reasons, the world's limited development funds could be better spent than in encouraging further fertility decline.

Douglas A. Sylva is senior fellow of the Catholic Family and Human Rights Institute, a think tank and lobbying group that consults governments on international social policy. He is also a regular columnist for thefactis.org, focusing on international affairs and development. He received a Ph.D. in political science from Columbia University.

Explore this issue further by accessing **Viewpoints** at www.aw-bc.com/withgott.

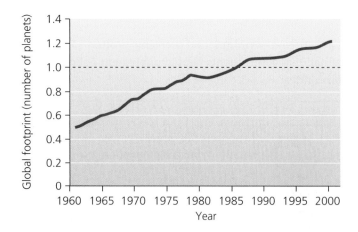

FIGURE 6.15 The global ecological footprint of the human population is 2.5 times larger than it was in 1961 and now exceeds what Earth can bear in the long run, scientists have calculated. The estimate shown here indicates that we have already overshot our carrying capacity by at least 20%. That is, we are using renewable natural resources 20% faster then they are being replenished. Data from WWF-World Wide Fund for Nature, 2004. *Living planet report.* Gland, Switzerland: WWF.

one-fifth of the world's people possesses over 80 times the income of the poorest one-fifth (Figure 6.16). The richest one-fifth also uses 86% of the world's resources. As the gap between rich and poor grows wider and as the sheer number of those living in poverty continues to increase, we will likely see increasing tensions between the "haves" and the "have-nots." This is why the inequitable distribution of wealth is one of the key factors the U.S. Departments of Defense and State take into account when assessing the potential for armed conflict around the world, whether it be conventional warfare or terrorism.

If today's developing nations are unable to overcome their mounting social, economic, and environmental challenges, and if the developed nations do not step in to help them, then these countries could fail to advance through the demographic transition. Instead, rising death rates could push birth rates back up, potentially causing these countries to fall back to the pre-industrial stage of the demographic transition model. Such an outcome would lead to greater population growth while economic and social conditions worsen. It would be a profoundly negative outcome, both for human welfare and for the well-being of the environment.

If one of humanity's goals is to generate a high standard of living and quality of life for all the world's people, then developing nations must find ways to reduce their population growth. However, those of us living in the industrialized world must also be willing to reduce our consumption. Earth does not hold enough resources to sustain all

(a) A family living in the United States

(b) A family living in Egypt

FIGURE 6.16 A typical U.S. family (**a**) may own a large house, keep numerous material possessions, and have enough money to afford luxuries such as vacation travel. A typical family in a developing nation such as Egypt (**b**) may live in a small, sparsely furnished dwelling with few material possessions and little money or time for luxuries.

6.5 billion of us at the current North American standard of living, nor can we go out and find extra planets; so, we must make the best of the one place that supports us all.

Conclusion

Today, several years after welcoming its six-billionth member, the human population is larger than at any time in the past. Our growing population, as well as our growing consumption, affects the environment and our ability to meet the needs of all the world's people. Approximately 90% of children born today are likely to live their lives in conditions far less healthy and prosperous than most of us in the industrialized world are accustomed to.

However, there are at least two major reasons to be optimistic. First, although global population is still rising, the *rate* of growth has decreased nearly everywhere, and some countries are even seeing population declines. Most developed nations have passed through the demographic transition, lowering death rates while stabilizing population and creating more prosperous societies. A second reason to feel optimistic is the progress in expanding rights for women worldwide. Although there is still a long way to go, women are slowly being treated more fairly, receiving better education, obtaining more economic independence, and gaining more ability to control their reproductive decisions. Aside from the clear ethical progress these developments entail, they are helping to slow population growth.

Human population cannot continue to rise forever. The question is how it will stop rising: through the gentle and benign process of the demographic transition, through restrictive governmental intervention such as China's one-child policy, or through the miserable Malthusian checks of disease and social conflict caused by overcrowding and competition for scarce resources. Moreover, sustainability demands a further challenge—that we stabilize our population size in time to avoid destroying the natural systems that support our societies. We are indeed a special species. We are the only one to come to such dominance as to alter so much of Earth's landscape, and even its climate system. We are also the only species with the intelligence needed to turn around an increase in our own numbers before we destroy the very systems on which we depend.

TESTING YOUR COMPREHENSION

1. What is the approximate current human global population? How many people are being added to the population each day?
2. Why has the human population continued to grow in spite of environmental limitations?
3. Contrast the views of environmental scientists with those of the libertarian writer Sheldon Richman and similar-thinking economists over whether population growth is a problem. Why does Richman think the concept of carrying capacity does not apply to human populations?
4. Explain the IPAT model. How can technology either increase or decrease environmental impact? Provide at least two examples.
5. What characteristics and measures do demographers use to study human populations? How do each of these help determine the impact of human population on the environment?
6. What is total fertility rate (TFR)? Can you explain why the replacement fertility for humans is approximately 2.1? How is Europe's TFR affecting its natural rate of population change?
7. Why have fertility rates fallen in many countries?
8. In the demographic transition model, why is the pre-industrial stage characterized by high birth and death rates, and the industrial stage by falling birth and death rates?
9. How does the demographic transition model explain the increase in population growth rates in recent centuries? How does it explain the decrease in population growth rates in recent decades?
10. Why do poorer societies have higher population growth rates than wealthier societies? How does poverty affect the environment? How does affluence affect the environment?

SEEKING SOLUTIONS

1. China's reduction in birth rates is significantly changing the nation's age structure. Review Figure 6.9 (▶ p. 129) and note how the population is growing older, as shown by the top-heavy age pyramid for the year 2030. What effects might this ultimately have on Chinese society? What impacts might the aging population of the United States have on U.S. society? How might we address these impacts?
2. Apply the IPAT model to the example of China provided in the chapter. How do population, affluence, and technology affect China's environment? Now consider your own country, state, or region. How do population, affluence, technology, and ecological sensitivity affect your environment? How can we regulate the relationship between population and its effects on the environment?
3. Do you think that all of today's developing nations will complete the demographic transition and come to enjoy a permanent state of low birth and death rates? Why or why not? What steps might we as a global society take to help ensure that they do? Now think about developed nations like the United States and

Canada. Do you think these nations will continue to lower and stabilize their birth and death rates in a state of prosperity? What factors might affect whether they do so?

4. Imagine that India's prime minister puts you in charge of that nation's population policy. India has a population growth rate of 1.7% per year, a TFR of 3.0, a 43% rate of contraceptive use, and a population that is 72% rural. What policy steps would you recommend, and why?

5. Now imagine that you have been tapped to design population policy for Germany. Germany is losing population at an annual rate of 0.1%, has a TFR of 1.3, a 72% rate of contraceptive use, and a population that is 88% urban. What policy steps would you recommend, and why?

CALCULATING ECOLOGICAL FOOTPRINTS

The equation I = PAT (Impact = Population × Affluence × Technology) suggests that a population's size and affluence are not the only determinants of its ecological impact, but that its technological choices also have an effect. Technologies can be either efficient or wasteful. One way of gauging the relative value of T is to calculate a per capita value of I/A (equivalent to I divided by A divided by P). The table presents per capita values of I (estimated ecological footprints) and A (income). Calculate the relative values of T by completing the blank column.

1. If the world average value of T were decreased (improved) to that of the United States, what per capita GNI PPP could be supported at the current average per capita ecological footprint of 6.9 acres?

2. What value of T would enable the world's population to live at its current affluence within the 4.9 acres per capita that Wackernagel et al. estimate are available? Do you think this is achievable?

3. Which country's technological choices would you choose to study if you were interested in learning how to maximize your standard of living while minimizing your ecological impact? Using the value of T for this country and the mid-2005 world population of 6,477,000,000, calculate the following:

   (a) The number of people the world could support at the current per capita impact of 6.9 acres and affluence of $8,540

   (b) The number of people the world could support sustainably on the available 4.9 acres per capita and at an affluence of $8,540

   (c) The per capita GNI PPP that the world's current population could achieve on 6.9 acres per capita

   (d) The per capita GNI PPP that the world's current population could achieve on 4.9 acres per capita

Country	Impact (ecological footprint, in acres per capita)	Affluence (per capita income, in GNI PPP*)	Technology (I/A) (footprint per $1,000 income)
Bangladesh	1.2	$1,980	0.61
Colombia	4.9	$6,820	
Mexico	6.4	$9,590	
Sweden	14.6	$29,770	
Thailand	6.9	$8,020	
United States	25.4	$39,710	
World average	6.9	$8,540	

* GNI PPP is "gross national income in purchasing power parity," a measure that standardizes income and makes it comparable among nations by converting income to "international" dollars using a conversion factor. International dollars indicate the amount of goods and services one could buy in the United States with a given amount of money.

Data sources: Population Reference Bureau. 2005. *World population data sheet 2005*; and Wackernagel, M., et al. 1999. National natural capital accounting with the ecological footprint concept. *Ecological Economics* 29: 375–390.

Take It Further

Go to www.aw-bc.com/withgott, where you'll find:

▶ Suggested answers to end-of-chapter questions

▶ Quizzes, animations, and flashcards to help you study

▶ *Research Navigator*™ database of credible and reliable sources to assist you with your research projects

▶ **GRAPHit!** tutorials to help you interpret graphs

▶ **INVESTIGATEit!** current news articles that link the topics that you study to case studies from your region to around the world

Soil, Agriculture, and the Future of Food

7

No-till farm
in Ceara, Brazil

Upon successfully completing this chapter, you will be able to:

▶ Explain the importance of soils to agriculture and describe the impacts of agriculture on soils

▶ Delineate the fundamentals of soil science, including soil formation and properties

▶ State the causes and predict the consequences of soil erosion and soil degradation

▶ Recite the history and explain the principles of soil conservation

▶ Identify the goals, methods, and environmental impacts of the "green revolution"

▶ Categorize the strategies of pest management

▶ Discuss the importance of pollination

▶ Describe the science behind genetically modified food

▶ Evaluate controversies and the debate over genetically modified food

▶ Identify approaches for preserving crop diversity

▶ Assess feedlot agriculture for livestock and poultry and weigh approaches in aquaculture

▶ Evaluate sustainable agriculture

Central Case: No-Till Agriculture in Southern Brazil

In southernmost Brazil, hundreds of thousands of people make their living farming. The warm climate and rich soils of this region's rolling highlands and coastal plains have historically made for bountiful harvests. However, repeated cycles of plowing and planting over many decades diminished the productivity of the soil. More and more topsoil—the valuable surface layer of soil richest in organic matter and nutrients—was being eroded away by water and wind. Meanwhile, the synthetic fertilizers used to restore nutrients were polluting area waterways. Yields were falling, and by 1990 farmers were looking for help.

As a result, many of southern Brazil's farmers abandoned the conventional practice of tilling the soil after harvests. In its place, they turned to *no-tillage*, or *no-till*, farming. Turning the earth by tilling (plowing, disking, harrowing, or chiseling) aerates the soil and works weeds and old crop residue into the soil to nourish it. However, tilling also leaves the surface bare, allowing wind and water to erode precious topsoil. Tilling historically boosted the productivity of agriculture in Europe, but in subtropical regions such as southern Brazil, heavy rainfall promotes erosion, causing tilled soils to lose organic matter and nutrients, and hot weather can overheat tilled soil.

Working with agricultural scientists and government extension agents, southern Brazil's farmers began leaving crop residues on their fields after harvesting, and planting "cover crops" to keep soil protected. When they went to plant the next crop, they merely cut a thin, shallow groove into the soil surface, dropped in seeds, and covered them. They did not invert the soil, but kept it covered with plants or their residues at all times, reducing erosion by 90%.

With less soil eroding away, and more organic material being added to it, the soil held more water and encouraged better plant growth. In the state of Santa Catarina, maize yields per hectare increased by 47% between 1991 and 1999, wheat yields rose by 82%, and soybean yields by 83%. In the states of Paraná and Rio Grande do Sul, maize yields were up 67% over 10 years, and soybean yields were up 68%.

No-till farming methods also reduced farmers' costs, because farmers now used less labor and fuel. And by enhancing soil conditions and reducing erosion, no-till techniques reduced air and water pollution. No-till agriculture spread quickly in the region as farmers saw their neighbors' successes and traded information through Friends of the Land clubs. In Paraná and Rio Grande do Sul, the area being farmed with no-till methods shot up from 700,000 ha (1.7 million acres) in 1990 to 10.5 million ha (25.9 million acres) in 1999, when it involved 200,000 farmers.

Reduced tillage is not a panacea for all areas of the world. But in regions suitable for reduced tillage, proponents say these approaches can help make agriculture sustainable. We will need sustainable agriculture if we are to feed the world's human population while protecting the natural environment, including the soils that vitally support our production of food.

Soil: The Foundation for Agriculture

As the human population has grown, so have the amounts of land and resources we devote to agriculture, which currently covers 38% of Earth's land surface. We can define **agriculture** as the practice of raising crops and livestock for human use and consumption. We obtain most of our food and fiber from **cropland**, land used to raise plants for human use, and **rangeland** or pasture, land used for grazing livestock.

For most of the 10,000 or so years that people have practiced agriculture, we have relied on human and animal muscle power, along with hand tools and simple machines. Such **traditional agriculture** is still widely practiced in the developing world. The industrial revolution (▸ p. 5) introduced large-scale mechanization and fossil fuel combustion, enabling farmers to replace horses and oxen with faster and more powerful means of working with crops and livestock. Such **industrialized agriculture** also boosted yields by intensifying irrigation and introducing synthetic fertilizers, while the advent of chemical pesticides reduced competition from weeds and herbivory by crop pests.

Modern industrialized agriculture has enabled us to feed more people than ever before, but it is exacting a high ecological price. When poorly managed, industrial agriculture can remove forests; destroy wetlands; turn grasslands to deserts; diminish biodiversity; encourage invasive species; pollute soil, air, and water with toxic chemicals, nutrients, and sediments; and allow fertile soil to be blown and washed away.

If we are to feed our growing population, we cannot simply keep expanding agriculture into new areas, because suitable land is running out. Instead, we must find ways to improve the efficiency of food production in areas that are already in agricultural use—and do so sustainably, without degrading soil, the very foundation of our terrestrial food supply.

Soil is complex and full of life

Most of us tend to think of soil as an inert, lifeless substance, but it is much more. Scientists define **soil** as a complex plant-supporting system consisting of disintegrated rock, organic matter, water, gases, nutrients, and microorganisms (Figure 7.1). By volume, soil consists

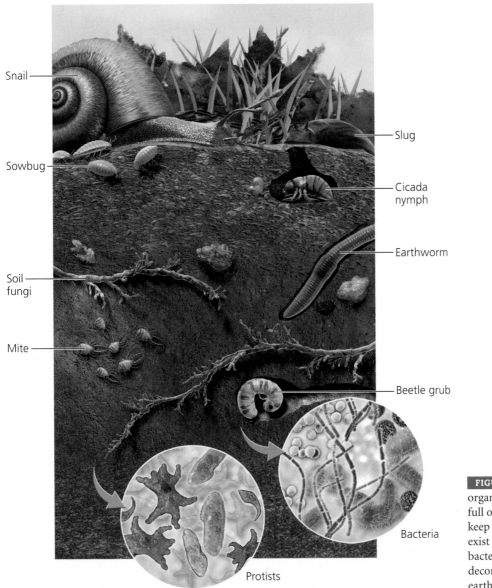

Snail

Sowbug

Soil fungi

Mite

Slug

Cicada nymph

Earthworm

Beetle grub

Bacteria

Protists

FIGURE 7.1 Soil is a complex mixture of organic and inorganic components and is full of living organisms whose actions help keep it fertile. In fact, entire ecosystems exist in soil. Most soil organisms, from bacteria to fungi to insects to earthworms, decompose organic matter. Many, such as earthworms, also help to aerate the soil.

very roughly of half mineral matter and up to 5% organic matter. The rest consists of pore space taken up by air or water. The organic matter in soil includes living and dead microorganisms as well as decaying material derived from plants and animals. A single teaspoon of soil can contain 100 million bacteria, 500,000 fungi, 100,000 algae, and 50,000 protists. Soil also provides habitat for earthworms, insects, mites, millipedes, centipedes, nematodes, sow bugs, and other animals. Because soil is composed of living and nonliving components that interact in complex ways, soil itself meets the definition of an ecosystem (▶ p. 60).

Soil formation begins with **parent material**, the base geological material in a particular location. This can include lava or volcanic ash; rock or sediment deposited by glaciers; wind-blown dunes; sediments deposited by rivers, in lakes, or in the ocean; or *bedrock*, the continuous mass of solid rock that makes up Earth's crust. Parent material is broken down by **weathering**, the physical, chemical, and biological processes that turn large rock particles into smaller particles.

Biological activity next contributes to soil formation through the deposition, decomposition, and accumulation of organic matter. As plants, animals, and microbes die or deposit waste, this material is incorporated into the substrate, mixing with minerals. Partial decomposition of organic matter creates *humus*, a dark, spongy, crumbly mass of material made up of complex organic compounds. Soils with high humus content hold moisture well and are productive for plant life.

A soil profile consists of distinct layers known as horizons

Once weathering has produced an abundance of small particles between the parent material and the atmosphere, then wind, water, and organisms begin to move and sort them. Eventually, distinct layers develop. Each layer of soil is known as a **horizon**, and the cross-section as a whole, from surface to bedrock, is known as a **soil profile**. Six major horizons found in a typical soil profile are the O, A, E, B, C, and R horizons (Figure 7.2).

Generally, the degree of weathering and the concentration of organic matter decrease as one moves downward in the soil profile. Minerals are generally transported downward as a result of **leaching**, the process whereby solid particles suspended or dissolved in liquid are transported to another location.

A crucial horizon for agriculture and ecosystems is the A horizon, or **topsoil**. Topsoil consists mostly of

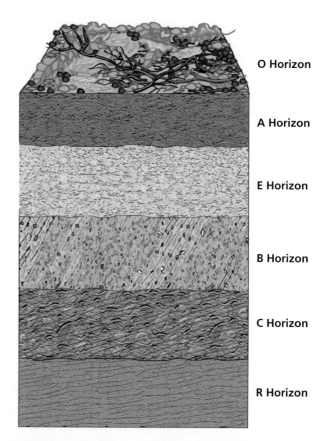

O Horizon

A Horizon

E Horizon

B Horizon

C Horizon

R Horizon

FIGURE 7.2 Mature soil consists of layers, or horizons, that have different compositions and characteristics. The uppermost layer (the **O horizon**, or litter layer) consists mostly of organic matter deposited by organisms. Below it lies the **A horizon**, or topsoil, consisting of some organic material mixed with mineral components. The **E horizon** is characterized by the loss of some minerals and organic matter through leaching. Minerals tend to leach out of the E horizon down into the **B horizon**, or subsoil, where they accumulate. The **C horizon** consists largely of weathered parent material unaltered or only slightly altered by the processes of soil formation. The C horizon may overlie an **R horizon** of pure parent material.

inorganic mineral components such as weathered substrate, with organic matter and humus from above mixed in. Topsoil is the portion of soil that is most nutritive for plants, and it takes its loose texture and dark coloration from its humus content. The O and A horizons are home to most of the countless organisms that give life to soil.

The six horizons presented in Figure 7.2 depict an idealized soil profile, but real soils display great variety. Soil scientists classify soils using properties such as color, texture, structure, and pH. Soil with a relatively even mixture of pore and particle sizes is known as *loam*. Scientists' studies of soil and the practical experience of farmers have shown that the most desirable soil for agriculture is loam with a pH close to neutral that is workable and capable of holding nutrients.

Soil Degradation and Conservation

Most of the world's soils are not ideal for agriculture, but as our planet gains over 70 million new people each year, we are pressing into cultivation many lands that are unsuitable for farming. This is causing considerable environmental damage, and we are losing 5–7 million ha (12–17 million acres) of productive cropland annually as a result.

Soil degradation around the globe results from roughly equal parts forest removal, cropland agriculture, and overgrazing of livestock (Figure 7.3). Scientists estimate that soil degradation over the past 50 years has reduced agricultural yields by 13% on cropland and by 4% on rangeland. Common problems affecting soil productivity include erosion, desertification, salinization, waterlogging, nutrient depletion, structural breakdown, and pollution.

Erosion carries soil away

Erosion is the removal of material from one place and its transport toward another by the action of wind or water. *Deposition* is the arrival of eroded material at its new location. Erosion and deposition are natural processes that in the long run help create soil. Flowing water can deposit eroded sediment in river valleys and deltas, producing rich and productive soils. This is why floodplains are excellent for farming and why flood-control measures can decrease farming productivity in the long run.

However, erosion often becomes a problem locally for ecosystems and agriculture because it nearly always occurs much more quickly than soil is formed. Furthermore, erosion tends to remove topsoil, the most valuable soil layer for living things. People have increased the vulnerability of fertile lands to erosion through three widespread practices:

▶ Overcultivating fields through poor planning or excessive tilling

▶ Overgrazing rangelands with more livestock than the land can support

▶ Clearing forests on steep slopes or with large clear-cuts (▶ pp. 209–210)

Grasslands, forests, and other plant communities protect soil from wind and water erosion. Vegetation breaks the wind and slows water flow, whereas plant roots hold

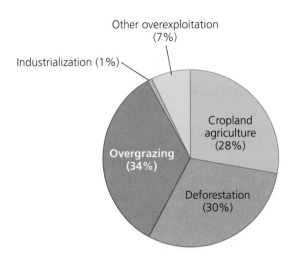

FIGURE 7.3 The great majority of the world's soil degradation results from cropland agriculture, overgrazing by livestock, and deforestation. Data from Wali, M. K. et al., 1999. Assessing terrestrial ecosystem sustainability: Usefulness of regional carbon and nitrogen models. *Nature and Resources* 35: 21–33.

soil in place and take up water. Removing plant cover nearly always accelerates erosion. In general, steeper slopes, greater precipitation intensities, and sparser vegetative cover all lead to greater water erosion.

Erosion and desertification are global problems

In today's world, humans are the primary cause of erosion, and we have accelerated it to unnaturally high rates. In a 2004 study, geologist Bruce Wilkinson analyzed prehistoric erosion rates from the geologic record and compared these with modern rates. He concluded that human activities move over 10 times more soil than all other natural processes on the surface of the planet combined.

More than 19 billion ha (47 billion acres) of the world's croplands suffer from erosion and other forms of soil degradation resulting from human activities. Between 1957 and 1990, China lost as much arable farmland as exists in Denmark, France, Germany, and the Netherlands combined. For Africa, projections indicate that soil degradation over the next 40 years could reduce crop yields by half. In the United States, erosion rates have declined from 9.1 tons/ha (3.7 tons/acre) in 1982 to 5.9 tons/ha (2.4 tons/acre) in 2001, thanks to soil conservation measures discussed below. Yet U.S. farmlands still lose 6 tons of soil for every ton of grain harvested.

Soil degradation is especially severe in arid environments, where **desertification** is a concern. This term describes a loss of more than 10% productivity due to erosion, soil compaction, forest removal, overgrazing, drought, salinization, climate change, water depletion, and other factors. By some estimates, desertification affects one-third of the planet's land area, costing people in over 100 countries many billions of dollars in income each year.

Severe desertification can expand desert areas and create new ones in once-fertile regions. This occurred historically in the "Fertile Crescent" of the Middle East—a region that hosted the very origin of agriculture but is now desert. Today, areas of northern Africa, central Asia, and China are most affected, and gigantic dust storms from denuded land blow soil across oceans to North America (see Figure 12.5a, ▸ p. 276). Such massive dust storms occurred in the United States in the early 20th century, when desertification shook American agriculture and society to their very roots.

The Dust Bowl inspired measures to slow soil degradation

Prior to cultivation of North America's Great Plains, native prairie grasses of this temperate grassland region held soils in place. In the late 19th and early 20th centuries, homesteading settlers spread through Oklahoma, Texas, Kansas, New Mexico, and Colorado. Farmers grew abundant wheat, and ranchers grazed many thousands of cattle, sometimes expanding onto unsuitable land and contributing to erosion by removing native grasses and breaking down soil structure.

In the early 1930s, a drought exacerbated the ongoing human impacts, and the region's strong winds began to erode millions of tons of topsoil (Figure 7.4). Dust storms traveled up to 2,000 km (1,250 mi), blackening rain and snow as far away as New York and Vermont. Some areas in the affected states lost as much as 10 cm (4 in.) of topsoil in a few years. The affected region in the Great Plains became known as the **Dust Bowl**, a term now also used for the historical event itself. The "black blizzards" of the Dust Bowl forced thousands of farmers off their land, and many who remained had to rely on government assistance programs to survive.

In response, the U.S. government, along with state and local governments, increased support of research into soil conservation measures. Congress passed the Soil Conservation Act of 1935, establishing the Soil Conservation Service (SCS), which worked closely with farmers to develop conservation plans for individual farms. Early SCS teams included soil scientists, forestry experts, engineers, economists, and biologists, and these were among

FIGURE 7.4 Drought combined with poor agricultural practices brought devastation and despair to millions of U.S. farmers in the 1930s, especially in the Dust Bowl region of the southern Great Plains. The tragedy spurred the development of soil conservation practices that have since been put into place in the United States and around the world.

the earliest examples of interdisciplinary approaches to environmental problem solving.

The SCS (today renamed the *Natural Resources Conservation Service*) served as a model for efforts elsewhere. Southern Brazil's no-till movement came about through local grass-roots organization by farmers, with the help of agronomists and government extension agents who provided them information and resources. In this model of collaboration between local farmers and trained experts, 8,000 Friends of the Land clubs now exist in Paraná and Rio Grande do Sul, and 7,700 in Santa Catarina.

Farmers can protect soil against degradation in various ways

Several farming techniques can reduce the impacts of cultivation on soil (Figure 7.5). In **crop rotation**, farmers alternate the type of crop grown in a given field from one season or year to the next (Figure 7.5a). Rotating crops can return nutrients to the soil, break cycles of disease associated with continuous cropping, and minimize the erosion that can come from letting fields lie fallow. Many U.S. farmers rotate their fields between wheat or corn and soybeans from one year to the next. Soybeans have specialized bacteria on their roots that fix nitrogen (▸ p. 65), revitalizing soil that the previous crop had partially depleted of nutrients. Crop rotation also reduces insect pests; if an insect is adapted to feed and lay eggs on one crop, planting a different type of crop will leave its offspring with nothing to eat.

(a) Crop rotation

(b) Contour farming

(c) Intercropping

(d) Terracing

(e) Shelterbelts

(f) No-till farming

FIGURE 7.5 The world's farmers have adopted various strategies to conserve soil. Rotating crops such as soybeans and corn (**a**) helps restore soil nutrients and reduce impacts of crop pests. Contour farming (**b**) reduces erosion on hillsides. Intercropping (**c**) can reduce soil loss while maintaining soil fertility. Terracing (**d**) minimizes erosion in steep mountainous areas. Shelterbelts (**e**) protect against wind erosion. In (**f**), corn grows up from amid the remnants of a "cover crop" used in no-till agriculture.

Farmers have developed several methods for cultivating slopes, where water easily erodes soil. **Contour farming** (Figure 7.5b) consists of plowing furrows sideways across a hillside, perpendicular to its slope and following the natural contours of the land, to help prevent formation of rills and gullies.

Farmers may also minimize erosion by **intercropping**, planting different types of crops in alternating bands or other spatially mixed arrangements (Figure 7.5c). Intercropping provides more ground cover and reduces vulnerability to insects and disease. Some southern Brazilian farmers intercrop food crops with cover crops intended to reduce erosion and nitrogen loss. Government extension agents worked with Santa Catarina farmers to test over 60 species, mixing these cover crops with primary crops such as maize, soybeans, wheat, onions, cassava, grapes, tomatoes, tobacco, and orchard fruit.

On extremely steep terrain, **terracing** (Figure 7.5d) is the most effective method for reducing erosion. Terraces are level platforms, sometimes with raised edges, that are cut into steep hillsides to contain water from irrigation and precipitation. Terracing transforms slopes into series of steps like a staircase, enabling farmers to cultivate hilly land without losing huge amounts of soil to water erosion.

A widespread technique to reduce erosion from wind is to establish **shelterbelts** or *windbreaks* (Figure 7.5e). These are rows of trees or other tall plants that are planted along the edges of fields to slow the wind. On the Great Plains, fast-growing species such as poplars are often used.

Finally, agriculture that bypasses plowing has gained popularity in recent years (Figure 7.5f). For instance, as the appeal of no-till farming spread in Brazil, it also spread in neighboring nations such as Argentina and Paraguay. These methods were pioneered in the United States and United Kingdom, where no-till is still rare but where reduced tillage has been slowly spreading for decades. Today nearly half of U.S. acreage is farmed with reduced-tillage methods.

Critics of no-till and reduced-tillage farming in the United States have noted that these techniques often require substantial use of chemical herbicides (because weeds are not physically removed from fields) and synthetic fertilizer (because other plants take up much of the soil's nutrients). In many industrialized countries, this has been the case. Proponents, however, point out that southern Brazil's farmers have departed somewhat from the industrialized model by relying more heavily on *green manures* (dead plants as fertilizer) and by rotating fields with cover crops, including nitrogen-fixing legumes. The manures and legumes nourish the soil, and cover crops reduce weeds by taking up space the weeds might occupy.

Critics maintain, however, that green manures are not practical for large-scale intensive agriculture. Reduced tillage methods seem to work well in some areas but not in others, and to work better with some crops than with others.

Weighing the issues:
How Would You Farm?

You are a farmer owning land on both sides of a steep ridge. You want to plant a sun-loving crop on the sunny, but very windy, south slope of the ridge and a crop that needs a great deal of irrigation on the north slope. What type of farming techniques might maximize conservation of your soil? What other factors might you want to know about before you decide to commit to one or more methods?

Irrigation boosts productivity but may also cause soil problems

Erosion is not the only threat to soil. Soil can also be degraded when we apply water to crops. The artificial provision of water to support agriculture is known as **irrigation**. Some crops, such as rice and cotton, require large amounts of water whereas others, such as beans and wheat, require relatively little. By irrigating crops, people have managed to turn dry and unproductive regions into fertile farmland.

Seventy percent of all freshwater that people withdraw is used for irrigation. Irrigated acreage has increased dramatically worldwide, reaching 276 million ha (683 million acres) in 2002, greater than the entire area of Mexico and Central America.

If some water is good for plants and soil, it might seem that more must be better. But we can supply too much water. Overirrigated soils saturated with water may experience **waterlogging** when the water table rises such that water bathes plant roots, depriving them of gases and essentially suffocating them.

A more frequent problem is **salinization**, the buildup of salts in surface soil layers. In arid areas, water evaporating from the soil's A horizon may pull water from lower horizons upward by capillary action. As this water rises through the soil, it carries dissolved salts, and when it evaporates, those salts precipitate and are left at the surface. Irrigation in arid areas generally hastens salinization, and irrigation water often contains some dissolved salt in the first place, introducing new salt to the soil. Salinization now inhibits agricultural production on one-fifth of all irrigated cropland globally, costing more than $11 billion each year.

The best way to prevent salinization is to avoid planting crops that require a great deal of water in areas that are prone to the problem. A second way is to irrigate with water low in salt content. A third way is to irrigate efficiently, supplying no more water than the crop requires. This minimizes the amount of water that evaporates and hence the amount of salt that accumulates in the topsoil. Currently, irrigation efficiency worldwide is low; only 43% of the water applied actually gets used by plants. Drip irrigation systems (Figure 7.6) that target water directly to plants provide more control over how much water is used, and they waste far less.

Fertilizers boost crop yields but can be over-applied

Salinization is not the only way we may chemically damage soil. Overapplying fertilizers can do so as well. Plants require nitrogen, phosphorus, and potassium to grow, as well as smaller amounts of over a dozen other nutrients. If soils contain too few nutrients, crop yields decline. Therefore, a great deal of effort has aimed to enhance nutrient-limited soils by adding **fertilizer**, any of various substances that contain essential nutrients.

There are two main types of fertilizers. **Inorganic fertilizers** are mined or synthetically manufactured mineral supplements. **Organic fertilizers** consist of natural materials and include animal manure; crop residues; fresh vegetation (*green manure*); and *compost*, a mixture produced when decomposers break down organic matter, including food and crop waste, in a controlled environment. Inorganic fertilizers are generally more susceptible to leaching and runoff, and somewhat more likely to cause unintended off-site impacts. Organic fertilizers can provide some benefits that inorganic fertilizers cannot. For instance, the proper use of compost improves soil structure, nutrient retention, and water-retaining capacity, helping to prevent erosion.

Inorganic fertilizer use grew globally by leaps and bounds between 1960 and 1990, and we now apply 150 million metric tons per year. Its use has greatly boosted the amount of food produced worldwide. But applying fertilizer to croplands has impacts far beyond the boundaries of the fields, and fertilizer overuse is causing increasingly severe pollution problems. We saw in Chapter 3 how nitrogen and phosphorus runoff from farms and other

(a) Conventional irrigation

(b) Drip irrigation

FIGURE 7.6 Currently, less than half the water we apply in irrigation actually gets taken up by plants. Conventional methods that lose a great deal of water to evaporation **(a)** are now being replaced by more efficient ones in which water is more precisely targeted to plants. In drip irrigation systems **(b)**, water drips from holes in hoses directly onto the plants.

sources in the Mississippi River basin each year spurs phytoplankton blooms in the Gulf of Mexico and creates an oxygen-depleted "dead zone" that kills fish and shrimp. Such eutrophication occurs at countless river mouths, lakes, and ponds throughout the world. Moreover, nitrates readily leach through soil and contaminate groundwater, posing health threats to people.

Grazing practices and policies can contribute to soil degradation

We have focused so far largely on the cultivation of crops as a source of impacts on soils and ecosystems, but raising livestock also has impacts. When sheep, goats, cattle, or other livestock graze on the open range, they feed primarily on grasses. As long as livestock populations do not exceed a range's carrying capacity (▸ p. 89) and do not consume grasses faster than they can be replaced, grazing may be sustainable. However, when too many animals eat too much plant cover, impeding plant regrowth and the replacement of biomass, the result is **overgrazing**.

Rangeland scientists have shown that overgrazing causes a number of impacts, some of which give rise to positive feedback cycles that exacerbate damage to soils, natural communities, and the land's productivity for grazing (Figure 7.7). When livestock remove too much plant cover, more soil is exposed and made vulnerable to erosion. Soil erosion makes it difficult for vegetation to regrow, perpetuating the lack of cover and giving rise to more erosion. Moreover, non-native weedy plants may invade denuded soils. These invasive plants are usually less palatable to livestock and can outcompete native vegetation in the new, modified environment. Overgrazing can also compact soils and alter their structure. Soil compaction makes it harder for water to infiltrate, harder for soils to be aerated, harder for plants' roots to expand, and harder for roots to conduct cellular respiration.

As a cause of soil degradation worldwide, overgrazing is equal to cropland agriculture, and it is a greater cause of desertification. Humans keep a total of 3.3 billion cattle, sheep, and goats, and rangeland degradation is estimated to cost $23.3 billion per year. In the American West, government policy has long facilitated overgrazing, allowing ranchers to graze cattle on public lands with little restriction and for inexpensive fees. However, today increasing numbers of ranchers are working cooperatively with government agencies, environmental scientists, and even environmental advocates to find ways to ranch more sustainably and safeguard the health of the land.

Recent U.S. laws promote soil conservation

In recent years, the U.S. Congress has enacted a number of laws promoting soil conservation. The Food Security Act of 1985 required farmers to adopt soil conservation plans and practices as a prerequisite for receiving price supports and other government benefits.

The Conservation Reserve Program, also enacted in 1985, pays farmers to stop cultivating highly erodible cropland and instead place it in conservation reserves planted with grasses and trees. The USDA estimates that for an annual cost of $1 billion, this program saves 700 million metric tons (771 million tons) of topsoil each year. Besides reducing erosion, the Conservation Reserve Program has generated income for farmers and has provided habitat for native wildlife. In 1996, Congress extended the program by passing the Federal Agricultural Improvement and Reform Act. Also known as the "Freedom to Farm Act," this law aimed to reduce subsidies and government influence over many farm products and to promote and pay for the adoption of conservation practices in agriculture.

In 1998, the USDA initiated the Low-Input Sustainable Agriculture Program to provide funding for individual farmers to develop and practice sustainable agriculture. Improved policies and practices in all regions of the world will be necessary if we are to continue to provide for our planet's growing human population.

FIGURE 7.7 When grazing by livestock exceeds the carrying capacity of rangelands and their soil, overgrazing can set in motion a series of consequences and positive feedback loops that degrade soils and grassland ecosystems.

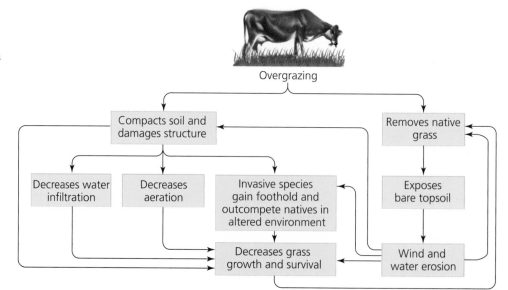

The Race to Feed the World

Although human population growth has slowed, we can still expect our numbers to swell to 9 billion by the middle of this century. For every two people living today, there will be three in 2050. The kind of world we live in then will depend on the choices we make now, and some of the most important choices involve our means of producing food.

We are producing more food per person

Over the past half-century, our ability to produce food has grown even faster than global population (Figure 7.8). However, because of poverty, political obstacles, and inefficiencies in distribution, today 850 million people in developing countries do not have enough to eat. These people suffer from *undernourishment* (receiving too few calories per day) and *malnutrition* (receiving too few nutrients in food; Figure 7.9). Every 5 seconds, somewhere in the world, a child starves to death. Meanwhile, in the developed world, where food is available in abundance, many of us are *overnourished* (receiving too many calories each day) and lead sedentary lives with little exercise. More than three out of five adults are technically overweight, and over one out of four are obese. Many U.S. children face a future of severe health problems, extensive health care costs, and reduced life expectancy.

Agricultural scientists and policymakers pursue a goal of **food security**, the guarantee of an adequate, reliable,

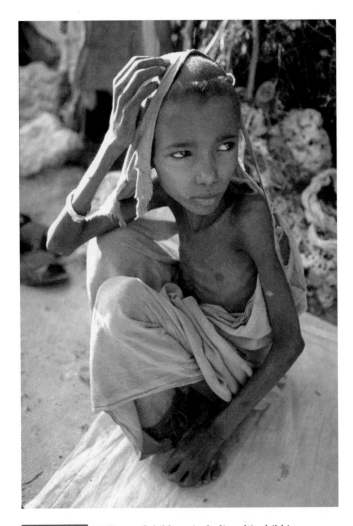

FIGURE 7.9 Millions of children, including this child in Somalia, suffer from conditions caused by malnutrition.

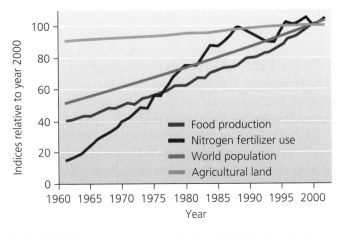

FIGURE 7.8 Over the past four decades, global food production rose by over two-and-a-half times, growing at a faster rate than world population. Additional land was converted to agricultural use at a slower rate during this time, and fertilizer use grew most rapidly. All data are graphed relative to year 2000 values, so that each line equals 100% in 2000. Data from Food and Agriculture Organization of the United Nations.

and available food supply to all people at all times. Making a food supply sustainable depends on maintaining healthy soil, water, and biodiversity.

Starting in the 1960s, a number of scientists such as Paul Ehrlich (▸ p. 5) predicted widespread starvation and a catastrophic failure of agricultural systems, arguing that the human population could not continue to grow without outstripping its food supply. However, the human population has continued to increase well past their predictions. In percentage terms, we have reduced hunger by half, from 26% of the population in 1970 to 13% today.

We have achieved this success, and a dramatic increase in our carrying capacity (▸ p. 89), largely by increasing our ability to produce food through the industrialization of agriculture. We now devote more energy (especially fossil fuel energy) to agriculture; plant and harvest more frequently; use more irrigation, fertilizer, and pesticides; cultivate more land area; and have developed (through crossbreeding and genetic engineering) more productive crop and livestock varieties. To gain efficiency of scale, industrialized agriculture demands that vast fields be planted with single types of crops. The uniform planting of a single crop, termed **monoculture**, differs from the *polyculture* approach of traditional agriculture such as Native American farming systems that mixed maize, beans, squash, and peppers in the same fields. Today, monocultural industrialized agriculture occupies about 25% of the world's cropland.

However, with grain crops, the world's staple foods, we are producing less food per person each year. World grain production per person peaked in 1985 and has since fallen slightly. Moreover, the world's soils are in decline, and nearly all the planet's arable land has already been claimed. Simply because agricultural production has outpaced population growth so far, there is no guarantee that it will continue to do so.

The "green revolution" led to dramatic increases in agricultural production

The desire for greater quantity and quality of food for the growing human population led people to introduce industrialized agriculture to the developing world in the mid- and late 20th century, in what is called the **green revolution**. Realizing that farmers could not go on indefinitely cultivating more land, agricultural scientists introduced methods and technologies to increase crop output per unit area of existing cultivated land.

The transfer of technology to the developing world that marked the green revolution began in the 1940s, when U.S. agricultural scientist Norman Borlaug introduced Mexico's farmers to a specially bred type of wheat (Figure 7.10). This strain of wheat produced large seed heads, was short in stature to resist wind, resisted diseases, and produced high yields. Within two decades of planting and harvesting this specially bred crop, Mexico tripled its wheat production and began exporting wheat. The stunning success of this program inspired others. Borlaug—who won the Nobel Peace Prize for his work—took his wheat to India and Pakistan and helped transform agriculture there. Soon many developing nations were doubling, tripling, or quadrupling their yields using selectively bred strains of wheat, rice, corn, and other crops.

Along with the new grains, developing nations began applying large amounts of synthetic fertilizers and chemical pesticides on their fields, irrigating crops with generous amounts of water, and using heavy equipment powered by fossil fuels. From 1900 to 2000, people expanded

FIGURE 7.10 Norman Borlaug holds examples of the wheat variety he bred that helped launch the green revolution. The high-yielding disease-resistant wheat helped increase agricultural productivity in many developing countries.

the world's total cultivated area by 33%, yet increased energy inputs into agriculture 80-fold.

This high-input agriculture succeeded dramatically in allowing farmers to harvest more corn, wheat, rice, and soybeans from each hectare of land. In India, intensive agriculture saved millions from starvation in the 1970s and eventually turned that nation into a net exporter of grain.

These developments had mixed effects on the environment. On the positive side, the intensified use of already-cultivated land reduced pressures to convert additional natural lands for new cultivation. Between 1961 and 2002, food production rose by 150% and population rose by 100%, while area converted for agriculture increased only 10% (see Figure 7.8). For this reason, the green revolution prevented some degree of deforestation and habitat conversion in many countries at the very time those countries were experiencing their fastest population growth.

On the negative side, the intensive use of water, fossil fuels, and chemical fertilizers and pesticides exacerbated pollution, salinization, and desertification. The monocultures that made planting and harvesting more efficient and thereby increased output also reduced biodiversity, because many fewer wild organisms are able to live in monocultures than in native habitats or in traditional small-scale polycultures. Moreover, when all plants in a field are genetically similar, as in monocultures, all are equally susceptible to viral diseases, fungal pathogens, or insect pests that spread quickly from plant to plant. For this reason, monocultures bring significant risks of catastrophic failure.

Weighing the Issues:
The Green Revolution and Population

In the 1960s, India's population was skyrocketing, and its traditional agriculture was not producing enough food to support the growth. By adopting green revolution agriculture, India sidestepped mass starvation. In the years since intensifying its agriculture, India has added several hundred million more people and continues to suffer widespread poverty and hunger. Norman Borlaug called his green revolution methods "a temporary success in man's war against hunger and deprivation," something to give us breathing room in which to deal with what he called "the Population Monster."

Do you think we can call the green revolution a success? Do you think the green revolution has solved problems, or delayed our resolution of problems, or created new ones? How sustainable are green revolution approaches? Consider our discussion of demographic transition theory from Chapter 6 (▸ pp. 131–132). Have we been dealing with the "Population Monster" during the breathing room that the green revolution has bought for us?

Pests and Pollinators

Throughout the history of agriculture, the insects, fungi, viruses, rats, and weeds that eat or compete with our crop plants have taken advantage of the ways we cluster food plants into agricultural fields. In monocultures, a pest adapted to specialize on the crop can easily move from one individual plant to many others of the same type.

What people term a *pest* is any organism that damages crops that are valuable to us. What we term a *weed* is any plant that competes with our crops. These are subjective categories that we define entirely by our own economic interests. There is nothing inherently malevolent in the behavior of a pest or a weed. These organisms are simply trying to survive and reproduce. From the viewpoint of an insect that happens to be adapted to feed on corn, grapes, or apples, a grain field, vineyard, or orchard represents an endless buffet.

We have developed thousands of chemical pesticides

To prevent pest outbreaks and to limit competition with weeds, people have developed thousands of artificial chemicals to kill insects (*insecticides*), plants (*herbicides*),

and fungi (*fungicides*). Such poisons that target pest organisms are collectively termed **pesticides**. Table 7.1 shows the 10 most widely used pesticides in U.S. agriculture. All told, roughly 400 million kg (900 million lb) of active ingredients from conventional pesticides are applied in the United States each year, three-quarters of this total to agricultural land. Since 1960, pesticide use has risen fourfold worldwide. Usage in industrialized nations has leveled off in the past two decades, but it continues to rise in the developing world. Today over $32 billion is expended annually on pesticides, with one-third of that total spent in the United States. We address the health consequences of synthetic pesticides for humans and other organisms in Chapter 10.

Pests evolve resistance to pesticides

Despite the toxicity of these chemicals (Chapter 10), their usefulness tends to decline with time as pests evolve resistance to them. Recall from our discussion of natural selection (▸ pp. 76–78) that organisms within populations vary in their traits. Because most insects and microbes occur in huge numbers, it is likely that a small fraction of individuals may by chance have genes that confer some degree of immunity to a given pesticide. Even if a pesticide application kills 99.99% of the insects

Table 7.1	Most Commonly Used Pesticides* in Agriculture in the United States	
Active ingredient[†]	**Type of pesticide**	**Millions of kg applied per year**
Glyphosate	Herbicide	39–41
Atrazine	Herbicide	34–36
Metam sodium	Fumigant	26–28
Acetochlor	Herbicide	14–16
2,4-D	Herbicide	13–15
Malathion	Insecticide	9–11
Methyl bromide	Fumigant	9–11
Dichloropropene	Fumigant	9–11
Metolachlor-s	Herbicide	9–11
Metolachlor	Herbicide	7–10

*Includes only "conventional pesticides" used in agriculture. Does not include many other types of pesticides, such as disinfectants and wood preservatives.
[†]Includes only active ingredients, not ingredients such as oil, sulfur, and sulfuric acid.
Data from Kiely, T., et al. 2004. *Pesticides industry sales and usage: 2000 and 2001 market estimates.* Washington, DC: U.S. Environmental Protection Agency.

in a field, 1 in 10,000 survives. If an insect survives by being genetically resistant to a pesticide, and if it mates with other resistant individuals of the same species, the insect population may grow. This new population will consist of individuals that are genetically resistant to the pesticide. As a result, pesticide applications will cease to be effective (Figure 7.11).

In many cases, industrial chemists are caught up in an "evolutionary arms race" (▶ p. 100) with the pests they battle, racing to increase or retarget the toxicity of their chemicals while the armies of pests evolve ever-stronger resistance to their efforts. The number of species known to have evolved resistance to pesticides has grown over the decades. As of 2000, there were nearly 2,700 known cases of resistance by 540 species to over 300 pesticides. Insects such as the green peach aphid, Colorado potato beetle, and diamondback moth have evolved resistance to multiple insecticides.

Biological control pits one organism against another

Because of pesticide resistance and the health risks from some synthetic chemicals, agricultural scientists increasingly battle pests and weeds with organisms that eat or infect them. This strategy, called **biological control**, or **biocontrol**, operates on the principle that "the enemy of one's enemy is one's friend." For example, parasitoid wasps (▶ p. 100) are natural enemies of many caterpillars. These wasps lay eggs on a caterpillar, and the larvae that hatch from the eggs feed on the caterpillar, eventually killing it. Some such efforts have succeeded in controlling pests and have led to steep reductions in chemical pesticide use.

A classic case of successful biological control was the introduction of the cactus moth, *Cactoblastis cactorum*, from Argentina to Australia in the 1920s to control invasive prickly pear cactus that was overrunning rangeland.

1 Outbreak of pests on crops

2 Application of pesticide

3 All pests except a few with innate resistance are killed

4 Survivors breed and produce a pesticide-resistant population

5 Pesticide is applied again

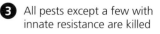
6 Pesticide has little effect and new, more toxic pesticides must be developed

FIGURE 7.11 Through the process of natural selection, crop pests frequently evolve resistance to the poisons we apply to kill them. When a pesticide is applied to an outbreak of insect pests, it may kill virtually all individuals except those few with an innate immunity to the poison. Those surviving individuals may found a population with genes for resistance to the poison. Future applications of the pesticide may then be ineffective, forcing us to develop a more potent poison or an alternative means of pest control.

(a) Before cactus moth introduction

(b) After cactus moth introduction

FIGURE 7.12 In one of the classic cases of biocontrol, larvae of the cactus moth, *Cactoblastis cactorum*, were used to clear non-native prickly pear cactus from millions of hectares of rangeland in Queensland, Australia. These photos from the 1920s show an Australian ranch before (**a**) and after (**b**) introduction of the moth.

Within just a few years, the moth managed to free millions of hectares of rangeland from the cactus (Figure 7.12). One widespread modern biocontrol effort has been the use of *Bacillus thuringiensis* (Bt), a naturally occurring soil bacterium that produces a protein that kills many caterpillars and the larvae of some flies and beetles.

However, biocontrol involves risks. Because most biocontrol agents are introduced from foreign ecosystems, no one can know for certain in advance what effects they might have. In some cases, biocontrol agents have become invasive and harmed nontarget organisms. Following the cactus moth's success in Australia, for example, it was introduced in other countries to control prickly pear. Moths introduced to Caribbean islands spread to Florida on their own and are now eating their way through rare native cacti

in Florida and spreading to other states. If these moths reach Mexico and the southwestern United States, they could decimate many native species of prickly pear there.

When biocontrol works as planned, it can be a permanent solution that requires no further maintenance and is environmentally benign. However, if the agent has nontarget effects, the harm done may also become permanent, because removing the agent from the system once it is established is far more difficult than simply halting a chemical pesticide application.

Integrated pest management combines methods

As it became clear that both chemical and biocontrol approaches have their drawbacks, agricultural scientists and farmers began developing more sophisticated strategies, trying to combine the best attributes of each approach. In **integrated pest management (IPM)**, numerous techniques are integrated, including biocontrol, use of chemicals, close monitoring of populations, habitat alteration, crop rotation, transgenic crops, alternative tillage methods, and mechanical pest removal.

Indonesia stands as an exemplary case of IPM's success. The nation had subsidized pesticide use heavily for years, but its scientists came to understand that pesticides were actually making pest problems worse. They were killing the natural enemies of the brown planthopper, which began to devastate rice fields as its populations exploded. Concluding that pesticide subsidies were costing money, causing pollution, and apparently decreasing yields, the Indonesian government in 1986 banned the importation of 57 pesticides, slashed pesticide subsidies, and encouraged IPM. Within 4 years, pesticide production fell to below half its 1986 level, imports fell to one-third, and subsidies were phased out (saving $179 million annually). Rice yields rose by 13%.

We depend on insects to pollinate crops

Managing insect pests is such a major issue in agriculture that it is easy to fall into a habit of thinking of all insects as somehow bad or threatening. But in fact, most insects are harmless to agriculture, and some are absolutely essential. The insects that pollinate agricultural crops are one of the most vital, yet least understood and appreciated, factors in cropland agriculture.

Pollination is the process by which male sex cells of a plant (pollen) fertilize female sex cells of a plant; it is the botanical version of sexual intercourse. The showy flowers of many plants are, in fact, evolutionary adaptations that function to attract pollinating animals such as

FIGURE 7.13 North American farmers hire beekeepers to bring hives of European honeybees to their crops when it is time for flowers to be pollinated. In recent years, certain parasitic mites have swept through honeybee populations, devastating hives and making it increasingly vital for us to conserve native species of pollinators.

insects, bats, and hummingbirds, which transfer pollen among plants. The sugary nectar and protein-rich pollen in flowers serve as rewards to lure these sexual intermediaries, and the sweet smells and bright colors of flowers are signals to advertise these rewards.

Our staple grain crops are derived from grasses and are wind-pollinated, but at least 800 cultivars, or types of cultivated plants, are known to rely on bees and other insects for pollination. An estimated 73% of cultivars are pollinated, at least in part, by bees; 19% by flies; 5% by wasps; 5% by beetles; and 4% by moths and butterflies.

Preserving the biodiversity of native pollinators is especially important today because the domesticated workhorse of pollination, the European honeybee (Figure 7.13), is being devastated by parasites. Moreover, research indicates that honeybees are sometimes less effective pollinators than many native species and often outcompete them, keeping the native species away from the plants. Farmers and homeowners alike can help maintain populations of pollinating insects by reducing or eliminating pesticide use on fields and lawns, providing nesting sites for bees, and planting gardens of flowering plants.

Biotechnology and the Future of Food

The green revolution enabled us to feed a greater number and proportion of the world's people, but relentless population growth demands still more. A new set of potential solutions began to arise in the 1980s and 1990s as advances in genetics enabled scientists to directly alter the genes of organisms, including crop plants and livestock. The genetic modification of organisms that provide us food holds promise for increasing nutrition and the efficiency of agriculture while lessening the impacts of agriculture on the planet's environmental systems. However, genetic modification may also pose risks that are not yet well understood, and this has given rise to widespread protest from consumer advocates, small farmers, opponents of big business, and environmental activists.

Food can be genetically modified

The genetic modification of crops and livestock is one type of genetic engineering. **Genetic engineering** is any process whereby scientists directly manipulate an organism's genetic material in the lab, by adding, deleting, or changing segments of its DNA (▸ p. 55). Using *recombinant DNA* technology, scientists break up DNA from multiple organisms and then splice segments together, trying to place genes that produce certain proteins and code for desirable traits (such as rapid growth, disease and pest resistance, or better nutrition) into the genomes of organisms lacking those traits. **Genetically modified (GM) organisms** are organisms that have been genetically engineered. An organism that contains DNA from another species is known as a *transgenic* organism, and the genes that have moved between them are called *transgenes*.

The creation of transgenic organisms is one type of **biotechnology**, the material application of biological science to create products derived from organisms. Biotechnology has helped us develop medicines, clean up pollution, understand the causes of cancer and other diseases, and dissolve blood clots after heart attacks. Figure 7.14 details several notable examples of GM foods, whose stories illustrate both the promises and pitfalls of food biotechnology.

In one sense, our genetic alteration of plants and animals is nothing new; through artificial selection (▸ pp. 78–79) we have been influencing the genetic makeup of livestock and crop plants for thousands of years. However, as biotech critics are quick to point out, the techniques geneticists use to create GM organisms differ from traditional selective breeding. Selective breeding generally mixes genes of individuals of the same species, whereas with recombinant DNA technology, scientists mix genes of different ones, even species as different as viruses and crops, or spiders and goats. Moreover,

Several Notable Examples of Genetically Modified Food Technology			
Food	Development	Food	Development
Golden rice 	Millions of people in the developing world get too little vitamin A in their diets, causing diarrhea, blindness, immune suppression, and even death. The problem is worst with children in east Asia, where the staple grain, white rice, contains no vitamin A. Researchers took genes from plants that produce vitamin A and spliced the genes into rice DNA to create more-nutritious "golden rice" (the vitamin precursor gives it a golden color). Critics charged that biotech companies over-hyped their product.	StarLink corn 	StarLink corn, a variety of Bt corn, had been approved and used in the United States for animal feed but not for human consumption. In 2000, StarLink corn DNA was discovered in taco shells and other corn products. These products were recalled amid fears that the corn might cause allergic reactions. No such health effects were confirmed, but the corn's French manufacturer, Aventis CropScience, chose to withdraw the product from the market.
Ice-minus strawberries 	Researchers removed a gene that facilitated the formation of ice crystals from the DNA of a bacterium, *Pseudomonas syringae*. The modified, frost-resistant bacteria could then serve as a kind of antifreeze when sprayed on the surface of frost-sensitive crops such as strawberries, protecting them from frost damage. However, news coverage showed scientists spraying plants while wearing face masks and protective clothing, an image that caused public alarm.	Sunflowers and superweeds 	Research on Bt sunflowers suggests that their transgenes might spread to other plants and turn them into vigorous weeds that compete with the crop. This is most likely to happen with crops like squash, canola, and sunflowers that can breed with their wild relatives. Researchers bred wild sunflowers with Bt sunflowers and found that hybrids with the Bt gene produced more seeds and suffered less herbivory than hybrids without it. They concluded that if Bt sunflowers were planted commercially, the Bt gene would spread into wild sunflowers, potentially turning them into superweeds.
Bt crops 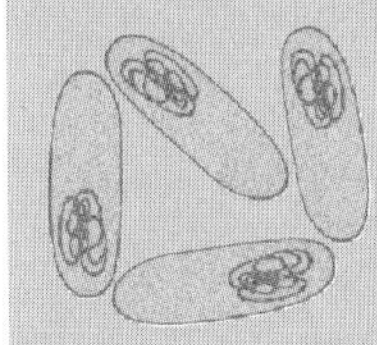	By equipping plants with the ability to produce their own pesticides, scientists hoped to boost crop yields by reducing losses to insects. Scientists working with *Bacillus thuringiensis* (Bt) pinpointed the genes responsible for producing that bacterium's toxic effects on insects, and inserted the genes into the DNA of crops. The USDA and EPA approved Bt versions of 18 crops for field testing, from apples to broccoli to cranberries. Corn and cotton are the most widely planted Bt crops today. Proponents say Bt crops reduce the need for chemical pesticides. Critics worry that they will induce insects to evolve resistance to the toxins, cause allergic reactions in humans, and harm nontarget species.	Roundup Ready crops 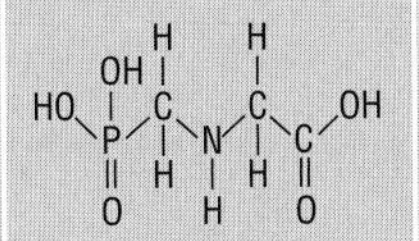	The Monsanto Company's widely used herbicide, Roundup, kills weeds, but it kills crops too, so farmers must apply it carefully. Thus, Monsanto engineered Roundup Ready crops, including soybeans, corn, cotton, and canola, that are immune to the effects of its herbicide. With these variants, farmers can spray Roundup on their fields without killing their crops. Of course, this also creates an incentive for farmers to use Monsanto's Roundup herbicide rather than a competing brand. Unfortunately, Roundup is not completely benign; its active ingredient, glyphosate, is the third-leading cause of illness for California farm workers.

FIGURE 7.14 The early development of genetically modified foods has been marked by a number of cases in which these products ran into trouble in the marketplace or were opposed by activists. A selection of these cases serves to illustrate some of the issues that proponents and opponents of GM foods have being debating.

selective breeding deals with whole organisms living in the field, whereas genetic engineering involves lab experiments dealing with genetic material apart from the organism. And whereas traditional breeding selects from among combinations of genes that come together on their own, genetic engineering creates the novel combinations directly. Thus, traditional breeding changes organisms through the process of selection, whereas genetic engineering is more akin to the process of mutation (▸ p. 77).

Transgenic Contamination of Native Maize?

Corn is a staple grain of the world's food supply. We can trace its ancestry back roughly 5,500 years, when people in the highland valleys of what is now Oaxaca in southern Mexico first domesticated that region's wild maize plants. The corn we eat today arose from some of the many varieties that evolved from the selective crop breeding conducted by the people of this region. Today Oaxaca remains a world center of diversity for maize, with numerous native varieties, or cultivars, growing in the rich, well-watered soil. Preserving such varieties of major crops in their ancestral homelands is important for securing the future of our food supply, scientists maintain.

Thus, it caused global consternation when, in 2001, Mexican scientists conducting routine genetic tests of Oaxacan farmers' maize turned up DNA that seemed to match genes from genetically modified corn. GM corn was widely grown in the

Ignacio Chapela (left) and David Quist (right) ignited a firestorm of controversy with their scientific paper reporting transgenic corn DNA in Mexican maize.

United States, but Mexico had banned its cultivation in 1998. The concern was that transgenic crops might crossbreed with native crops and thereby "contaminate" the genetic makeup of native crops.

Two researchers at the University of California at Berkeley, Ignacio Chapela and his postdoctoral associate David Quist, shared these concerns and ventured to Oaxaca to test samples of native maize. Their analyses seemed to

confirm the Mexican scientists' findings, revealing what they argued were traces of DNA from GM corn in the genes of native maize plants. They also argued that the invading genes had split up and spread throughout the maize genome. Quist and Chapela published their findings in the scientific journal *Nature* in 2001. Activists opposed to GM food trumpeted the news and urged a ban on imports of transgenic crops from producer nations such as the United States into developing nations. The agrobiotech industry defended the safety of its crops and questioned the validity of the research—as did many of Quist and Chapela's peers.

One of Quist and Chapela's claims in particular came under fire: the idea that a transgenic promoter, which determines where to begin transcribing a gene, had been jumping at random throughout the genome. Soon after publication, several geneticists pointed out flaws in

Biotechnology is transforming the products around us

In just three decades, GM foods have gone from science fiction to big business. Since the 1980s, industry has developed hundreds of applications, from improved medicines (such as hepatitis B vaccine and insulin for diabetes) to designer plants and animals. Traits engineered into crops, such as built-in pest resistance and herbicide resistance, made it efficient, and in some cases more economical, for large-scale commercial farmers to do their jobs. As a result, sales of GM seeds to these farmers in the United States and other nations rose quickly.

Today well over two-thirds of the U.S. harvests of soybeans, corn, and cotton consists of genetically modified strains. Worldwide in 2004, it was estimated that GM

crops grew on 81 million ha (200 million acres) of farmland—an area the size of Kansas, Nebraska, and California combined. Because many of the 17 nations growing GM crops in 2004 (Figure 7.15) are major food exporters, much of the produce on the world market is transgenic for crops such as soybeans, corn, cotton, and canola. The global area planted in GM crops has jumped by more than 10% every year since 1996. The market value of GM crops in 2004 was estimated at $4.7 billion.

Scientists and citizens debate the impacts of GM crops

As GM crops were adopted, as research proceeded, and as biotech business expanded, many citizens, scientists, and policymakers became concerned. Some feared the new

the methods Quist and Chapela used to determine the location of the transgene. Interpreting results of an experimental technique known as inverse polymerase chain reaction (i-PCR), Quist and Chapela had reported that the transgene was surrounded by essentially random sequences of DNA. Critics argued that i-PCR was unreliable and that Quist and Chapela had used insufficient controls. Their results, the critics suggested, could have arisen from similarities between the transgene and stretches of maize DNA.

Several geneticists also attacked Quist and Chapela's more fundamental claim that transgenes had been integrated into the genome of native maize. Some argued that the low levels of transgenic DNA detected indicated, at most, first-generation hybrids between local cultivars and transgenic varieties. Other critics, including the editor of the journal *Transgenic Research*, noted that i-PCR was a highly sensi-

tive technique and that all of Quist and Chapela's findings could be unreliable if laboratory practices had been careless.

On April 11, 2002, *Nature* published two letters from scientists critical of the study, along with an editorial note concluding that "the evidence available is not sufficient to justify the publication of the original paper." Quist and Chapela responded by acknowledging that some of their initial findings, particularly those based on the i-PCR technique, were probably invalid. But they also presented new analyses to support their fundamental claim and pointed to a Mexican government study that also found high rates of transgenic contamination. However, this did little to convince skeptics, especially because the Mexican government's results remained unpublished.

A broader debate raged on the editorial pages of scientific journals, in newspapers and magazines, over

the Internet, and within the halls of academe. The debate sometimes turned personal. GM supporters pointed out that Chapela and Quist had long opposed transgenic crops and biotechnology corporations and that Chapela had spoken out against his university's plan to enter into a $25 million partnership with a biotechnology firm. GM opponents countered that Quist and Chapela's most vocal critics received funding from biotechnology corporations. Chapela was denied tenure by Berkeley's administration and gained it only after a high-profile three-year struggle.

The twists and turns of the still-unresolved debate illustrate that there can be more to the scientific process than merely the scientific method. Almost overlooked, however, was the fact that nearly all researchers agreed that gene flow between transgenic corn and Mexico's native landraces was bound to happen at some point.

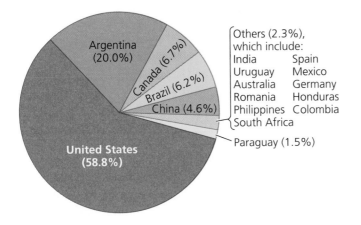

FIGURE 7.15 Five nations (the United States, Argentina, Canada, Brazil, and China) account for 96% of GM crops, with the United States alone accounting for three-fifths of the global total. Data from International Service for the Acquisition of Agri-Biotech Application (ISAAA), 2004.

foods might be dangerous for people to eat. Some were concerned that transgenes might escape and pollute ecosystems and damage nontarget organisms. Others worried that pests would evolve resistance to the super-crops and become "superpests" or that transgenes would move from crops to other plants and turn them into "superweeds." Still others worried that transgenes might ruin the integrity of native ancestral races of crops (see "The Science behind the Story," above).

Because the technology is new and its large-scale introduction into the environment is newer still, there remains a lot scientists don't know about how transgenic crops behave in the field. Certainly, millions of Americans eat GM foods every day without outwardly obvious signs of harm, and evidence for negative ecological effects is limited so far. However, it is too early to dismiss the concerns cited above without further scientific research. Therefore, critics argue that we should proceed with caution, adopting the **precautionary principle**, the idea that one should not undertake a new action until the ramifications of that action are well understood. Scientific studies conducted so far suggest that any environmental or health impacts of GM crops could be complex and will likely vary with the conditions under which different crops are grown.

Much more than science has been involved in the debate over GM foods. Ethical issues have played a large role. For many people, the idea of "tinkering" with the food supply seems dangerous or morally wrong. The perceived lack of control over one's own food has also driven widespread concern about domination of the global food supply by a few large agrobiotech corporations. Whereas the green revolution was a largely public venture, the "gene revolution" promised by GM crops has been largely driven by market considerations of companies selling proprietary products.

People of different cultures have reacted differently to these issues. European consumers have expressed widespread unease about possible risks of GM technologies, and European governments now demand that GM food be labeled as such. U.S. consumers, meanwhile, have largely accepted the GM crops approved by U.S. agencies. The future of GM foods seems likely to hinge on social, economic, legal, and political factors as well as on scientific ones.

Preserving crop diversity provides insurance against failure

Although biotechnology may greatly influence the future of our food supply, it is at least as important that we preserve the natural genetic diversity of our crop plants. Modern industrial agriculture relies on a small number of plant types, but its foundation lies in the diverse ancestral varieties that still exist in places like Oaxaca (see "The Science behind the Story," ▸ pp. 158–159). These varieties contain genes that, through conventional crossbreeding or genetic engineering, might confer resistance to disease, pests, inbreeding, and other pressures that challenge modern agriculture. Monocultures essentially place all our eggs in one basket, such that any single catastrophic cause could potentially wipe out entire crops. Having available the wild relatives of crop plants or the domesticated cultivars of crop plants provides us genetic diversity that may include ready-made solutions to unforeseen prob-

FIGURE 7.16 Seed banks preserve genetic diversity of traditional crop plants. Native Seeds/SEARCH of Tucson, Arizona, preserves seeds of beans, chiles, squashes, and other food plants important to traditional diets of Native Americans of the Southwest. At the farm where seeds are grown, care is taken to pollinate varieties by hand to protect their genetic distinctiveness.

lems. Because accidental interbreeding can decrease the diversity among local variants, many scientists argue that we need to protect areas like Oaxaca from contact with GM crops to retain important repositories of crop diversity.

Another way to preserve genetic assets for agriculture is to collect and store seeds from crop varieties and periodically plant and harvest them to maintain a diversity of cultivars. This is the work of **seed banks** or gene banks, institutions that preserve seed types as a kind of living museum of genetic diversity (Figure 7.16). In total, such facilities worldwide hold roughly 6 million seed samples.

Feedlot Agriculture: Livestock and Poultry

Food from cropland agriculture makes up a large portion of the human diet, but most people also eat animal products, such as meat, fish, milk, and eggs. Doing so has significant environmental, social, agricultural, and economic impacts. As wealth and global commerce have increased, so has our consumption of meat, milk, eggs, and other animal products (Figure 7.17). The world population of domesticated animals raised for food rose from 7.3 billion

VIEWPOINTS

Genetically Modified Foods

Proponents of GM foods say these products can alleviate hunger and malnutrition while posing no known threats to human health or the environment. Opponents say that these foods help only the large corporations that sell them and that they do pose risks to human health, wild organisms, and ecosystems. **What do you think? Should we encourage the continued development of GM foods? If so, what, if any, restrictions should we put on their dissemination?**

A Global Experiment without Controls

Genetic engineering, specifically transgenesis, gives us the unprecedented capacity to move DNA. In so doing, this technology breaches boundaries established through millions of years of evolution. As such, we should expect fundamental alterations in ecosystems with the release of transgenic crops, fish, insects, microbes, and so forth into uncontrolled areas. These alterations are similar in nature to those caused by the introduction of exotic species into new environments. Both processes are unpredictable and could have serious consequences.

Yet because of political and short-term economic imperatives, releases of transgenic organisms have continued unabated for at least a decade. Science has barely started to imagine the ecological and evolutionary consequences of releasing transgenic crops. We not only are experiencing a global experiment without controls, but also lack the tools to document it. Serious research, although extremely scarce, has already confirmed some of the theoretical fears concerning transgenesis aired by scientists a quarter century ago.

Today, we have cataclysmic world hunger paired with food surpluses. The claim that transgenesis can solve this problem is merely a diversion tailored to conceal how transgenesis manipulates the biosphere. Molecular biology might one day become part of the solution to world hunger, but it is certainly not the science most relevant to address the problem today.

What checks should be placed on the release of transgenic organisms? Every check. Through the unaccountable releases so far, we have seen enough, and possibly caused enough, environmental insult for me to say today that we should stop. We need to take stock of the consequences of transgenesis and continue researching under strictly regulated conditions.

Ignacio H. Chapela is associate professor (microbial ecology) in the Department of Environmental Science, Policy, and Management at the University of California at Berkeley. He helped found the Mycological Facility: Oaxaca, Mexico, where he serves as scientific director.

The United States Should Begin a Phased Deregulation of Biotech Crops

During the past two decades the international scientific community, biotechnology industry, and regulatory agencies in many countries have accumulated and critically evaluated a wealth of information about the production and use of biotech crops and products. Biotech crops have been planted since 1996 on more than 1 billion acres of farmland in nearly 20 countries. More than 1 billion humans and hundreds of millions of farm animals have consumed biotech foods and products. Yet there is not a single instance in which biotech crops and foods have been shown to cause illness in humans or animals or to damage the environment.

In spite of this exemplary safety record, a small but well-organized, well-financed, and vocal anti-biotechnology lobby has alleged that biotech crops and products are unsafe for humans and a danger to the environment, demanding a moratorium or outright ban on biotech crops. The rhetoric of the anti-biotechnology groups is alarming, confusing, and frightening to the public, but it is devoid of any substance, because they have never provided any credible scientific evidence to support their allegations.

Any further delay in combining the power of biotechnology with conventional breeding will seriously endanger future food security, political and economic stability, and the environment. Plant biotechnology is still the best hope for meeting the food needs of the ever-growing world population. Biotech crops are already helping to conserve valuable natural resources, reduce the use of harmful agro-chemicals, produce more nutritious foods, and promote economic development.

Twenty years ago, the United States set the precedent by developing regulations for the development and use of biotech crops. Now, as the world leader in plant biotechnology, it is imperative that it lead again by phasing out these redundant regulations in an organized and responsible manner.

Indra K. Vasil is graduate research professor emeritus at the University of Florida (Gainesville, Florida). His research focuses on the biotechnology of cereal crops, and he has been recognized as one of the world's most highly cited authors in the plant and animal sciences.

Explore this issue further by accessing **Viewpoints** at www.aw-bc.com/withgott.

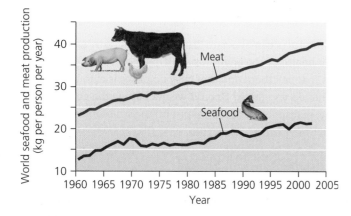

animals to 20.6 billion animals between 1961 and 2000. Global meat production has increased fivefold since 1950, and per capita meat consumption has nearly doubled.

Increased consumption of animal products has led to feedlot agriculture

In traditional agriculture, livestock were kept by farming families near their homes or were grazed on open grasslands by nomadic herders or sedentary ranchers. These traditions have survived, but the advent of industrial agriculture has added a new method. **Feedlots**, also known as *factory farms* or *concentrated animal feeding operations (CAFOs)*, are essentially huge warehouses or pens designed to deliver energy-rich food to animals living at extremely high densities (Figure 7.18). Today over half of the world's pork and poultry come from feedlots, as does much of its beef.

Feedlot operations allow for greater production of food and are probably necessary for a country with a level of meat consumption like that of the United States. Feedlots have one overarching benefit for environmental quality: Taking livestock off the land and concentrating them in feedlots reduces the impact they would otherwise exert on large portions of the landscape.

Of course, feedlots are not without impact, and environmental advocates have attacked them for their water and air pollution. Waste from feedlots can emit strong odors and can pollute surface water and groundwater, because livestock produce prodigious amounts of feces and urine. One dairy cow can produce about 20,400 kg (44,975 lb) of waste in a single year. Poor waste containment practices at feedlots in North Carolina, Maryland,

FIGURE 7.18 These chickens at a Pennsylvania factory farm are housed several to a cage and have been "debeaked," the tips of their beaks cut off to prevent them from pecking one another. The hens cannot leave the cages and essentially spend their lives eating, defecating, and laying eggs, which roll down slanted floors to collection trays. The largest U.S. chicken farms house hundreds of thousands of individuals.

and other states have been linked to outbreaks of disease, including virulent strains of *Pfiesteria*, a microbe that poisons fish. The crowded and dirty conditions under which animals are often kept necessitate heavy use of antibiotics to control disease. These chemicals can be transferred up the food chain, and their overuse can cause microbes to evolve resistance to them.

Feedlot impacts can be minimized when properly managed, and both the EPA and the states regulate U.S. feedlots. Most feedlot manure is applied to farm fields as fertilizer, reducing the need for chemical fertilizers.

Our food choices are also energy choices

What we choose to eat has ramifications for how we use energy and the land that supports agriculture. Recall our discussions of thermodynamics and trophic levels (▸ pp. 58–59 and ▸ pp. 101–102). Every time energy moves from one trophic level to the next, roughly 90% is lost. For example, if we feed grain to a cow and then eat beef from the cow, we lose most of the grain's energy to the cow's digestion and metabolism. Energy is used as the cow converts the grain to tissue as it grows, and as the cow uses its muscle mass on a daily basis to maintain itself. For this reason, eating meat is far less energy-efficient than relying on a vegetarian diet. The lower in the food chain we take our food, the greater the proportion of the sun's energy we utilize, and the more people Earth can support.

Some animals convert grain feed into milk, eggs, or meat more efficiently than others (Figure 7.19). Scientists have calculated relative energy-conversion efficiencies for different types of animals. Such energy efficiencies have ramifications for land use, because land and water are required to raise food for the animals, and some animals require more than others. Figure 7.20 shows the area of land and weight of water required to

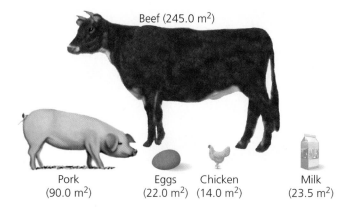

(a) Land required to produce 1 kg of protein

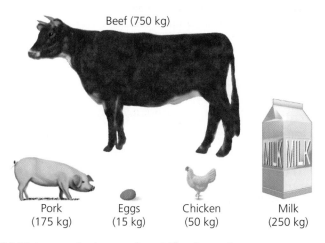

(b) Water required to produce 1 kg of protein

FIGURE 7.20 Producing different types of animal products requires different amounts of land and water. Raising cattle for beef requires by far the most land and water of all animal products. Go to **GRAPHit!** at www.aw-bc.com/withgott. Data from Smil, V. 2001. *Feeding the world: A challenge for the twenty-first century*, Cambridge, MA: MIT Press.

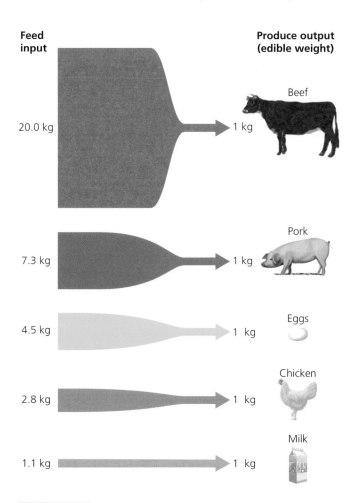

FIGURE 7.19 Different animal food products require different amounts of input of animal feed. Chickens must be fed 2.8 kg of feed for each 1 kg of resulting chicken meat, for instance, whereas 20 kg of feed must be provided to cattle to produce 1 kg of beef.

Go to **GRAPHit!** at www.aw-bc.com/withgott. Data from Smil, V. 2001. *Feeding the world: A challenge for the twenty-first century*. Cambridge, MA: MIT Press.

produce 1 kg (2.2 lb) of food protein for milk, eggs, chicken, pork, and beef. Producing eggs and chicken meat requires the least space and water, whereas producing beef requires the most. Such differences make clear that when we choose what to eat, we are also indirectly choosing how to make use of resources such as land and water.

Aquaculture is the fastest-growing type of food production

We also rely on aquatic organisms for food. Wild fish populations are plummeting throughout the world's oceans as increased demand and new technologies have led us to

FIGURE 7.21 Efforts to genetically modify important food fish have resulted in the creation of transgenic salmon (top), which can be considerably larger than wild salmon of the same species.

overharvest most marine fisheries (▶ pp. 263–265). This means that raising fish and shellfish on "fish farms" may be the only way to meet our growing demand for these foods.

We call the raising of aquatic organisms for food in controlled environments **aquaculture**. Many aquatic species are grown in open water in large, floating net-pens. Others are raised in land-based ponds or holding tanks. People pursue both freshwater and marine aquaculture. Aquaculture is the fastest-growing type of food production; in the past 20 years, global output has increased sevenfold. Aquaculture today provides a third of the world's fish for human consumption, is most common in Asia, and involves over 220 species. Some, such as carp, are grown for local consumption, whereas others, such as salmon and shrimp, are exported to affluent countries.

Aquaculture can provide substantial economic, social, and environmental benefits. Small-scale fish-farming helps ensure people a reliable protein source, can be sustainable, and is compatible with other activities. For instance, uneaten fish scraps make excellent fertilizers for crops. Aquaculture on larger scales can help improve a region's or nation's food security and reduce fishing pressure on wild stocks. Fish farming can also be remarkably energy-efficient, producing up to 10 to 1,000 times more fish per unit area than can be harvested from marine waters.

Aquaculture also has drawbacks. Dense concentrations of farmed animals can increase the incidence of disease, which reduces food security, necessitates antibiotic treatment, and results in additional expense. A virus outbreak wiped out half a billion dollars in shrimp in Ecuador in 1999, for instance. Aquaculture can also

produce prodigious amounts of waste, from farmed organisms and feed that goes uneaten and decomposes in the water column. Farmed fish often are fed grain or fish meal, reducing the overall energy efficiency of food production. Moreover, fish meal often comes from wild ocean fish such as herring and anchovies, whose harvest may place additional stress on wild populations.

If farmed aquatic organisms escape into ecosystems where they are not native, they may spread disease to native stocks or may outcompete native organisms for food or habitat. Genetically modified salmon, for instance, may outcompete their non-GM wild cousins while also interbreeding with them and spreading disease to them (Figure 7.21).

Sustainable Agriculture

Industrialized agriculture involves many adverse environmental impacts, from soil degradation to reliance on fossil fuels to problems arising from pesticide use, genetic modification, and intensive feedlot and aquaculture operations. Although many of these developments in intensive commercial agriculture have alleviated some environmental pressures, they have often exacerbated others. Industrial agriculture in some form seems necessary to feed our planet's 6.5 billion people, but many feel we will be better off in the long run by practicing less-intensive methods of raising animals and crops.

Farmers, ranchers, and researchers have made great advances recently toward **sustainable agriculture**, farming and ranching that does not deplete soils faster than they form and that maintains the clean water and genetic diversity essential to long-term crop and livestock production. It is, simply, agriculture that can be practiced in the same way far into the future. The no-till agriculture practiced in southern Brazil appears to be sustainable, as does the traditional Chinese practice of carp aquaculture in small ponds. Sustainable agriculture is closely related to *low-input agriculture*, agriculture that uses smaller amounts of pesticides, fertilizers, growth hormones, water, and fossil fuel energy than are used in industrial agriculture. Food-growing practices that use no synthetic fertilizers, insecticides, fungicides, or herbicides—but instead rely on biological approaches such as composting and biocontrol—are termed **organic agriculture**.

Organic agriculture is on the increase

Citizens, government officials, farmers, and agricultural industry representatives have debated the meaning of the word *organic* for many years. In 1990, Congress passed the

Organic Food Production Act, establishing national standards for organic products and facilitating the sale of organic food. In 2000, the USDA issued criteria by which crops and livestock could be officially certified as organic. Today 17 U.S. states have laws spelling out standards for organic products.

Weighing the **issues:**
Do You Want Your Food Labeled?

The USDA issues labels to certify that produce claiming to be organic has met the government's standards. Increasingly, critics of GM products want them to be labeled as well. Given that 70% of processed food currently contains GM ingredients, labeling would cause added—and many people think, unnecessary—costs. But the European Union currently labels such foods. Do you want your food to be labeled? Would you choose among foods based on whether they are organic or genetically modified? Do you feel your food choices have environmental impacts, good or bad? Is purchasing power an effective way to make your views heard?

Long viewed as a small niche market, the market for organic foods is on the increase. Although it accounts for only 1% of food expenditures in the United States and Canada, sales of organic products increased by 20% annually from 1989 to 2002, when global sales of organic products reached $25 billion. Today farmers in more than 130 nations practice organic farming commercially to some extent.

These trends have been fueled by the desire of many consumers to reduce health risks and to improve environmental quality by reducing chemical pollution. Government initiatives have also spurred the growth of organic farming. For example, several million hectares of land has undergone conversion from conventional to organic farming in Europe since the European Union adopted a policy in 1993 to support farmers financially during the first years of conversion. Such support is important, because conversion often means a temporary loss in income for farmers. More and more studies, however, suggest that reduced inputs and higher market prices can, in the long run, make organic farming more profitable for the farmer than conventional methods.

Locally supported agriculture is also growing

Increasing numbers of farmers and consumers are also supporting local small-scale agriculture. Farmers' markets (Figure 7.22) are becoming more numerous throughout North America as consumers rediscover the joys of fresh, locally grown produce. The average food product sold in U.S. supermarkets travels at least 2,300 km (1,400 mi)

FIGURE 7.22 Farmers' markets, like this one in San Francisco, have become more widespread as consumers have rediscovered the benefits of buying fresh, locally grown produce.

between the farm and the shelf, and supermarket produce is often chemically treated to preserve freshness and color. At farmers' markets, consumers buy fresh produce in season from local farmers and often have a wide choice of organic items and unique local varieties.

In the end, consumer choice will largely determine the future of our agriculture. Current trends suggest that local agriculture and organic agriculture will continue to increase. To ensure our long-term well-being, sustainable agriculture will sooner or later need to become the rule rather than the exception.

Conclusion

Many of the intensive commercial agricultural practices we have discussed have substantial negative environmental impacts. At the same time, many aspects of industrialized agriculture have had positive environmental effects by relieving pressures on land or resources. Whether Earth's natural systems would be under more pressure from 6.5 billion people practicing traditional agriculture or from 6.5 billion people living with the industrialized agriculture model is a very complicated question.

What is certain is that if our planet is to support 9 billion people by mid-century without further degradation of the soil, water, pollinators, and other resources and ecosystem services that support our food production, we must find ways to shift to sustainable agriculture. Approaches such as biological pest control, organic agriculture, pollinator conservation, preservation of native crop diversity, sustainable aquaculture, and likely some degree of careful and responsible genetic modification of food may all be parts of the game plan we will need to set in motion.

TESTING YOUR COMPREHENSION

1. How do scientists define soil? Compare and contrast the major horizons in a typical soil profile.
2. Why is erosion generally considered a destructive process? Name three human activities that can promote soil erosion. What factors affect the intensity of erosion by water?
3. Describe several farming techniques (such as terracing and no-till farming) that can help reduce the risk of erosion. How does each of them achieve this goal?
4. Describe the effects of overgrazing on soil. What conditions characterize sustainable grazing practices?
5. What kinds of techniques have people employed to increase agricultural food production? How did agricultural scientist Norman Borlaug help inaugurate the green revolution?
6. Explain how pesticide resistance occurs. Now explain the concept of biocontrol. List several components of a system of integrated pest management (IPM).

7. Roughly how many and what types of cultivated plants are known to rely on insects for pollination? Why is it important to preserve the biodiversity of native pollinators?
8. How is a transgenic organism created? How is genetic engineering different from traditional selective breeding? How is it similar?
9. Describe several reasons why many people support the development of genetically modified organisms, and name several uses of such organisms that have been developed so far. Now describe the concerns of those opposed to genetically modified crops.
10. Name several positive and negative environmental effects of feedlot operations. Why is beef an inefficient food from the perspective of energy consumption?

SEEKING SOLUTIONS

1. How do you think a farmer can best help to conserve soil? How do you think a scientist can best help to conserve soil? How do you think a national government can best help to conserve soil?
2. Imagine that you are the head of an international granting agency that assists farmers with soil conservation and sustainable agriculture. You have $10 million to disburse. Your agency's staff has decided that the funding should go to (1) farmers in an arid area of Africa prone to salinization, (2) farmers in southern Brazil practicing no-till agriculture, and (3) farmers in a dryland area of Mongolia undergoing desertification. What types of projects would you recommend funding in each of these areas, how would you apportion your funding among them, and why?

3. Assess several ways in which high-input agriculture can be beneficial for the environment and several ways in which it can be detrimental to the environment. Now suggest several strategies we might follow to modify industrial agriculture to lessen its environmental impact.

4. What factors make for an effective biological control strategy of pest management? What risks are involved in biocontrol? If you had to decide whether to use biocontrol against a particular pest, what questions would you want to have answered before you decide?

5. Imagine it is your job to make the regulatory decision whether to allow the planting of a new genetically modified strain of cabbage that produces its own pesticide and has twice the vitamin content of regular cabbage. What questions would you ask of scientists before deciding whether to approve the new crop? What scientific data would you want to see, and how much would be enough? Would you also consult non-scientists or take ethical, economic, and social factors into consideration?

CALCULATING ECOLOGICAL FOOTPRINTS

 As food production became more industrialized during the 20th century, we began expending increasing amounts of energy to store food and ship it to market. In the United States today, food travels an average of 1,400 miles from the field to your table. The price you pay for the food covers the cost of this long-distance transportation, which in 2004 was approximately one dollar per ton per mile. Assuming that the average person eats 2 pounds of food per day, calculate the food transportation costs for each category in the table below.

Consumer	Daily Cost	Annual Cost
You	$1.40	$511
Your class		
Your hometown		
Your state		
United States		

1. What specific challenges to environmental sustainability are imposed by a food production and distribution system that relies on long-range transportation to bring food to market?

2. One recent study noted that locally produced food traveled only about 50 miles to market, saving 96% of the transportation costs. Locally grown foods may be fresher and cause less environmental impact as they are brought to market, but what are the disadvantages to you as a consumer in relying on local food production? Do you think the advantages outweigh those disadvantages?

3. What has happened to gasoline prices recently? Would future increases in the price of gas affect your answers to the preceding questions?

Take It Further

Go to www.aw-bc.com/withgott, where you'll find:

▶ Suggested answers to end-of-chapter questions
▶ Quizzes, animations, and flashcards to help you study
▶ *Research Navigator*™ database of credible and reliable sources to assist you with your research projects

▶ **GRAPHit!** tutorials to help you interpret graphs
▶ **INVESTIGATEit!** current news articles that link the topics that you study to case studies from your region to around the world

8

Biodiversity and Conservation Biology

A Siberian tiger in the Sikhote-Alin Mountains

Upon successfully completing this chapter, you will be able to:

▶ Characterize the scope of biodiversity on Earth

▶ Describe ways to measure biodiversity

▶ Contrast background extinction rates and periods of mass extinction

▶ Evaluate the primary causes of biodiversity loss

▶ Specify the benefits of biodiversity

▶ Assess conservation biology and its practice

▶ Explain island biogeography theory and its application to conservation biology

▶ Compare and contrast traditional and more innovative biodiversity conservation efforts

Central Case: Saving the Siberian Tiger

"Future generations would be truly saddened that this century had so little foresight, so little compassion, such lack of generosity of spirit for the future that it would eliminate one of the most dramatic and beautiful animals this world has ever seen."
—GEORGE SCHALLER, WILDLIFE BIOLOGIST, ON THE TIGER

"Except in pockets of ignorance and malice, there is no longer an ideological war between conservationists and developers. Both share the perception that health and prosperity decline in a deteriorating environment. They also understand that useful products cannot be harvested from extinct species."
—EDWARD O. WILSON, HARVARD UNIVERSITY BIODIVERSITY EXPERT

International conservation groups began to get involved, working with Russian biologists to try to save the dwindling tiger population. One such group was the Hornocker Wildlife Institute, now part of the Wildlife Conservation Society. In 1991 the group helped launch the Siberian Tiger Project, devoted to studying the tiger and its habitat. The team put together a plan to protect the tiger, began educating people regarding the tiger's importance and value, and worked closely with those who live in proximity to the big cats.

Thanks to such efforts by conservation biologists, today Siberian tigers in the wild number roughly 330–370, and about 600 more survive in zoos and captive breeding programs around the world. The outlook for the species' survival still looks daunting, but many people are trying to save this endangered animal. It is one of many efforts around the world today to stem the loss of our planet's priceless biological diversity.

Historically, tigers roamed widely across Asia from Turkey to northeast Russia to Indonesia. Within the past 200 years, however, people have driven the majestic striped cats from most of their historic range. Today, tigers are exceedingly rare and are creeping toward extinction.

Of the tigers that still survive, those of the subspecies known as the Siberian tiger are the largest cats in the world. Males reach 3.66 m (12 ft) in length and weigh up to 363 kg (800 lb). Also named Amur tigers for the watershed they occupied along the Amur River, which divides Siberian Russia from Manchurian China, these cats now find their last refuge in the forests of the remote Sikhote-Alin Mountains of the Russian Far East.

For thousands of years the Siberian tiger coexisted with the region's native people and held a prominent place in native language and lore. These people referred to the tiger as "Old Man" or "Grandfather" and equated it with royalty or viewed it as a guardian of the mountains and forests. Indigenous people of the region rarely killed a tiger unless it had preyed on a person.

The Russians who moved into the region and exerted control in the early 20th century had no such cultural traditions. They hunted tigers for sport and hides, and some Russians reported killing as many as 10 tigers in a single hunt. In addition, poachers began killing tigers to sell their body parts to China and other Asian countries, where they are used in traditional medicine and as aphrodisiacs. Meanwhile, road building, logging, and agriculture began to fragment tiger habitat and provide easy access for well-armed hunters. The tiger population dipped to perhaps 20–30 animals.

Our Planet of Life

Rising human population and resource consumption are placing ever-greater pressure on the flora and fauna of our planet, from tigers to tiger beetles. We are diminishing Earth's diversity of life, the very quality that makes our planet so special.

Biodiversity encompasses several levels

In Chapter 4 (▸ p. 79) we introduced *biological diversity*, or *biodiversity*, as the sum total of all organisms in an area, taking into account the diversity of species, their genes, their populations, and their communities. Biodiversity is a concept as multifaceted as life itself, and different biologists employ different working definitions according to their own aims, interests, and values.

Nonetheless, there is broad agreement that the concept applies across several major levels in the organization of life (Figure 8.1).

Species diversity As you will recall (▸ p. 79), a *species* is a distinct type of organism, a set of individuals that uniquely share certain characteristics and can breed with one another and produce fertile offspring. Biologists may use somewhat differing criteria to delineate species boundaries; some emphasize characteristics shared because of common ancestry, whereas others emphasize ability to interbreed. In practice, however, scientists broadly agree on species identities. We can express **species diversity** in terms of the number or variety of species in the world or in a particular region. One component of species diversity is *species richness*, the number of species. Another is *evenness* or *relative abundance*, the extent to which numbers of individuals of different species are equal or skewed.

As we saw in Chapter 4 (▸ pp. 80–81), speciation generates new species, adding to species richness, whereas extinction decreases species richness. Although immigration, emigration, and local extinction may increase or decrease species richness locally, only speciation and extinction change it globally.

Taxonomists, the scientists who classify species, use an organism's physical appearance and genetic makeup to determine its species. Taxonomists also group species by their similarity into a hierarchy of categories meant to reflect evolutionary relationships. Related species are grouped together into *genera* (singular, *genus*), related genera are grouped into families, and so on (Figure 8.2). Every species is given a two-part Latin or Latinized scientific name denoting its genus and species. The tiger, *Panthera tigris*, differs from the world's other species of large cats such as the jaguar *(Panthera onca)*, the leopard *(Panthera pardus)*, and the African lion *(Panthera leo)*. These four species are closely related in evolutionary terms, as indicated by the genus name they share, *Panthera*. They are more distantly related to cats in other genera such as the cheetah *(Acinonyx jubatus)* and the bobcat *(Felis rufus)*, although all cats are classified together in the family Felidae.

Biodiversity exists below the species level in the form of *subspecies*, populations of a species that occur in different geographic areas and vary from one another in some characteristics. Subspecies are formed by the same processes that drive speciation, but they result when divergence does not proceed far enough to create separate species. Scientists denote subspecies

Ecosystem diversity

Species diversity

Genetic diversity

FIGURE 8.1 The concept of biodiversity encompasses several levels in the hierarchy of life. Species diversity (middle frame of figure) refers to the number or variety of species. Genetic diversity (bottom frame) refers to variation in DNA composition among individuals within a species. Ecosystem diversity (top frame) and related concepts refer to variety at levels above the species level, such as ecosystems, communities, habitats, or landscapes.

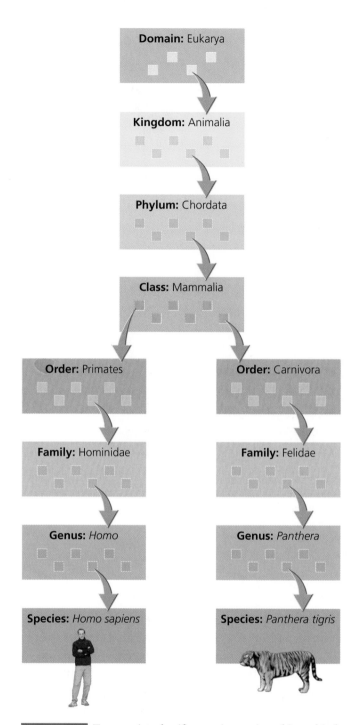

FIGURE 8.2 Taxonomists classify organisms using a hierarchical system meant to reflect evolutionary relationships. Species that are similar in their appearance, behavior, and genetics (because of recent common ancestry) are placed in the same genus. Organisms of similar genera are placed within the same family. Families are placed within orders, orders within classes, classes within phyla, phyla within kingdoms, and kingdoms within domains. For instance, humans (*Homo sapiens*, a species in the genus *Homo*) and tigers (*Panthera tigris*, a species in the genus *Panthera*) are both within the class Mammalia. However, the differences between our two species, which have evolved over millions of years, are great enough that we are placed in different orders and families.

with a third part of the scientific name. The Siberian tiger, *Panthera tigris altaica*, is one of five subspecies of tiger still surviving (Figure 8.3). Tiger subspecies differ in color, coat thickness, stripe patterns, and size. For example, *Panthera tigris altaica* is 5–10 cm (2–4 in.) taller at the shoulder than the Bengal tiger *(Panthera tigris tigris)* of India and Nepal, and it has a thicker coat and larger paws.

Genetic diversity Scientists designate subspecies when they recognize substantial, genetically based differences among individuals from different populations of a species. However, all species consist of individuals that vary genetically from one another to some degree, and this genetic diversity is an important component of biodiversity. **Genetic diversity** encompasses the differences in DNA composition among individuals within species and populations.

Genetic diversity provides the raw material for adaptation to local conditions. A diversity of genes for coat thickness in tigers allowed natural selection to favor genes for thin coats of fur in Bengal tigers living in warm regions, and to favor genes for thick coats of fur for Siberian tigers living in cold regions. In the long term, populations with more genetic diversity may stand better chances of persisting, because their variation better enables them to cope with environmental change.

Populations with little genetic diversity are vulnerable to environmental change for which they are not genetically prepared. Populations with depressed genetic diversity may also be more vulnerable to disease and may suffer *inbreeding depression*, which occurs when genetically similar parents mate and produce weak or defective offspring. Scientists have sounded warnings over low genetic diversity in species that have dropped to low population sizes in the past, including cheetahs, bison, and elephant seals, but the full consequences of reduced diversity in these species remain to be seen. Diminishing genetic diversity in our crop plants also is a prime concern (▸ p. 160).

Ecosystem diversity Biodiversity also encompasses levels above the species level. *Ecosystem diversity* refers to the number and variety of ecosystems, but biologists may also refer to the diversity of biotic community types or habitats within some specified area. If the area is large, scientists may also consider the geographic arrangement of habitats, communities, or ecosystems at the landscape level, including the sizes, shapes, and interconnectedness of patches of these entities. Under any of these concepts, a seashore of rocky and sandy beaches, forested cliffs, offshore coral

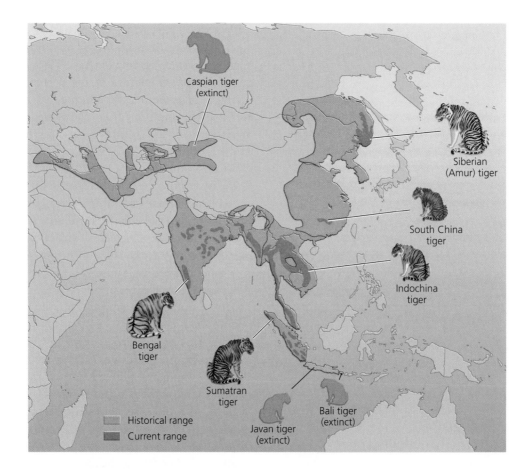

FIGURE 8.3 Three of the eight subspecies of tiger became extinct during the 20th century. The Bali, Javan, and Caspian tigers are extinct. Today only the Siberian (Amur), Bengal, Indochina, Sumatran, and South China tigers persist, and the Chinese government estimates that less than 30 individuals of the South China tiger remain. Deforestation, hunting, and other pressures from people have caused tigers of all subspecies to disappear from most of the geographic range they historically occupied. This map contrasts the ranges of the eight subspecies in the years 1800 (orange) and 2000 (red). Data from the Tiger Information Center.

reefs, and ocean waters would hold far more biodiversity than the same acreage of a monocultural cornfield. A mountain slope whose vegetation changes from desert to hardwood forest to coniferous forest to alpine meadow would hold more biodiversity than an area the same size consisting of only desert, forest, or meadow.

Measuring biodiversity is not easy

Coming up with precise quantitative measurements to express a region's biodiversity is difficult. Scientists often express biodiversity in terms of its most easily measured component, species diversity, and in particular, species richness. Species richness is a good gauge for overall biodiversity, but we still are profoundly ignorant of the number of species that exist worldwide. So far, scientists have

identified and described 1.7–2.0 million species of plants, animals, and microorganisms. However, estimates for the total number that actually exist range from 3 million to 100 million, with our best educated guesses ranging from 5 million to 30 million. Our knowledge of species numbers is incomplete because some areas of Earth remain little explored, many species are tiny and easily overlooked, and many organisms are difficult to identify.

Species are not evenly distributed among taxonomic groups. In terms of number of species, insects show a staggering predominance over all other forms of life (Figure 8.4). Within insects, about 40% are beetles, which outnumber all noninsect animals and all plants. No wonder the 20th-century British biologist J. B. S. Haldane famously quipped that God must have had "an inordinate fondness for beetles."

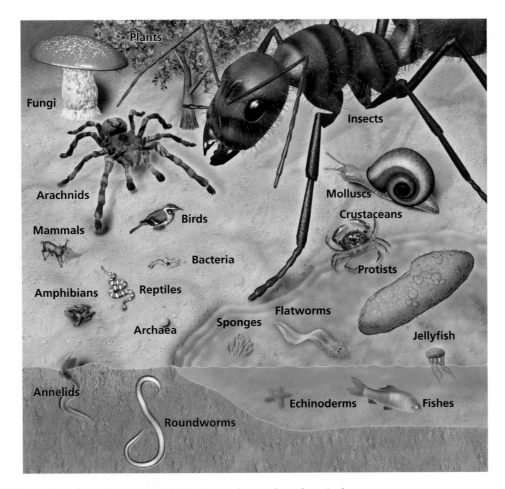

FIGURE 8.4 This illustration shows organisms scaled in size to the number of species known so far from each major taxonomic group, giving a visual sense of the disparity in species richness among groups. However, because most species are not yet discovered or described, some groups (such as bacteria, archaea, insects, nematodes, protists, fungi, and others) may contain far more species than we now know of. Data from Groombridge, B., and M. D. Jenkins. 2002. *Global biodiversity: Earth's living resources in the 21st century.* UNEP-World Conservation Monitoring Centre. Cambridge, U.K.: Hoechst Foundation.

Biodiversity is unevenly distributed

Numbers of species tell only part of the story of Earth's biodiversity. Living things are distributed across our planet unevenly, and scientists have long sought to explain the distributional patterns they see. For example, as we have noted, some groups of organisms include only one or a few species, whereas other groups contain many. Some groups have given rise to many species in a relatively short period of time through the process of adaptive radiation (see Figure 4.2, ► p. 78).

Species diversity also varies according to biome (► pp. 111–118). For instance, tropical dry forests and rainforests tend to support more species than tundra and boreal forests. The variation by biome is related to one of the planet's most striking patterns of species diversity: the fact that species richness generally increases as one approaches the equator (Figure 8.5). This pattern of variation with latitude, called the *latitudinal gradient*, has been one of the most obvious patterns in ecology, but one of the most difficult ones for scientists to explain.

Hypotheses abound for the cause of this pattern, but it seems likely that plant productivity and climate stability play key roles in the phenomenon. Greater amounts of solar energy, heat, and humidity at tropical latitudes lead to more plant growth, making areas nearer the equator more productive and able to support larger numbers of animals. Moreover, the relatively stable climates of equatorial regions—their similar temperatures and rainfall from day to day and season to season—help ensure that single species won't dominate ecosystems, but that instead numerous species can coexist. Whereas varying

FIGURE 8.5 For many types of organisms, number of species per unit area tends to increase as one moves toward the equator. This trend, the latitudinal gradient in species richness, is one of the most readily apparent—yet least understood—patterns in ecology. One example is bird species in North and Central America: In any one spot in arctic Canada and Alaska, 30 to 100 species can be counted; in areas of Costa Rica and Panama, the number rises to over 600. Adapted from Cook, R. E. 1969. Variation in species density in North American birds. *Systematic Zoology* 18: 63–84.

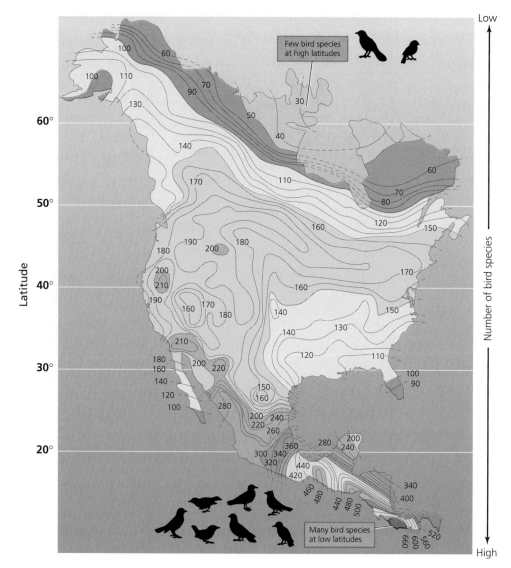

environmental conditions favor generalists—species that can deal with a wide range of circumstances but that do no single thing very well—stable conditions favor specialists, organisms with narrow niches that do particular things very well. In addition, polar and temperate regions may be relatively lacking in species because glaciation events repeatedly forced organisms out of these regions and toward more tropical latitudes.

Biodiversity Loss and Species Extinction

Biodiversity at all levels is being lost to human impact — irretrievably when species become extinct. Once vanished, a species can never return. *Extinction* (▶ p. 81) occurs when the last member of a species dies and the species ceases to exist. The disappearance of a particular population from a given area, but not the entire species globally, is referred to as **extirpation**. The tiger has been extirpated from most of its historic range, but it is not yet extinct. Extirpation is an erosive process that can, over time, lead to extinction.

Extirpation and extinction occur naturally. If organisms did not naturally go extinct, we would be up to our ears in dinosaurs, trilobites, ammonites, and the millions of other types of creatures that vanished from Earth long before humans appeared. Paleontologists estimate that roughly 99% of all species that have ever lived are now extinct. Most extinctions preceding the appearance of humans occurred one by one for independent reasons, at a rate that paleontologists refer to as the **background rate of extinction** (▶ p. 83). For example, the fossil record indicates that for mammals and marine animals, one species out of 1,000 would typically become extinct every 1,000 to 10,000 years. This translates to an annual rate of one extinction per 1 to 10 million species.

Earth has experienced five mass extinction episodes

Extinction rates have risen far above this background rate during several discrete time periods in Earth's history. In the past 440 million years, our planet has experienced five major episodes of **mass extinction** (▶ p. 83; Figure 8.6). Each mass extinction event has eliminated more than one-fifth of life's families and at least half its species. The most severe episode occurred at the end of the Permian period, 248 million years ago, when close to 54% of all families, 90% of all species, and 95% of marine species went extinct. The best-known episode occurred at the end of the Cretaceous period, 65 million years ago, when an apparent asteroid impact brought an end to the dinosaurs and many other groups.

If current trends continue, the modern era, known as the Quaternary period, may see the extinction of more than half of all species. Although similar in scale to previous mass extinctions, today's ongoing mass extinction is different in two primary respects. First, humans are causing it. Second, humans will suffer as a result of it.

Humans have set the sixth mass extinction in motion

We have recorded hundreds of instances of human-induced species extinction over the past few centuries. Among North American birds alone, we have indis-

putably driven into extinction the Carolina parakeet, great auk, Labrador duck, and passenger pigeon (▶ p. 85); likely the Bachman's warbler and Eskimo curlew; and possibly the ivory-billed woodpecker. Several more species, including the whooping crane, Kirtland's warbler, and California condor (Figure 8.14, ▶ p. 186), teeter on the brink of extinction.

However, species extinctions caused by humans precede written history. Indeed, people may have been hunting species to extinction for thousands of years. Archaeological evidence shows that in case after case, a wave of extinctions followed close on the heels of human arrival on islands and continents (Figure 8.7). After Polynesians reached Hawaii, half its birds went extinct. Birds, mammals, and reptiles vanished following human arrival on many other oceanic islands, including large island masses such as New Zealand and Madagascar. Dozens of species of large vertebrates died off in Australia after people arrived roughly 50,000 years ago. North America lost 33 genera of large mammals after people arrived on the continent at least 10,000 years ago.

Today, species loss is accelerating as our population growth and resource consumption put increasing strain on habitats and wildlife. In 2005, scientists with the Millennium Ecosystem Assessment (▶ p. 20) calculated that the current global extinction rate is 100 to 1,000 times greater than the background rate. They projected that the rate would increase tenfold or more in future decades.

To keep track of the current status of endangered species, the World Conservation Union maintains the **Red List**, an updated list of species facing high risks of extinction. The 2006 Red List reported that 23% (1,093) of mammal species and 12% (1,206) of bird species are

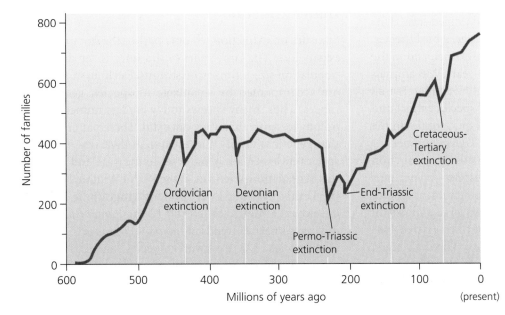

FIGURE 8.6 The fossil record shows evidence of five episodes of mass extinction during the past half-billion years of Earth history. At the end of the Ordovician, Devonian, Permian, Triassic, and Cretaceous periods, 50–95% of the world's species appear to have gone extinct. Each time, biodiversity later rebounded to equal or higher levels, but the rebound required millions of years in each case. Data from Raup, D. M., and J. J. Sepkoski. 1982. Mass extinctions in the marine fossil record. *Science* 215: 1501–1503.

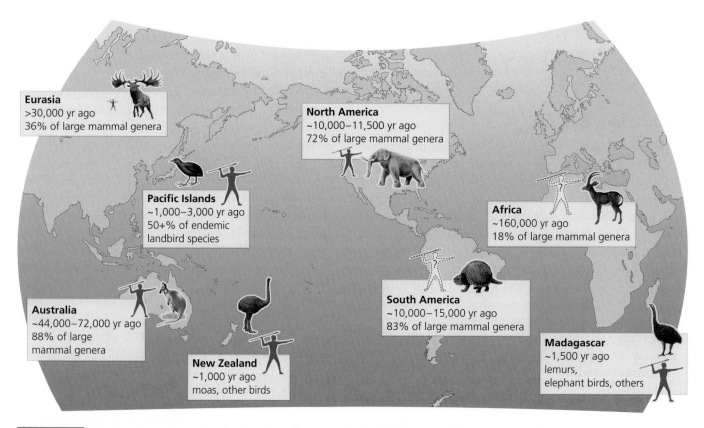

Eurasia
>30,000 yr ago
36% of large mammal genera

North America
~10,000–11,500 yr ago
72% of large mammal genera

Pacific Islands
~1,000–3,000 yr ago
50+% of endemic
landbird species

Africa
~160,000 yr ago
18% of large mammal genera

Australia
~44,000–72,000 yr ago
88% of large
mammal genera

South America
~10,000–15,000 yr ago
83% of large mammal genera

Madagascar
~1,500 yr ago
lemurs,
elephant birds, others

New Zealand
~1,000 yr ago
moas, other birds

FIGURE 8.7 This map shows for each region the time of human arrival and the extent of the recent extinction wave. Illustrated are representative extinct megafauna from each region. The human hunter icons are sized according to the degree of evidence that human hunting was a cause of extinctions; larger icons indicate more certainty that humans (as opposed to climate change or other forces) were the cause. Data for South America and Africa are so far too sparse to be conclusive, and future archaeological and paleontological research could well alter these interpretations. Adapted from Barnosky, A. D., et al. 2004. Assessing the causes of late Pleistocene extinctions on the continents. *Science* 306: 70–75; and Wilson, E. O. 1992. *The diversity of life.* Cambridge, MA: Belknap Press.

threatened with extinction. Among other major groups (for which assessments are not fully complete), estimates of the percentage of species threatened ranged from 31% to 86%. Since 1970, at least 58 fish species, 9 bird species, and one mammal species have become extinct, and in the United States alone over the past 500 years, 236 animals and 17 plants are confirmed to have gone extinct. For all of these figures, the *actual* numbers of species extinct and threatened—like the actual number of total species in the world—are doubtless greater than the *known* numbers.

Among the 1,093 mammals facing possible extinction on the Red List is the tiger, which despite—or perhaps because of—its tremendous size and reputation as a fierce predator, is one of the most endangered large animals on the planet. In 1950, eight tiger subspecies existed (see Figure 8.3). Today, three are extinct. The Bali tiger, *Panthera tigris balica*, went extinct in the 1940s; the Caspian tiger, *Panthera tigris virgata*, during the 1970s; and the Javan tiger, *Panthera tigris sondaica*, during the 1980s.

Biodiversity loss involves more than extinction

Statistics on extinction tell only part of the story of biodiversity loss. The larger part of the story is the decline in population sizes of many organisms. Declines in numbers are accompanied by shrinkage of species' geographic ranges. Thus, many species today are less numerous and occupy less area than they once did. These patterns mean that genetic diversity and ecosystem diversity, as well as species diversity, are being lost. To measure and quantify this degradation, scientists at the World Wildlife Fund and the United Nations Environment Programme (UNEP) developed a metric called the *Living Planet Index*. This index summarizes trends in the populations of 555 terrestrial species, 323 freshwater species, and 267 marine species that are well enough monitored to provide reliable data. Between 1970 and 2000, the Living Planet Index fell by roughly 40% (Figure 8.8).

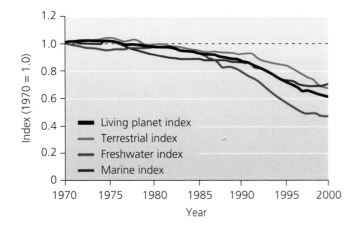

FIGURE 8.8 The Living Planet Index serves as an indicator of the state of global biodiversity. Index values summarize population trends for 1,145 species. Between 1970 and 2000, the Living Planet Index fell by roughly 40%. The indices for terrestrial and marine species fell 30%, and the index for freshwater species fell 50%. Data from World Wide Fund for Nature and U.N. Environment Programme. 2004. *The Living Planet Report, 2004*. Gland, Switzerland: WWF.

Several major causes of biodiversity loss stand out

Reasons for the decline of any given species are often multifaceted, complex, and difficult to determine. Nonetheless, scientists have identified four primary causes of population decline and species extinction: habitat alteration, invasive species, pollution, and overharvesting. Global climate change (Chapter 12) now threatens to become the fifth.

Habitat alteration Nearly every human activity can alter the habitat of the organisms around us. Farming replaces diverse natural communities with simplified ones of only one or a few plant species. Grazing modifies the structure and species composition of grasslands. Either type of agriculture can lead to desertification. Clearing forests removes the food and shelter that forest-dwelling organisms need to survive. Hydroelectric dams turn rivers into reservoirs upstream and affect water conditions and floodplain communities downstream. Urbanization and suburban sprawl supplant diverse natural communities with simplified human-made ones, driving many species from their homes. Because organisms are adapted to the habitats in which they live, any major change is likely to render the habitat less suitable for them.

Habitat alteration is by far the greatest cause of biodiversity loss today. It is the primary source of population declines for 83% of threatened mammals and 85% of threatened birds, according to UNEP data. As just

one example, the prairies of North America's Great Plains have been almost entirely converted to agriculture. Area of prairie habitat has been reduced by more than 99%. As a result, grassland bird populations have declined by an estimated 82–99%. Many grassland species have been extirpated from large areas, and the two species of prairie chickens still persisting could soon go extinct. Habitat destruction has occurred widely in every biome. Estimates by UNEP in 2002 reported that 45% of Earth's forests, 50% of its mangrove ecosystems, and 10% of its coral reefs had been destroyed by recent human activity.

Invasive species Our introduction of non-native species to new environments, where some may become invasive (Figure 8.9), has also pushed native species toward extinction. Some introductions have been accidental. Examples include aquatic organisms (such as zebra mussels; Chapter 5) that have been transported among continents in the ballast water of ships, animals that have escaped from the pet trade, and the weed seeds that cling to our socks as we travel from place to place. Other introductions have been intentional. People have brought with them food crops, domesticated animals, and other organisms as they colonized new places, generally unaware of the ecological consequences that could result.

Most organisms introduced to new areas perish, but the few types that survive may do very well, especially if they find themselves without the predators and parasites that attacked them back home or without the competitors that had limited their access to resources. Once released from the limiting factors (▶ pp. 89) of predation, parasitism, and competition, an introduced species may increase rapidly, spread, and displace native species. Moreover, invasive species cause billions of dollars in economic damage each year.

Pollution Pollution can harm organisms in many ways. Air pollution (Chapter 12) can degrade forest ecosystems. Water pollution (Chapter 11) can adversely affect fish and amphibians. Agricultural runoff (including fertilizers, pesticides, and sediments; Chapters 3 and 7) can harm many terrestrial and aquatic species. Heavy metals, PCBs, and other toxic chemicals can poison people and wildlife (Chapter 10), and the effects of oil and chemical spills on wildlife are dramatic and well known. However, although pollution is a substantial threat, it tends to be less significant than public perception holds it to be, and less severe than the damage caused by habitat alteration or invasive species.

Invasive Species			
Species	**Native to...**	**Invasive in...**	**Effects**
Gypsy moth (*Lymantria dispar*) 	Eurasia	Northeastern United States	In the 1860s, a scientist introduced the gypsy moth to Massachusetts in the mistaken belief that it might be bred with others to produce a commercial-quality silk. The gypsy moth failed to start a silk industry, and instead spread through the northeastern United States and beyond, where its outbreaks defoliate trees over large regions every few years.
European starling (*Sturnus vulgaris*) 	Europe	North America	The bird was first introduced to New York City in the late 19th century by Shakespeare devotees intent on bringing every bird mentioned in Shakespeare's plays to the new continent. It only took 75 years for the birds to spread to the Pacific coast, Alaska, and Mexico, becoming one of the most abundant birds on the continent. Starlings are thought to outcompete native birds for nest sites.
Cheatgrass (*Bromus tectorum*) 	Eurasia	Western United States	In just 30 years after its introduction to Washington state in the 1890s, cheatgrass spread across much of the western United States. Its secret: fire. Its thick patches that choke out other plants and use up the soil's nitrogen burn readily. Fire kills many of the native plants, but not cheatgrass, which grows back even stronger amid the lack of competition.
Brown tree snake (*Boiga irregularis*) 	Southeast Asia	Guam	Nearly all native forest bird species on the South Pacific island of Guam have disappeared. The culprit is the brown tree snake. The snakes were likely brought to the island inadvertantly as stowaways in cargo bays of military planes in World War II. Guam's birds had not evolved with tree snakes, and so had no defenses against the snake's nighttime predation. The snakes also cause numerous power outages each year on Guam and have spread to other islands where they are repeating their ecological devastation. The arrival of this snake is the greatest fear of conservation biologists in Hawaii.
Kudzu (*Pueraria montana*) 	Japan	Southeastern United States	A vine that can grow 30 m (100 ft) in a single season. The U.S. Soil Conservation Service introduced kudzu in the 1930s to help control erosion. Adaptable and extra-ordinarily fast-growing, kudzu has taken over thousands of hectares of forests, fields, and roadsides in the south-eastern United States.
Asian long-horned beetle (*Anoplophora glabripennis*) 	Asia	United States	Having first arrived in the United States in imported lumber in the 1990s, these beetles burrow into hardwood trees and interfere with the trees' ability to absorb and process water and nutrients. They may wipe out the majority of hardwood trees in an area. Several U.S. cities, including Chicago in 1999 and Seattle in 2002, have cleared thousands of trees after detecting these invaders.
Rosy wolfsnail (*Euglandina rosea*) 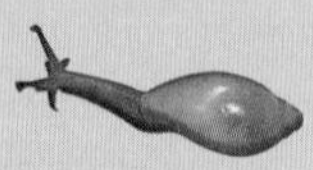	Southeastern United States and Latin America	Hawaii	In the 1950s, well-meaning scientists introduced the rosy wolfsnail to Hawaii to prey upon and reduce the popula-tion of another invasive species, the giant African land snail (*Achatina fulica*), which had been introduced early in the 20th century as an ornamental garden animal. Within a few decades, however, the carnivorous rosy wolfsnail had instead driven more than half of Hawaii's native species of banded tree snails to extinction.

FIGURE 8.9 Invasive species are species that thrive in areas where they are introduced, outcompeting, preying on, or otherwise harming native species. Of the many thousands of invasive species, this chart shows a few of the best known.

Overharvesting For most species, hunting or harvesting by people does not *in itself* pose a threat of extinction, but for some species it can. The Siberian tiger is one such species. Large in size, few in number, long-lived, and raising few young in its lifetime—a classic K-selected species (▶ pp. 90–91)—the Siberian tiger is just the type of animal to be vulnerable to population reduction by hunting. The political freedom that came with the Soviet Union's breakup in 1989 brought with it a freedom to harvest Siberia's natural resources, the tiger included, without restriction. This coincided with an economic expansion in many Asian countries, where tiger penises are traditionally used to try to boost human sexual performance and where tiger bones, claws, whiskers, and other body parts are used to treat a wide variety of maladies. Thus, the early 1990s brought a boom in poaching (poachers killed at least 180 Siberian tigers between 1991 and 1996).

Over the past century, hunting has led to steep population declines of many other K-selected animals. The Atlantic gray whale has gone extinct, and several other whales remain threatened or endangered. Gorillas and other primates that are killed for their meat could face extinction soon. Thousands of sharks are killed each year solely for their fins, which are used in soup. Today the oceans contain only 10% of the large animals they once did (▶ p. 264).

Climate change As we will explore in Chapter 12, our emissions of carbon dioxide and other "greenhouse gases" are causing temperatures to warm worldwide, modifying global weather patterns and increasing the frequency of extreme weather events. Extreme weather events such as droughts and hurricanes stress populations, and warming temperatures force species to move toward the poles and upward in elevation. Some species will cope, but others will not. Like the cloud-forest fauna at Monteverde (Chapter 4), mountaintop organisms cannot move further upslope to escape warming temperatures, and they will likely perish. Trees may not be able to spread poleward fast enough. Animals and plants may find themselves among altered communities of prey, predators, and parasites to which they are not adapted.

All five of these primary causes of population decline are exacerbated by human population growth and rising per capita consumption. More people and more consumption mean more habitat alteration, more invasive species, more pollution, more overharvesting, and more climate change. Growth in population and growth in consumption are the ultimate reasons behind the proximate threats to biodiversity.

Benefits of Biodiversity

Scientists worldwide are presenting us with data that confirm what any naturalist who has watched the habitat change in his or her hometown already knows: Biodiversity is being lost rapidly and visibly within our lifetimes. This suggests the question, "Does it matter?" There are many ways to answer this question, but we can begin by considering the ways that biodiversity benefits people.

Biodiversity provides ecosystem services free of charge

Contrary to popular opinion, some things in life can indeed be free, as long as we choose to protect the ecological systems that provide them. Intact forests provide clean air and buffer hydrologic systems against flooding and drought. Native crop varieties provide insurance against disease and drought. Abundant wildlife can attract tourists and boost the economies of developing nations. Intact ecosystems provide these and other valuable processes, known as *ecosystem services* (▶ pp. 25, 27), for all of us, free of charge.

Maintaining ecosystem services is one clear benefit of protecting biodiversity. According to UNEP, biodiversity:

▶ Provides food, fuel, and fiber
▶ Provides shelter and building materials
▶ Purifies air and water
▶ Detoxifies and decomposes wastes
▶ Stabilizes and moderates Earth's climate
▶ Moderates floods, droughts, wind, and temperature extremes
▶ Generates and renews soil fertility and cycles nutrients
▶ Pollinates plants, including many crops
▶ Controls pests and diseases
▶ Maintains genetic resources as inputs to crop varieties, livestock breeds, and medicines
▶ Provides cultural and aesthetic benefits
▶ Gives us the means to adapt to change

Organisms and ecosystems support a vast number of vital processes that humans could not replicate or would need to pay for if nature did not provide them. As we have seen (▶ pp. 30–31), the annual value of just 17 of these ecosystem services may be in the neighborhood of $16–54 trillion per year.

Biodiversity enhances food security

Biodiversity benefits our agriculture. As our discussion of native landraces of corn in Oaxaca, Mexico, in Chapter 7 showed, genetic diversity within crop species and their

Food Security and Biodiversity: Potential New Food Sources		
Species	**Native to...**	**Potential uses and benefits**
Amaranths (three species of *Amaranthus*) 	Tropical and Andean America	Grain and leafy vegetable; livestock feed; rapid growth, drought resistant
Buriti palm (*Mauritia flexuosa*) 	Amazon lowlands	"Tree of life" to Amerindians; vitamin-rich fruit; pith as source for bread; palm heart from shoots
Maca (*Lepidium meyenii*) 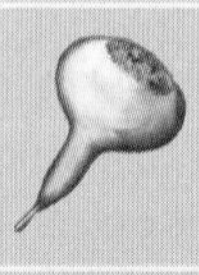	Andes Mountains	Cold-resistant root vegetable resembling radish, with distinctive flavor; near extinction
Babirusa (*Babyrousa babyrussa*) 	Indonesia: Moluccas and Sulawesi	A deep-forest pig; thrives on vegetation high in cellulose and hence less dependent on grain
Capybara (*Hydrochoeris hydrochoeris*) 	South America	World's largest rodent; meat esteemed; easily ranched in open habitats near water
Vicuna (*Lama vicugna*) 	Central Andes	Threatened species related to llama; valuable source of meat, fur, and hides; can be profitably ranched
Chachalacas (*Ortalis*, many species) 	South and Central America	Birds, potentially tropical chickens; thrive in dense populations; adaptable to human habitations; fast-growing

FIGURE 8.10 By protecting biodiversity, we can enhance food security. The wild species shown here are a tiny fraction of the many plants and animals that could someday supplement our food supply. Adapted from Wilson, E. O. 1992. *The diversity of life.* Cambridge, MA: Belknap Press.

ancestors is enormously valuable. In 1995, Turkey's wheat crops received at least $50 billion worth of disease resistance from wild wheat strains. California's barley crops annually receive $160 million in disease resistance benefits from Ethiopian strains of barley.

In addition, new potential food crops are waiting to be used (Figure 8.10). The babassu palm of the Amazon produces more vegetable oil than any other plant. The serendipity berry produces a sweetener 3,000 times sweeter than table sugar. Several species of salt-tolerant grasses and trees are so hardy that farmers can irrigate them with saltwater. These same plants also produce animal feed, a substitute for conventional vegetable oil, and other economically important products.

Organisms provide drugs and medicines

People have made medicines from plants for centuries, and many of today's widely used drugs were discovered when researchers studied chemical compounds present in wild plants, animals, and microbes (Figure 8.11). Each year pharmaceutical products owing their origin to wild species generate up to $150 billion in sales.

It can truly be said that every species that goes extinct represents one lost opportunity to find a cure for cancer or AIDS. The rosy periwinkle produces compounds that treat Hodgkin's disease and a particularly deadly form of leukemia. Had this native plant of Madagascar become extinct before researchers discovered it, two deadly diseases

Medicines and Biodiversity: Natural Sources of Pharmaceuticals			
Plant		**Drug**	**Medical application**
Pineapple (*Ananas comosus*) 		Bromelain	Controls tissue inflammation
Autumn crocus (*Colchicum autumnale*) 		Colchicine	Anticancer agent
Yellow cinchona (*Cinchona ledgeriana*) 		Quinine	Antimalarial
Common thyme (*Thymus vulgaris*) 		Thymol	Cures fungal infection
Pacific yew (*Taxus brevifolia*) 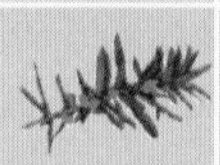		Taxol	Anticancer (especially ovarian cancer)
Velvet bean (*Mucuna deeringiana*) 		L-Dopa	Parkinson's disease suppressant
Common foxglove (*Digitalis purpurea*) 		Digitoxin	Cardiac stimulant

FIGURE 8.11 By protecting biodiversity, we can enhance our ability to treat illness. Shown here are just a few of the plants that have so far been found to provide chemical compounds of medical benefit. Adapted from Wilson, E. O. 1992. *The diversity of life.* Cambridge, MA: Belknap Press.

would have claimed far more victims than they have. In Australia, the government has prioritized research into products from rare and endangered species. There, a rare species of cork, *Duboisia leichhardtii*, now provides hyoscine, a compound that physicians use to treat cancer, stomach disorders, and motion sickness. Another Australian plant, *Tylophora*, provides a drug that treats lymphoid leukemia. Now researchers are testing the compound prostaglandin E2 for treating gastric ulcers. This compound was first discovered in two frog species unique to the rainforest of Queensland, Australia. Scientists believe that both species are now extinct.

Biodiversity generates economic benefits through tourism and recreation

Besides providing for our food and health, biodiversity can represent a direct source of income through tourism, particularly for developing countries in the tropics that have impressive species diversity. Many people like to travel to experience protected natural areas, and in so doing they create economic opportunity for residents living near those natural areas. Visitors spend money at local businesses, hire local people as guides, and support parks that employ local residents. Such **ecotourism** can bring jobs and income to areas that otherwise might be poverty-stricken.

Ecotourism has become a vital source of income for nations such as Costa Rica, with its rainforests; Australia, with its Great Barrier Reef; Belize, with its reefs, caves, and rainforests; and Kenya and Tanzania, with their savanna wildlife. The United States, too, benefits from ecotourism; U.S. national parks draw hundreds of millions of visitors domestically and from around the world. Ecotourism serves as a powerful financial incentive for nations, states, and local communities to preserve natural areas and reduce impacts on the landscape and on native species.

People value and seek out connections with nature

Not all of the benefits of biodiversity to people can be expressed in the hard numbers of economics or the day-to-day practicalities of food and medicine. Some scientists and philosophers argue that there is a deeper importance to biodiversity. Harvard biologist E. O. Wilson has described a phenomenon he calls **biophilia**, "the connections that human beings subconsciously seek with the rest of life." Wilson and others have cited as evidence of biophilia our affinity for parks and wildlife, our keeping of pets, the high value of real estate with a view of natural landscapes, and

our interest—despite being far removed from a hunter-gatherer lifestyle—in hiking, bird-watching, fishing, hunting, backpacking, and similar outdoor pursuits.

In a 2005 book, writer Richard Louv added that as today's children are increasingly deprived of outdoor experiences and direct contact with wild organisms, they suffer what he calls "nature-deficit disorder." Although it is not a medical condition, this alienation from biodiversity and the natural environment, Louv argues, may damage childhood development and may lie behind many of the emotional and physical problems young people struggle with today.

Weighing the **Issues:**
Biophilia and Nature-Deficit Disorder

What do you think of the concepts of biophilia and "nature-deficit disorder"? Have you ever felt a connection to other living things that you couldn't explain in scientific or economic terms? Do you think that an affinity for other living things is innately human?

Aside from all of biodiversity's pragmatic benefits, many people feel that living organisms have an innate right to exist. In this view, biodiversity conservation is justified on ethical grounds alone. Despite our ethical convictions, however, and despite biodiversity's many benefits—from the pragmatic and economic to the philosophical and spiritual—the future of biodiversity is far from secure. The search for solutions to today's biodiversity crisis is an exciting and active one, and scientists are playing a leading role in developing innovative approaches to maintaining the diversity of life on Earth.

Conservation Biology: The Search for Solutions

Today, more and more scientists and citizens perceive a need to do something to stem the loss of biodiversity. In his 1994 autobiography, *Naturalist*, E. O. Wilson wrote:

> When the [20th] century began, people still thought of the planet as infinite in its bounty. . . . In one lifetime exploding human populations have reduced wildernesses to threatened nature reserves. Ecosystems and species are vanishing at the fastest rate in 65 million years. Troubled by what we have wrought, we have begun to turn in our role from local conqueror to global steward.

The urge to act as responsible stewards of natural systems, and to use science as a tool in that endeavor, helped spark the rise of conservation biology. **Conservation biology** is a scientific discipline devoted to understanding the factors, forces, and processes that influence the loss, protection, and restoration of biological diversity. It arose as biologists became increasingly alarmed at the degradation of the natural systems they had spent their lives studying.

Conservation biologists choose questions and pursue research with the aim of developing solutions to such problems as habitat degradation and species loss. Conservation biology is thus an applied and goal-oriented science, with implicit values and ethical standards. Conservation biologists integrate an understanding of evolution and extinction with ecology and the dynamic nature of environmental systems. They use field data, lab data, theory, and experiments to study the impacts of humans on other organisms. They also attempt to design, test, and implement ways to mitigate human impact.

Island biogeography theory is a key component of conservation biology

Safeguarding habitat for species and conserving communities and ecosystems requires thinking and working at the landscape level. One key conceptual tool for doing so is the **equilibrium theory of island biogeography**. This theory, introduced by E. O. Wilson and ecologist Robert MacArthur in 1963, explains how species come to be distributed among oceanic islands. Since then, researchers have also applied it to "habitat islands"—patches of one habitat type isolated within "seas" of others. The Sikhote-Alin Mountains, last refuge of the Siberian tiger, are an island of habitat, isolated from other mountains by deforested regions, a seacoast, and populated lowlands.

Island biogeography theory predicts the number of species on an island based on the island's size and its distance from the nearest mainland. The number of species on an island results from a balance between the number added by immigration and the number lost through extinction (or more precisely, extirpation from the particular island). Immigration and extinction are ongoing dynamic processes, so the balance between them represents an equilibrium state.

The farther an island is located from a continent, the fewer species tend to find and colonize it. Thus, remote islands host fewer species because of lower immigration rates. Large islands have higher immigration rates because they present fatter targets for wandering or dispersing organisms to encounter. Large islands have lower extinction rates because more space allows for larger populations, which are less vulnerable to dropping to zero by

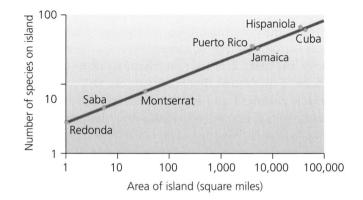

FIGURE 8.12 The larger the island, the greater the number of species—a prediction borne out by data from around the world. By plotting the number of amphibians and reptile species on Caribbean islands (**a**) as a function of the areas of these islands, the species-area curve shows that species richness increases with area (**b**). The increase is not linear, but logarithmic; note the scales of the axes. Go to **GRAPHIt!** at www.aw-bc.com/withgott. Data from MacArthur, R. H., and E. O. Wilson. 1967. *The theory of island biogeography*. Princeton University Press.

chance. Together, these trends give large islands more species at equilibrium than small islands. Very roughly, the number of species on an island is expected to double as island size increases tenfold. This effect can be illustrated with *species-area curves* (Figure 8.12). Large islands also tend to contain more species because they generally possess more habitats than do smaller islands, providing suitable environments for a wider variety of arriving species.

These theoretical patterns have been widely supported by empirical data from the study of species on islands (see "The Science behind the Story," ► pp. 184–185). The patterns hold up for terrestrial habitat islands, such as forests fragmented by logging and road building (Figure 8.13). Small islands of forest lose their diversity fastest, starting with large species that were few in number to begin with. In a landscape of fragmented habitat, species requiring the habitat will gradually disappear, winking out from one fragment after another over time. We will examine a few strategies conservation biologists have designed to deal with fragmentation in our discussion of parks and preserves in Chapter 9 (► pp. 214–215).

Should endangered species be the focus of conservation efforts?

The primary legislation for protecting biodiversity in the United States is the **Endangered Species Act (ESA)**. Passed in 1973, the ESA forbids the government and private citizens from taking actions (such as developing land) that destroy endangered species or their habitats. The ESA also forbids trade in products made from

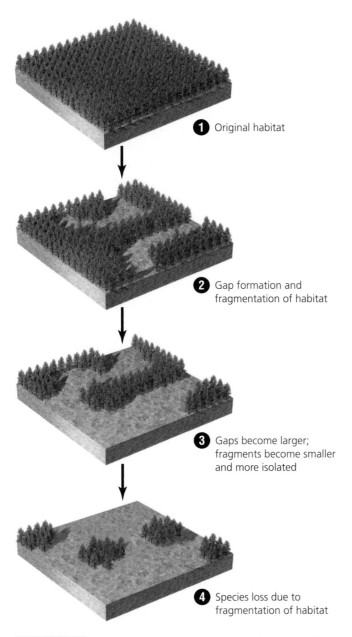

① Original habitat

② Gap formation and fragmentation of habitat

③ Gaps become larger; fragments become smaller and more isolated

④ Species loss due to fragmentation of habitat

FIGURE 8.13 Forest clearing, farming, road building, and other types of human land use and development can fragment natural habitats. Habitat fragmentation usually begins when gaps are created within a natural habitat. As development proceeds, these gaps expand, join together, and eventually dominate the landscape, stranding islands of habitat in their midst. As habitat becomes fragmented, fewer populations can persist, and numbers of species in the fragments decrease with time.

endangered species. The aim is to prevent extinctions, stabilize declining populations, and enable populations to recover. As of 2006, there were 1,311 species in the United States listed as "endangered" or as "threatened," the latter status considered one notch less severe than endangered.

The ESA has had a number of notable successes. Following the ban on the pesticide DDT and years of intensive effort by wildlife managers, the peregrine falcon, brown pelican, bald eagle, and other birds have recovered and have been removed from the endangered list (▸ p. 228). Intensive management programs with other species, such as the red-cockaded woodpecker, have held stable roughly 40% of formerly declining populations, despite continued pressure on habitat.

Although most Americans support protecting endangered species, some resent provisions of the ESA. Many opponents feel that the ESA places more value on the life of an endangered organism than it does on the livelihood of a person. This was a common perception in the Pacific Northwest in the 1990s, when protection for the northern spotted owl slowed logging in old-growth rainforest, causing many loggers to fear for their jobs. Moreover, many landowners worry that federal officials will restrict the use of private land if threatened or endangered species are found on it. This has led in many cases to a practice described as "Shoot, shovel, and shut up," among landowners who want to conceal the presence of such species on their land.

ESA supporters maintain that such fears are overblown, pointing out that the ESA has stopped few development projects. Moreover, a number of provisions of the ESA and its amendments promote cooperation with landowners. *Habitat conservation plans* and *safe harbor agreements* allow landowners to harm species in some ways if they voluntarily improve habitat for the species in others.

Nonetheless, efforts to reauthorize the ESA have faced stiff opposition in the U.S. Congress since the 1990s. As this book went to press, Republican Congressman Richard Pombo was leading efforts to weaken the ESA, including stripping it of its crucial ability to safeguard habitat.

Captive breeding, reintroduction, and cloning are single-species approaches

To save threatened and endangered species, zoos and botanical gardens have become centers for **captive breeding**, in which individuals are bred and raised in controlled conditions and then reintroduced into the wild. One example is the program to save the California condor, North America's largest bird (Figure 8.14). Other reintroduction programs have been more controversial. Reintroducing wolves to Yellowstone National Park has proven popular with the public, but reintroducing them to sites in Arizona and New Mexico met stiff resistance from ranchers who feared the wolves would attack their livestock. The program is making slow headway; a number of the wolves have been shot.

The newest idea for saving species from extinction is to create more individuals by cloning them. In this

Testing and Applying Island Biogeography Theory

The researchers who first experimentally tested the equilibrium theory of island biogeography and applied it to conservation biology leaned on their own resourcefulness, along with some plastic, some pesticides, and a small station wagon.

Robert MacArthur and Edward O. Wilson had originally developed island biogeography theory by using observational data from oceanic islands, correlating numbers of species found on islands with island size and distance between landmasses. Yet as of 1966, no one had tested its precepts in the field with a manipulative experiment. Wilson decided to remedy that.

Wilson began looking for islands in the United States where he could run an experimental test: to remove all animal life from islands and then observe and measure recolonization.

To be suitable, the islands would need to be small, contain few forms of life, and be situated close to the mainland to ensure an influx of immigrating species. Wilson found his research sites off the tip of Florida: six small mangrove islands 11–18 m (36–59 ft) in diameter, home only to trees and a few dozen species of insects, spiders, centipedes, and other arthropods.

Daniel Simberloff, Wilson's graduate student at the time, painstakingly counted each island's arthropods, breaking up bark and poking under branches to find every mite, midge, and millipede. Then with the help of professional exterminators, Wilson and Simberloff wrapped each island in a plastic tarpaulin and gassed the interior with the pesticide methyl bromide. After removing the tarpaulins, Simberloff checked to make

sure no creatures were left alive. The researchers then waited to see how life on the islands would return.

Over the next 2 years, Simberloff scrambled up trees and turned over leaves, looking for newly arrived organisms. His monitoring showed that life recovered on most islands within a year, regaining about the same number of species and total number of arthropods the islands had sheltered originally. Larger islands once again became home to a greater number of species than smaller islands. Outlying islands recovered more slowly and reached lower species diversities than did islands near the mainland. These results provided the first evidence from a manipulative experiment for the predictions of island biogeography theory.

Published in the journal *Ecology* in 1969 and 1970, Simberloff and Wilson's research gave new empirical rigor to a set of ideas that was increasingly helping scientists understand geographic patterns of biodiversity. Their research also fueled a question of pressing concern for conservation biology: Could island biogeography theory also be applied to isolated "islands" of habitat on continents? At the University of Michigan, biology graduate student William Newmark set out to address this question.

Newmark had learned that many North American national parks kept records that documented

Daniel Simberloff and E. O. Wilson used mangrove islands in the Florida Keys to test island biogeography theory.

technique, DNA from an endangered species is inserted into a cultured egg without a nucleus, and the egg is implanted into a closely related species that can act as a surrogate mother. So far two Eurasian mammals have

been cloned in this way. With future genetic technology, some scientists even talk of recreating extinct species from DNA recovered from preserved body parts. However, even if cloning can succeed from a technical standpoint, most

sightings of wildlife over the course of the parks' existence. He surmised that by examining these historical records, he could infer which species had vanished from parks, which were new arrivals, and roughly when these changes occurred. The parks, increasingly surrounded by development, were islands of natural habitat isolated by farms, roads, towns, and cities, so Newmark hypothesized that island biogeography theory would apply.

In 1983, Newmark drove his Toyota station wagon to 24 parks in the western United States and Canada. At each park, he studied the wildlife sighting records, focusing on larger mammals such as bear, lynx, and river otter (but not species, such as wolves, that had been deliberately eradicated by hunting). Newmark found a few species missing from many parks, and they added up to a troubling total. Forty-two species had disappeared in all, and not as a result of direct human action. The red fox and river otter had vanished from Sequoia and Kings Canyon National Parks, for example, and the white-tailed jackrabbit and spotted skunk no longer lived in Bryce Canyon National Park. As theory predicted, the smallest parks showed the greatest number of losses, and the largest parks retained a greater number of species.

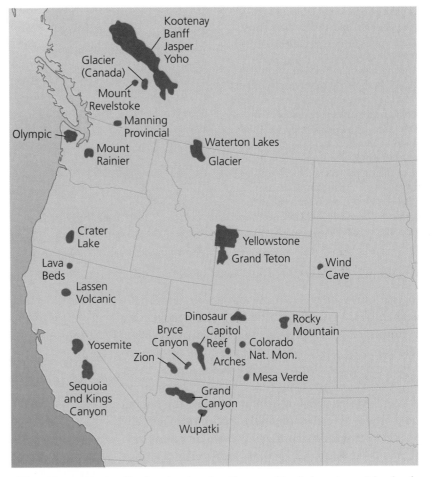

William Newmark visualized national parks of western North America as islands of natural habitat in a sea of development. His data showed that mammal species were disappearing from these recently created terrestrial "islands," in accordance with island biogeography theory. Adapted from Quammen, D. 1997. *The song of the dodo.* New York: Simon & Schuster.

These species disappeared because the parks, Newmark concluded, were too small to sustain their populations in the long term. Moreover, the parks had become island habitats that were too isolated to be recolonized by new arrivals. Newmark's findings, published in the journal *Nature* in 1987, placed island biogeography theory squarely on the mainland. Today, this theory helps inform national park policy and biodiversity conservation plans around the world.

biologists agree that such efforts are not an adequate response to biodiversity loss. Without ample habitat and protection in the wild, having cloned animals in a zoo does little good.

Some species act as "umbrellas" for protecting habitat and communities

Protecting habitat and conserving communities, ecosystems, and landscapes are the goals of many conservation

FIGURE 8.14 In efforts to save the California condor from extinction, biologists have raised hundreds of chicks in captivity with the help of hand puppets designed to look and feel like the heads of adult condors. Using these puppets, biologists feed the growing chicks in an enclosure and shield them from all contact with humans, so that when the chick is grown it does not feel an attachment to people. Condors were persecuted in the early 20th century, collided with electrical wires, and succumbed to lead poisoning from scavenging carcasses of animals killed with lead shot. By 1982, only 22 condors remained, and biologists decided to take all the birds into captivity, in hopes of boosting their numbers and then releasing them. The ongoing program is succeeding. So far, over 100 of the 250 birds raised in captivity have been released into the wild at sites in California and Arizona, where a few pairs have begun nesting.

biologists. Often, they use particular species essentially as tools to conserve communities and ecosystems, for a very simple reason: Although the ESA provides legal justification and resources for species conservation, no such law exists for communities or ecosystems. Large species that roam great distances, such as the Siberian tiger, require large areas of habitat. Meeting the habitat needs of these so-called *umbrella species* automatically helps meet those of thousands of less charismatic animals, plants, and fungi that would never elicit as much public interest.

Environmental advocacy organizations have found that using large and charismatic vertebrates as spearheads for biodiversity conservation has been an effective strategy. This approach of promoting particular *flagship species* is evident in the longtime symbol of the World Wide Fund for Nature (World Wildlife Fund in North America), the panda. The panda is a large endangered animal requiring sizeable stands of undisturbed bamboo forest. Its lovable appearance has made it a favorite with the public—and an effective tool for soliciting funding for conservation efforts that protect far more than just the panda.

Weighing the **Issues:**
Single-Species Conservation?

What would you say are some advantages of focusing on conserving single species, versus trying to conserve broader communities, ecosystems, or landscapes? What might be some of the disadvantages? Which do you think is the better approach, or should we use both?

International conservation efforts include widely signed treaties

At the international level, biodiversity protection has been pursued with several treaties facilitated by the United Nations. The 1973 **Convention on International Trade in Endangered Species of Wild Fauna and Flora (CITES)** protects endangered species by banning the international transport of their body parts. When nations enforce it, CITES can protect the tiger and other rare species whose body parts are traded internationally.

In 1992, leaders of many nations agreed to the **Convention on Biological Diversity.** This treaty aims to conserve biodiversity, use biodiversity in a sustainable manner, and ensure the fair distribution of biodiversity's benefits. Its many accomplishments so far include ensuring that Ugandan people share in the economic benefits of wildlife preserves, increasing global markets for "shade-grown" coffee and other crops grown without removing forests, and replacing pesticide-intensive farming practices with sustainable ones in some rice-producing Asian nations. As of 2006, 188 nations had become parties to the Convention on Biological Diversity. Nations that have chosen *not* to do so include Iraq, Somalia, the Vatican, and the United States. This decision is just one example of why the U.S. government is no longer widely regarded as a leader in biodiversity conservation efforts.

Biodiversity hotspots pinpoint areas of high diversity

One international approach oriented around geographic regions, rather than single species, has been the effort to map **biodiversity hotspots**. The concept of biodiversity hotspots was introduced in 1988 by British ecologist Norman Myers (▸ p. 188) as a way to prioritize regions that are most important globally for biodiversity conservation. A hotspot is an area that supports an especially great number of species that are **endemic** (▸ p. 81) to the area, that is, found nowhere else in the world (Figure 8.15). To qualify as a hotspot, a location must harbor at least 1,500 endemic plant species, or 0.5% of the world total. In addition, a hotspot must have already lost 70% of its habitat as a result of human impact and be in danger of losing more.

The nonprofit group Conservation International maintains a list of 34 biodiversity hotspots (Figure 8.16). The ecosystems of these areas together once covered 15.7% of the planet's land surface, but today, because of habitat loss, cover only 2.3%. This small amount of land is the exclusive home for 50% of the world's plant species and 42% of all terrestrial vertebrate species. The hotspot concept gives incentive to focus on these areas of endemism, where the greatest number of unique species can be protected with the least amount of effort.

FIGURE 8.15 The golden lion tamarin *(Leontopithecus rosalia)*, a species endemic to Brazil's Atlantic rainforest, is one of the world's most endangered primates. Captive breeding programs have produced roughly 500 individuals in zoos, but the tamarin's habitat is fast disappearing.

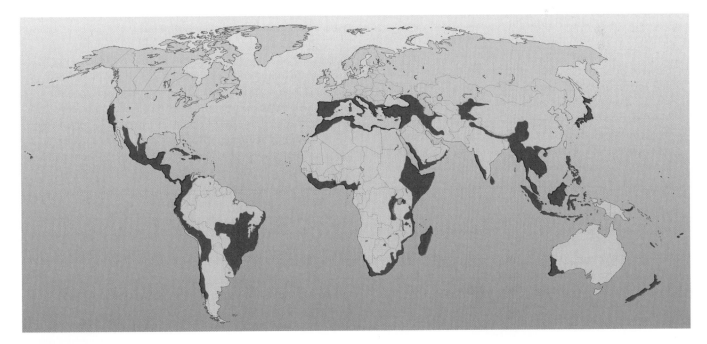

FIGURE 8.16 Some areas of the world possess exceptionally high numbers of species found nowhere else. Some conservation biologists have suggested prioritizing habitat preservation in these areas, dubbed *biodiversity hotspots*. Shown in red are the 34 biodiversity hotspots mapped by Conservation International in 2005. Data from Conservation International, 2005.

Biodiversity

VIEWPOINTS

Biodiversity is being lost worldwide at an accelerating pace. **How should we respond to biodiversity loss? What solutions should we seek, and what strategies should we prioritize?**

Mainstreaming Conservation through Ecosystem Services

Parks and nature reserves are the jewels of conservation, but they are not, and never will be, enough. Even the most ambitious conservation plans aspire to having no more than 20–30% of the world's lands protected as nature reserves, and the reality is that closer to 5–10% of land is currently under protection. If we put a fence around our parks to keep humans out, our nature preserves will still end up deteriorating because of how we treat the remaining 70–90% of Earth. Nature cannot be sequestered in some small portion of the globe while the bulk of land and water are left at the mercy of humans. Conservation will never be attainable unless humans learn to live in, work in, exploit, and harvest nature in a manner that does not ravage biodiversity.

An additional 3 billion people are expected in the next 50 years, bringing the total population to 9 billion. Many of the most impoverished people in the world depend on nature for their livelihood. All of us depend on nature for flood control, protection against storm damage, and other regulating services. To protect biodiversity and meet human needs, we must align biodiversity protection with these ecosystem services. As we make clear the links between ecosystem services and biodiversity protection, institutions and business should increasingly be willing to pay for nature's protection.

The disastrous tsunami of 2004 did minimal damage in coastal areas sheltered by natural mangrove forests. If only more of the Asia-Pacific region had been willing to invest in the insurance provided by mangroves, we would have saved human lives and biodiversity. The solution to biodiversity loss lies in clearly recognizing and valuing ecosystem services and in setting priorities that focus simultaneously on biodiversity and ecosystem services.

Peter Kareiva is lead scientist for The Nature Conservancy and an adjunct professor at the University of California at Santa Barbara and at Santa Clara University. Dr. Kareiva has also served as director of the Division for Conservation Biology at NOAA Fisheries, at the Northwest Fisheries Science Center, Seattle. His research has concerned organisms as diverse as whales, owls, Antarctic seabirds, ladybug beetles, butterflies, wildebeest, and genetically engineered microbes and crops.

Parks: The Best Way to Protect Biodiversity?

Although protected areas cover 12% of Earth's land surface, we need many more of them. Consider 34 "biodiversity hotspots" containing the last habitats of 50% of Earth's vascular plant species and 42% of vertebrate species (excluding fish). Once covering 16% of Earth's land surface, their habitats have since lost 86% of their expanse and now cover just 2.3%. If we could safeguard these relatively small areas, we could reduce the number of extinct species by at least one-third. Protection of the hotspots (parks, reserves, and so on) would cost roughly $1.5 billion per year for 5 years—just one-seventh of all conservation funding worldwide. What a massive need and massive opportunity for protected areas!

In the tropics, however, where most biodiversity is found and where it is most threatened, no island is an island, so to speak. The Kruger Park in South Africa is drying out because rivers arising outside the park are losing their waters to ranches and other development works. The park is also being overtaken by acid rain from South Africa's main industrial region upwind.

Moreover, global warming will shift temperature bands, and consequently vegetation communities, away from the equator and toward the poles. Many plants and animals in Hawaii will have little place to go but into the sea, as will those in the southern tip of Africa, northern Philippines, and dozens of other unfortunate locales. Even if they were to be turned into giant parks and perfectly protected on the ground, they would still be vulnerable to global warming—half of which is caused by carbon dioxide emissions from fossil fuels.

Biodiversity enthusiasts, here's a key question for you: Which country has less than one-twentieth of the world's population but causes one-quarter of the world's carbon dioxide emissions?

Norman Myers is a Fellow of Oxford University and a member of the U.S. National Academy of Sciences. He works as an independent scientist, advising the United Nations, the World Bank, the World Conservation Union, the World Wildlife Fund, and dozens of other conservation bodies around the world.

Explore this issue further by accessing **Viewpoints** at www.aw-bc.com/withgott.

Community-based conservation is growing

Taking a global perspective and prioritizing optimal locations to set aside as parks and reserves makes good sense, but setting aside land for preservation affects the people that live in and near these areas. In past decades, many conservationists from developed nations, in their zeal to preserve ecosystems in other nations, too often neglected the needs of people in the areas they wanted to protect. Many developing nations came to view this international environmentalism as a kind of neocolonialism.

Today this has largely changed, and many conservation biologists actively engage local people in efforts to protect land and wildlife in their own backyards, in an approach sometimes called **community-based conservation**. Setting aside land for preservation deprives local people of access to natural resources, but it can also guarantee that these resources will not be used up or sold to foreign corporations and can instead be sustainably managed. Moreover, parks and reserves draw ecotourism, which can support local economies. Community-based conservation efforts have not always been successful, but in a world of increasing human population, locally based management that meets people's needs sustainably will likely be essential.

Conclusion

The erosion of biological diversity on our planet threatens to result in a mass extinction event equivalent to the mass extinctions of the geological past. Human-induced habitat alteration, invasive species, pollution, and overharvesting of biotic resources are the primary causes of biodiversity loss. This loss matters, because human society could not function without biodiversity's pragmatic benefits. As a result, conservation biologists are rising to the challenge of conducting science aimed at saving endangered species, preserving their habitats, restoring populations, and keeping natural ecosystems intact. The innovative strategies of these scientists hold promise to slow the erosion of biodiversity that threatens life on Earth.

TESTING YOUR COMPREHENSION

1. What is biodiversity? List and describe three levels of biodiversity.
2. What are the five primary causes of biodiversity loss? Can you give a specific example of each?
3. List and describe five invasive species and the adverse effects they have had.
4. Define the term *ecosystem services*. Give five examples of ecosystem services that humans would have a hard time replacing if their natural sources were eliminated.
5. What is the relationship between biodiversity and food security? Between biodiversity and pharmaceuticals? Give three examples of potential benefits of biodiversity conservation for food security and medicine.
6. Describe four reasons why people suggest biodiversity conservation is important.
7. Explain the theory of island biogeography. Use the example of the Siberian tiger to describe how this theory can be applied to fragmented terrestrial landscapes.
8. Name two successful accomplishments of the U.S. Endangered Species Act. Now name two reasons some people have criticized it.
9. What is the difference between an umbrella species and a flagship species? Could one species be both?
10. What is a biodiversity hotspot? Define community-based conservation.

SEEKING SOLUTIONS

1. Many arguments have been advanced for the importance of preserving biodiversity. Which argument do you think is most compelling, and why? Which argument do you think is least compelling, and why?
2. Some people argue that we shouldn't worry about endangered species because extinction has always occurred. How would you respond to this view?
3. Imagine that you are an influential legislator in a country that has no endangered species act and that you want to introduce legislation to protect your country's vanishing biodiversity. Consider the U.S. Endangered Species Act, as well as international efforts such as CITES and the Convention on Biological Diversity. What strategies would you write into your legislation? How would your law be similar to and different from the U.S. and international efforts?

4. Environmental advocates from developed nations who want to preserve biodiversity globally have long argued for setting aside land in biodiversity-rich regions of developing nations. Many leaders of developing nations have responded by accusing the advocates of neocolonialism. "Your nations attained their prosperity and power by overexploiting their environments decades or centuries ago," these leaders asked, "so why should we now sacrifice our own development by setting aside our land and resources?" What would you say to these leaders of developing countries? What would you say to the environmental advocates? Do you see ways that both preservation and development goals might be reached?

5. Compare the biodiversity hotspot approach to the approach of community-based conservation. What are the advantages and disadvantages of each? Can we—and should we—follow both approaches?

CALCULATING ECOLOGICAL FOOTPRINTS

 Of the five major causes of biodiversity loss discussed in this chapter, habitat alteration arguably has the greatest impact. In their 1996 book introducing the ecological footprint concept, authors Mathis Wackernagel and William Rees present a consumption/land-use matrix for an average North American. Each cell in the matrix lists the number of hectares of land of that type required to provide for the different categories of a person's consumption (food, housing, transportation, consumer goods, and services). Of the 4.27 hectares required to support this average person, 0.59 hectares are forest, with most (0.40 hectares) being used to meet the housing demand. Using this information, calculate the missing values in the table.

	Hectares of forest used for housing	Total forest hectares used
One person	0.40	0.59
Your class		
Your state		
United States		

Data from Wackernagel, M., and W. Rees. 1996. *Our ecological footprint: reducing human impact on the earth.* British Columbia, Canada: New Society Publishers.

1. Approximately two-thirds of the forests' productivity is consumed for housing. To what use(s) would you speculate that most of the other third is put?

2. If the harvesting of forest products exceeds the sustainable harvest rate, what will be the likely consequence for the forest?

3. What will be the impact of deforestation, or of the loss of old-growth forests and their replacement with plantations of young trees, on the species diversity of the forest community? In your answer, discuss the possibilities of both extirpation and extinction.

Take It Further

Go to www.aw-bc.com/withgott, where you'll find:

▶ Suggested answers to end-of-chapter questions
▶ Quizzes, animations, and flashcards to help you study
▶ *Research Navigator™* database of credible and reliable sources to assist you with your research projects

▶ **GRAPHit!** tutorials to help you interpret graphs

▶ **INVESTIGATEit!** current news articles that link the topics that you study to case studies from your region to around the world

Cities, Forests, and Parks: Land Use and Resource Management

9

Tom McCall Waterfront
Park in Portland, Oregon

Upon successfully completing this chapter, you will be able to:

▶ Describe the scope of urbanization and assess urban and suburban sprawl

▶ Outline city and regional planning and land use strategies

▶ Evaluate transportation options, the roles of urban parks, and approaches such as smart growth and the new urbanism

▶ Analyze environmental impacts and advantages of urban centers and assess the pursuit of sustainable cities

▶ Identify major approaches in resource management

▶ Summarize the ecological roles and economic contributions of forests and outline the history and extent of forest loss.

▶ Explain aspects of forest management and describe major methods of harvesting timber

▶ Identify major U.S. land management agencies and the lands they manage

▶ Recognize types of parks and reserves and evaluate issues involved in their design

Central Case: Managing Growth in Portland, Oregon

> "Sagebrush subdivisions, coastal condomania, and the ravenous rampage of suburbia in the Willamette Valley all threaten to mock Oregon's status as the environmental model for the nation."
> —OREGON GOVERNOR TOM McCALL, 1973

> "We have planning boards. We have zoning regulations. We have urban growth boundaries and 'smart growth' and sprawl conferences. And we still have sprawl."
> —ENVIRONMENTAL SCIENTIST DONELLA MEADOWS, 1999

With the fighting words above, Oregon governor Tom McCall challenged his state's legislature in 1973 to take action against runaway urbanization, which many Oregon residents feared would ruin the communities and landscapes they had come to love. McCall echoed the growing concerns of state residents that farms, forests, and open space were being gobbled up for development, including housing for people moving in from California and elsewhere. Foreseeing a future of subdivisions, strip malls, and traffic jams engulfing the pastoral Willamette Valley, Oregon acted. With Senate Bill 100, the state legislature in 1973 passed a sweeping land use law that would become the focus of acclaim, criticism, and careful study for years afterward by other states and communities trying to manage their own urban and suburban growth.

Oregon's law required every city and county to draw up a comprehensive land use plan, in line with statewide guidelines that had gained popular support from the state's electorate. As part of each land use plan, each metropolitan area had to establish an **urban growth boundary (UGB)**, a line on a map intended to separate areas desired to be urban from areas desired to remain rural. Development for housing, commerce, and industry would be encouraged within these urban growth boundaries, but severely restricted beyond them. The intent was to revitalize city centers, prevent suburban sprawl, and protect farmland, forests, and open landscapes around the edges of urbanized areas.

Residents of the area around Portland, the state's largest city, established a new regional entity to help plan how land would be apportioned in their region. The Metropolitan Service District, or Metro, represents 25 municipalities and three counties. Metro adopted the Portland-area UGB in 1979 and has tried to focus growth on existing urban centers and to build communities where people can walk or take mass transit between home, work, and shopping. These policies have largely worked as intended; Portland's downtown and older neighborhoods have thrived, regional urban centers are becoming denser and more community oriented, mass transit has expanded, and development has been limited on land beyond the UGB. Portland began attracting international attention for its "livability."

To many Portlanders today, the UGB remains the key to maintaining quality of life in city and countryside alike. In the view of its critics, however, the "Great Wall of Portland" is an elitist and intrusive government regulatory tool. Ironically, the Portland area's successes may one day prove its undoing. A continuing influx of people attracted by the city's allure has meant rapid development and rising housing prices. Still, most citizens have supported Oregon's land use rules for the past 30 years, and the system survived three state referenda and many legal challenges.

In November 2004, however, Oregon voters approved a ballot measure that threatens to eviscerate the very land use reforms they had backed for three decades. Ballot Measure 37 requires the state to compensate certain landowners if government regulation has decreased the value of their land. For example, regulations prevent landowners outside UGBs from subdividing their lots and selling them for housing development. Under Measure 37, the state now has to pay these landowners to make up for theoretically lost income—or else allow them to ignore the regulations. Because the state does not have enough money to pay such claims, the measure could effectively gut Oregon's zoning, planning, and land use rules.

Lawsuits delayed Measure 37's implementation until February 2006, and opponents may introduce a ballot measure to reverse it. Meanwhile, the state legislature is debating whether to intervene, and a task force is convening to review the state's overall land use vision. Events in Oregon over the next few years may tell us much about how our urban areas—and the landscapes around them that provide them the resources they need—may change in the future.

Our Urbanizing World

We live at a turning point. Beginning about the year 2007, for the first time in human history, more people will live in urban areas than in rural areas. This shift from the countryside into towns and cities—**urbanization**—is arguably the single greatest change our society has undergone since its transition from a nomadic hunter-gatherer lifestyle to a sedentary agricultural one.

As we design our new urban-centered world, and as human population and resource consumption increase, the ways we apportion land for different uses and the ways we manage natural resources take on increasing importance. Urban areas rely on natural resources imported from surrounding regions. Cities and suburbs are sinks for resources, and cannot function without a steady supply of goods and raw materials from areas beyond their borders.

Urban centers depend on the food and fiber that agricultural lands (Chapter 7) supply. They rely on mineral resources from land where minerals are extracted. They need timber and other forest products from forested land where forests are managed. And as we manage these types of lands and the extraction of resources from them, we also require that some land be left undeveloped, so that natural ecosystems continue to function, providing vital ecosystem services (▸ pp. 25–27, 30–31), homes for wildlife, and areas of wilderness.

In this chapter we begin by exploring our urbanizing world and the issues that our urban areas face today. We then go on to explore how land is used and resources are managed in some of the regions beyond our urban areas—in forests and in protected parks and reserves.

Industrialization has driven the move to urban centers

Since 1950, the world's urban population has quadrupled. Urban populations are growing for two reasons: (1) the human population overall is growing (Chapter 6), and (2) more people are moving from farms to cities than are moving from cities to farms. Industrialization has decreased the need for farm labor and has increased commerce and jobs in cities. In a process of positive feedback, industrialization and urbanization have bred further technological advances that increase production efficiencies, both on the farm and in the city.

Worldwide, the proportion of population that is urban rose from 30% half a century ago to 48% today. Between 1975 and 2000, the global urban population increased each year by 2.53%, while the rural population rose only by 0.92%. From 2000 to 2030, the United Nations projects that the urban population will grow by 1.83% annually, while the rural population will decline by 0.03% each year.

In developed nations urbanization has slowed, because roughly three of every four people already live in cities, towns, and **suburbs**, the smaller communities that ring cities. In 1850, the U.S. Census Bureau classified only 15% of U.S. citizens as urban dwellers. That percentage passed 50% shortly before 1920 and now stands at 80%. Most U.S. urban dwellers reside in suburbs; fully 50% of the U.S. population today is suburban.

In contrast, today's developing nations, where many people still reside on farms, are urbanizing rapidly right now (Figure 9.1). In nations such as China, India, and Nigeria, rural people are streaming to cities in search of jobs and urban lifestyles.

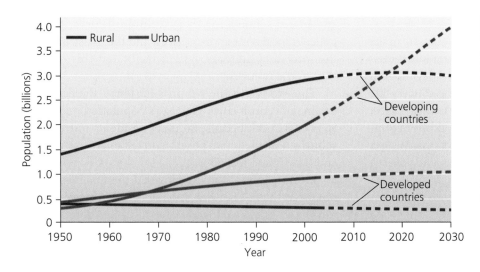

FIGURE 9.1 In developing countries today, urban populations are growing quickly, whereas rural populations are leveling off and may soon begin to decline. Developed countries are already largely urbanized, so in these countries urban populations are growing more slowly, whereas rural populations are falling. Solid lines in the graph indicate past data, and dashed lines indicate projections of future trends. Data from United Nations Population Division. 2004. *World urbanization prospects: The 2003 revision.* New York: UNPD.

Today, 46 metropolitan areas hold more than 5 million people, and 20 are home to over 10 million residents (Table 9.1). However, less than 5% of urban dwellers live in cities of greater than 10 million. Most live in smaller cities, such as Portland, Omaha, Winnipeg, Raleigh, Austin, and their still-smaller suburbs.

Well-situated cities have grown rapidly

Environmental variables such as climate, topography, and the configuration of waterways help determine whether a small settlement will become a large city. Successful cities tend to be located in places that give them economic advantages. Because cities rely on outlying regions for their resources, it is generally advantageous to be located along a major river, seacoast, railroad, highway, or other corridor for trade. Many well-located cities have acted as linchpins in trading networks, funneling in resources from agricultural regions, processing them and manufacturing products, and shipping those products to other markets.

Portland got its start in the mid-19th century as pioneers arriving by the Oregon Trail settled where the Willamette River flowed into the Columbia River. The city grew as it shipped farm products from the fertile Willamette Valley overseas and accepted products shipped in from other North American ports and from Asia. Cities from ancient times to the present day have supported themselves by drawing in resources from outlying rural areas through trade, persuasion, or conquest. In turn, cities influence how people use land and harvest or manage resources in surrounding areas.

In recent years, many southern and western U.S. cities have experienced growth spurts, as people (particularly retirees) have moved south and west in search of warmer weather or more space. Between 1990 and 2000, the population of the Atlanta metropolitan area grew by 39%, that of the Phoenix region grew by 45%, and that of the Las Vegas metropolitan area grew by 83%.

Internationally, most fast-growing cities today are in the developing world, where many nations are rapidly industrializing. Sadly, wars, conflict, and ecological degradation are also driving millions of people out of the countryside and into cities. Cities like Mumbai (Bombay), India; Lagos, Nigeria; and Cairo, Egypt are growing dramatically in population, but without the economic growth to match. As a result, they are facing overcrowding, pollution, and poverty.

People have moved to suburbs

In the United States, Canada, and other developed nations, many cities by the mid-20th century had accumulated more people than these cities had jobs to offer.

Table 9.1 Metropolitan Areas with 10 Million Inhabitants or More	
City	**Millions of people**
Tokyo, Japan	35.0
Mexico City, Mexico	18.7
New York, United States	18.3
Sao Paulo, Brazil	17.9
Mumbai (Bombay), India	17.4
Delhi, India	14.1
Calcutta, India	13.8
Buenos Aires, Argentina	13.0
Shanghai, China	12.8
Jakarta, Indonesia	12.3
Los Angeles, United States	12.0
Dhaka, Bangladesh	11.6
Osaka-Kobe, Japan	11.2
Rio de Janeiro, Brazil	11.2
Karachi, Pakistan	11.1
Beijing, China	10.8
Cairo, Egypt	10.8
Moscow, Russian Federation	10.5
Metro Manila, Philippines	10.4
Lagos, Nigeria	10.1

Source: United Nations Population Division. 2004. *World urbanization prospects: The 2003 revision.* New York: UNPD

Unemployment rose, and crowded inner-city areas began to suffer increasing poverty and crime. As inner cities declined economically from the 1950s onward, many affluent city dwellers chose to move outward to the cleaner, less crowded, and more parklike suburban communities. These people were pursuing more space, better economic opportunities, cheaper real estate, less crime, and better schools for their children. Federal housing policies, local zoning ordinances, and the construction of the interstate highway system facilitated this outward shift to the suburbs.

The exodus to the suburbs hastened the economic decline of central cities. Chicago's population declined to 80% of its peak because so many residents moved to its suburbs. Philadelphia's population fell to 76% of its peak, and Detroit's to just 55%.

In most ways, suburbs delivered the qualities people sought in them. The wide spacing of houses gave families room and privacy. However, by allotting more space to each person, suburban growth spread human impact across the landscape. Natural areas disappeared as housing developments were constructed. Extensive road networks eased travel, but suburbanites found themselves needing to drive everywhere, commuting longer distances to work and spending more time in congested traffic. The expanding rings of suburbs surrounding cities grew larger than the cities themselves, and towns ran into one another. These aspects of suburban growth inspired a new term: *sprawl*.

Today's urban areas spread outward

Although the term *sprawl* has become laden with meanings and connotes different things to different people, we can assign **sprawl** a simple nonjudgmental definition: The spread of low-density urban or suburban development outward from an urban center. The spatial growth of urban and suburban areas is obvious from maps and satellite images of rapidly spreading cities such as Las Vegas (Figure 9.2). Houses and roads supplant over 1 million ha (2.5 million acres) of U.S. rural land each year—over 2,700 ha (6,700 acres) every day.

Because suburban growth entails allotting more space per person than in cities, in most cases this outward spatial growth across the landscape has outpaced the growth in numbers of people. In fact, many researchers define sprawl as the physical spread of development at a rate greater than the rate of population growth. For instance, from 1950 to 2002, Phoenix's population grew 12 times larger, but its land area grew 27 times larger. Chicago's metropolitan area now spreads over an area 40 times the size of the city proper, and on average each suburban resident takes up 11 times more space than a resident of the city proper. Between 1950 and 1990, the population of 58 major U.S. metropolitan areas rose by 80%, but the land area they covered rose by 305%. Even in 11 metro areas where population declined between 1970 and 1990 (for instance, rust-belt cities such as Detroit, Cleveland, and Pittsburgh), the amount of land covered increased. Several types of development approaches can lead to sprawl (Figure 9.3).

Economists, politicians, and city boosters have traditionally encouraged the unbridled spatial expansion of cities and suburbs. The conventional assumption has been that growth is good and that attracting business, industry, and residents will unfailingly increase a community's economic well-being, political power, and cultural influence. Today, as negative impacts of sprawl on citizens' lifestyles accumulate, this assumption is increasingly being challenged.

(a) Las Vegas, Nevada, 1972　　　　**(b) Las Vegas, Nevada, 2002**

FIGURE 9.2 Satellite images show the type of rapid urban and suburban expansion that many people have dubbed *sprawl*. Las Vegas, Nevada, is currently one of the fastest-growing cities in North America. Between 1972 **(a)** and 2002 **(b)**, its population increased more than fivefold, and its developed area rose more than threefold.

(a) Uncentered commercial strip development

(b) Low-density single-use development

(c) Scattered, or leapfrog, development

(d) Sparse street network

FIGURE 9.3 Several standard approaches to development can result in sprawl. In uncentered commercial strip development (**a**), businesses are arrayed in a long strip along a roadway, and no attempt is made to create a centralized community with easy access for consumers. In low-density, single-use residential development (**b**), homes are located on large lots in residential tracts far away from commercial amenities. In scattered or leapfrog development (**c**), developments are created at great distances from a city center and are not integrated. In developments with a sparse street network (**d**), roads are far enough apart that moderate-sized areas go undeveloped, but not far enough apart for these areas to function as natural areas or sites for recreation. All these development approaches necessitate frequent automobile use.

What is wrong with sprawl?

To some people, the word *sprawl* evokes strip malls, homogenous commercial development, and tracts of cookie-cutter houses encroaching on farmland and ranchland. It may suggest traffic jams, destruction of wildlife habitat, and loss of natural land around cities. However, for other people, sprawl represents the collective result of choices made by millions of well-meaning individuals trying to make a better life for themselves and their families. In this view, those who criticize sprawl are elitist and fail to appreciate the good things about suburban life. Let us try, then, to leave the emotional debate aside and assess the impacts of sprawl (see "The Science behind the Story," ▶ pp. 198–199).

Transportation Most studies show that sprawl constrains transportation options, essentially forcing people to drive cars. Across the United States, during the 1980s and 1990s the average length of work trips rose by 36%, and total vehicle miles driven increased at three times the rate of population growth. An automobile-oriented culture also increases dependence on petroleum, with its attendant economic and environmental consequences (▶ pp. 312–317).

Pollution Sprawl's effects on transportation give rise to increased pollution. Carbon dioxide emissions from vehicles exacerbate global climate change (Chapter 12) while nitrogen- and sulfur-containing air pollutants contribute

Sprawl

Do you see problems with urban and suburban sprawl? If so, what are the best solutions? What strategies should we pursue in making our communities more livable?

The Myth of Urban Sprawl

Russians say Americans have no real problems, so they make them up. Urban sprawl is one of those made-up problems. Yet the proposed remedy to sprawl—sometimes called "smart growth," though it is anything but smart—will cause far more problems than it solves.

The U.S. Department of Agriculture says that urban development does not threaten American farm productivity. Nor is it a threat to rural open space: All of the cities, suburbs, and towns in the United States occupy less than 4% of the nation's land area.

University of Southern California planning professor Peter Gordon points out that low-density development is a remedy for, not the cause of, congestion, air pollution, and many other problems. Studies claiming that suburbs cause obesity, crime, and other social problems are little more than junk science, being based on inadequate data with little statistical significance and usually confusing cause and effect.

So-called smart growth says more people should live in high-density, mixed-use developments. These developments can be attractive to some people, mainly young adults with no children. But the market for them is limited. Most Americans still find a single-family home with a large yard to be their American dream.

Attempting to impose smart growth on more people has many unfortunate effects. Because it doesn't significantly reduce the miles people drive, it increases congestion; and because cars pollute more in stop-and-go traffic, it increases air pollution. Smart growth makes housing unaffordable to low- and even middle-income families, and makes neighborhoods more vulnerable to crime.

Instead of attempting to impose their lifestyle preferences on others, city officials should simply ensure that people pay the full costs of whatever lifestyle they prefer. Once that happens, people can be free to choose to live in high densities or low, and to drive, walk, bicycle, or ride transit.

Randal O'Toole is an economist and the director of the American Dream Coalition, which seeks to solve urban problems without reducing people's personal freedom. He is the author of *The Vanishing Automobile and Other Urban Myths: How Smart Growth Will Harm American Cities.* He has taught environmental economics at Yale, the University of California at Berkeley, and Utah State University.

The Real Problem with Sprawl

The most visible problem associated with sprawl is one of livability—sprawl's effect on our quality of life. However, the bigger but less visible problem is sprawl's contribution to global warming.

Until the mid-20th century, every village, town, and city in the world was made up of mixed-use, pedestrian-friendly neighborhoods. But after World War II, the time-tested *neighborhood* was discarded in favor of an untested invention, now known as *sprawl*.

Sprawl is based on two simple premises: first, that each land use be separated from every other; and second, that these now distant land uses be connected by a massive automotive infrastructure. In sprawl, walking not only serves limited purposes but also can often be dangerous. For this reason, a typical suburban household generates more than 12 one-way car trips per day.

The ecological impact is profound. Motor vehicles, the lifeblood of sprawl, are the single greatest contributor to global warming. Over the past 60 years, we have created a built environment that requires most adults to own a car and drive it every day. Unless we quickly make the change to nonpolluting vehicles, global warming will remain the strongest argument against sprawl.

But what if cars did not pollute; would sprawl then offer a satisfactory solution? This is where quality of life enters the picture. In a society in which cars are a prerequisite to social and economic viability, those who can't drive—one-third of the population—become second-class citizens. And because everyone who can drive must drive, we spend inordinate amounts of time stuck in traffic, time that would be better spent in less stressful pursuits. The frustration with this situation—the hours that we spend trying to reconnect our artificially disassociated lives—is the reason why most people who argue against sprawl cite concerns over quality of life.

Jeff Speck is director of design at the National Endowment for the Arts. Prior to joining the Endowment in 2003, he spent 10 years as director of town planning at the firm of Duany Plater-Zyberk and Co., Architects and Town Planners (DPZ). With Andres Duany and Elizabeth Plater-Zyberk, he is the co-author of the book *Suburban Nation: The Rise of Sprawl and the Decline of the American Dream*, published in March 2000 by North Point/Farrar Straus Giroux.

Explore this issue further by accessing **Viewpoints** at www.aw-bc.com/withgott.

Measuring the Impacts of Sprawl

Critics of sprawl have blamed it for so many societal ills that it can make a person feel guilty just for being born in the suburbs or shopping at a mall. But what does scientific research tell us are the actual consequences of sprawl?

When Reid Ewing of Rutgers University and his team set out to measure the impacts of sprawl, they discovered that researchers studying sprawl have been hard-pressed even to agree on a definition of the term or on how to measure it. Surveying the literature, Ewing's team found that researchers using different criteria ranked cities in very different ways. For instance, in most studies Los Angeles was deemed more sprawling than Portland, but in some, Portland was judged to suffer worse sprawl than L.A.

So Ewing, Rolf Pendall of Cornell University, and Don Chen of the nonprofit group Smart Growth America tried to define *sprawl* in terms as simple as possible, without mixing sprawl's consequences into the definition. They decided that sprawl occurs when the spread of development across the landscape far outpaces population growth.

Traffic congestion is one of the most recognized impacts of sprawl.

Ewing, Pendall, and Chen then devised four criteria by which to rank 83 of the largest U.S. metropolitan areas in terms of sprawl. Sprawling cities would show:

▶ Low residential density
▶ Distant separation of homes, employment, shopping, and schools
▶ Lack of "centeredness," i.e., lack of activity in community centers and downtown areas
▶ Street networks that make many streets hard to access

For each criterion, Ewing's team decided to measure multiple factors—22 variables in all. They then devised a way to analyze the vari-

ables and arrive at a cumulative index of sprawl. Finally, they obtained data from municipalities throughout the country.

Ewing's team's rankings showed that the most sprawling area in the nation was the Riverside–San Bernardino, California, region, which has expanded quickly in recent decades. The area with the least sprawl was New York City, whose historically dense population and vibrant neighborhoods kept it geographically compact, relative to its number of inhabitants. The accompanying tables list the 10 most- and least-sprawling areas.

The researchers next correlated those scores with a number of transportation variables. They found that people in the 10 most-sprawling metros owned more cars (180 per 100 households) than people in the 10 least-sprawling metros (162 per 100 households). They also found that residents of the most-sprawling metros drove an average of 43 km (27 mi) per day, whereas those of the least-sprawling metros drove only 34 km (21 mi). In addition, people in the most-sprawling metros used public transit far less and suffered 67% more traffic fatalities than those in the least-sprawling metros.

to urban smog and acid precipitation (▶ pp. 280, 282–285). Motor oil and road salt from roads and parking lots pollute waterways, and runoff of polluted water from paved areas is estimated to be about 16 times greater than from naturally vegetated areas.

Health Aside from the health impacts of pollution, research suggests that sprawl promotes physical inactivity and obesity, because driving cars takes the place of walking during daily errands. One 2003 study found

that people from the most-sprawling U.S. counties weigh 2.7 kg (6 lb) more for their height than people from the least-sprawling U.S. counties and that slightly more people from the most-sprawling counties have high blood pressure.

Land use The spread of low-density development means that more land is developed while less is left as forests, fields, farmland, or ranchland. Of the estimated 1 million ha (2.5 million acres) of U.S. land converted each

Strikingly, the study found no significant difference in commute time from home to work for people of sprawling versus less-sprawling metro areas. Critics of sprawl have long blamed sprawl for traffic congestion and commute delays. Advocates of suburban spread have argued that more streets ease commutes and that regions can sprawl their way out of congestion. The Ewing team's results seem to suggest that each side may have a point and that the effects may cancel one another out.

Because vehicle emissions cause air pollution, the researchers also measured levels of tropospheric ozone (▶ p. 277). Sprawling areas had worse air pollution, they found; ozone levels were 40% higher in the most-sprawling metros.

These results were published in 2003 in the *Transportation Research Record* and in a 2002 report published by Smart Growth America. None of the results prove that sprawl *causes* these impacts, because statistical correlation alone does not imply causation. However, taken together, the results suggest that spatial patterns of development may influence people's options, impacts, and behavior relative to transportation.

The 10 Most-Sprawling American Urban Areas	
Rank	**Metropolitan Region**
1	Riverside–San Bernardino, CA
2	Greensboro–Winston-Salem–High Point, NC
3	Raleigh–Durham, NC
4	Atlanta, GA
5	Greenville–Spartanburg, SC
6	West Palm Beach–Boca Raton–Delray Beach, FL
7	Bridgeport–Stamford–Norwalk–Danbury, CT
8	Knoxville, TN
9	Oxnard–Ventura, CA
10	Fort Worth–Arlington, TX

1 = most-sprawling, 10 = less-sprawling.

The 10 Least-Sprawling American Urban Areas	
Rank	**Metropolitan Region**
83	New York, NY
82	Jersey City, NJ
81	Providence–Pawtucket–Woonsocket, RI
80	San Francisco, CA
79	Honolulu, HI
78	Omaha, NE-IA
77	Boston–Lawrence–Salem–Lowell–Brockton, MA
76	Portland, OR
75	Miami–Hialeah, FL
74	New Orleans, LA

83 = least-sprawling, 74 = more-sprawling.
Source: Ewing, R., et al. 2002. *Measuring sprawl and its impact.* Washington, D.C.: Smart Growth America.

year, roughly 60% is agricultural land and 40% is forest. These lands provide resource production, recreation, aesthetic beauty, wildlife habitat, water purification, and many other ecosystem services (▶ p. 25–27, 30–31). Sprawl generally diminishes all these amenities.

Economics Sprawl drains tax dollars from existing communities and funnels them into infrastructure for new development on the fringes of those communities. Money that could be spent maintaining and improving downtown centers is instead spent extending the road system, water and sewer system, electricity grid, and telephone lines to distant developments, and extending police and fire service, schools, and libraries. Advocates for sprawling development argue that taxes on new development eventually pay back the investment made in infrastructure, but studies have found that in most cases taxpayers continue to subsidize new development, especially if municipalities do not pass on infrastructure costs to developers.

Creating Livable Cities

To respond to the challenges that urban and suburban sprawl present, architects, planners, developers, and policymakers across North America are trying to restore the vitality of city centers and to plan and manage how urbanizing areas develop.

City and regional planning are means for creating livable urban areas

City planning is the professional pursuit that attempts to design cities so as to maximize their efficiency, functionality, and beauty. City planners advise policymakers on development options, transportation needs, public parks, and other matters. City planning in North America came into its own in the early 20th century as urban leaders sought to improve neighborhood living conditions, streamline traffic systems, and establish outdoor public spaces.

In Portland in 1912, planner Edward Bennett's *Greater Portland Plan* recommended rebuilding the harbor; dredging the river channel; constructing new docks, bridges, tunnels, and a waterfront railroad; superimposing wide radial boulevards on the old city street grid; establishing civic centers downtown; and greatly expanding the number of parks. As the century progressed, several other major planning efforts were conducted, and some ideas, such as establishing a downtown public square, came to fruition.

City planning grew in importance throughout the 20th century as urban populations expanded, inner cities decayed, and wealthier residents fled to the suburbs. In today's world of sprawling metropolitan areas, **regional planning** has become just as important. Regional planners deal with the same issues as city planners, but they work on broader geographic scales and must coordinate their work with multiple municipal governments. In some places, regional planning has been institutionalized in formal governmental bodies; the Portland area's Metro is the epitome of such a regional planning entity.

Zoning is a key tool for planning

One tool that planners use is **zoning**, the practice of classifying areas for different types of development and land use (Figure 9.4). For instance, to preserve the cleanliness and tranquility of residential neighborhoods, industrial plants may be kept out of districts zoned for residential

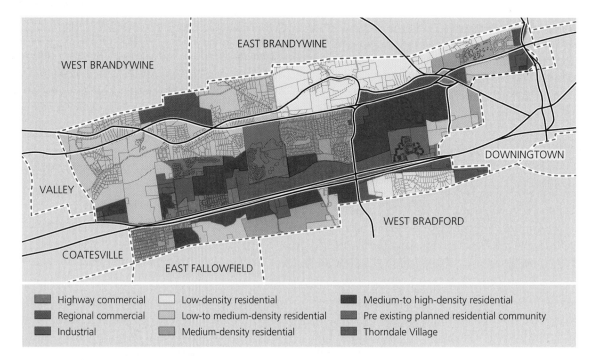

FIGURE 9.4 By zoning areas of a city for different uses, planners can guide how an urban area develops. Zoning puts restrictions on what private landowners can do with their land, but it is intended to maximize prosperity, efficiency, and quality of life for the community. This zoning map for Caln Township, Pennsylvania, shows several patterns common to modern zoning practice. Public and institutional uses are clustered together in a downtown area. Industrial uses are clustered together, away from most residential areas. Commercial uses are clustered along major roadways, and residential zones generally are higher in density toward the center of town.

use. The specification of zones for different types of development gives planners a powerful means of guiding what gets built where. Zoning also gives home buyers and business owners security; they know in advance what types of development can and cannot be located nearby.

Zoning involves government restriction on the use of private land, so some people view zoning as a top-down constraint on personal property rights. Others defend zoning, saying that government has a proper role in setting certain limitations on property rights for the good of the community. Similar debates arise with endangered species management (▸ p. 183) and other environmental issues.

Oregon voters sided with private property rights in 2004 when they passed Ballot Measure 37, which shackled government's ability to enforce zoning regulations with landowners who had owned their land before the regulations were enacted. However, opponents of the measure predicted that the state's voters would change their minds once they begin seeing new development they do not wish to occur. For the most part, people have supported zoning over the years because the public good it produces for communities is widely felt to outweigh the restrictions on private use.

Urban growth boundaries are now widely used

Planners intended Oregon's urban growth boundaries to limit sprawl by containing growth largely within existing urbanized areas. The UGBs aimed to revitalize downtowns; protect farms, forests, and their industries; and ensure urban dwellers some access to open space near cities. Since Oregon began its experiment, other states, regions, and cities have adopted UGBs—from Boulder, Colorado, to Lancaster, Pennsylvania, to many California communities. In their own ways, all the UGBs aim to concentrate development, prevent sprawl, and preserve working farms, orchards, ranches, and forests.

UGBs also appear to reduce the amounts municipalities have to pay for infrastructure, compared to sprawl. The best estimate is that UGBs save taxpayers about 20% on infrastructure costs. However, UGBs also seem to increase housing prices within their boundaries. In the Portland area, housing is becoming less affordable, but in most other ways its UGB is working as intended. It has lowered prices for land outside the UGB while increasing prices within it. It has restricted development outside the UGB. It has increased the density of new housing inside the UGB by over 50% as homes and multistory apartments are built on smaller lots. Downtown employment

rose by 73% between 1970 and 1995 as businesses and residents invested in the central city. However, urbanized area still increased by 101 km^2 (39 mi^2) in the decade after the UGB was established, because 146,000 people joined the population. This suggests that relentless population growth may thwart even the best anti-sprawl efforts. Indeed, Metro has incrementally enlarged the Portland-area UGB, and population projections for the region suggest there will be still more expansion.

"Smart growth" and the "new urbanism" aim to counter sprawl

As more people have begun to feel negative effects of sprawl on their everyday lives, states, regions, and cities throughout North America have adopted land use policies to manage growth. Urban growth boundaries and many other ideas from these policies have coalesced under the concept of **smart growth** (Table 9.2). Proponents of smart growth seek to rejuvenate the older existing communities that so often are drained and impoverished by sprawl. Smart growth often means "building up, not out"—focusing development and economic investment in existing urban centers and favoring multistory shop-houses and high-rises over one-story homes spreading outward.

A related movement among many architects, planners, and developers is labeled the **new urbanism**. This approach seeks to design neighborhoods on a walkable scale, with homes, businesses, schools, and other amenities all close together for convenience. The aim is to create

Table 9.2   Ten Principles of "Smart Growth"
▸ Mix land uses
▸ Take advantage of compact building design
▸ Create a range of housing opportunities and choices
▸ Create walkable neighborhoods
▸ Foster distinctive, attractive communities with a strong sense of place
▸ Preserve open space, farmland, natural beauty, and critical environmental areas
▸ Strengthen and direct development towards existing communities
▸ Provide a variety of transportation choices
▸ Make development decisions predictable, fair, and cost-effective
▸ Encourage community and stakeholder collaboration in development decisions

Source: U.S. Environmental Protection Agency, 2005.

functional neighborhoods in which most of a family's needs can be met close to home without using a car. Green spaces, trees, a mix of architectural styles, and creative street layouts add to the visual interest and pleasantness of new urbanist developments (Figure 9.5). These developments mimic the traditional urban neighborhoods that existed until the advent of suburbs. Moreover, new urbanist neighborhoods are generally connected to public transit systems, enabling people to travel most places they need to go by train and foot alone.

To develop urban centers in these ways, zoning rules must allow it. Traditionally, many zoning rules have limited the density of development. Although well intentioned, such rules encouraged sprawl, so some are now being rethought in communities that desire new urbanism and smart growth.

Transportation options are vital to livable cities

A key ingredient in any planner's recipe for improving the quality of urban life is making multiple transportation options available to citizens. These options include public buses, trains and subways, and light rail (smaller rail systems powered by electricity). As long as an urban center is large enough to support the infrastructure necessary, these mass transit options are cheaper, more energy-efficient, and cleaner than roadways choked with cars (Figure 9.6).

The fuel and productivity lost to traffic jams have been estimated to cost the U.S. economy $74 billion each year. Some cities with severe traffic problems, such as Bangkok,

FIGURE 9.5 This plaza at Mizner Park, Boca Raton, Florida, is part of a planned community built in the style of the "new urbanism." Homes, schools, and businesses are mixed close together in a centered neighborhood so that most amenities are within walking distance.

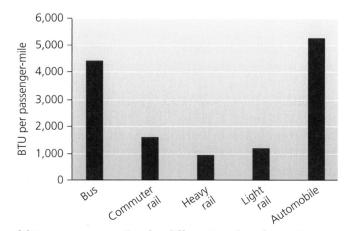

(a) Energy consumption for different modes of transit

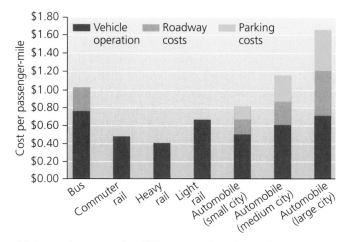

(b) Operating costs for different modes of transit

FIGURE 9.6 Rail transit tends to be cheaper and more energy-efficient than road-based transportation. Rail transit consumes far less energy per passenger mile (**a**) than bus or automobile transit. Rail transit involves fewer costs per passenger mile (**b**) than bus or automobile transit. *Source:* Litman, T. 2005. *Rail transit in America: A comprehensive evaluation of benefits.* Victoria, B.C.: Victoria Transport Policy Institute.

Thailand, and Athens, Greece, have recently built new rail systems that carry hundreds of thousands of commuters each day. Curitiba, Brazil is famous for its outstanding bus system that is used daily by three-quarters of the city's 2.5 million citizens. In 1986 Portland introduced a light rail system called MAX (Figure 9.7), and Metro policy encouraged the development of self-sufficient neighborhood communities in the new urbanist style along the rail lines. Light rail ridership has steadily increased as this system has expanded.

To make urban transportation more efficient, governments can also raise fuel taxes, tax inefficient modes of transport, reward carpoolers with carpool lanes, encourage bicycle use and bus ridership, and charge trucks for road damage. They can choose to minimize investment in

FIGURE 9.7 Portland's light rail system is one component of an urban planning strategy that has helped make it one of North America's most livable cities.

FIGURE 9.8 City parks were developed in many urban areas in the late 19th century to provide citizens aesthetic pleasure, recreation, and relief from the stresses of the city. Manhattan's Central Park, shown here, was one of the first.

infrastructure that encourages sprawl, and stimulate investment in renewed urban centers.

Parks and open space are key elements of livable cities

City dwellers often desire some sense of escape from the noise, commotion, and stress of urban life. Natural lands, public parks, and open space provide greenery, scenic beauty, freedom of movement, and places for recreation. These lands also keep ecological processes functioning by regulating climate, producing oxygen, purifying air and water, and providing habitat for wildlife.

In the late 1800s, public parks began to be established in eastern U.S. cities, using aesthetic ideals borrowed from European parks, gardens, and royal hunting grounds. The lawns, shaded groves, curved pathways, and pastoral vistas we see today in many American city parks and cemeteries originated with these European ideals, as interpreted by the leading American landscape architect, Frederick Law Olmsted. Olmsted designed New York City's Central Park in 1853 and a host of urban park systems afterwards (Figure 9.8).

Portland's quest for urban parks began in 1900, when city leaders created a parks commission and then hired Olmsted's son, John Olmsted, to design a park system. His 1904 plan recommended acquiring land to ring the city generously with parks, but no action was taken. A full 44 years later, citizen pressure finally resulted in the creation of Forest Park along a large forested ridge on the northwest side of the city. At 11 km (7 mi) long, it is today the largest city park in the United States.

Although large city parks are vital to a healthy urban environment, even small spaces can make a big difference. Playgrounds provide places where children can be active

outdoors and interact with their peers. Community gardens allow people to grow their own vegetables and flowers in a neighborhood setting (Figure 9.9). Greenways along rivers, streams, or canals can protect water quality, boost property values, provide hiking trails, and serve as corridors for the movement of birds and wildlife.

Weighing the **issues:**
Nature in the City

How important do you feel it is for urban residents to have access to parks and natural areas within their cities or towns? Would you be willing to pay extra taxes to establish more parks in your community? What types of parks or natural areas do you think are most helpful, and why?

Urban centers have mixed consequences for the environment

Urbanization and urban centers exert both positive and negative environmental impacts. These impacts depend strongly on how we utilize resources, produce goods, transport materials, and deal with waste.

Resource consumption Cities and towns are sinks for resources, having to import from widespread regions beyond their borders nearly all the food, water, timber, metal, and mined fuels they need to feed, clothe, and house their inhabitants and power their commerce. The long-distance transportation of resources and goods

FIGURE 9.9 Urban community gardens like this one in Seattle provide city residents with a place to grow vegetables and also serve as greenspaces that beautify cities.

requires a great deal of fossil fuel use. For this reason, the centralization of resource use that urbanization entails may seem a bad thing for the environment.

However, imagine that the world's 3 billion urban residents were instead spread evenly across the landscape. What would the transportation requirements be, then, to move all those resources and goods around to all those people? A world without cities would likely require *more* transportation to provide people the same level of access to diverse resources and goods. Once resources have arrived at an urban center, cities should be able to maximize the efficiency of resource use and delivery of goods and services. The density of cities facilitates the provision of many services that improve quality of life, including electricity, medical services, education, water and sewer systems, waste disposal, and public transportation.

Land preservation Because cities draw resources from afar, their ecological footprints (▸ pp. 6–7) are much larger than their actual land areas. However, because people are packed densely together in cities, more land outside cities is left undeveloped. Indeed, this is the idea behind urban

growth boundaries. If cities did not exist, and if instead all 6.5 billion of us were evenly spread across the planet's land area, we would have much less room for agriculture, wilderness, biodiversity, or privacy. There would be no large blocks of land left uninhabited by people. The fact that half the human population is concentrated in discrete locations helps allow room for natural ecosystems to continue functioning and provide the ecosystem services on which all of us, urban and rural, depend.

Pollution Just as cities import resources, they export wastes, either passively through pollution or actively through trade. In so doing, urban centers transfer the costs of their activities to other regions—and mask the costs from their own residents. Citizens of Toronto may not recognize that pollution from coal-fired power plants in their region exacerbates acid precipitation hundreds of miles to the east. Citizens of New York City may not realize how much garbage their city produces if it is shipped to other states or nations for disposal.

However, not all waste and pollution leaves the city. Urban residents are exposed to high levels of toxic industrial compounds, photochemical and industrial smog, noise pollution, and light pollution. These forms of pollution and the health threats they pose are not shared evenly among urban residents. As environmental justice advocates point out (▸ p. 19), those who receive the brunt of the pollution are most often those who are too poor to live in cleaner areas.

Innovation Cities promote a flourishing cultural life and, by mixing together diverse people and influences, spark innovation and creativity. The urban environment can promote education and scientific research, and cities have long been viewed as engines of technological and artistic inventiveness. This inventiveness can lead to solutions to societal problems, including ways to reduce environmental impacts.

Some seek sustainability for cities

Modern cities that import all their resources and export all their wastes have a linear, one-way metabolism. Such linear models of production and consumption tend to destabilize environmental systems. Proponents of *urban ecology* urge us to develop circular systems, akin to systems found in nature that recycle materials and use renewable sources of energy. Advocates of urban sustainability suggest that cities:

▸ Maximize efficient use of resources
▸ Recycle as much as possible (▸ pp. 361–362)

▶ Develop environmentally friendly technologies
▶ Account fully for external costs (▶ pp. 28–29)
▶ Offer tax incentives to encourage sustainable practices
▶ Use locally produced resources
▶ Use organic waste and wastewater to restore soil fertility
▶ Encourage urban agriculture

More and more cities are adopting these strategies. Curitiba, Brazil shows the kind of success that can result when a city invests in well-planned infrastructure. Besides its highly effective bus network, the city provides recycling, environmental education, job training for the poor, and free health care. Surveys show that its citizens are unusually happy and better off economically than people of other Brazilian cities. Successes in places from Portland to Curitiba suggest that urban sustainability is feasible. Indeed, because they affect the environment in some positive ways and have the potential for efficient resource use, cities can be a key element in achieving environmental progress.

Urban areas will always depend on an influx of resources from rural and undeveloped lands around them, however. The ways we use these lands and the ways we manage the resources from them are crucial components in the quest for sustainability. Thus we will next examine how we manage forest lands and the timber resources we need for modern life.

Forestry and Resource Management

We need to manage the resources we take from the natural world because many of them are limited. We have seen (▶ pp. 3–4) how some resources, such as fossil fuels, are nonrenewable on human time scales, whereas other resources, such as the sun's energy, are perpetually renewable. Between these extremes lie resources that are renewable if they are not exploited too rapidly or carelessly—resources such as soils, freshwater, wildlife and fisheries, rangeland, and timber. All these resources also serve functions in the ecosystems of which they are a part.

Resource management is the practice of harvesting potentially renewable resources in ways that do not deplete them. Resource managers are guided in their decision making by available research in the natural sciences, as well as by political, economic, and social factors. The management of timber and other forest resources is a clear and representative example of resource management. Foresters, those professionals who manage forests through the practice of forestry, must balance the central importance of forests as ecosystems with civilization's demand for wood products.

Resource managers follow several strategies

A key question in managing resources is whether to focus narrowly on the resource of interest or to look more broadly at the environmental system of which the resource is a part. Taking a broader view can often help avoid degrading the system and may thereby help sustain the resource in the long term.

Maximum sustainable yield One traditional guiding principle in resource management has been **maximum sustainable yield**. The aim is to achieve a maximum amount of resource extraction without depleting the resource from one harvest to the next. Recall the logistic growth curve (Figure 4.11, ▶ p. 89), which shows that a population grows most quickly when it is at an intermediate size. A fisheries manager aiming for maximum sustainable yield will prefer to keep fish populations at intermediate levels so that they rebound quickly after each harvest. Doing so should result in the greatest amount of fish harvested over time, while the population sustains itself (Figure 9.10).

This management approach, however, keeps the fish population at only about half its carrying capacity (▶ p. 89)—well below the level it would attain in the absence of

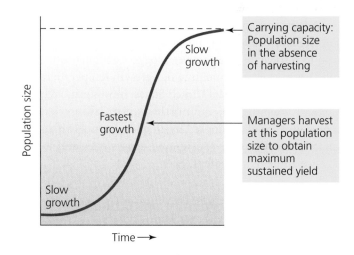

FIGURE 9.10 Using the concept of maximum sustainable yield, resource managers attempt to maximize the amount of resource harvested, as long as the harvest is sustainable in perpetuity. In the case of a wildlife population or fisheries stock that grows according to a logistic growth curve, managers aim to keep the population at an intermediate level, well below the carrying capacity, because populations grow fastest at intermediate sizes.

fishing. Reducing one population in this way will likely have effects on other species and on the food web dynamics of the community. From an ecological point of view, management for maximum sustainable yield may thereby set in motion complex and unpredictable changes.

Ecosystem-based management Increasing numbers of managers today espouse **ecosystem-based management**, which attempts to manage the harvesting of resources in ways that minimize impact on the ecosystems and ecological processes that provide the resource. Some modern forestry plans protect certain forested areas, restore ecologically important habitats such as riparian corridors, and consider patterns at the landscape level, allowing timber harvesting while preserving the functional integrity of the forest ecosystem.

It can be challenging to determine how best to implement this type of management. Ecosystems are complex, and our understanding of how they operate is limited. Thus, ecosystem-based management has often come to mean different things to different people.

Adaptive management Some management actions will succeed, and some will fail. **Adaptive management** involves systematically testing different approaches with the aim of improving methods through time. For managers, it entails monitoring the results of one's practices and adjusting them as needed, based on what is learned. Adaptive management is intended as a true fusion of science and management, because hypotheses about how best to manage resources are explicitly tested.

Adaptive management was featured in the Northwest Forest Plan, a 1994 plan crafted by the Clinton administration to resolve disputes between loggers and environmentalists over the last remaining old-growth temperate rainforest in the continental United States. In western Oregon, western Washington, and northwestern California, this plan sought to allow limited logging to continue with adequate protections for species and ecosystems, and to let science guide management.

Forests are ecologically valuable, but forest products are economically valued

Because of their structural complexity and their ability to provide many niches for organisms, forests comprise some of the richest ecosystems for biodiversity. Trees furnish food and shelter for an immense diversity of vertebrate and invertebrate animals, while understory shrubs and groundcover plants provide further habitat. The leaves, stems, and roots of forest plants are colonized by an extensive array of fungi and microbes, while on the forest floor, leaf litter nourishes the soil and myriad organisms decompose plant material and cycle nutrients.

In general, forests with a greater diversity of plants host a greater diversity of organisms. And, in general, old-growth forests host more biodiversity than younger forests, because older forests contain more structural diversity and thus more microhabitats and resources for more species.

Forests also provide vital ecosystem services. Forest vegetation—like that of grasslands and other plant communities—stabilizes soil and prevents erosion. Plants help regulate the hydrologic cycle (▸ pp. 67–68, 70), slowing runoff, lessening flooding, and purifying water as they take it up from the soil and release it to the atmosphere. Forest plants also store carbon, release oxygen, and moderate climate. By performing such ecological functions, forests are indispensable for our survival.

In addition, forests provide people with economically valuable wood products. For millennia, wood from forests has housed us, and built the ships that carried people and cultures from one continent to another. It allowed us to produce paper, the medium of the first information revolution. In recent decades, industrial harvesting has allowed us to extract more timber than ever before. The exploitation of forest resources has been instrumental in helping our society achieve the standard of living we enjoy today.

Demand for wood has led to deforestation

People clear forests to exploit timber resources, and also to make way for agriculture. **Deforestation**, the clearing and loss of forests, has altered the landscapes and ecosystems of much of our planet. It has caused soil degradation, population declines, species extinctions, and, as we saw with Easter Island (▸ pp. 8–9), has in some cases helped bring whole civilizations to ruin. Forest loss also adds carbon dioxide (CO_2) to the atmosphere: CO_2 is released when plant matter is burned or decomposed, and thereafter less vegetation remains to soak up CO_2. Deforestation thus is one contributor to global climate change (Chapter 12).

Today forests are being felled at the fastest rates in the tropical rainforests of Latin America and Africa (Figure 9.11). Developing countries in these regions are striving to expand areas of settlement for their burgeoning populations and to boost their economies by extracting natural resources and selling them abroad. Moreover, most people in these societies cut trees for fuelwood for their daily cooking and heating needs (▸ p. 332). In contrast, many areas of Europe and eastern North America are now slowly gaining forest as they recover from severe deforestation of the past.

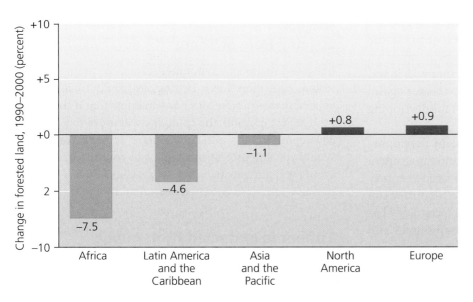

FIGURE 9.11 Nations in Africa and in Latin America are experiencing rapid deforestation as they attempt to develop, extract resources, and provide new agricultural land for their growing populations. In parts of Europe and North America, meanwhile, forested area is slowly increasing as some former farmed areas are abandoned and allowed to grow back into forest. Go to **GRAPHIt!** at www.aw-bc.com/ withgott. Data from United Nations Environmental Programme. 2002. *Global environmental outlook 3.* London: Earthscan Publications.

Deforestation for timber and farmland propelled the growth of the United States and Canada throughout their phenomenal expansion westward across the North American continent over the past 400 years. The vast deciduous forests of the East were virtually stripped of their trees by the mid-19th century, making way for countless small farms and supplying timber to build the cities of the Atlantic seaboard and the upper Midwest. As a farming economy shifted to an industrial one, wood was used to stoke the furnaces of industry. Logging operations then moved south, where vast pine and bottomland hardwood forests were logged and converted to pine plantations.

Once most mature trees were removed from these areas, timber companies moved west, cutting the continent's biggest trees in the Rocky Mountains, the Sierra Nevada, the Cascade Mountains, and the Pacific Coast ranges.

By the early 20th century, very little virgin timber was left in the lower 48 U.S. states (Figure 9.12). Today, the largest oaks and maples found in eastern North America, and even most redwoods of the California coast, are merely *second-growth* trees, trees that have sprouted and grown to partial maturity after old-growth timber has been cut.

The fortunes of loggers have risen and fallen with the availability of big trees. As each region was denuded of

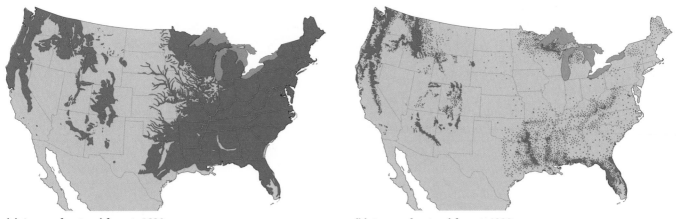

(a) Areas of natural forest, 1620

(b) Areas of natural forest, 1920

FIGURE 9.12 When Europeans first were colonizing North America, the entire eastern half of the continent and substantial portions of the western half were covered in forest (**a**). By the early 20th century, the vast majority of this forest had been cut, replaced mostly by agriculture and other human land uses (**b**). Since that time, most of the remaining original forest has been cut, but much of the landscape has also become reforested with second-growth forest. *Source:* Williams, M. 1989. *Americans and their forests.* Cambridge: Cambridge University Press. As adapted by Goudie, A., 2000. *The human impact.* Cambridge, MA: MIT Press.

timber, the industry declined and the timber companies moved on, while local loggers lost their jobs. If the remaining ancient trees of North America—most in British Columbia and Alaska—are cut, many U.S. and Canadian loggers will likely be out of jobs once again. Their employers will move on to nations of the developing world, as many already have.

Uncut tropical forests still remain in many developing countries, and nations such as Brazil and Indonesia are in the position the United States and Canada faced a century or two ago: having a vast forested frontier that they can develop for human use. Today's advanced technology has enabled these countries to exploit their resources and push back their frontiers even faster than occurred in North America.

Developing nations are often desperate enough for economic development that they impose few restrictions on logging. Often their timber is extracted by foreign multinational corporations, which pay fees to the developing nation's government for a *concession*, or right to extract the resource. In such cases, the foreign corporation has little or no incentive to manage forest resources sustainably. Many of the short-term economic benefits are reaped not by local residents but by the corporations that log the timber and export it elsewhere. Local people may or may not receive temporary employment from the corporation, but once the timber is harvested they no longer have the forest and the ecosystem services it once provided.

Weighing the Issues:
Logging Here or There

Imagine you are an environmental activist protesting a logging operation that is cutting old-growth trees near your hometown. Now let's say you know that if the protest is successful, the company will move to a developing country and cut its virgin timber instead. Would you still protest the logging in your hometown? Would you pursue any other approaches?

Timber is extracted from public and private lands

In the United States, the depletion of the eastern forests and widespread fear of a "timber famine" spurred the formation of a system of forest reserves: public lands set aside to grow trees, produce timber, protect watersheds, and serve as insurance against future scarcities of lumber. Today the U.S. **national forest** system consists of 77 million ha (191 million acres) managed by the U.S. Forest Service and covering over 8% of the nation's land area (Figure 9.13).

Most timber harvesting in the United States now takes place on private land, including land owned by timber companies (Figure 9.14). Private companies also extract timber from the publicly held national forests. U.S. Forest

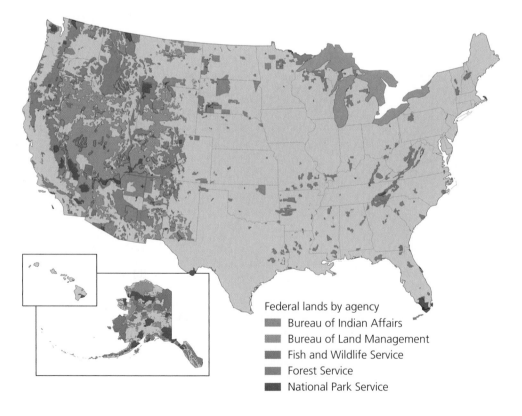

FIGURE 9.13 Federal agencies own and manage well over 250 million ha (600 million acres) of land in the United States, particularly in the western states. These include national forests, national parks, national wildlife refuges, Native American reservations, and Bureau of Land Management lands. *Source:* United States Geological Survey.

Federal lands by agency
- Bureau of Indian Affairs
- Bureau of Land Management
- Fish and Wildlife Service
- Forest Service
- National Park Service

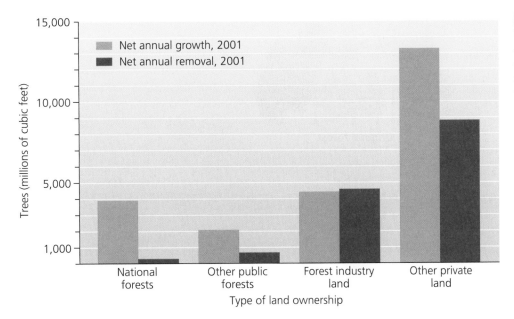

FIGURE 9.14 As the United States begins to recover from deforestation, trees (measured in cubic feet of wood biomass) are now growing at a faster rate than they are being removed. The exception is on land owned by the timber industry, where equivalent rates of growth and removal reflect attempts by timber companies to manage their resources in accordance with the maximum sustainable yield approach. On public lands, rates of growth and removal reflect not only economic forces but social and political ones as well. Go to **GRAPHit!** at www.aw-bc.com/withgott. Data are for 2001, from United States Forest Service, 2001.

Service employees plan and manage timber sales and build roads to provide access for logging companies, which sell the timber they harvest for profit. Timber extraction from U.S. national forests began to increase in the 1950s as the country experienced a postwar economic boom, consumption of paper products rose, and the population expanded into newly built suburban homes. During the 1980s and 1990s, harvests from public lands decreased as economic trends shifted, public concern over clear-cutting grew, and forest management philosophy evolved. By 2001, tree regrowth outpaced tree removal on national forests by nearly 12 to 1.

Timber is harvested by several methods

Most timber this century has been harvested by **clear-cutting**, in which all trees in an area are cut, leaving only stumps (Figure 9.15). Clear-cutting is the most cost-efficient harvest method in the short term, but it exerts the greatest impacts on forest ecosystems. Widespread clear-cutting occurred at a time when public awareness of environmental problems was blossoming. The combination produced public outrage toward the timber industry and public forest managers. Eventually the industry integrated other harvesting methods (Figure 9.16).

Today the North American timber industry focuses on production in the Northwest and the South from plantations of fast-growing tree species that are single-species monocultures (▶ p. 152). Because all trees in a given stand are planted at the same time, the stands are *even-aged*, with virtually all trees the same age (see

Figures 9.16a and 9.16b). Stands are cut after a certain number of years (called the *rotation time*), and the land is replanted with seedlings. Most ecologists and foresters view these plantations more as crop agriculture than as ecologically functional forests. Because there are few tree species and little variation in tree age, plantations do not offer many forest organisms the habitat they need. However, some harvesting methods (see Figure 9.16c) aim to maintain *uneven-aged* stands, where a mix of ages (and often a mix of tree species) makes the stand more similar to a natural forest.

FIGURE 9.15 Clear-cutting is the most cost-efficient method for timber companies, but it can have severe ecological consequences, including soil erosion, species loss, and community change. Although some species do use clear-cuts as they regrow, most people find these areas aesthetically unappealing, and public reaction to clear-cutting has driven changes in forestry methods.

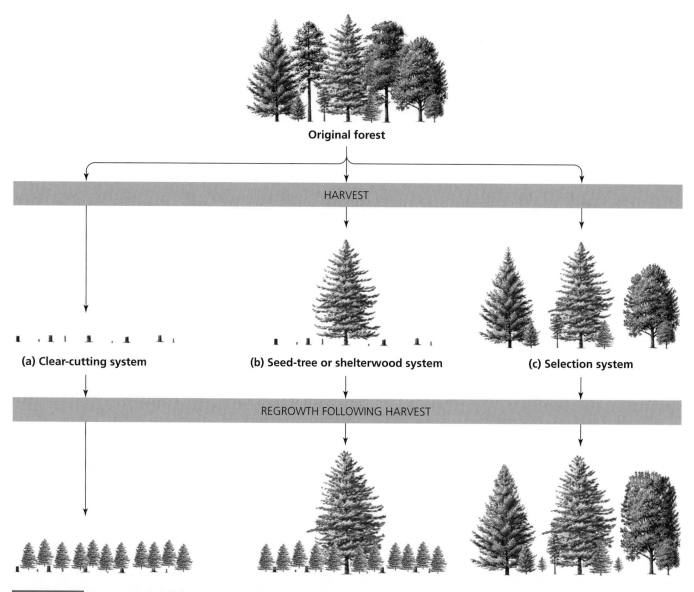

Original forest

HARVEST

(a) Clear-cutting system   **(b) Seed-tree or shelterwood system**   **(c) Selection system**

REGROWTH FOLLOWING HARVEST

FIGURE 9.16 Foresters and timber companies have devised various methods of harvesting timber from forests. In clear-cutting (**a**), all trees in an area are cut, extracting a great deal of timber inexpensively but leaving a vastly altered landscape. In seed-tree systems and shelterwood systems (**b**), small numbers of large trees are left in clear-cuts, to help reseed the area or provide shelter for growing seedlings. In selection systems (**c**), only some trees are removed at any one time, while most are left standing. These latter methods involve less environmental impact than clear-cutting, but all methods can cause significant changes to the structure and function of natural forest communities.

Public forests may be managed for recreation and ecosystems

All methods of logging result in habitat disturbance, which invariably affects the plants and animals inhabiting an area. All methods change forest structure and composition. Most methods increase soil erosion, leading to siltation of waterways, which can degrade habitat and affect drinking water quality. Most methods also speed runoff, sometimes causing flooding. In extreme cases, as when steep hillsides are clear-cut, landslides can result. In recent decades, increased awareness of these problems has prompted many citizens to protest the way public forests are managed in the United States and Canada. These citizens have urged that public forests be managed for recreation, wildlife, and ecosystem integrity, rather than for timber.

In 1976 the U.S. Congress passed the *National Forest Management Act*, which mandated that plans for renewable resource management be drawn up for every national forest. These plans were to be subject to broad public participation under the National Environmental Policy Act (▸ p. 38). Following passage of the National Forest Management Act, the U.S. Forest Service began developing new programs to manage wildlife, non-game animals, and endangered species. It pushed for ecosystem-based management and even ran programs of ecological restoration, attempting to recover whole plant and animal communities that had been lost or degraded. Timber harvesting methods were brought more in line with ecosystem-based management goals.

In 2004, however, the George W. Bush administration freed local forest managers from requirements imposed by the National Forest Management Act, granting them more flexibility in managing forests, but loosening environmental protections and restricting public oversight. Then in 2005, the Bush administration repealed the Clinton administration's roadless rule, by which 23.7 million ha (58.5 million acres)—31% of national forest land and 2% of total U.S. land—were in 2001 put off limits to further road construction or maintenance. Although the roadless rule had been supported by a record 4.2 million public comments, the Bush administration overturned it and required state governors to petition the federal government if they want to keep areas in their states roadless. The states of California, Oregon, and New Mexico responded by suing the federal government, asking that the roadless rule be reinstated.

Fire policy has also stirred controversy

Some ecosystem management efforts, ironically, run counter to the U.S. Forest Service's best-known symbol, Smokey Bear. The cartoon bear wearing a ranger's hat who urges us to fight forest fires is widely recognized, but unfortunately Smokey's message is badly outdated, and many scientists assert that it has done great harm to American forests.

For over a century, the Forest Service and other land management agencies suppressed fire whenever and wherever it broke out. Yet ecological research now clearly shows that many ecosystems depend on fire. Certain plants have seeds that germinate only in response to fire, and researchers studying tree rings have documented that many ecosystems historically experienced frequent fire. (Burn marks in a tree's rings reveal past fires, giving scientists an accurate history of fire events extending back hundreds or even thousands of years.) Ecosystems dependent on fire are adversely affected by its suppression; pine woodlands become cluttered with hardwood understory, and animal diversity and abundance decline.

In the long term, fire suppression can lead to catastrophic fires that truly do damage forests, destroy human property, and threaten human lives. Fire suppression allows limbs, logs, sticks, and leaf litter to accumulate on the forest floor, producing kindling for a catastrophic fire.

To reduce fuel load and improve the health and safety of forests, the Forest Service and other agencies have in recent years been burning areas of forest under carefully controlled conditions. These **prescribed burns** have worked effectively, but all too often, these efforts have been impeded by public misunderstanding and by interference from politicians who have not taken time to understand the science behind the approach.

In the wake of the southern California fires of 2003, the U.S. Congress, intending to make forests less fire-prone, passed the Bush administration's Healthy Forests Restoration Act. Although this legislation encourages some prescribed burning, it primarily promotes the physical removal of small trees, underbrush, and dead trees by timber companies. The removal of dead trees, or snags, following a natural disturbance is called **salvage logging**. From an economic standpoint, salvage logging may seem to make good sense. However, ecologically, snags have immense value; the insects that decay them provide food for wildlife, and many birds, mammals, and reptiles depend on holes in snags for nesting and roosting sites. Removing timber from recently burned land can also cause severe erosion and soil damage. In fact, a 2006 study showed that salvage logging at the site of Oregon's Biscuit Fire decreased the number of seedlings that resprouted and also created debris that could fuel further fires, thus likely hindering the recovery of the forest.

Sustainable forestry is gaining ground

Any company can claim that its timber harvesting practices are sustainable, but how is the purchaser of wood products to know whether they really are? In the last several years, a consumer movement has grown that is making informed consumer choice possible. The Forest Stewardship Council and several other organizations now examine the practices of timber companies and certify products produced using methods they consider sustainable. Consumers can look for the logos of these organizations on forest products they purchase. Consumer demand for sustainable wood has been great enough that Home Depot and other major retail businesses have begun selling sustainable wood. Sustainable forestry is more costly for the timber industry, but if certification standards can be kept adequately strong, then consumer choice in the marketplace can be a powerful driver for good forestry practices for the future.

Our need for sustainable forestry and resource management is growing. As resources dwindle, as forests and soils are degraded, and as the landscape fills with more people, the arguments for conservation of resources—for their sustainable use—have grown stronger. Also growing stronger is the argument for preservation of land—setting aside tracts of relatively undisturbed land intended to remain forever undeveloped.

Parks and Reserves

Preservation has been part of the American psyche ever since John Muir rallied support for saving scenic lands in the Sierras (▶ p. 18). For ethical reasons as well as pragmatic ecological and economic ones, U.S. citizens and many other people worldwide have chosen to set aside tracts of land in perpetuity to be preserved and protected from development.

Why have we created parks and reserves?

The historian Alfred Runte has cited four traditional reasons that parks and protected areas have been established:

1. Enormous, beautiful, or unusual features such as the Grand Canyon, Mount Rainier, or Yosemite Valley inspire people to protect them—an impulse termed *monumentalism* (Figure 9.17).
2. Protected areas offer recreational value to tourists, hikers, fishers, hunters, and others.
3. Protected areas offer utilitarian benefits. For example, undeveloped watersheds provide cities with clean drinking water and a buffer against floods.
4. Parks make use of sites that lack economically valuable material resources or that are hard to develop; land with little monetary value is easy to set aside.

To these four traditional reasons, we have added a fifth in recent years: the preservation of biodiversity. As we saw in Chapter 8, human impact alters habitats and has led to countless population declines and species extinctions. A park or reserve is widely viewed as a kind of Noah's Ark, an island of habitat that can, scientists hope, maintain species that might otherwise disappear.

Public parks and reserves began in the United States

The striking scenery of the American West impelled the U.S. government to create the world's first **national parks**, publicly held lands protected from resource extraction and development but open to nature appreciation and various forms of recreation. In 1872, Yellowstone National Park was established as "a public park or pleasuring-ground for the benefit and enjoyment of the

FIGURE 9.17 The awe-inspiring beauty of some regions of the western United States was one reason for the establishment of national parks. Images of scenic vistas such as this one of Bridal Veil Falls in Yosemite National Park, portrayed by the landscape painter Albert Bierstadt, have inspired millions of people from North America and abroad to visit these parks.

people." Yosemite, General Grant, Sequoia, and Mount Rainier National Parks followed after 1890.

The National Park Service (NPS) was created in 1916 to administer the growing system of parks and monuments, which today numbers 388 sites totaling 32 million ha (79 million acres) and includes national historic sites, national recreation areas, national wild and scenic rivers, and other types of areas (see Figure 9.13). This most widely used park system in the world received 277 million reported recreation visits in 2004—about as many visits as there are U.S. residents.

Another type of federal protected area in the United States is the **national wildlife refuge**. The system of national wildlife refuges, begun in 1903, now totals 37 million ha (91 million acres) spread over 541 sites (see Figure 9.13). The U.S. Fish and Wildlife Service (FWS) administers the refuges, which serve not only as havens for wildlife, but also in many cases encourage hunting, fishing, wildlife observation, photography, environmental education, and other public uses.

In response to the public's desire for undeveloped areas of land, in 1964 the U.S. Congress passed the Wilderness Act, which allowed some areas within existing federal lands to be designated as **wilderness areas**. These areas are off-limits to private development but are open to hiking, nature study, and other low-impact public recreation. Congress declared that wilderness areas were necessary "to assure that an increasing population, accompanied by expanding settlement and growing mechanization, does not occupy and modify all areas within the United States and its possessions, leaving no lands designated for preservation and protection in their natural condition."

Not everyone supports land set-asides

The restriction of activities in wilderness areas has helped generate opposition to U.S. land protection policies. Sources of such opposition include the governments of some western states, where large portions of land are federally owned. When those states came into existence, the federal government retained ownership of much of the acreage inside their borders. Idaho, Oregon, and Utah own less than 50% of the land within their borders, and in Nevada 80% of the land is federally owned. Western state governments have traditionally sought to obtain land from the federal government and encourage resource extraction and development on it.

The drive to extract more resources, secure local control of lands, and expand recreational access to public lands is epitomized by the **wise-use movement**, a loose confederation of individuals and groups that coalesced in the 1980s and 1990s in response to the increasing success of environmental advocacy. Wise-use advocates are dedicated to protecting private property rights; opposing government regulation; transferring federal lands to state, local, or private hands; and promoting more motorized recreation on public lands. Wise-use advocates include farmers, ranchers, trappers, and mineral prospectors at the grassroots level who live off the land, as well as groups representing the large corporations of industries that extract timber, mineral, and fossil fuel resources. Under the Bush administration, wilderness protection policies have been weakened, and administrative agencies have generally shifted policies and enforcement away from preservation and conservation, and toward recreation and extractive uses.

Nonfederal entities also protect land

Efforts to set aside land—and the debates over such efforts—at the federal level are paralleled at regional and local levels. Each U.S. state and Canadian province has agencies that manage resources on state or provincial lands, as do many counties and municipalities.

Private nonprofit groups also preserve land. **Land trusts** are local or regional organizations that purchase land with the aim of preserving it in its natural condition. The Nature Conservancy is the world's largest land trust, but over 900 local and regional land trusts in the United States together own 177,000 ha (437,000 acres) and have helped preserve an additional 930,000 ha (2.3 million acres), including scenic areas such as Big Sur on the California coast, Jackson Hole in Wyoming, and Maine's Mount Desert Island.

Parks and reserves are increasing internationally

Many nations have established national park systems and are benefiting from ecotourism as a result—from Costa Rica (Chapter 4) to Ecuador to Thailand to Tanzania. The total worldwide area in protected parks and reserves increased more than fourfold from 1970 to 2000, and in 2003 the world's 38,536 protected areas covered 1.3 billion ha (3.2 billion acres), or 9.6% of the planet's land area. However, parks in developing countries do not always receive the funding they need to manage resources, provide for recreation, and protect wildlife from poaching and timber from logging. Thus many of the world's protected areas are merely "paper parks"—protected on paper but not in reality.

Some types of protected areas are designated or partly managed internationally by the United Nations. *World heritage sites* are an example; currently over 560 sites across 125 countries are listed for their cultural value and nearly 150 for their natural value. *Biosphere reserves*, another type of internationally protected area, are tracts of land with exceptional biodiversity that couple preservation with sustainable development to benefit local people. (Figure 9.18).

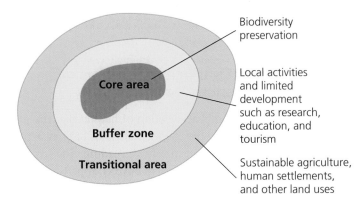

FIGURE 9.18 Biosphere reserves are international efforts that couple preservation with sustainable development to benefit local residents. Each reserve includes a core area that preserves biodiversity, a buffer zone that allows limited development, and a transition zone that permits various uses.

(a) Clear-cuts in Mount Hood National Forest, Oregon

(b) Wood thrush

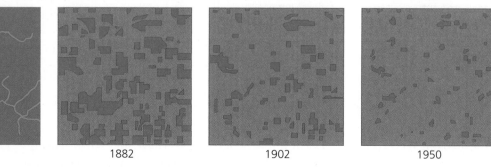

1831   1882   1902   1950

(c) Fragmentation of wooded area (green) in Cadiz Township, Wisconsin

FIGURE 9.19 As human populations have grown and human impacts have increased, most large expanses of natural habitat have become fragmented into smaller disconnected areas. Forest fragmentation from timber harvesting, for example, is evident on the Mount Hood National Forest, Oregon, where clearcuts show up as snowy patches in this photo **(a)**. Fragmentation has significant impacts on forest-dwelling species such as the wood thrush, *Hylocichla mustelina* **(b)**, a migrant songbird of eastern North America. In forest fragments, wood thrush nests are parasitized by cowbirds that thrive in open country and edge habitats. Forest fragmentation has been extreme in the eastern and midwestern United States; shown in **(c)** are historical changes in forested area in Cadiz Township, Wisconsin, between 1831 and 1950. *Source for (c):* Curtis, J. T. 1956. The modification of mid-latitude grasslands and forests by man. In Thomas, W. L. Jr., ed., *Man's role in changing the face of the earth*. Chicago: Univ. of Chicago Press.

The design of parks and reserves has consequences for biodiversity

Often it is not outright destruction of habitat that threatens species, but rather the fragmentation of habitat (▶ pp. 182–183). Expanding agriculture, spreading cities, highways, logging, and other impacts have chopped up large contiguous expanses of habitat into small disconnected ones (Figure 9.19a and 9.19c). When this happens, many species suffer. Bears, mountain lions, and other animals that need large areas may disappear. Bird species that thrive in the interior of forests may fail to reproduce when forced near the edge of a fragment (Figure 9.19b). Their nests often are attacked by predators and parasites that favor open habitats surrounding the fragment or that travel along habitat edges. Avian ecologists judge forest fragmentation to be a main reason why populations of many songbirds of eastern North America are declining.

Because habitat fragmentation is such a central issue in biodiversity conservation, and because there are limits on how much land can be set aside, conservation biologists have argued heatedly about whether it is better to make reserves large in size and few in number, or many in number but small in size. Nicknamed the **SLOSS dilemma**, for "**s**ingle **l**arge **o**r **s**everal **s**mall," this debate is ongoing and complex, but it seems clear that large species that roam great distances, such as the Siberian tiger (Chapter 8), benefit more from the "single large" approach to reserve design. In contrast, creatures such as insects that live as larvae in small areas may do just fine in a number of small isolated reserves, if they can disperse as adults by flying from one reserve to another.

A related issue is whether **corridors** of protected land are important for allowing animals to travel between islands of protected habitat. In theory, connections between fragments provide animals access to more habitat and help enable gene flow to maintain populations in the long term. Many land management agencies and environmental groups try, when possible, to join new reserves to existing reserves for these reasons. It is clear that we will need to think on the landscape level if we are to preserve a great deal of our natural heritage.

Conclusion

As half the human population has shifted from rural to urban lifestyles, the nature of our impact on the environment has changed. As urban and suburban dwellers, our impacts are less direct but more far-reaching. Resources must be delivered to us over long distances, requiring the use of still more resources. Limiting the waste of those resources by making our urban areas more sustainable will be vital for our future.

Key components for creating more livable and sustainable cities include land use planning, expanding transportation options, and ensuring access to parklands. Proponents of smart growth and the new urbanism believe they have solutions to the challenges posed by urban and suburban sprawl, although free-market theorists maintain that if people desire suburban lifestyles, governments should not stand in the way of sprawl. Continuing experimentation will help us determine how best to ensure that urban growth improves our quality of life and does not degrade the quality of our environment.

On the lands beyond our cities, prudent resource management will be essential. In forest management, early emphasis on resource extraction evolved into policies on sustainable yield and eventually more holistic strategies for ecosystem-based management—shifts that occurred as land and resource availability declined and as the public became more aware of environmental degradation. Meanwhile, public support for preservation of natural lands has resulted in parks, wilderness areas, and other reserves, in North America and abroad.

TESTING YOUR COMPREHENSION

1. What factors lie behind the shift of population from rural areas to urban areas? What types of cities and countries are experiencing the fastest urban growth today, and why?
2. Give two definitions of *sprawl*. Describe five negative impacts that critics suggest result from sprawl.
3. What are city planning and regional planning? Contrast planning with zoning.
4. Describe several key elements of "smart growth." What effects, positive and negative, do urban growth boundaries tend to have? What are the roles of mass transit and of public parks?
5. Describe some ways in which urban centers can exert beneficial environmental effects.
6. Compare and contrast maximum sustainable yield, adaptive management, and ecosystem management. Why may pursuing maximum sustainable yield sometimes conflict with what is ecologically desirable?
7. Name several major causes of deforestation. Describe some consequences of forest loss. Where is deforestation most severe today, and why?
8. Compare and contrast the major methods of timber harvesting. What are some ecological effects of logging?
9. Name five reasons that parks and reserves have been created. How do wilderness areas differ from national parks and national wildlife refuges?
10. Roughly what percentage of Earth's land is protected in parks or reserves? What types of protected areas have been established in countries outside the United States?

SEEKING SOLUTIONS

1. Evaluate the causes of the spread of suburbs and of the environmental, social, and economic impacts of sprawl. Overall, do you think the spread of urban and suburban development that many people label *sprawl* is predominantly a good thing or a bad thing? Do you think it is inevitable? Give reasons for your answers.

2. Would you personally want to live in a neighborhood developed in the style of the new urbanism? Would you like to live in a city or region with an urban growth boundary? Why or why not?

3. Environmentalists in developed countries are fond of warning people in developing countries to stop destroying rainforest. Leaders of developing countries often respond that this is hypocritical, because the developed nations became wealthy by deforesting their land and exploiting its resources in the past. What would you say to the president of a developing nation, such as Brazil or Indonesia, that is clearing its forest in pursuit of economic development?

4. A century of fire suppression has left millions of hectares of forested lands in North America in danger of catastrophic wildfires. Yet we will probably never have adequate resources to conduct careful prescribed burning over all these lands. Can you suggest any solutions that might help protect people's homes near forests while improving the ecological condition of some forested lands?

5. Which is more ecologically sustainable: living in a high-rise apartment in a big city, or living on a 40-acre ranch abutting a national forest? Why?

CALCULATING ECOLOGICAL FOOTPRINTS

 We all depend on forest resources. How much paper do you think you use? The average North American uses over 300 kg (660 lb) of paper and paperboard per year. Using the estimates of paper and paperboard consumption and of population for each region of the world for the year 2000, calculate the per capita consumption of paper and paperboard for each region.

	Population (millions)*	Total paper consumed in 2000 (millions of tons)	Per capita paper consumed in 2000 (lbs)
Africa	840	6	14
Asia	3,766	115	
Europe	728	99	
Latin America	531	21	
North America	319	110	
Oceania	32	5	
World	6,216		~114

Data source: Population Reference Bureau.

1. How much paper would North Americans save each year if we consumed paper at the rate of Europeans?

2. How much paper would be consumed if everyone in the world used as much paper as the average European? As the average North American?

3. Why do you think people in other regions consume less paper, per capita, than North Americans? Name three things you could do to reduce your paper consumption.

Take It Further

Go to www.aw-bc.com/withgott, where you'll find:

▶ Suggested answers to end-of-chapter questions

▶ Quizzes, animations, and flashcards to help you study

▶ *Research Navigator*™ database of credible and reliable sources to assist you with your research projects

▶ **GRAPH it!** tutorials to help you interpret graphs

▶ **INVESTIGATE it!** current news articles that link the topics that you study to case studies from your region to around the world

Environmental Health and Toxicology

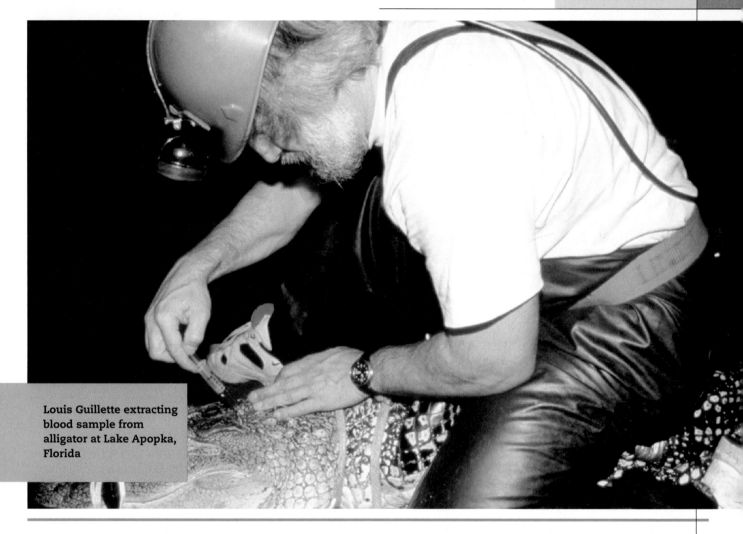

Louis Guillette extracting blood sample from alligator at Lake Apopka, Florida

Upon successfully completing this chapter, you will be able to:

▶ Identify the major types of environmental health hazards and explain the goals of environmental health

▶ Describe the types, abundance, distribution, and movement of synthetic and natural toxicants in the environment

▶ Discuss the study of hazards and their effects, including epidemiology, animal testing, and dose-response analysis

▶ Assess risk assessment and risk management

▶ Compare philosophical approaches to risk, and how these relate to policy

Central Case: Alligators and Endocrine Disruptors at Lake Apopka, Florida

"Over the past 50 years we have all been unwitting participants in a vast, uncontrolled, worldwide chemistry experiment involving the oceans, air, soils, plants, animals, and human beings."
—UNITED NATIONS ENVIRONMENT PROGRAMME, IN A GUIDE TO THE STOCKHOLM CONVENTION ON PERSISTENT ORGANIC POLLUTANTS

"Every man in this room is half the man his grandfather was."
REPRODUCTIVE BIOLOGIST LOUIS GUILLETTE TO A U.S. CONGRESSIONAL COMMITTEE IN 1995, REGARDING DECLINING SPERM COUNTS IN MEN WORLDWIDE OVER THE PREVIOUS HALF CENTURY

When biologist Louis Guillette began studying the reproductive biology of alligators in Florida lakes in 1985, he soon discovered that not all was well. Alligators from one lake in particular, Lake Apopka northwest of Orlando, showed a number of bizarre reproductive problems.

Females were having trouble producing viable eggs. Male hatchling alligators had severely depressed levels of the male sex hormone testosterone, and female hatchlings showed greatly elevated levels of the female sex hormone estrogen. The young animals had abnormal gonads, too. Testes of the males produced sperm at a premature age, and ovaries of the females contained multiple eggs per follicle instead of the expected one egg per follicle.

As Guillette and his co-workers tested and rejected hypotheses for the reproductive abnormalities one by one, they grew to suspect that environmental contaminants might be responsible. Lake Apopka had suffered a major spill of the pesticides dicofol and DDT in 1980, and yearly surveys thereafter showed a precipitous decline in the number of juvenile alligators in the lake. In addition, the lake received high levels of chemical runoff from agriculture and was experiencing eutrophication (▸ pp. 52 and 252–253) from nutrient input from fertilizers.

The puzzle pieces fell together when a colleague of Guillette's told him that recent research was showing that some environmental contaminants mimic hormones and interfere with the functioning of animal endocrine (hormone) systems. At very low doses, these endocrine disruptors can affect animals, mimicking estrogen and feminizing males. Guillette realized that his alligator observations were similar to results from tests in which rodents were exposed to estrogen during embryonic development. He hypothesized that certain chemical contaminants in Lake Apopka were disrupting the endocrine systems of alligators during development in the egg.

To test his hypothesis, Guillette and his co-workers compared alligators from heavily polluted Lake Apopka with those from cleaner lakes nearby. They found that Lake Apopka alligators had abnormally low hatching rates in the years after the pesticide spill and, even as hatching rates recovered in the 1990s, continued to show aberrant hormone levels and bizarre gonad abnormalities. In addition, the penises of Lake Apopka male alligators were 25% smaller than those of male alligators from surrounding lakes.

Similar problems began cropping up in other lakes that experienced runoff of chemical pesticides. In the lab, researchers found that several contaminants detected in alligator eggs and young could bind to receptors for estrogen and exist in concentrations great enough to cause sex reversal of male embryos. One chemical in particular, atrazine—the most widely used herbicide in the United States—appeared to disrupt hormones by inducing production of aromatase, an enzyme that converts testosterone to estrogen. In 2003, Guillette reported that nitrate from fertilizer runoff may also act as an endocrine disruptor; when nitrate concentrations in lakes were above the standard for drinking water, juvenile male alligators had smaller penises and 50% lower testosterone levels.

Guillette's results have raised concern for human health. Because alligators and humans share many of the same hormones, many scientists suspect that chemical contaminants could be affecting people, just as they have affected alligators.

Environmental Health

Examining impacts of human-made chemicals on wildlife and people comprises one aspect of the broad field of environmental health. The study and practice of **environmental health** assesses environmental factors that influence human health and quality of life, and seeks to prevent adverse effects on human health and ecological systems.

Environmental hazards can be physical, chemical, biological, or cultural

For each type of environmental health threat, or hazard, in the world around us, there is some amount of risk that we cannot avoid—but there is also some amount of risk that we *can* avoid by taking precautions. Much of environmental health consists of taking steps to minimize the impacts of hazards and the risks of encountering them.

Physical hazards *Physical* processes that occur naturally in our environment and pose health hazards include discrete events such as earthquakes, volcanic eruptions, fires, floods, blizzards, landslides, hurricanes, and droughts. They also include ongoing natural phenomena, such as ultraviolet (UV) radiation from sunlight (Figure 10.1a), which at excessive exposure damages DNA and

has been tied to skin cancer, cataracts, and immune suppression. We can do little to predict the timing of a natural disaster, and nothing to prevent one. However, we can take steps to reduce our vulnerability to them. For instance, deforesting slopes makes landslides more likely, and diking rivers may make flooding more likely in some areas. We can reduce risk from such hazards by improving our forestry and flood control practices and by choosing not to build in areas prone to floods, landslides, fires, and coastal waves. For hazards such as exposure to UV light, we can reduce our exposure and risk by using clothing and sunscreen to shield our skin from intense sunlight.

Chemical hazards *Chemical* hazards include many of the synthetic chemicals that our society produces, such as disinfectants, pesticides (Figure 10.1b), and the compounds that contributed to reproductive problems for the alligators at Lake Apopka. Chemicals produced naturally by organisms also can be hazardous. Following our overview of environmental health, much of our chapter

(a) Physical hazard

(b) Chemical hazard

(c) Biological hazard

(d) Cultural hazard

FIGURE 10.1 Environmental health hazards can be divided into four types. The sun's ultraviolet radiation is an example of a physical hazard (**a**). Excessive exposure increases the risk of skin cancer. Chemical hazards (**b**) include both artificial and natural chemicals. Much of our exposure comes from household chemical products, such as pesticides that some people apply to their lawns. Biological hazards (**c**) include other organisms that transmit disease. Some mosquitoes are vectors for certain pathogenic microbes, including those that cause malaria. Cultural or lifestyle hazards (**d**) include the decisions we make about how to behave, as well as the constraints forced on us by socioeconomic factors. Smoking is a lifestyle choice that raises one's risk of lung cancer and other disease considerably.

will focus on chemical health hazards and the ways people study and regulate them.

Biological hazards *Biological* hazards result from ecological interactions among organisms (Figure 10.1c). When we become sick from a virus, bacterial infection, or other pathogen, we are suffering parasitism by other species that are simply fulfilling their ecological roles. This is **infectious disease**, also called *communicable* or *transmissible disease*. Infectious diseases such as malaria, cholera, tuberculosis, and influenza (flu) all are considered environmental health hazards. As with physical and chemical hazards, it is impossible for us to avoid risk from biological agents completely, but we can take steps to reduce the likelihood of infection.

Cultural hazards Hazards that result from the place we live, our socioeconomic status, our occupation, or our behavioral choices can be thought of as *cultural* or *lifestyle* hazards. For instance, choosing to smoke cigarettes, or living or working with people who do, can greatly increase our risk of lung cancer (Figure 10.1d). Choosing to smoke is a personal behavioral decision, but exposure to secondhand smoke in the home or workplace may not be under one's control. Much the same might be said for

drug use, diet and nutrition, crime, and mode of transportation. As advocates of environmental justice (▸ p. 19) argue, such health factors as living in proximity to toxic waste sites or working unprotected with pesticides are often correlated with socioeconomic deprivation.

Disease is a major focus of environmental health

Among the hazards people face, disease stands preeminent. Despite all our technological advances, we still find ourselves battling disease, which causes the vast majority of human deaths worldwide (Figure 10.2a). Many major killers such as cancer, heart disease, and respiratory disorders have genetic bases but are also influenced by environmental factors. For instance, whether a person develops asthma is influenced not only by the genes, but also by environmental conditions. Pollutants from fossil fuel combustion worsen asthma, and children raised on farms suffer less asthma than children raised in cities, studies have shown. Malnutrition (▸ p. 151) can foster a wide variety of illnesses, as can poverty and poor hygiene. Moreover, lifestyle choices can affect risks of acquiring some noninfectious diseases: Smoking can lead to lung cancer, and lack of exercise to heart disease, for example.

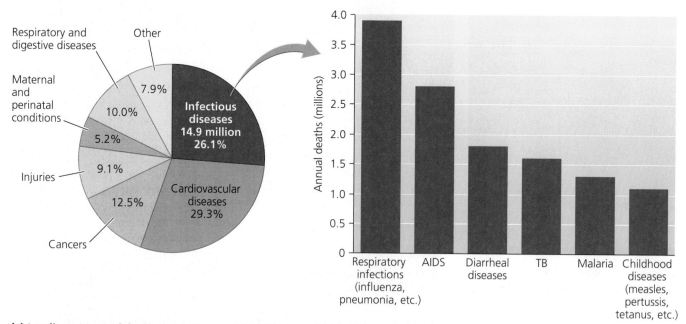

(a) Leading causes of death across the world, 2004

(b) Leading causes of death by infectious disease, 2004

FIGURE 10.2 Infectious diseases are the second-leading cause of death worldwide, accounting for over one-quarter of all deaths per year (**a**). Six types of diseases—respiratory infections, AIDS, diarrhea, tuberculosis (TB), malaria, and childhood diseases such as measles—account for 80% of all deaths from infectious disease (**b**). Data from World Health Organization, 2004.

Infectious diseases account for 26% of deaths that occur worldwide each year—nearly 15 million people (Figure 10.2b). Some pathogenic microbes attack us directly, and sometimes infection occurs through a *vector*, an organism that transfers the pathogen to the host. Infectious diseases account for close to half of all deaths in developing countries, but for very few deaths in developed nations. This discrepancy is due to differences in hygiene conditions and access to medicine, which are tightly correlated with wealth. Public health efforts have lessened the impact of infectious disease in developed nations and even have eradicated some diseases. Other diseases, among them tuberculosis, acquired immunodeficiency syndrome (AIDS), and West Nile virus, are increasing. Still others, such as the avian flu that began spreading worldwide in 2005–2006, remain as threats for a possible global epidemic.

Environmental health hazards exist indoors as well as outdoors

Outdoor hazards are generally more familiar to us, but we spend most of our lives indoors. Therefore, we must consider the spaces inside our homes and workplaces to be part of our environment and, as such, the sources of potential environmental hazards (Table 10.1). Radon is

Table 10.1 Selected Environmental Hazards

Air

▶ Smoking and secondhand smoke

▶ Chemicals from automotive exhaust

▶ Chemicals from industrial pollution

▶ Tropospheric ozone

▶ Pesticide drift

▶ Dust and particulate matter

Water

▶ Pesticide and herbicide runoff

▶ Nitrates and fertilizer runoff

▶ Mercury, arsenic, and other heavy metals in groundwater and surface water

Food

▶ Natural toxins

▶ Pesticide and herbicide residues

Indoors

▶ Asbestos

▶ Lead in paint and pipes

▶ Toxicants in plastics and consumer products

one indoor hazard (▶ p. 286). *Radon* is a highly toxic radioactive gas that is colorless and undetectable without specialized kits. Radon seeps up from the ground in areas with certain types of bedrock and can build up inside basements and homes with poor air circulation. The U.S. Environmental Protection Agency (EPA) estimates that slightly less than 1 person in 1,000 may contract lung cancer as a result of a lifetime of radon exposure at average levels for U.S. homes.

Lead poisoning represents another indoor health hazard. When ingested, lead, a heavy metal, can cause damage to the brain, liver, kidney, and stomach; learning problems and behavioral abnormalities; anemia; hearing loss; and even death. Lead poisoning can result from drinking water that has passed through the lead pipes common in older houses. Even in newer pipes, lead solder was widely used into the 1980s and is still sold in stores. Until 1978, most paints contained lead, and interiors in most houses were painted with lead-based paint. Babies and young children often take an interest in peeling paint from walls and may ingest or inhale some of it. Today lead poisoning is thought to affect 3–4 million young children, fully one in six children under age 6.

Another indoor hazard is *asbestos*. There are several types of asbestos, each a mineral that forms long, thin microscopic fibers. This structure allows asbestos to insulate buildings for heat, muffle sound, and resist fire. When inhaled and lodged in lung tissue, asbestos induces the body to produce acid to combat it. The acid scars the lung tissue but does little to dislodge or dissolve the asbestos. Consequently, within a few decades the scarred lungs may cease to function, a disorder called *asbestosis*. Asbestos can also cause types of lung cancer. Because of these risks, asbestos has been removed from many schools and offices, although the dangers of exposure from asbestos removal may sometimes exceed the dangers of leaving it in place.

One recently recognized hazard is a group of chemicals known as polybrominated diphenyl ethers (PBDEs), fire-retardant compounds used in a diverse array of consumer products, including computers, televisions, plastics, and furniture. PBDEs appear to be released during production and disposal of products and also to evaporate at very slow rates throughout the lifetime of products. These chemicals persist and accumulate in living tissue, and their abundance in the environment and in people in the United States is doubling every few years. Lab testing with animals shows that PBDEs affect thyroid hormones and may cause cancer and affect brain and nervous system development. The European Union decided in 2003 to ban PBDEs, but in the United States there has so far been little movement to address the issue.

Toxicology is the study of poisonous substances

Studying the health effects of chemical agents suspected to be harmful, such as PBDEs, is the focus of the field of **toxicology**, the science that examines the effects of poisonous substances on humans and other organisms. Toxicologists assess and compare substances to determine their *toxicity*, the degree of harm a chemical substance can inflict. Chemical hazards differ in their capacity to endanger us, but any chemical substance may exert negative effects if it is ingested in great enough quantities or if exposure is extensive enough. Conversely, a toxic agent, or **toxicant**, in a minute enough quantity may pose no health risk at all. These facts are often summarized in the catchphrase, "The dose makes the poison." In other words, a substance's toxicity depends not only on its chemical identity, but also on its quantity.

During the past century, our ability to produce new chemicals has expanded, concentrations of chemical contaminants in the environment have increased, and public concern for health and the environment has grown. These trends have driven the rise of **environmental toxicology**, which deals specifically with toxic substances that come from or are discharged into our environment. Environmental toxicology includes the study of health effects on humans, other animals, and ecosystems, and it represents one approach within the broader scope of environmental health.

Toxicologists generally focus on human health, using other organisms as models and test subjects. In environmental toxicology, animals are also studied out of concern for their welfare and because—like canaries in a coal mine—animals can serve as indicators of health threats that could soon affect humans. For example, reproductive abnormalities like those seen in Lake Apopka's alligators have also been found in some Taiwanese boys. Studies showed that these boys were born to mothers who used cooking oil contaminated with toxicants called polychlorinated biphenyls (PCBs), which are by-products of chemicals used in transformers and other electrical equipment.

As we review the sometimes poisonous effects of human-made chemicals throughout this chapter, it is important to keep in mind that artificially produced chemicals have played a crucial role in giving us the standard of living we enjoy today. These chemicals have helped create the industrial agriculture that produces our food, the medical advances that protect our health and prolong our lives, and many of the modern materials and conveniences we use every day. It is appropriate to remember these benefits as we examine some of the unfortunate side effects of these advances and as we search for better alternatives.

Toxic Agents in the Environment

The environment contains countless natural chemical substances that may pose health risks. These substances include oil oozing naturally from the ground; radon gas seeping up from bedrock; and toxic chemicals stored or manufactured in the tissues of living organisms—for example, *toxins* that plants use to ward off herbivores or that insects use to deter predators. In addition, we are exposed to many synthetic (artificial, or human-made) chemicals.

Synthetic chemicals are ubiquitous in our environment

Synthetic chemicals are all around us in our daily lives. Tens of thousands have been manufactured, and many have found their way into soil, air, and water. A 2002 study by the U.S. Geological Survey found that 80% of U.S. streams contain at least trace amounts of 82 wastewater contaminants, including antibiotics, detergents, drugs, steroids, plasticizers, disinfectants, solvents, perfumes, and other substances. The pesticides we use to kill insects and weeds on farms, lawns, and golf courses are some of our most widespread synthetic chemicals (▶ pp. 153–154). As a result of all this exposure, every one of us carries traces of numerous industrial chemicals in our bodies.

This should not *necessarily* be cause for alarm. Not all synthetic chemicals pose health risks, and relatively few are known with certainty to be toxicants. However, of the roughly 100,000 synthetic chemicals on the market today, very few have been thoroughly tested for harmful effects. For the vast majority, we simply do not know what effects, if any, they may have.

Why are there so many synthetic chemicals around us? Let's consider pesticides and herbicides, made widespread by advances in chemistry and production capacity during and following World War II. As material prosperity grew in Westernized nations following the war, people began using pesticides for agriculture, to improve the look of lawns and golf courses, and to kill insects inside their homes and offices. Pesticides were viewed as means toward a better quality of life.

It was not until the 1960s that people began to learn about the risks of exposure to pesticides. The key event was the publication of Rachel Carson's 1962 book *Silent Spring*

(▶ pp. 36–38), which brought the pesticide dichloro-diphenyl-trichloroethane (DDT) to the public's attention.

Silent Spring began the public debate over synthetic chemicals

Rachel Carson was a naturalist, author, and government scientist. In *Silent Spring*, she brought together a diverse collection of scientific studies, medical case histories, and other data that no one had previously synthesized and presented to the general public. Her message was that DDT in particular and artificial pesticides in general were hazardous to people's health, the health of wildlife, and the well-being of ecosystems. Carson wrote at a time when large amounts of pesticides virtually untested for health effects were indiscriminately sprayed over residential neighborhoods and public areas, on the assumption that the chemicals would do no harm to people (Figure 10.3). Most consumers had no idea that the store-bought chemicals they used in their houses, gardens, and crops might be toxic.

Although challenged vigorously by spokespeople for the chemical industry, who attempted to discredit both the author's science and her personal reputation, Carson's book was a best-seller. Carson suffered from cancer as she finished *Silent Spring*, and she lived only briefly after its publication. However, the book helped generate significant social change in views and actions toward the environment. The use of DDT was banned in the United States in 1973 and has been banned in a number of other nations. The United States still manufactures and exports DDT to countries that do use it, however. Many developing countries use DDT to control human disease vectors, such as mosquitoes that transmit malaria, which in these countries may represent a greater health threat than the toxic effects of the pesticide.

Weighing the Issues:
A Circle of Poison?

It has been called the "circle of poison." Although the United States has banned the use of DDT, U.S. companies still manufacture and export the compound to many developing nations. Thus, it is possible that pesticide-laden food can be imported back into the United States. How do you feel about this? Is it unethical for one country to sell to others a substance that it has deemed toxic? Are there factors or circumstances that might change the view you take?

Toxicants come in several different types

Toxicants can be classified based on their particular effects on health. The best-known are **carcinogens**, which are chemicals or types of radiation that cause cancer. In cancer, malignant cells grow uncontrollably, creating tumors, damaging the body's functioning, and often leading to death. In our society today, the greatest number of cancer cases is thought to result from carcinogens contained in cigarette smoke. Carcinogens can be difficult to identify because there may be a long lag time between exposure to the agent and the detectable onset of cancer.

Mutagens are chemicals that cause mutations in the DNA of organisms (▶ p. 57). Although most mutations have little or no effect, some can lead to severe problems, including cancer and other disorders. If mutations occur in an individual's sperm or egg cells, then the individual's offspring suffer the effects.

Chemicals that cause harm to the unborn are called **teratogens**. Teratogens that affect the development of

FIGURE 10.3 Before the 1960s, the environmental and health effects of potent pesticides such as DDT were not widely studied or publicly known. Public areas such as parks, neighborhoods, and beaches were regularly sprayed for insect control without safeguards against excessive human exposure. Here children on a Long Island, New York, beach are fogged with DDT from a pesticide spray machine being tested in 1945.

human embryos in the womb can cause birth defects. One example involves the drug thalidomide, developed in the 1950s as a sleeping pill and to prevent nausea during pregnancy. Tragically, the drug turned out to be a powerful teratogen, and its use caused birth defects in thousands of babies. Thalidomide was banned in the 1960s once scientists recognized this connection.

The human immune system protects our bodies from disease. Some toxicants weaken the immune system, reducing the body's ability to defend itself against bacteria, viruses, allergy-causing agents, and other attackers. Other toxicants, called **allergens**, overactivate the immune system, causing an immune response when one is not necessary. One hypothesis for the increase in asthma in recent years is that allergenic synthetic chemicals have become more prevalent in our environment.

Still other chemical toxicants, **neurotoxins**, assault the nervous system. Neurotoxins include various heavy metals such as lead, mercury, and cadmium, as well as pesticides and some chemical weapons developed for use in war. A famous case of neurotoxin poisoning occurred in Japan, where a chemical factory dumped mercury waste into Minamata Bay between the 1930s and 1960s. Thousands of people in and around the town on the bay ate fish contaminated with the mercury. These victims of poisoning showed odd symptoms, including slurred speech, loss of muscle control, sudden fits of laughter, and in some cases death. The company and the government eventually paid out about $5,000 in compensation to each affected resident.

Most recently, scientists have recognized **endocrine disruptors**, toxicants that interfere with the *endocrine system*, or hormone system. The endocrine system consists of chemical messengers (*hormones*) that travel though the body. Sent through the bloodstream at extremely low concentrations, these messenger molecules have many vital functions. They stimulate growth, development, and sexual maturity, and they regulate brain function, appetite, sexual drive, and many other aspects of our physiology and behavior. Hormone-disrupting toxicants can block the action of hormones or accelerate their breakdown. Many endocrine disruptors are so similar to some hormones in their molecular structure and chemistry that they "mimic" the hormone by interacting with receptor molecules just as the hormone would (Figure 10.4). One type of endocrine disruption involves the feminization of male animals, as Louis Guillette found with alligators, and as other studies have indicated with fish, frogs, and other organisms. Feminization may occur widely because a number of chemicals appear to mimic the female sex hormone estrogen and bind to estrogen receptors.

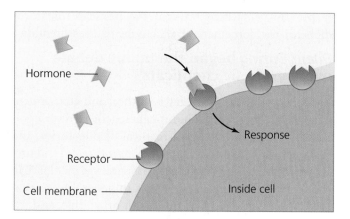

(a) Normal hormone binding

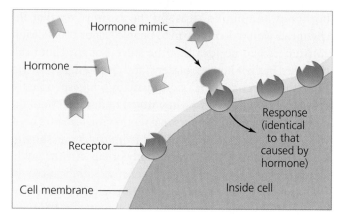

(b) Hormone mimicry

FIGURE 10.4 Many endocrine-disrupting substances act by mimicking the structure of hormone molecules. Like a key similar enough to fit into another key's lock, the hormone mimic binds to a cellular receptor for the hormone, causing the cell to react as though it had encountered the hormone.

Endocrine disruption may be widespread

Scientists first noted endocrine-disrupting effects in the 1960s with DDT, but the idea that synthetic chemicals might be altering our hormones was not widely appreciated until the 1996 publication of the book *Our Stolen Future*, by Theo Colburn, Dianne Dumanoski, and J. P. Myers. Like *Silent Spring*, this book (and its Web page that provides updates on new research) integrated scientific work from toxicologists, medical doctors, and wildlife biologists. And like *Silent Spring, Our Stolen Future* presented a unified picture that shocked many readers—and brought criticism from some scientists and from the chemical industry.

One line of research that has sparked debate is that of scientist Tyrone Hayes of the University of California at Berkeley (Figure 10.5), who has found in frogs gonadal abnormalities similar to those of Guillette's alligators and

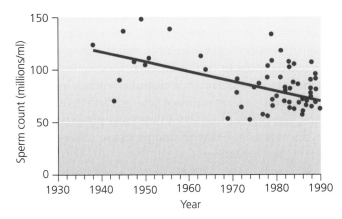

(a) Declining sperm count in humans, based on 61 studies

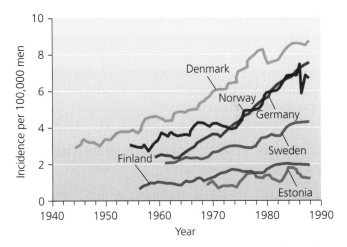

(b) Increasing incidence of testicular cancer

FIGURE 10.6 Research in 1992 synthesized the results of 61 studies that had reported sperm counts in men from various localities since 1938. The data were highly variable but showed a significant decrease in human sperm counts over time. Many scientists have hypothesized that this decrease may result from exposure to endocrine-disrupting chemicals in the environment. Data from Carlsen, E., et al. 1992. Evidence for decreasing quality of semen during the last 50 years. *British Medical Journal* 305: 609–613, as adapted by Toppari, J., et al. 1996. Male reproductive health and environmental xenoestrogens. *Environmental Health Perspectives* 104(Suppl 4): 741–803.

FIGURE 10.5 According to research by Tyrone Hayes of the University of California at Berkeley, frogs may suffer reproductive abnormalities from exposure to the best-selling herbicide atrazine. Industry-backed scientists have disputed the findings.

has attributed them to atrazine. In lab experiments, male frogs raised in water containing very low doses of this herbicide became feminized and hermaphroditic, developing both testes and ovaries. In field studies, leopard frogs across North America showed similar problems in areas of atrazine use. Hayes found these effects at atrazine concentrations well below EPA standards for human health and at concentrations lower than commonly exist in U.S. drinking water.

Many scientists attribute the striking drop in sperm counts among men worldwide to endocrine disruptors. In the most noted work, Danish researchers reviewed 61 studies and reported in 1992 that the number and motility of sperm in men's semen had declined by 50% since 1938 (Figure 10.6). Subsequent studies by other researchers—including some who set out to disprove the findings—have largely confirmed the results using other methods and populations (although there is much geographic variation that remains unexplained). Still other researchers have suggested links with rising rates of testicular cancer, undescended testicles, and genital birth defects in men, as well as rising breast cancer rates in women.

Much of the research into hormone disruption has brought about strident debate. This is partly because a great deal of scientific uncertainty is inherent in any young and developing field. Another reason is that negative findings about chemicals pose an economic threat to the manufacturers of those chemicals. For instance, Tyrone Hayes's work has met with fierce criticism from scientists associated with atrazine's manufacturer, which stands to lose many millions of dollars if its top-selling herbicide were to be banned or restricted in the United States.

Indeed, our society has invested heavily in some chemicals that now are suspect. The estrogen mimic bisphenol-A occurs in a great variety of plastic products we use daily, from baby bottles to food cans to eating utensils to auto parts. This chemical leaches into water and food, and recent evidence ties it to birth defects in lab mice. The plastics industry vehemently protests that the chemical is safe, however, and points to other research backing its contention.

Research results with bisphenol-A and mice, atrazine and frogs, and other known or purported endocrine disruptors show effects at extremely low levels of the chemical. The likely reason is that the endocrine system is geared to respond to minute concentrations of substances (normally, hormones in the bloodstream). Because the endocrine system responds to minuscule amounts of chemicals, it may be especially vulnerable to effects from environmental contaminants that are dispersed and diluted through the environment and that reach our bodies in very low concentrations.

Toxicants may concentrate in surface water or groundwater

Toxicants are not evenly distributed in the environment, and they move about in specific ways (Figure 10.7). For instance, water often carries toxicants from large areas of land and concentrates them in small volumes of surface water. If chemicals can persist in soil, they can leach down into groundwater and contaminate drinking water supplies.

Many chemicals are soluble in water, and these are often the ones that are most accessible to organisms, entering organisms' tissues through drinking or absorption. For this reason, aquatic animals such as fish, frogs, and stream invertebrates are effective indicators of pollution. This is why many scientists see findings that show impacts of low concentrations of pesticides on frogs, fish, and invertebrates as a warning that humans could be next. The pesticides that wash into streams and rivers also flow and seep into the water we drink and drift through the air we breathe.

Airborne toxicants can travel widely

Because many chemical substances can be transported by air, the toxicological effects of chemical use can occur far from the site of direct chemical use. For instance, airborne transport of pesticides is sometimes termed *pesticide drift*. The Central Valley of California is widely considered the most productive agricultural region in the world. But because it is a naturally arid area, food production depends

on intensive use of irrigation, fertilizers, and pesticides. The region's frequent winds often blow the airborne spray—and dust particles containing pesticide residue—for long distances. Families living in towns in the Central Valley have suffered health impacts, and activists for farmworkers maintain that hundreds of thousands of the state's residents are at risk. In the mountains of the Sierra Nevada, research has associated pesticide drift from the Central Valley with population declines in four species of frogs.

Synthetic chemical contaminants are ubiquitous worldwide. Despite being manufactured and applied mainly in the temperate and tropical zones, contaminants appear in substantial quantities in the tissues of Arctic polar bears, Antarctic penguins, and people living in Greenland. Scientists can travel to the most remote and seemingly pristine alpine lakes in British Columbia and find them contaminated with foreign toxicants, such as PCBs. The surprisingly high concentrations in polar regions result from patterns of global atmospheric circulation that move airborne chemicals systematically toward the poles (▸ pp. 274–275).

Some toxicants persist for a long time

Once toxic agents arrive somewhere, they may degrade quickly and become harmless, or they may remain unaltered and persist for many months, years, or decades. The rate at which chemicals degrade depends on factors such as temperature, moisture, and sun exposure, and on how these factors interact with the chemistry of the toxicant. Toxicants that persist in the environment have the greatest potential to harm many organisms over long periods of time. A major reason people have been so concerned about toxic chemicals such as DDT and PCBs is that they have long persistence times. In contrast, the Bt toxin used in biocontrol and in genetically modified (GM) crops (▸ p. 155; Figure 7.14, ▸ p. 157) has a very short persistence time. The herbicide atrazine varies in its persistence, depending on environmental conditions.

Most toxicants eventually degrade into simpler compounds called *breakdown products*. Often these are less harmful than the original substance, but sometimes they are just as toxic as the original chemical, or more so. For instance, DDT breaks down into DDE, a highly persistent and toxic compound in its own right.

Toxicants may accumulate and move up the food chain

Of the toxicants that organisms absorb, breathe, or consume, some are quickly excreted and some are degraded into harmless breakdown products, but others remain

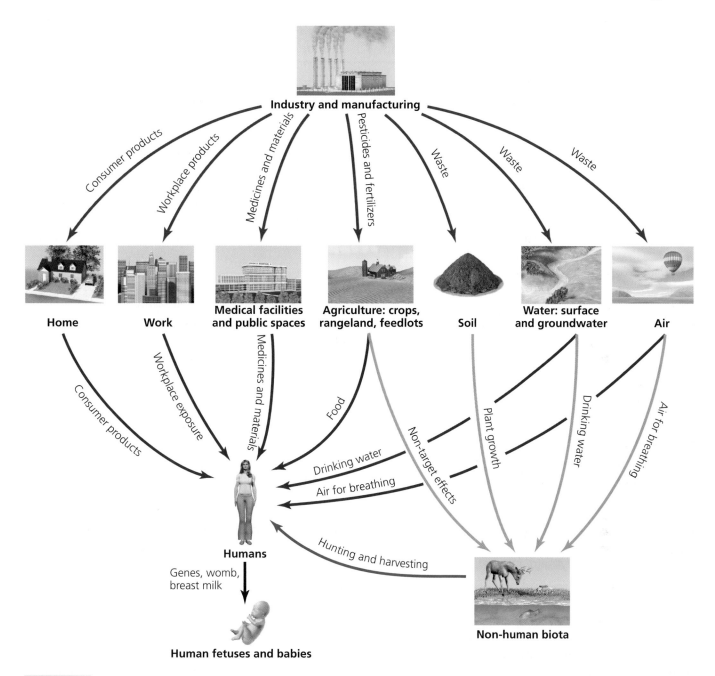

FIGURE 10.7 Synthetic chemicals take many routes in traveling through the environment. Although humans take in only a tiny proportion of these compounds, and although many compounds are harmless, humans—particularly babies—receive small amounts of toxicants from many sources.

intact within the body. Toxicants that are fat-soluble or oil-soluble (many of which are organic compounds such as DDT and DDE) are absorbed and stored in fatty tissues. Others, such as methyl mercury, may be stored in muscle tissue. Such toxicants may build up in an animal in a process termed **bioaccumulation**.

Toxicants that bioaccumulate in the tissues of one organism may be transferred to other organisms as predators consume prey. When one organism consumes another, it takes in any stored toxicants and stores them itself, along with the toxicants it has received from eating other prey. Thus with each step up the food chain, from producer to primary consumer to secondary consumer and so on, concentrations of toxicants can be greatly magnified. This process, called **biomagnification**, occurred most famously with DDT. Top predators, such as birds of prey, ended up with high concentrations of the pesticide because concentrations were magnified as

DDT moved from water to algae to plankton to small fish to bigger fish and finally to fish-eating birds (Figure 10.8).

Biomagnification caused populations of many North American birds of prey to decline precipitously from the 1950s to the 1970s. The peregrine falcon was almost totally wiped out in the eastern United States, and the bald eagle, the U.S. national bird, was virtually eliminated from the lower 48 states. The brown pelican vanished from its Atlantic Coast range, remaining only in Florida, and the osprey and other hawks saw substantial population declines. Eventually scientists determined that DDT was causing these birds' eggshells to grow thinner, so that eggs were breaking while in the nest.

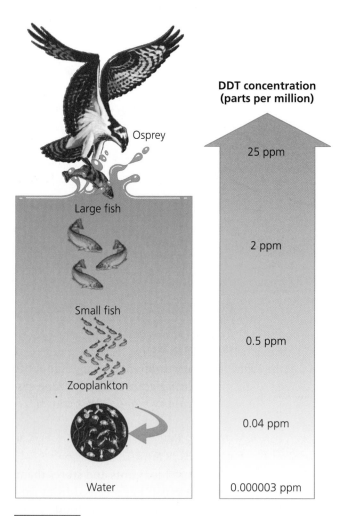

FIGURE 10.8 Fat-soluble compounds such as DDT bioaccumulate in the tissues of organisms. As animals at higher trophic levels eat organisms lower on the food chain, their load of toxicants passes up to each consumer. DDT moves from zooplankton through various types of fish, finally becoming highly concentrated in fish-eating birds such as ospreys.

Such scenarios are by no means a thing of the past. The polar bears of Svalbard Island in Arctic Norway are at the top of the food chain and feed on seals that have biomagnified toxicants. Despite their remote Arctic location, Svalbard Island's polar bears show some of the highest levels of PCB contamination of any wild animals tested, as a result of atmospheric currents that blow airborne chemicals toward the poles. The contaminants are likely responsible for the immune suppression, hormone disruption, and high cub mortality that the bears seem to be suffering. Cubs that survive receive PCBs in their mothers' milk, so that contamination persists and accumulates over generations.

Not all toxicants are synthetic

Although we have focused on synthetic chemicals thus far, chemical toxicants also exist naturally in the environment around us and in the foods we eat. We have good reason as citizens and consumers to insist on being informed about risks synthetic chemicals may pose, but it is a mistake to assume that all artificial chemicals are unhealthy and that all natural chemicals are healthy. In fact, the plants and animals we eat contain many chemicals that can cause us harm. Recall that plants produce toxins to ward off animals that eat them. In domesticating crop plants, we have selected for strains with reduced toxin content, but we have not eliminated these dangers. Furthermore, when we consume animal meat, we take in toxins the animals have ingested from plants or animals they have eaten. Scientists are actively debating just how much risk natural toxicants pose.

Studying Effects of Hazards

Determining health effects of particular environmental hazards is a challenging job, particularly because any given person or organism likely has a complex history of exposure to many hazards throughout life. Scientists rely on several different methods.

Wildlife toxicologists observe animals, determining the cause of death or sickness in an effort to understand the impacts of poisons. In studies of human health, medical professionals study and treat sickened individuals directly.

Environmental health researchers also conduct **epidemiological studies**. These involve large-scale comparisons among groups of people, usually contrasting a group known to have been exposed to some hazard and a group that has not. Epidemiologists track the fate of all people in the study, generally for a long period of time (often years or decades) and measure the rate at which deaths, cancers, or other health problems occur in each group. The epidemiologist then analyzes the data, looking for observable differences between the groups, and statistically tests hypotheses accounting for differences. When a group that has been exposed to a hazard shows a significantly greater degree of harm, it suggests that the hazard may be responsible. This process is akin to a natural experiment (▸ p. 13), in which the experimenter takes advantage of the presence of groups of subjects made possible by some event that has already occurred. A slightly different type of natural experiment was conducted by anthropologist Elizabeth Guillette (see "The Science behind the Story," ▸ pp. 230–231).

Manipulative experiments are generally not used in human toxicology because subjecting people to massive doses of toxicants in a lab experiment would clearly be unethical. So researchers have traditionally used other animals as subjects to test toxicity. Foremost among these animal models have been laboratory strains of rats, mice, and other mammals.

Dose-response analysis is a mainstay of toxicology

The standard method of testing with lab animals in toxicology is dose-response analysis. Scientists quantify the toxicity of a given substance by measuring how much effect a toxicant produces at different doses or how many animals are affected by different doses of the toxic agent. The *dose* is the amount of toxicant the test animal receives, and the *response* is the type or magnitude of negative effects the animal exhibits as a result. The response is generally quantified by measuring the proportion of animals exhibiting negative effects. The data are plotted on a graph, with dose on the *x* axis and response on the *y* axis (Figure 10.9). The resulting curve is called a **dose-response curve**.

Once they have plotted a dose-response curve, toxicologists can calculate a convenient shorthand gauge of a substance's toxicity: the amount of toxicant it takes to kill half the population of study animals used. This lethal dose for 50% of individuals is termed the **LD_{50}**. A high LD_{50} indicates low toxicity, and a low LD_{50} indicates high toxicity. Of course, the experimenter may

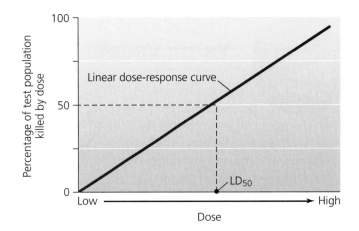

FIGURE 10.9 In a classic linear dose-response curve, the percentage of animals killed or otherwise affected by the substance rises with the dose. The point at which 50% of the animals are killed is labeled the lethal-dose-50, or LD_{50}. Go to **GRAPHit!** at www.aw-bc.com/withgott.

instead be interested in nonlethal health effects. A researcher often will want to document the level of toxicant at which 50% of the population of test animals is affected in whatever way is of interest in the study (for instance, what level of toxicant causes 50% of lab mice to lose their hair?). Such a level is called the effective-dose-50%, or **ED_{50}**.

Sometimes responses occur only above a certain dose. Such a *threshold* dose might be expected if the body's organs can fully metabolize or excrete a toxicant at low doses but become overwhelmed at higher concentrations. It might also occur if cells can repair damage to their DNA only up to a certain point.

Sometimes responses *decrease* with increased dose. Toxicologists are finding that some dose-response curves are U-shaped or J-shaped, or shaped like an inverted U. Such counterintuitive curves often appear to apply to endocrine disruptors (likely because the hormone system is geared to function with extremely low concentrations of hormones and is thus vulnerable to disruption by toxicants at extremely low concentrations). Inverted dose-response curves present a challenge for policymakers attempting to set safe environmental levels for toxicants. If many chemicals behave in these ways, we may have underestimated the dangers of these compounds, because many chemicals exist in very low concentrations over wide areas. Knowing the shape of dose-response curves is crucial if one is planning to extrapolate from them to predict responses at doses below those that have been tested.

Pesticides and Child Development in Mexico's Yaqui Valley

With spindly arms and big, round eyes, one set of pictures shows the sorts of stick figures drawn by young children everywhere. Next to them is another group of drawings, mostly disconnected squiggles and lines, resembling nothing. Both sets of pictures are intended to depict people. The main difference identified between the two groups of young artists: long-term pesticide exposure.

Children's drawings are not a typical tool of toxicology, but Elizabeth Guillette, an anthropologist married to Louis Guillette, wanted to try new methods. Guillette was interested in the effects of pesticides on children. She devised tests to measure childhood development based on techniques from anthropology and medicine. Searching for a study site, Guillette found the Yaqui Valley region of northwestern Mexico.

The Yaqui Valley is farming country, worked for generations by the indigenous group that gives the region its name. Synthetic pesticides arrived in the area in the 1940s. Some Yaqui embraced the agricultural innovations, spraying their farms in the valley to increase their yields. Yaqui farmers in the surrounding foothills, however, generally chose to bypass the chemicals and to continue following more traditional farming practices.

Drawings by children in the foothills

4-year-olds

5-year-olds

Drawings by children in the valley

4-year-olds

5-year-olds

Elizabeth Guillette's study in Mexico's Yaqui Valley offers a startling example of apparent neurological effects of pesticide poisoning. Young children from foothills areas where pesticides were not commonly used drew recognizable figures of people. Children the same age from valley areas where pesticides were used heavily in industrialized agriculture could draw only scribbles. Adapted from Guillette, E. A., et al. 1998. *Environmental Health Perspectives* 106: 347–353.

Although differing in farming techniques, Yaqui in the valley and foothills continued to share the same culture, diet, education system, income levels, and family structure.

At the time of the study, in 1994, valley farmers planted crops twice a year, applying pesticides up to 45 times from planting to harvest. A previous study conducted in the valley in 1990, focusing on areas with the largest farms, had indicated high levels of multiple pesticides in the breast milk of mothers and in the umbilical cord blood of newborn babies. In contrast, foothill families avoided

Because extrapolations go beyond the actual data obtained, they introduce uncertainty into the interpretation of what doses are acceptable for humans. As a result, to be on the safe side, regulatory agencies set standards for maximum allowable levels of toxicants that are well below the minimum toxicity levels estimated from lab studies.

Individuals vary in their responses to hazards

Different individuals may respond quite differently to identical exposures to hazards. These differences can be genetically based, or they can be due to a person's current

chemical pesticides in their gardens and homes.

To understand how pesticide exposure affects childhood development, Guillette and fellow researchers studied 50 preschoolers aged four to five, of whom 33 were from the valley and 17 from the foothills. Each child underwent a half-hour exam, during which researchers showed a red balloon, promising to give the balloon later as a gift. Researchers used the promise to evaluate long-term memory. Each child was then put through a series of physical and mental tests:

▶ Catching a ball from distances of up to 3 m (10 ft) away, to test overall coordination
▶ Jumping in place for as long as possible, to assess endurance
▶ Drawing a picture of a person, as a measure of perception
▶ Repeating a short string of numbers, to test short-term memory
▶ Dropping raisins into a bottle cap from a height of about 13 cm (5 in.), to gauge fine-motor skills

The researchers measured each child's height and weight but, because of lack of time and money, stopped short of taking blood or tissue samples to check for pesticides or other toxins. When all tests were completed, researchers asked each child what he or she had been promised and then gave the child a red balloon.

Although the two groups of children were not significantly different in height and weight, they differed markedly in other areas of development. Valley children lagged far behind the foothill children developmentally in coordination, physical endurance, long-term memory, and fine-motor skills:

▶ From a distance of 3 m (10 ft), valley children had great difficulty catching the ball.
▶ Valley children could jump for an average of 52 seconds, compared to 88 seconds for foothill children.
▶ Most valley children missed the bottle cap when dropping their raisins, whereas foothill children dropped them into the caps far more often.
▶ Valley children showed poor long-term memory. At the end of the test, all but one of the foothill children remembered that they had been promised a balloon, and 59% remembered it was red. However, of the valley children only 27% remembered the color of the balloon, only 55% remembered they'd be getting a balloon, and 18% were unable to remember anything about the balloon.

The children's drawings exhibited the most dramatic difference between valley and foothill children (see the figure). The researchers determined each drawing could earn 5 points, with 1 point each for a recognizable feature: head, body, arms, legs, and facial features. Foothill children drew pictures that looked like people, averaging about 4.5 points per drawing. Valley children, in contrast, averaged 1.6 points per drawing; their scribbles resembled little that looked like a person. By the standards of developmental medicine, the four- and five-year-old valley children drew at the level of a two-year-old.

Some scientists greeted Guillette's study skeptically, pointing out that its sample size was too small to be meaningful. Others said that factors the researchers missed, such as different parenting styles or unknown health problems, could account for the differences between groups. Prominent toxicologists argued that without blood or tissue tests on the children, the study results couldn't be tied to agricultural chemicals. Regardless of these criticisms, Guillette maintains that her findings show that nontraditional study methods are a valid way to track the effects of environmental toxins, and that pesticides present a complex long-term risk to human growth and health.

condition. People in poorer health are often more sensitive to biological and chemical hazards. Sensitivity also can vary with sex, age, and weight. Because of their smaller size and rapidly developing organ systems, fetuses, infants, and young children tend to be much more sensitive to toxicants than are adults. Regulatory agencies such as the EPA traditionally set standards for adults and extrapolated downward for infants and children. However, they have subsequently found that in many cases their linear extrapolations did not lower standards enough to protect babies adequately. Many critics today contend that despite improvements, regulatory agencies still do not account explicitly enough for risks to fetuses, infants, and children.

The type of exposure can affect the response

The risk posed by a hazard often varies according to whether a person experiences high exposure for short periods of time, known as **acute exposure**, or lower exposure over long periods of time, known as **chronic exposure**. Incidences of acute exposure are easier to recognize, because they often stem from discrete events, such as accidental ingestion, an oil spill, a chemical spill, or a nuclear accident. Lab tests and LD_{50} values generally reflect acute toxicity effects. However, chronic exposure is more common—and more difficult to detect and diagnose. Chronic exposure often affects organs gradually, as when smoking causes lung cancer, or when alcohol abuse induces liver or kidney damage. Pesticide residues on food or low levels of arsenic in drinking water also pose chronic risk. Because of the long time periods involved, relationships between cause and effect may not be readily apparent.

Mixes may be more than the sum of their parts

It is difficult enough to determine the impact of a single hazard on an organism, but the task becomes astronomically more difficult when multiple hazards interact. For instance, chemical substances, when mixed, may act in concert in ways that cannot be predicted from the effects of each in isolation. Mixed toxicants may sum each other's effects, cancel out each other's effects, or multiply each other's effects. Whole new types of impacts may arise when toxicants are mixed together. Such interactive impacts—those that are more than or different from the simple sum of their constituent effects—are called **synergistic effects**.

With Florida's alligators, lab experiments have indicated that DDE can either help cause or inhibit sex reversal, depending on the presence of other chemicals. Mice exposed to a mixture of nitrate, atrazine, and aldicarb have been found to show immune, hormone, and nervous-system effects that were not evident from exposure to each of these chemicals alone.

Traditionally, environmental health has tackled effects of single hazards one at a time. In toxicology, the complex experimental designs required to test interactions, and the sheer number of chemical combinations, have meant that single-substance tests have received priority. This approach is changing, but scientists in environmental health and toxicology will never be able to test all possible combinations. There are simply too many hazards in the environment.

Risk Assessment and Risk Management

Policy decisions to ban chemicals or restrict their use generally follow years of rigorous testing for toxicity. Likewise, strategies for combating disease and other environmental health threats are often based on extensive research. Policy and management decisions also reach beyond the scientific results on health to incorporate considerations about economics and ethics. And all too often, they are influenced by political pressure from powerful interests. The steps between the collection and interpretation of scientific data and the formulation of policy involve assessing and managing risk.

Risk is expressed in terms of probability

Exposure to an environmental health threat does not invariably produce some given effect. Rather, it causes some probability of harm, some statistical chance that damage will result. To understand the impact of an environmental health threat, a scientist must know more than just its identity and strength. He or she must also know the chance that an organism will encounter it, the frequency with which the organism may encounter it, the amount of substance or degree of threat to which the organism is exposed, and the organism's sensitivity to the threat. Such factors help determine the overall risk posed. Risk can be measured in terms of *probability*, a quantitative description of the likelihood of a certain outcome. The mathematical probability that some harmful outcome (for instance, injury, death, environmental damage, or economic loss) will result from a given action, event, or substance expresses the **risk** posed by that phenomenon.

Our perception of risk may not match reality

Every action we take and every decision we make involves some element of risk, some (generally small) probability that things will go wrong. Interestingly, our perceptions of risk do not always match statistical reality (Figure 10.10). People often worry unduly about negligibly small risks but happily engage in other activities that pose high risks. For instance, most people perceive flying in an airplane as a riskier activity than driving a car, but driving a car is statistically far more dangerous.

Psychologists agree that this difference between risk perception and reality stems from the fact that we feel more at risk when we are not controlling a situation and more safe when we are "at the wheel"—regardless

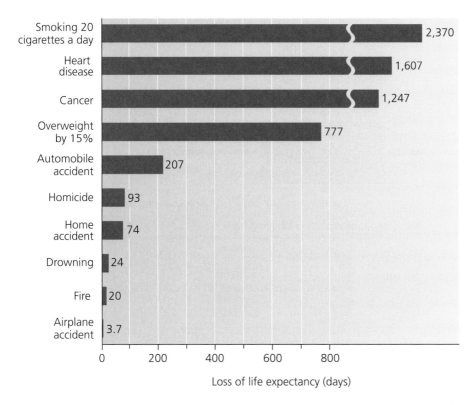

FIGURE 10.10 Our perceptions of risk do not always match the reality of risk. Listed here are several leading causes of death in the United States, along with a measure of the risk each poses. Risk is measured in days of lost life expectancy, or the number of days of life lost by people suffering the hazard, spread across the entire population—a measure commonly used by insurance companies. By this measure, one common source of anxiety, airplane accidents, poses 20 times less risk than home accidents, over 50 times less risk than auto accidents, and over 200 times less risk than being overweight. Data from Cohen, B. 1991. Catalog of risks extended and updated. *Health Physics* 61: 317–335.

of the actual risk involved. When we drive a car, we feel we are in control, even though statistics show we are at greater risk than as a passenger in an airplane. This psychology can account for people's great fear of nuclear power, toxic waste, and pesticide residues on foods—environmental hazards that are invisible or little understood, and whose presence in their lives is largely outside their personal control. In contrast, people are more ready to accept and ignore the risks of smoking cigarettes, overeating, and not exercising, all voluntary activities statistically shown to pose far greater risks to health.

Risk assessment analyzes risk quantitatively

The quantitative measurement of risk and the comparison of risks involved in different activities or substances together are termed **risk assessment**. Risk assessment is a way of identifying and outlining problems. In environmental health, it helps ascertain which substances and activities pose health threats to people or wildlife and which are largely safe. Assessing risk for a chemical substance involves several steps. The first steps involve the scientific study of toxicity outlined above—determining whether a given substance has toxic effects and, through dose-response analysis, measuring how effects on an organism vary with the degree of toxicant exposure. Subsequent steps involve assessing the individual's or population's likely extent of exposure to the substance, including the frequency of contact, the concentrations likely encountered, and the length of time the substance is likely to be encountered.

Risk management combines science and other social factors

Accurate risk assessment is a vital step toward effective **risk management**, which consists of decisions and strategies to minimize risk (Figure 10.11). In most developed nations, risk management is handled largely by federal agencies, such as the EPA, the Centers for Disease Control and Prevention (CDC), and the Food and Drug Administration (FDA) in the United States. In risk management, scientific assessments of risk are considered in light of economic, social, and political needs and values. The costs and benefits of addressing risk in various ways are assessed with regard to both scientific and nonscientific concerns. Decisions whether to reduce or eliminate risk are then made.

In environmental health and toxicology, comparing costs and benefits (▸ p. 28) is often difficult because the benefits are often economic and the costs often pertain to health. Moreover, economic benefits are generally known, easily quantified, and of a discrete and stable amount, whereas health risks are hard-to-measure probabilities, often involving a very small percentage of people likely to suffer greatly and a large majority likely to experience little effect. When a government agency bans a pesticide, it generally means measurable economic loss for the manufacturer

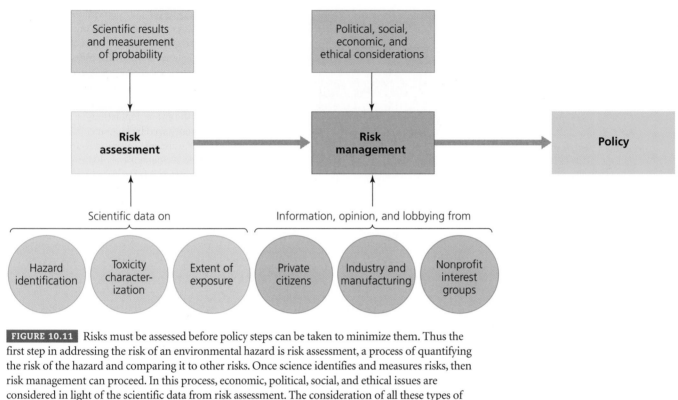

FIGURE 10.11 Risks must be assessed before policy steps can be taken to minimize them. Thus the first step in addressing the risk of an environmental hazard is risk assessment, a process of quantifying the risk of the hazard and comparing it to other risks. Once science identifies and measures risks, then risk management can proceed. In this process, economic, political, social, and ethical issues are considered in light of the scientific data from risk assessment. The consideration of all these types of information is designed to result in policy decisions that minimize the risk of the environmental hazard.

and the farmer, whereas the benefits accrue less predictably over the long term to some percentage of factory workers, farmers, and the general public. Because of the lack of equivalence in the way costs and benefits are measured, risk management frequently tends to stir up debate.

Philosophical and Policy Approaches

Because we do not know a substance's toxicity until we measure and test it, and because there are so many untested chemicals and combinations, science will never eliminate the many uncertainties that accompany risk assessment. In such a world of uncertainty, there are two basic philosophical approaches to categorizing substances as safe or dangerous (Figure 10.12).

Two approaches exist for determining safety

One approach is to assume that substances are harmless until shown to be harmful. We might nickname this the *innocent-until-proven-guilty approach.* Because thoroughly testing every existing substance (and combination of substances) for its effects is a hopelessly long, compli-

cated, and expensive pursuit, the innocent-until-proven-guilty approach has the benefit of not slowing down technological innovation and economic advancement. However, it has the disadvantage of putting into wide use some substances that may later turn out to be dangerous.

The other approach is to assume that substances are harmful until they are shown to be harmless. This approach follows the precautionary principle (▶ p. 160). This more cautious approach should enable us to identify troublesome toxicants before they are released into the environment, but it may also significantly impede the pace of technological and economic advance.

These two approaches are actually two ends of a continuum of possible approaches. The two endpoints differ mainly in where they lay the burden of proof—specifically, whether product manufacturers are required to prove safety or whether government, scientists, or citizens are required to prove danger.

- -
Weighing the **Issues:**
The Precautionary Principle

Given the substantial costs of testing chemicals for safety and the increasing concerns about their spread through the environment, should proof of safety be required by government prior to a chemical's release

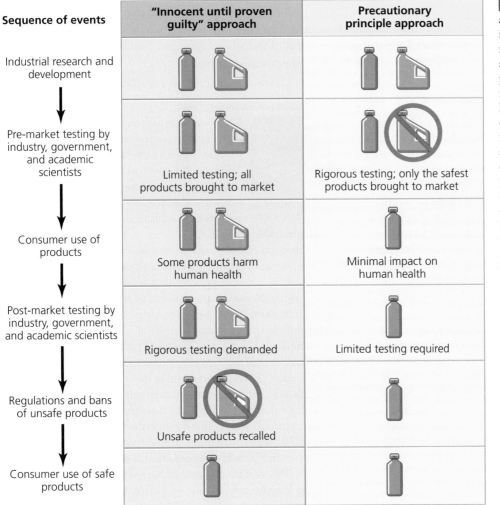

Sequence of events	"Innocent until proven guilty" approach	Precautionary principle approach
Industrial research and development		
Pre-market testing by industry, government, and academic scientists	Limited testing; all products brought to market	Rigorous testing; only the safest products brought to market
Consumer use of products	Some products harm human health	Minimal impact on human health
Post-market testing by industry, government, and academic scientists	Rigorous testing demanded	Limited testing required
Regulations and bans of unsafe products	Unsafe products recalled	
Consumer use of safe products		

FIGURE 10.12 Two main approaches can be taken to introducing new substances to the market. In one approach, substances are innocent until proven guilty; they are brought to market relatively quickly after limited testing. Products reach consumers more quickly, but some fraction of them may cause harm to some fraction of people. In the other approach, the precautionary principle is adopted, and substances are brought to market cautiously, only after extensive testing. Products that reach the market should be safe, but many perfectly safe products will be delayed in reaching consumers.

into the environment? Should the burden of proof fall to the company that stands to make a profit from a product's release? How do you think adopting the precautionary principle would affect the number of chemicals on the market?

Philosophical approaches are reflected in policy

One's philosophical approach has immediate and far-reaching impact on policy decisions, directly affecting what materials are allowed into our environment. Most nations follow a blend of the two approaches, but there is marked variation among countries. At present, European nations are to some extent following the precautionary principle regarding the regulation of synthetic chemicals, whereas the United States is not. Although industry frequently complains that government regulation is cumbersome, environmental and consumer advocates criticize U.S. policies for largely following an innocent-until-proven-guilty approach. For instance,

compounds involved in cosmetics require no FDA review or approval before being sold to the public.

In the United States, several federal agencies apportion responsibility for tracking and regulating synthetic chemicals. The FDA, under the Food, Drug, and Cosmetic Act of 1938 and its subsequent amendments, regulates foods and food additives, cosmetics, and drugs and medical devices. The EPA regulates pesticides under the Federal Insecticide, Fungicide, and Rodenticide Act of 1947 (FIFRA) and its amendments. The Occupational Safety and Health Administration (OSHA) regulates workplace hazards under a 1970 act. Several other agencies regulate other substances. Synthetic chemicals not covered by other laws are regulated by the EPA under the 1976 Toxic Substances Control Act (TSCA).

TSCA directed the EPA to monitor some 75,000 industrial chemicals manufactured in or imported into the United States, and gave the agency power to regulate these chemicals and ban them if they are found to pose excessive risk. However, many public health and environmental

Chemical Product Testing: Industry or Government?

The testing of chemical products for safety can take the so-called innocent-until-proven-guilty approach or the precautionary approach. **Should manufacturers be held responsible for comprehensive testing of new chemical products before they are introduced to the public? What would be some of the advantages and disadvantages? How extensive a role should government play in the testing process?**

Testing Must Ensure Public Health

Like most things in life, the controversy over product testing arises because both approaches have valid advantages and disadvantages.

Allowing industry to follow the innocent-until-proven-guilty approach, with limited testing, reduces development costs for new chemical products and may lead to greater economic activity. If industry were required to comprehensively test chemical product safety before introduction to the public, chemical industry profits could fall and result in job loss and costly new product development. Consumer prices might rise to cover these costs.

However, comprehensive testing would lower the number of chemicals that adversely affect biological species.

Our definition of *innocent* is often too narrow. In the innocent-until-proven-guilty approach, *what the consumer actually buys is never tested*, because only ultra-pure active ingredients are tested. Surfactants and organic soaps ("other ingredients") are added to improve the active ingredients' lipid or water solubility, and these other ingredients are frequently very active biologically. Also, production contaminants are not tested and registered. Therefore, a so-called innocent product can cause cancer and reproductive defects.

The assumption of a linear dose response is also coming under increased scrutiny, especially at very low physiological doses, where hormonal, immune, and neurological processes respond. At much higher pharmacological doses, where toxicity testing is typically done, the responses of physiological systems to the same chemical can be very different.

Given the inherent inadequacies of the testing process and the uncertainty of the economic impacts, both government and industry should share the responsibility of testing to ensure public safety.

Warren Porter is a toxicologist and physiological ecologist at the University of Wisconsin–Madison. He evaluates the connections among climate, animal energetics, and behavior using statistical experimental design.

An Industry Perspective

Chemical manufacturers already take an active role in testing new chemicals. This process is part of current EPA regulations. Additionally, government often tests chemicals to elucidate either hazard or exposure. Generation of these data by government adds to the body of knowledge generated by industry.

Chemical risk depends on two factors: hazard (toxicity) and exposure. To evaluate a chemical, manufacturers typically start by conducting screening-level toxicological and environmental studies and proceed to more or higher-tier studies as warranted. There is no single comprehensive testing program that is appropriate for all industrial chemicals.

The Toxic Substances Control Act requires almost all new commercial substances to undergo Premanufacture Notification (PMN) review, and to describe this preliminary process as an innocent-until-proven-guilty approach is an oversimplification. When the EPA reviews a PMN, it considers the physical and chemical properties of the substance, structural similarity to other compounds of known toxicity, and potential for human exposure and environmental release. If there is no evidence of harm from preliminary testing, longer-term or more specialized testing may not be conducted. In some cases, the EPA may require additional testing to determine whether the chemical poses an unreasonable risk to human health or the environment. If the EPA finds that risk can be addressed by reducing exposure, it may enter into a binding agreement to require the manufacturer to take steps to reduce the public's exposure to the substance, rather than requiring additional laboratory testing.

Manufacturers often voluntarily conduct new studies to support the continued safe use of their chemicals. The 150 member companies of the American Chemistry Council represent about 90% of U.S. chemical production. These companies are committed to Responsible Care®, under which chemical manufacturers, as good stewards of their products, continue to test as new data and methodologies become available. There is a role for both government and industry in chemical testing, and it is important that the EPA and manufacturers work together in evaluating chemicals to improve health, safety, and the environment.

Marian K. Stanley is senior director for the American Chemistry Council, which she joined in 1990. She is responsible for managing chemical-specific issue groups in the Council's self-funded CHEMSTAR Department.

Explore this issue further by accessing **Viewpoints** at www.aw-bc.com/withgott.

advocates view TSCA as being far too weak. They note that the screening required of industry is minimal and that to mandate more extensive and meaningful testing, the EPA must show proof of the chemical's toxicity. In other words, the agency is trapped in a Catch-22: To push for studies looking for toxicity, it must have proof of toxicity already. The result, these advocates say, is that most synthetic chemicals are not thoroughly tested before being put on the market. Of those that fall under TSCA, only 10% have been thoroughly tested for toxicity; only 2% have been screened for carcinogenicity, mutagenicity, or teratogenicity; fewer than 1% are government-regulated; and almost none have been tested for endocrine, nervous, or immune system damage, according to the U.S. National Academy of Sciences.

Internationally, action regarding chemical toxicants has been taken in the form of stricter policies of the European Union and in international treaties. The Stockholm Convention on Persistent Organic Pollutants (POPs), introduced in 2001, appears on its way to ratification. POPs are toxic chemicals that persist in the environment, bioaccumulate in the food chain, and often travel long distances. The PCBs and other contaminants found in polar bears are a prime example. Because these contaminants so often cross international boundaries, an international treaty seemed the best way of dealing fairly with such transboundary pollution. The Stockholm Convention aims first to end the use and release of 12 of the POPs shown to be most dangerous, a group nicknamed the "dirty dozen." It sets guidelines for phasing out these chemicals and encourages transition to safer alternatives.

Conclusion

International agreements such as the Stockholm Convention represent a hopeful sign that governments will act to protect the world's people, wildlife, and ecosystems from toxic chemicals and other environmental hazards. At the same time, solutions can often come more easily when they do not arise from government regulation alone. To many minds, consumer choice, exercised through the market, may be the best way to influence industry's decision making. Consumers of products can make decisions that influence industry when they have full information from scientific research regarding the risks involved. Once scientific results are in, a society's philosophical approach to risk management will determine what policy decisions are made.

Whether the burden of proof is laid at the door of industry or of government, it is important to realize that we will never attain complete scientific knowledge of any risk. Rather, we must make choices based on the information available. Synthetic chemicals have brought us innumerable modern conveniences, a larger food supply, and medical advances that save and extend human lives. Human society would be very different without them. Yet a safer and happier future, one that safeguards the well-being of both humans and the environment, depends on knowing the risks that some hazards pose and on having means in place to phase out harmful substances and replace them with safer ones.

TESTING YOUR COMPREHENSION

1. What four major types of health hazards does research in the field of environmental health encompass?
2. In what way is disease the greatest hazard that humans face? What kinds of interrelationships must environmental health experts study to learn about how diseases affect human health?
3. Where does most exposure to lead, asbestos, radon, and PBDEs occur? How has each of these exposure problems been addressed?
4. When did concern over the effects of pesticides start to grow in the United States? Describe the argument presented by Rachel Carson in *Silent Spring*. What policy resulted from the book's publication? Is DDT still used?
5. List and describe the six types or general categories of toxicants described in this chapter.
6. How do toxicants travel through the environment, and where are they most likely to be found? What are the life spans of toxic agents? Describe the processes of bioaccumulation and biomagnification.
7. What are epidemiological studies, and how are they most often conducted?
8. Why are animals used in laboratory experiments in toxicology? Explain the dose-response curve. Why are high LD_{50} and ED_{50} levels considered safe and low LD_{50} and ED_{50} levels considered unsafe for humans?
9. What factors may affect an individual's response to a toxic substance? Why is chronic exposure to toxic agents often more difficult to measure and diagnose than acute exposure? What are synergistic effects, and why are they difficult to measure and diagnose?
10. How do scientists identify and assess risks from substances or activities that may pose health threats?

SEEKING SOLUTIONS

1. Describe some environmental hazards that you think you may be living with indoors. How do you think you may have been affected by indoor or outdoor environmental hazards in the past? What philosophical approach do you plan to take in dealing with these toxicants in your own life?

2. Why is it that research on endocrine disruption has spurred so much debate? What steps do you think could be taken to help establish consensus among scientists, industry, regulators, policymakers, and the public?

3. Do you feel that laboratory-bred animals should be used in experiments in toxicology? Why or why not?

4. Discuss ways that we may cope with the uncertainty of risk assessment for synthetic chemicals in environmental health. Can you think of alternatives to taking one of the two philosophical approaches discussed in the chapter? Should these approaches apply to natural toxicants as well?

5. Describe what you have learned from this chapter regarding the policies of the United States toward the study and management of the risks of synthetic chemicals. Do you believe these policies are adequate, too weak, or too strong, and why?

CALCULATING ECOLOGICAL FOOTPRINTS

 In 2001, the population of the United States was approximately 285 million, and the world's population totaled 6.16 billion. In that same year, pesticide use in the United States was approximately 1.20 billion pounds of active ingredient, and world pesticide use totaled 5.05 billion pounds of active ingredient. Pesticides include hundreds of chemicals used as insecticides, fungicides, herbicides, rodenticides, repellants, and disinfectants by farmers, governments, industries, and individuals. In the table, calculate your share of pesticide use as a U.S. citizen in 2001 and the amount used by (or on behalf of) the average citizen of the world.

	Annual pesticide use (pounds of active ingredient)
You	4.21
Your class	
Your state	
United States	
World (total)	
World (per capita)	

1. What is the ratio of your annual pesticide use to the world's per capita average? Refer back to the "Calculating Ecological Footprints" question for Chapter 1 (▶ p. 22), and find the ecological footprints of the average U.S. citizen and the average world citizen. Compare the ratio of pesticide usage with the ratio of the overall ecological footprints. How would you explain the difference?

2. Does the figure for per capita pesticide use for you as a U.S. citizen seem reasonable for you personally? Why or why not? Do you find this figure alarming or of little concern? What else would you like to know to assess the risk associated with this level of pesticide use?

Take It Further

Go to www.aw-bc.com/withgott, where you'll find:

▶ Suggested answers to end-of-chapter questions

▶ Quizzes, animations, and flashcards to help you study

▶ *Research Navigator*™ database of credible and reliable sources to assist you with your research projects

▶ **GRAPHit!** tutorials to help you interpret graphs

▶ **INVESTIGATEit!** current news articles that link the topics that you study to case studies from your region to around the world

Freshwater and Marine Systems and Resources

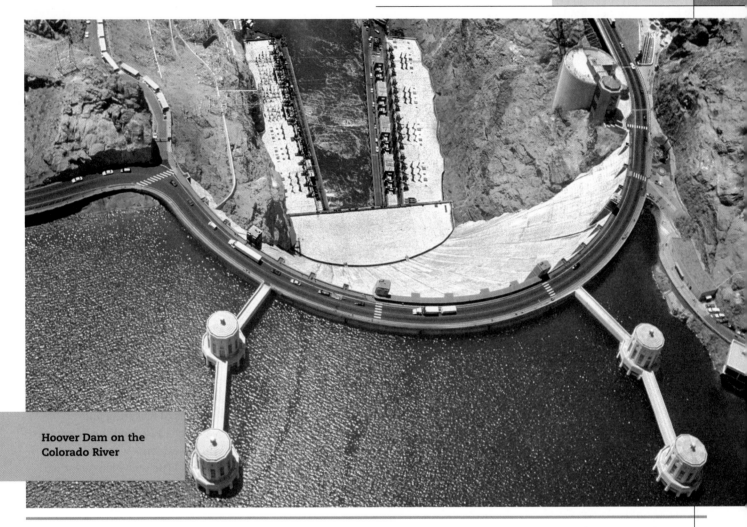

Hoover Dam on the Colorado River

Upon successfully completing this chapter, you will be able to:

▶ Explain the importance of water and the hydrologic cycle to ecosystems, human health, and economic pursuits

▶ Delineate freshwater distribution on Earth

▶ Describe major types of freshwater and marine ecosystems

▶ Discuss how we use water and alter aquatic systems

▶ Assess problems of water supply and propose solutions to address freshwater depletion

▶ Assess problems of water quality and propose solutions to address water pollution

▶ Explain how wastewater is treated

▶ Identify physical, geographical, chemical, and biological aspects of the marine environment

▶ Review the current state of ocean fisheries and reasons for their decline

▶ Evaluate marine protected areas and reserves as innovative solutions to overfishing

Central Case: Plumbing the Colorado River

"We've gone from being assured that we lived in this magical place where the rules of water didn't apply to [a] wake-up call about the fact that we do live in the California desert. People have lived in this false water utopia."
—BUFORD CRITES, CITY COUNCILOR, PALM DESERT CITY, CALIFORNIA, JUNE 2003

"Water promises to be to the 21st century what oil was to the 20th century: the precious commodity that determines the wealth of nations."
—FORTUNE MAGAZINE, MAY 2000

As the clock struck midnight on New Year's Eve, millions of Californians toasted the arrival of 2003 with champagne. But some people in the state that night had another liquid on their minds: water. Their fears were borne out the next day when the U.S. government followed through on its threat to cut off 15% of the water that California takes from the Colorado River.

Water is the lifeline for any civilization in an arid environment. Without generous supplies of freshwater delivered from elsewhere, southern California society as we know it could simply not exist. In ordering the New Year's Day cutoff, U.S. Interior Secretary Gale Norton was simply holding up her end of a deal that an irrigation district in California had scuttled. It may seem bizarre that a 3–2 vote of one county irrigation district could block enough water for 1.6 million households in Los Angeles and San Diego, but it was just the latest episode in the colorful history of California water politics and the battles among seven states jockeying for rights to what was once the West's wildest river.

The Colorado River begins in the high peaks of the Rocky Mountains, charges through the Grand Canyon, crosses the border into Mexico, and empties into the Gulf of California, draining 637,000 km² (246,000 mi²) of southwestern North America. Its raging waters have chiseled through thousands of feet of bedrock, creating the Grand Canyon and leaving extraordinary scenery along its 2,330-km (1,450-mi) length. Today, however, only a small amount of water—often none at all—reaches the river's mouth. Instead, the waters of the Colorado River irrigate 7% of U.S. cropland, quench the thirst of over 20 million people, keep hundreds of golf courses green in the desert, and fill the swimming pools and fountains of Las Vegas casinos. The Colorado provides vital water to the rapidly growing metropolitan areas of the arid U.S.

Southwest—Phoenix, Tucson, Las Vegas, San Diego, Los Angeles, and many others. The massive dams built across the river provide flood control and recreation, produce 12 billion kilowatt-hours of electricity from hydroelectric power each year, and provide irrigation that makes agriculture possible in this arid region.

The seven states along the Colorado divide the river's water among themselves, guided by the Colorado River Compact they signed in 1922, which apportioned water to each state. California had long been permitted to exceed its allotment because Colorado, Wyoming, Utah, Nevada, New Mexico, and Arizona were not using all of their portions. With the populations of these states booming, however, Interior Secretary Bruce Babbitt in 2000 pressured California to reduce its withdrawals by roughly 15% over 15 years.

California worked hard to get its agricultural districts, which controlled most water distribution in the state, to agree. At the last minute, however, the Imperial Irrigation District backed out of the agreement. The New Year's deadline passed, and 2003 saw the federal cutoff implemented. After 10 months of bickering and negotiation, the deal was patched up, and the cutoff was ended. Southern California's residents were able to continue living—at least for a little while longer—their mirage in the desert.

Freshwater Systems

"Water, water, everywhere, nor any drop to drink." The well-known line from Coleridge's poem *The Rime of the Ancient Mariner* describes the situation on our planet quite well. Water may seem abundant to us, but water that we can drink is actually quite rare (Figure 11.1). Roughly 97.5% of Earth's water resides in the oceans and is too salty to drink or use to water crops. Only 2.5% is considered **freshwater**, water with few dissolved salts. Because most freshwater is tied up in glaciers, icecaps, and underground aquifers (▶ p. 68), just over 1 part in 10,000 of Earth's water is easily accessible for human use.

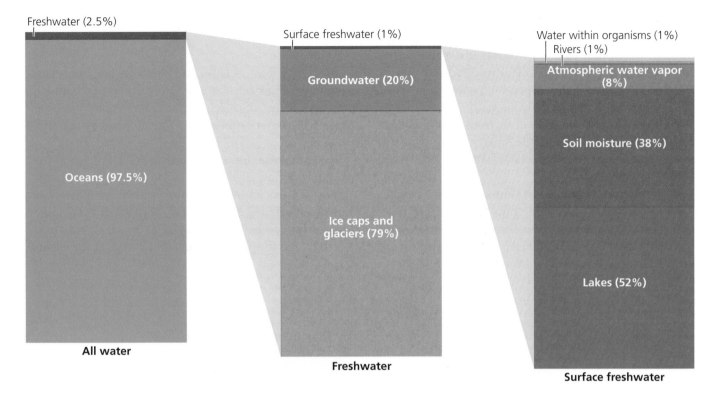

FIGURE 11.1 Only 2.5% of Earth's water is freshwater. Of that 2.5%, most is tied up in glaciers and ice caps. Of the 1% that is surface water, most is in lakes and soil moisture. Data from United Nations Environment Programme (UNEP) and World Resources Institute.

Water constantly moves among the reservoirs specified in Figure 11.1 via the *hydrologic cycle* (see Figure 3.16, ▶ pp. 67–68, 70). As water moves, it redistributes heat, erodes mountain ranges, builds river deltas, maintains organisms and ecosystems, shapes civilizations, and gives rise to political conflicts. Let's first examine the portions of the hydrologic cycle that are most conspicuous to us—surface water bodies—and take stock of the ecological systems they support.

Freshwater ecosystems are rich in life

Water physically shapes Earth's landscapes, and freshwater systems support a disproportionately large fraction of our planet's organisms. Several freshwater ecosystem types can be thought of as the aquatic equivalents of biomes (▶ pp. 111–118).

Rivers and streams Water from rain, snowmelt, or springs runs downhill and converges where the topography dips lowest, forming streams, creeks, or brooks. These watercourses merge into rivers (Figure 11.2), whose water eventually reaches the ocean or a landlocked water body. A smaller river flowing into a larger one is a *tributary*, and

FIGURE 11.2 Rivers and streams flow downhill, shaping landscapes. Water rounding a river's bend gradually eats away at the outer shore, eroding soil from the bank. Meanwhile, sediment is deposited along the inside of the bend, where water currents are weaker. In this way, over time, river bends become exaggerated in shape.

the area of land drained by a river and its tributaries is that river's *watershed* (▸ p. 24). Rivers shape the landscape through which they run. Over thousands or millions of years, a river may shift from one course to another, back and forth over a large area, carving out a flat valley. Areas that a river floods periodically are said to be within the river's **floodplain**. Deposition of silt from flooding makes floodplain soils especially fertile, so agriculture thrives in floodplains and *riparian* (riverside) forests are productive and species-rich.

The waters of rivers and streams host diverse ecological communities. Algae and detritus support many types of invertebrates, from water beetles to crayfish to dragonflies. Fish and amphibians consume aquatic insects, and birds such as kingfishers, herons, and ospreys dine on fish and amphibians.

Lakes and ponds Lakes and ponds are bodies of standing water. Their physical conditions and types of life vary with depth and the distance from shore. As a result, scientists have described several zones typical of lakes and ponds (Figure 11.3).

Around the nutrient-rich edges of a water body, the water is shallow enough that aquatic plants grow from the mud and reach above the water's surface. This region, named the *littoral zone*, abounds in invertebrates—such as insect larvae, snails, and crayfish—on which fish, birds, turtles, and amphibians feed. The *benthic zone* extends

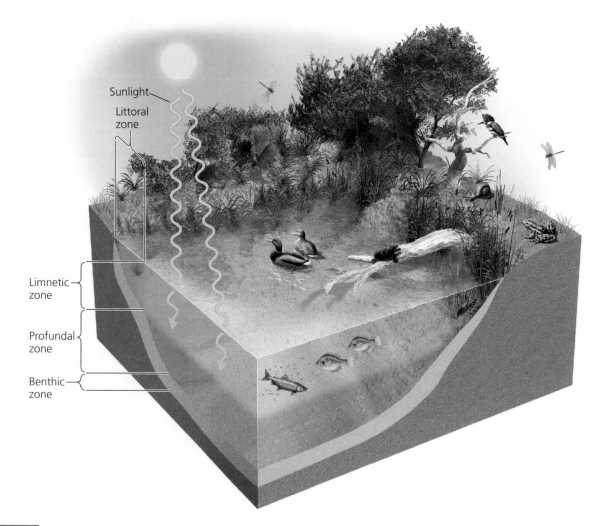

FIGURE 11.3 Lakes and ponds are open, still bodies of water consisting of different zones. Emergent plants grow around the shoreline in the littoral zone. The limnetic zone is the layer of open, sunlit water where photosynthesis takes place. Sunlight does not reach the deeper profundal zone. The benthic zone, which is the bottom of the water body, often is muddy, rich in detritus and nutrients, and low in oxygen.

along the bottom of the lake or pond, from shore to the deepest point. In the open portion of a lake or pond, away from shore, sunlight penetrates shallow waters of the *limnetic zone*. Because light enables photosynthesis, the limnetic zone supports phytoplankton (algae, protists, and cyanobacteria; ▸ p. 52) which in turn support zooplankton (▸ p. 52), both of which fish consume. Within the limnetic zone, sunlight intensity (and therefore water temperature) decreases with depth. Below the limnetic zone lies the *profundal zone*, the volume of open water that sunlight does not reach. This zone lacks photosynthetic life and thus is lower in dissolved oxygen and supports fewer animals.

Ponds and lakes change over time naturally as streams and runoff bring them sediment and nutrients. *Oligotrophic* lakes and ponds, which have low-nutrient and high-oxygen conditions, slowly give way to the high-nutrient, low-oxygen conditions of *eutrophic* water bodies. Eventually, water bodies may fill in completely by the process of aquatic succession (see Figure 5.8, ▸ p. 108).

Marshes, swamps, and bogs Systems that combine elements of freshwater and dry land—often lumped under the term **wetlands**—are enormously rich and pro-ductive. In *freshwater marshes*, shallow water allows plants such as cattails to grow above the water surface. *Swamps* also consist of shallow water rich with vegetation, but they occur in forested areas. *Bogs* are ponds covered with thick floating mats of vegetation, and they can represent a stage in aquatic succession.

Wetlands are extremely valuable as habitat for wildlife. They also provide ecosystem services by slowing runoff, reducing flooding, recharging aquifers, and filtering pollutants. People have drained and filled wetlands extensively, largely for agriculture. Southern Canada and the United States have lost well over half their wetlands since European colonization.

Groundwater plays key roles in the hydrologic cycle

Any precipitation reaching Earth's land surface that does not evaporate, flow into waterways, or get taken up by organisms infiltrates the surface. Most percolates downward through the soil to become **groundwater** (Figure 11.4).

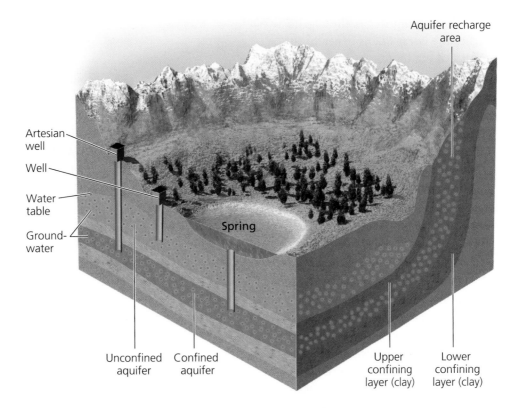

FIGURE 11.4 Groundwater may occur in *unconfined aquifers* above impermeable layers or in *confined aquifers* under pressure between impermeable layers. Water may rise naturally to the surface at springs and through the wells we dig. Artesian wells tap into confined aquifers to mine water under pressure.

As we first saw in Chapter 3 (▶ p. 68), groundwater is contained within **aquifers**, porous, spongelike formations of rock, sand, or gravel that hold water. An aquifer's upper layer, or zone of aeration, contains pore spaces partly filled with water. In the lower layer, or zone of saturation, the spaces are completely filled with water. The boundary between these two zones is the **water table** (▶ p. 68). Picture a sponge resting partly submerged in a tray of water; the lower part of the sponge is saturated, whereas the upper portion contains plenty of air in its pores.

Groundwater flows downhill and from areas of high pressure to areas of low pressure. A typical rate of flow might be only 1 m (3 ft) per day, so groundwater can remain in an aquifer for a long time. When we pump groundwater through wells, we are drawing up ancient water. The average age of groundwater has been estimated at 1,400 years, and some is tens of thousands of years old.

Nonetheless, volumes of groundwater are large enough that each day in the United States alone, aquifers release 1.9 trillion L (492 billion gal) into bodies of surface water—nearly as much as the daily flow of the Mississippi River.

The world's largest known aquifer is the Ogallala Aquifer, which underlies the Great Plains beneath eight U.S. states from South Dakota to Texas (Figure 11.5). Water from this massive aquifer has enabled American farmers to create the most bountiful grain-producing region in the world. The Ogallala Aquifer spans 453,000 km^2 (176,700 mi^2), is 370 m (1,200 ft) deep at its thickest point, and held 3,700 km^3 (881 mi^3) of water before pumping began. Overpumping for irrigation has reduced this aquifer's volume by nearly 10% so far—a volume equal to 18 years' worth of the entire flow of the Colorado River.

Water is unequally distributed across Earth's surface

Different regions of the world possess vastly different amounts of groundwater, surface water, and precipitation. Precipitation ranges from about 1,200 cm (470 in.) per year at Mount Waialeale on the Hawaiian island of Kauai to virtually zero in Chile's Atacama Desert.

People are not distributed across the globe in accordance with water availability (Figure 11.6). For example, Canada has 20 times more water per citizen than does China. The Amazon River carries 15% of the world's runoff, but its watershed holds less than half a percent of the world's human population. Many densely populated nations, such as Pakistan, Iran, and Egypt, face serious water shortages. Asia possesses the most water of any continent but has the least water available per person, whereas Australia, with the least amount of water, boasts the most water available per person.

Freshwater is distributed unevenly in time as well as space. In India, monsoon storms can dump half of a region's annual rain in just a few hours. Northwest China receives three-fifths of its annual precipitation during 3 months when crops do not need it. Uneven distribution of water across time is one reason people have erected dams to store water, so that it may be distributed when needed.

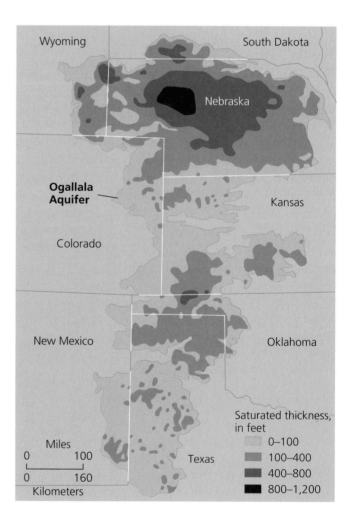

FIGURE 11.5 The Ogallala Aquifer is the world's largest. It underlies the Great Plains beneath portions of eight U.S. states from South Dakota to Texas. Overpumping for irrigation is currently reducing the volume and extent of this vital groundwater resource.

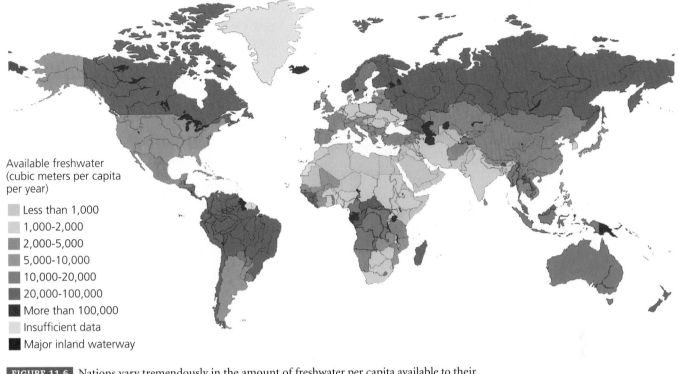

Available freshwater
(cubic meters per capita
per year)

- Less than 1,000
- 1,000-2,000
- 2,000-5,000
- 5,000-10,000
- 10,000-20,000
- 20,000-100,000
- More than 100,000
- Insufficient data
- Major inland waterway

FIGURE 11.6 Nations vary tremendously in the amount of freshwater per capita available to their citizens. For example, Iceland, Papua New Guinea, Gabon, and Guyana (colored red in this map) have more than 100 times as much water per capita as do many Middle Eastern and North African countries. Go to **GRAPH***it!* at www.aw-bc.com/withgott. Data from United Nations Environment Programme and World Resources Institute, as presented by Harrison, P., and F. Pearce. 2000. *AAAS atlas of population and the environment.* Berkeley, CA: University of California Press.

How We Use Water

In our attempts to harness freshwater sources for countless purposes, we have achieved impressive engineering accomplishments. In so doing, we have altered many environmental systems. It is estimated that 60% of the world's largest 227 rivers (and 77% of those in North America and Europe) have been strongly or moderately affected by artificial dams, canals, and diversions.

We are also using too much water. Data indicate that our present freshwater consumption in much of the world is unsustainable and that we are depleting many sources of surface water and groundwater. One-third of the world's people are already affected by water scarcity (with less than 1,000 m³ (35,000 ft³) of water per person per year), according to a comprehensive global assessment presented in late 2006.

Water supplies our households, agriculture, and industry

We all use water at home for drinking, cooking, and cleaning. Farmers and ranchers use water to irrigate crops

and water livestock. Most manufacturing and industrial processes require water. Globally, we spend about 70% of our annual freshwater allotment on agriculture. Industry accounts for roughly 20%, and residential and municipal uses for only 10%.

When we remove water from an aquifer or surface water body and do not return it, this is called **consumptive use**. A large portion of agricultural irrigation and of many industrial and residential uses is consumptive. **Nonconsumptive use** of water does not remove, or only temporarily removes, water from its source. Using water to generate electricity at hydroelectric dams is an example of nonconsumptive use; water is taken in, passed through dam machinery to turn turbines, and released downstream.

We have erected thousands of dams

A **dam** is any obstruction placed in a river or stream to block the flow of water so that water can be stored in a reservoir. We build dams to prevent floods, provide drinking water, facilitate irrigation, and generate electricity (Figure 11.7). Power generation with hydroelectric dams is discussed in Chapter 14 (▶ pp. 335–337). Worldwide, we

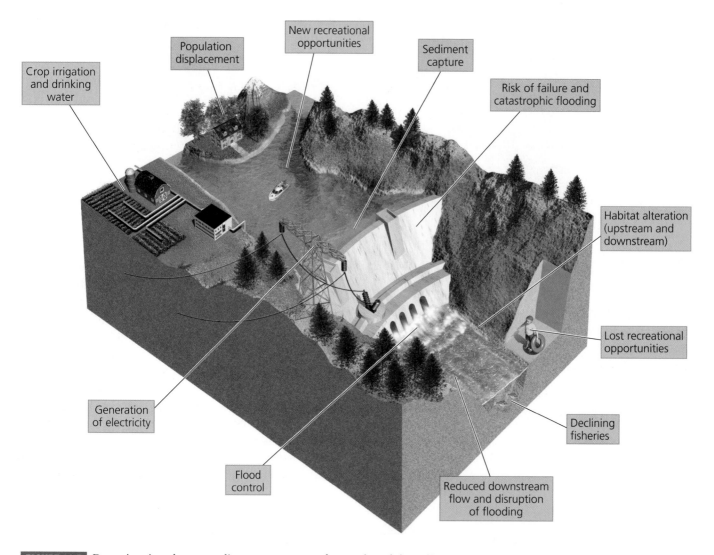

FIGURE 11.7 Damming rivers has many diverse consequences for people and the environment. The generation of clean and renewable electricity is one of several major benefits (green boxes) of hydroelectric dams. Habitat alteration is one of several negative impacts (red boxes).

have erected more than 45,000 large dams (>15 m, or 49 ft, high) across rivers in over 140 nations. We have built tens of thousands of smaller dams. Virtually the only major rivers in the world that remain undammed run through the tundra and taiga of Canada, Alaska, and Russia, and in remote regions of Latin America and Africa.

Our largest dams comprise some of humanity's greatest engineering feats. The two behemoths of the Colorado River, Hoover Dam and Glen Canyon Dam, stand 221 m (726 ft) and 216 m (710 ft) high, respectively, stretch for 379 m (1,244 ft) and 476 m (1,560 ft) across, and consist of 6.6 and 7.3 million tons of concrete and steel.

The complex mix of benefits and costs that dams produce (see Figure 11.7) is exemplified by the world's largest dam, the Three Gorges Dam on China's Yangtze River (Figure 11.8a). This dam will generate hydroelectric power,

enable boats and barges to travel farther upstream, and provide flood control. The power it generates may replace dozens of large coal or nuclear plants. However, the Three Gorges Dam has cost $25 billion to build, and its reservoir is flooding 22 cities and the homes of 1.13 million people, requiring the largest resettlement project in China's history (Figure 11.8b). The reservoir is also submerging 10,000-year-old archaeological sites, productive farmlands, and wildlife habitat. Moreover, as water flow slows, suspended sediment is settling and filling the reservoir.

People who feel that the costs of some dams have outweighed their benefits have pushed for such dams to be dismantled. By removing dams and letting rivers flow free, these people say, we can restore riparian ecosystems, reestablish economically valuable fisheries, and revive river recreation such as fly-fishing and rafting. Increasingly,

(a) The Three Gorges Dam in Yichang, China

(b) Displaced people in Sichuan Province, China

FIGURE 11.8 China's Three Gorges Dam (**a**), completed in 2003, is the world's largest dam, at 186 m (610 ft) high and 2 km (1.3 mi) wide. Once it fills in 2009, the resulting reservoir should be 616 km (385 mi) long, as long as Lake Superior. The reservoir will hold over 38 trillion L (10 trillion gal) of water. Well over a million people were displaced and whole cities were leveled for its construction (**b**).

private dam owners and the Federal Energy Regulatory Commission (FERC), the U.S. government agency charged with renewing licenses for dams, have agreed. Roughly 500 dams have been removed in the United States, nearly 200 of them in the past decade. One reason is that many aging

dams are in need of costly repairs or have outlived their economic usefulness.

Dikes and levees are meant to control floods

Flood prevention ranks high among reasons we control the movement of freshwater. People have always been attracted to riverbanks for their water supply, for their beauty, and for the flat topography and fertile soil of floodplains. Flooding is a normal, natural process due to snowmelt or heavy rain, and floodwaters spread nutrient-rich sediments over large areas, benefiting both natural systems and human agriculture.

In the short term, however, floods can damage farms, homes, and property. To protect against floods, individuals and governments have built *dikes* and *levees* (long raised mounds of earth) along the banks of rivers to hold rising water in main channels. Many dikes are small and locally built, but in the United States the Army Corps of Engineers has constructed thousands of miles of massive levees along the banks of major waterways (those that failed in New Orleans after Hurricane Katrina are examples). Although these structures prevent flooding at most times and places, they can sometimes worsen flooding because they force water to stay in channels and accumulate, leading to occasional catastrophic overflow events.

We divert—and deplete—surface water to suit our needs

People have long diverted water from rivers, streams, lakes, and ponds to homes, cities, and farm fields. The Colorado River's water is heavily diverted and utilized (Figure 11.9). What water is left after all the diversions comprises just a trickle making its way to the Gulf of California. On some days, water does not reach the Gulf at all. This flow reduction has drastically altered the ecology of the lower river and the once-rich delta, changing vegetative communities, wiping out whole populations of fish and aquatic invertebrates, and devastating fisheries.

The Colorado's plight is not unique. Several hundred miles to the east, the Rio Grande also frequently runs dry, the victim of overextraction by Mexican and U.S. farmers in times of drought. The situation is even worse for China's Yellow River, so the Chinese government wants to build a massive aqueduct to supplement its flow with water from the Yangtze River. Even the river that has nurtured human civilization as long as any other, the Nile, now peters out before reaching its mouth.

VIEWPOINTS

Dam Removal

Dams bring us many benefits, but they also exert ecological and social impacts. **Have some dams outlived their usefulness, and if so, should we dismantle them?**

Dams for Today and Tomorrow

Our need for dams today is greater than ever because water is a finite resource. Over the past century, global water use has increased at twice the rate of population growth. Currently, world population is expected to grow by 50%, to 9 billion people in total, by 2050. Dams and reservoirs address the needs of a growing world by efficiently storing and regulating water for multiple uses. Our world relies on the benefits they bring—drinking water, flood control, power generation, irrigation, and recreation.

Ninety percent of the dams in the United States are small, local projects that lack controversy. Consider what dams do every day for millions of people. Along the Mississippi, 70% of America's grain exports are barged to the Gulf of Mexico. Dams support 55 million irrigated acres of crop and pasture land (mostly in the arid West). Dams and reservoirs carry water to millions of people via canals and aqueducts. And dams help communities avoid billions of dollars in flood damage.

With dams, however, come environmental concerns such as fish passage, changes to water quality, and altered habitats. The challenge for any community is to balance economic, environmental, and social considerations. Today's choices often reflect the values, needs, wealth, and options of different communities and countries. It should be no surprise that in a world where 1.7 billion people are without electricity, hydropower is being developed in 80 countries.

Our challenge is to make decisions that embrace what research, sound science, technological innovation, and engineering prowess offers. We can embrace these things while also staying true to our historic and evolving cultural, environmental, and economic values. We owe it to future generations to make thoughtful, responsible policy choices about dams that affect not only our way of life, but also theirs.

Thomas Flint is a fifth-generation farmer, actively farming in Grant County, Washington. He was elected to the Grant County Public Utility District board of commissioners in 2000, is founder and director of the public education effort known as AgFARMation, is a grassroots activist, and holds director's positions on the Black Sands Irrigation District and the Columbia Basin Development League.

The Case for Dam Removal

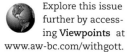

The dams currently in existence in the United States were built to provide a variety of services, including flood control, water supply, and hydropower, which runs mills and generates electricity. Although many dams continue to provide a useful service, large numbers are considered obsolete, providing no direct economic, safety, or social function. For example, many mill dams continue to stand across streams and rivers 100 to 200 years after the mill they powered went out of operation or was torn down. These dams should be considered for removal.

Regardless of size, all dams harm riparian environments. Dams block the free flow of water down a stream corridor and create a pool, or impoundment, behind them—an artificial lake in the middle of a stream community. Impounded waters often divide into layers by temperature and depth, with heated waters in the upper layer and oxygen-poor cooler water in the lower layer. The macroinvertebrates that fish depend on for food cannot survive under these lake conditions. Carp and non-native lake fish that can survive in hotter and oxygen-poor waters often displace trout and other cold-water stream species.

Dams block the movement of migratory fish and other aquatic species, preventing them from reaching upstream areas to feed, spawn, and successfully reproduce. Dams also block river sediments that would normally travel downstream and replenish beaches or gravel stream bottoms, where most macroinvertebrates live and where fish spawn.

Rivers are dynamic systems. They move within floodplains, exchanging nutrients, sediments, and interacting on many levels. When dams interrupt that exchange, river functions are impaired, and the fish and wildlife dependent on free-flowing river systems do not thrive as well. Once a dam has outlived its utility, it makes great sense to restore the river back to its original condition.

Sara Nicholas is associate director of dam programs for American Rivers and works out of their mid-Atlantic office in Harrisburg, Pennsylvania. She has a master of science degree in environmental science from the Yale School of Forestry.

Explore this issue further by accessing **Viewpoints** at www.aw-bc.com/withgott.

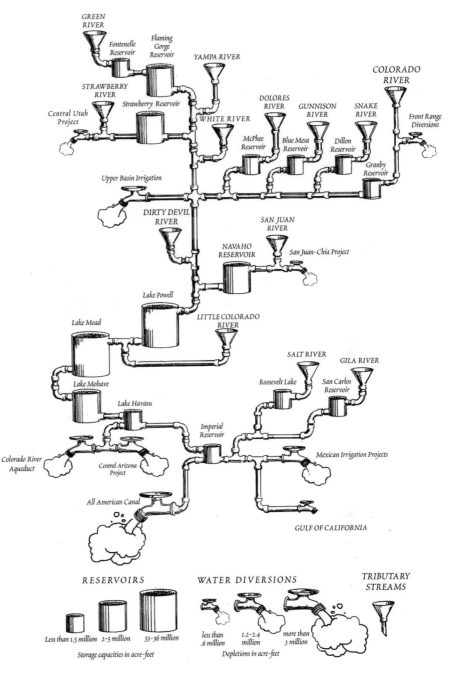

FIGURE 11.9 The 2,330-km (1,450-mile) Colorado River drains a watershed of 637,000 km² (246,000 mi²). The degree to which this once-wild river has been engineered led cartoonist Lester Dore to portray the Colorado and its tributaries as an immense plumbing system. Early in its course, some Colorado River water is piped through a mountain tunnel and down the Rockies' eastern slope to supply Denver. More is removed for Las Vegas and other cities and for farmland as the water proceeds downriver. When it reaches Parker Dam on the California-Arizona state line, large amounts are diverted into the Colorado River Aqueduct, which brings water to millions in Los Angeles and San Diego in a long open-air canal. From Parker Dam, Arizona also draws water, transporting it in the canals of the Central Arizona Project. Further south at Imperial Dam, water is diverted into the Coachella and All-American Canals, destined for agriculture in California's arid Imperial Valley. At the Colorado River's mouth, often there is no water left. *Source: High Country News, 10 November 1997.*

Nowhere are the effects of surface water depletion so evident as at the Aral Sea. Once the fourth-largest freshwater body on Earth, just larger than Lake Huron, it has lost four-fifths of its volume in 40 years (Figure 11.10). This dying inland sea, on the border of present-day Uzbekistan and Kazakhstan, is the victim of irrigation practices. The former Soviet Union instituted large-scale cotton farming in this region by flooding the dry land with water from the two rivers leading into the Aral Sea. For a few decades this practice boosted Soviet cotton production, but today 60,000 fishing jobs are gone, and

winds blow pesticide-laden dust from the dry lakebed. What cotton grows on the blighted soil cannot bring the region's economy back. Scientists, engineers, and local people are struggling to save the Aral Sea and its damaged ecosystems, and they now may have finally begun reversing its decline.

We are depleting groundwater

Groundwater is more easily depleted than surface water because most aquifers recharge very slowly. If we compare

(a) Ships stranded by the Aral Sea's fast-receding waters

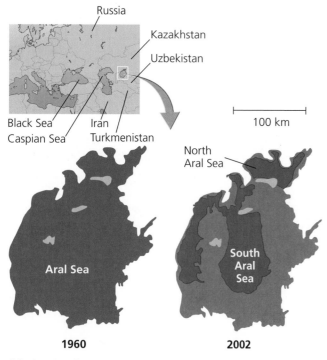

(b) The shrinking Aral Sea, then and now

FIGURE 11.10 Ships lie stranded in the sand (**a**) because the waters of Central Asia's Aral Sea have receded so far and so quickly. The Aral Sea was once the world's fourth-largest freshwater body. It has been shrinking for the past four decades (**b**) and could disappear completely in the near future. The primary cause has been overwithdrawal of water to irrigate cotton crops.

an aquifer to a bank account, we are making more withdrawals than deposits, and the balance is dwindling.

Most groundwater use goes toward agriculture. We withdraw 60% more water for irrigation today than in 1960, and since then the amount of land under irrigation

has doubled. Expansion of irrigated agriculture has kept pace with population growth; irrigated area per capita has remained for four decades at around 460 m^2 (4,900 ft^2)—about half a football field for each of us. Irrigation can more than double crop yields by allowing farmers to apply water when and where it is needed. The world's 274 million ha (677 million acres) of irrigated cropland make up only 18% of world farmland but yield fully 40% of world agricultural produce, including 60% of the global grain crop. Still, most irrigation remains highly inefficient (see Figure 7.6a, ▶ p. 149), and overirrigation leads to waterlogging and salinization (▶ pp. 148–149).

Worldwide today, 15–35% of irrigation withdrawals are thought to be unsustainable, and, like the Ogallala, many aquifers are being drained as water is "mined" at rates faster than it is recharged. As aquifers are depleted, water tables drop. Groundwater becomes more difficult and expensive to extract, and eventually it may run out. In parts of Mexico, India, China, and other nations in Asia and the Middle East, water tables are falling 1–3 m (3–10 ft) per year. When groundwater is overpumped in coastal areas, salt water can intrude into aquifers, making water undrinkable. Moreover, as aquifers lose water, the land surface above may subside. For this reason, cities from Venice to Bangkok to Mexico City are slowly sinking, while streets buckle, buildings flood, and pipes break. Falling water tables also do vast ecological harm by drying up wetlands vital for wildlife and ecosystem services.

Will we see a future of water wars?

Freshwater depletion leads to shortages, and resource scarcity can lead to conflict. On the Colorado River in 1933, the governor of Arizona foresaw that California's water diversion might endanger Arizona's future allotment, so he sent the state's National Guard to threaten the construction of Parker Dam. After a long standoff, the U.S. interior secretary halted the project to avoid hostilities while the issue was mediated in court. Arizona won the court case, but California got the dam authorized by Congress, and Arizona chose not to tackle the U.S. Army troops sent to protect the dam's construction.

Many predict that water's role in regional conflicts will increase as populations continue to grow in water-poor areas. World Water Commission chairman Ismail Serageldin has remarked that "the wars of the twenty-first century will be fought over water." A total of 261 major rivers cross national borders, and transboundary disagreements are common. Water is already a key element in the hostilities among Israel, the Palestinian people, and neighboring nations.

On the positive side, many nations have cooperated with neighbors to resolve water disputes. India has struck cooperative agreements over management of transboundary rivers with Pakistan, Bangladesh, Bhutan, and Nepal. In Europe, treaties have been signed by multiple nations along the Rhine and Danube rivers. Such progress gives reason to hope that future water wars may be few and far between.

Solutions to Freshwater Depletion

To address freshwater shortages, we can aim either to increase supply or to reduce demand. Lowering demand (such as through conservation and efficiency measures) is more difficult politically but will likely be necessary in the long term. In the developing world, international aid agencies are increasingly funding demand-based solutions over supply-based solutions, because demand-based solutions offer better economic returns and cause less ecological and social damage.

Some solutions involve increasing supply

To increase supply in a given area, people have transported water through pipes and aqueducts from areas where it is more plentiful or accessible. In many instances, water-poor regions have forcibly appropriated water from communities too weak to keep it for themselves. For instance, Los Angeles built itself up with water it appropriated from the Owens Valley, Mono Lake, and other rural regions of California. Transporting water from place to place, however, is not a just solution if it harms the region that loses the water.

Weighing the **ISSUES:**
Reaching for Water

In 1941, the burgeoning metropolis of Los Angeles needed water and decided to divert streams feeding into Mono Lake, over 565 km (350 mi) away in northern California. As the lake level fell 14 m (45 ft) in 40 years, salt concentrations doubled, and aquatic communities suffered. Other desert cities—such as Las Vegas, Phoenix, and Denver—are expected to double in population in coming decades. Where might they go for water to sustain their communities? What challenges might they face in trying to pipe in water from distant sources? How might people living in these source areas be affected? How else could these cities meet their future water needs?

Another strategy is to "make" more freshwater by removing salt from seawater or other water of marginal quality. This technological approach is called **desalination**, or *desalinization*. One method of desalination mimics the hydrologic cycle by hastening evaporation from allotments of ocean water with heat and then condensing the vapor— essentially distilling freshwater. Another method forces water through membranes to filter out salts; the most common such process is reverse osmosis. Unfortunately, desalination is quite expensive and requires substantial inputs of fossil fuel energy. Over 7,500 desalination facilities are operating worldwide. Most are in the arid Middle East, where water is scarce and oil is cheap (Figure 11.11).

Other solutions involve reducing demand

Because most water use is for agriculture, it makes sense to look first to agriculture for ways to decrease demand. Farmers can improve efficiency by lining irrigation canals to prevent leaks, leveling fields to minimize runoff, and adopting efficient irrigation methods. Low-pressure spray irrigation sprays water downward toward plants, and drip irrigation systems target individual plants and introduce water directly onto the soil (see Figure 7.6a, ▸ p. 148). Both methods reduce water lost to evaporation and surface runoff. Experts have estimated that drip irrigation, which has efficiencies as high as 90%, could cut water use in half while raising yields by 20–90% and giving farmers of the developing world $3 billion in extra annual income.

Choosing crops to match the land and climate in which they are being farmed can save huge amounts of water. Currently, crops that require a great deal of water, such as cotton, rice, and alfalfa, are often planted in arid areas with government-subsidized irrigation. As a result, the true cost of water is not part of the costs of growing

FIGURE 11.11 Kuwaiti engineers walk along water intake pipes for a desalination plant on the Persian Gulf.

the crop. Eliminating subsidies and growing crops in climates with adequate rainfall could greatly reduce water use. Finally, selective breeding (▶ pp. 78–79, 152) and genetic modification (▶ pp. 156–160) can result in crop varieties that require less water.

As individuals, we each can help reduce agricultural water use by decreasing the amount of meat we eat, because producing meat requires far greater water inputs than producing grain or vegetables (▶ p. 163). In our households, we can reduce water use by installing low-flow faucets, showerheads, washing machines, and toilets. Automatic dishwashers, studies show, use less water than does washing dishes by hand. We can water lawns at night, when water loss from evaporation is minimal. Better yet, we can replace water-intensive lawns with native plants adapted to our region's natural precipitation patterns. *Xeriscaping*, landscaping using plants adapted to arid conditions, has become popular in much of the U.S. Southwest.

Industry and municipalities can take water-saving steps as well. Manufacturers can shift to processes that use less water and, in doing so, can reduce their costs. In many areas, municipal wastewater is being recycled for irrigation and industrial uses. Finding and patching leaks in pipes has alone saved some cities and companies large amounts of water—and money. Boston and its suburbs reduced water demand by 31% between 1987 and 2004 by patching leaks, retrofitting homes with efficient plumbing, auditing industry, and promoting conservation to the public. This program enabled Massachusetts to avoid an unpopular $500-million river diversion scheme.

Economists who want to use market-based strategies to achieve sustainable water use have suggested ending government subsidies of inefficient practices and letting water become a commodity whose price reflects the true costs of its extraction. Others worry that making water a fully priced commodity can make this life-giving substance unavailable to the poor. In the past two decades, the ownership or operation of some public water systems has been transferred to private companies. This was done in hope of increasing the systems' efficiency, but companies have little incentive to allow equitable access to water for rich and poor alike. Already in some developing countries, rural residents without access to public water supplies, who are forced to buy water from private vendors, end up paying on average 12 times more than those connected to public supplies.

Regardless of how demand is addressed, the ongoing shift from supply-side to demand-side solutions is beginning to pay dividends. The United States decreased its total water consumption by 10% from 1980 to 1995, thanks to conservation measures, even while its population grew 16%.

Freshwater Pollution and Its Control

The quantity and distribution of freshwater poses one set of environmental and social challenges. Safeguarding the *quality* of water involves another collection of environmental and human health dilemmas. Developed nations have made admirable advances in cleaning up water pollution over the past few decades. Still, the World Commission on Water recently concluded that over half the world's major rivers are "seriously depleted and polluted, degrading and poisoning the surrounding ecosystems, threatening the health and livelihood of people who depend on them." Meanwhile, the largely invisible pollution of groundwater has been termed a "covert crisis."

Water pollution comes from point and non-point sources

The term **pollution** describes the release of matter or energy into the environment that causes undesirable impacts on the health and well-being of people or other organisms. Pollution can be physical, chemical, or biological, and it can affect water, air, or soil.

Some water pollution is emitted from **point sources**—discrete locations, such as a factory or sewer pipe. In contrast, **non-point-source** pollution arises from multiple cumulative inputs over larger areas, such as farms, city streets, and residential neighborhoods (Figure 11.12). The U.S. Clean Water Act (▶ p. 39), by targeting industrial discharges, addressed point-source pollution with some success, such that non-point-source pollution has a greater impact on water quality in the United States today. Many common activities give rise to non-point-source water pollution, such as applying fertilizers and pesticides to lawns, applying salt to roads in winter, and changing automobile oil. To minimize non-point-source pollution of drinking water, governments limit development on watershed land surrounding reservoirs.

Water pollution comes in many forms and can cause diverse impacts on aquatic ecosystems and human health.

Nutrient pollution We saw in Chapter 3 with the Gulf of Mexico's dead zone how nutrient pollution from fertilizers and other sources can lead to eutrophication and hypoxia in coastal marine areas (see Figure 3.3, ▶ p. 52). Eutrophication proceeds in a similar fashion in freshwater

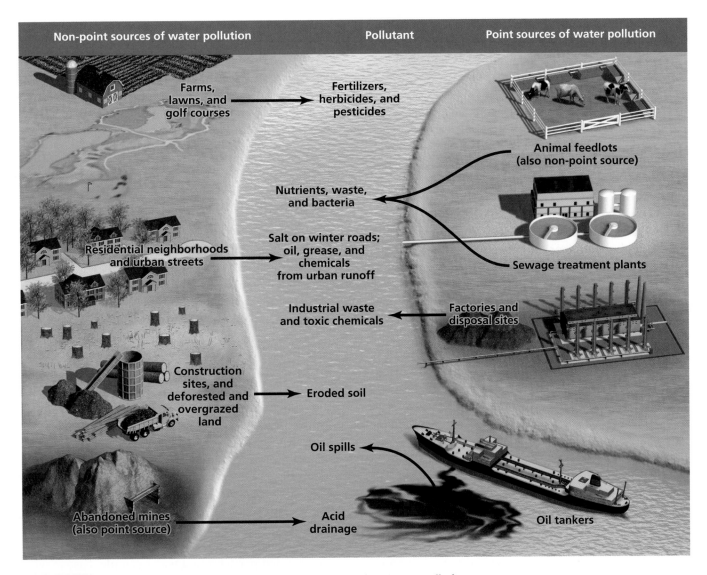

FIGURE 11.12 Point-source pollution comes from discrete facilities or locations, usually from single outflow pipes. Non-point-source pollution (such as runoff from streets, residential neighborhoods, lawns, and farms) originates from numerous sources spread over large areas.

systems, where phosphorus is usually the nutrient that spurs growth (▸ p. 61). When excess phosphorus enters surface waters, it boosts the growth of algae, which cover the water's surface, depriving deeper-water plants of sunlight. As algae die off, they provide food for decomposing bacteria. Decomposition requires oxygen, so the increased bacterial activity drives down levels of dissolved oxygen. These levels can drop too low to support fish and shellfish, leading to dramatic changes in aquatic ecosystems.

Eutrophication is a natural process, but excess nutrient input from runoff from farms, golf courses, lawns, and sewage can dramatically increase the rate at which it occurs. We can reduce nutrient pollution by treating wastewater, reducing fertilizer application, planting vegetation to increase nutrient uptake, and purchasing phosphate-free detergents.

Pathogens and waterborne diseases Disease-causing organisms (pathogenic viruses, protists, and bacteria) can enter drinking water supplies when these are contaminated with human waste from inadequately treated sewage or animal waste from feedlots (▸ p. 162). Biological pollution by pathogens causes more human health problems than any other type of water pollution. An international study in 2000 showed that despite many advances, over 1.1 billion people are still without safe water supplies, and 2.4 billion people lack sewer or sanitation facilities. Most of these people are Asians and Africans, and four-fifths of those without sanitation live in rural areas. These conditions contribute to widespread health impacts and 5 million deaths per year.

Treating sewage (▸ pp. 255–257) constitutes one approach for reducing health risks. Another is using chemical or other means to disinfect drinking water. Personal hygiene is vital, as is government enforcement of regulations to ensure the cleanliness of food production, processing, and distribution.

Toxic chemicals Our waterways have become polluted with toxic organic substances of our own making, including pesticides, petroleum products, and other synthetic chemicals. Many of these can poison animals and plants, alter aquatic ecosystems, and cause an array of human health problems, including cancer. In addition, toxic metals such as arsenic, lead, and mercury, as well as acids from acid precipitation and from acid drainage from mining sites (▸ pp. 313–314), also cause negative impacts on human health and the environment. We discussed health impacts of toxic chemicals in Chapter 10. Legislating and enforcing stricter regulations of industry can help reduce releases of toxic chemicals. Better yet, we can modify our industrial processes and our purchasing decisions to rely less on these substances.

Sediment Although floods build fertile farmland, sediment that rivers transport can also impair aquatic ecosystems. Mining, clear-cutting, overgrazing, land clearing for real estate development, and tilling of farm fields all expose soil to wind and water erosion (▸ p. 145). Some water bodies, such as the Colorado River and China's Yellow River, are naturally sediment-rich, but many others are not. When a clear-water river receives a heavy influx of eroded sediment, aquatic habitat can change dramatically, and fish adapted to clear-water environments may not be able to handle the change. We can reduce sediment pollution by better managing farms and forests and by avoiding large-scale disturbance of vegetation.

Thermal pollution Water's ability to hold dissolved oxygen decreases as temperature rises, so some aquatic organisms may not survive when human activities raise water temperatures. When we withdraw water from a river and use it to cool an industrial facility, we transfer heat energy from the facility back into the river where the water is returned. People also raise surface water temperatures by removing streamside vegetation that shades water.

Too little heat can also cause problems. On the Colorado and other dammed rivers, water at the bottoms of reservoirs is colder than water at the surface. When dam operators release water from the depths of a reservoir, downstream water temperatures drop suddenly. In the Colorado River system, these low water temperatures have favored cold-loving invasive trout over an endangered native species of suckerfish.

Scientists use several indicators of water quality

Scientists and technicians measure certain physical, chemical, and biological properties of water to characterize its quality. Biological properties include the presence of fecal coliform bacteria, which indicate contamination by human waste and suggest the presence of other disease-causing organisms. Algae and aquatic invertebrates are also commonly used as biological indicators of water quality.

Chemical properties include nutrient concentrations, pH (▸ pp. 56–57), taste and odor, and hardness. Hard water contains high concentrations of calcium and magnesium ions, prevents soap from lathering, and leaves chalky deposits behind when heated or boiled. An important chemical characteristic is dissolved oxygen content. Dissolved oxygen is an indicator of aquatic ecosystem health because surface waters low in dissolved oxygen support less aquatic life.

Among physical characteristics, scientists use temperature, color, and turbidity. Turbidity measures the density of suspended particles in a water sample. Fast-moving rivers that cut through arid or eroded landscapes, such as the Colorado River, carry a great deal of sediment and are turbid and muddy-looking as a result. If scientists can measure only one parameter, they will often choose turbidity, because it tends to be correlated with many others and is thus a good indicator of overall water quality.

Groundwater pollution is a serious problem

Many of these types of pollution threaten groundwater as well as surface water. Groundwater pollution is largely hidden from view and is extremely difficult to monitor; it can be out-of-sight, out-of-mind for decades until widespread contamination of drinking supplies is discovered. Groundwater pollution is also more difficult to manage than surface water pollution. Rivers flush their pollutants fairly quickly, but groundwater retains its contaminants until they decompose, which can sometimes be many years or decades. The long-lived pesticide DDT, for instance, is still found widely in U.S. aquifers even though it was banned 35 years ago. Moreover, because groundwater generally contains less dissolved oxygen, microbes, minerals, and organic matter than do surface water or soil water, chemicals are broken down more slowly. For instance, concentrations of the herbicide alachlor are reduced by half after 20 days in soil, but in groundwater this reduction takes almost 4 years.

Some chemicals that are toxic at high concentrations, including aluminum, fluoride, nitrates, and sulfates, occur naturally in groundwater. For instance, thousands of wells dug for drinking water in Bangladesh by international aid workers in the 1970s were later found to contain water naturally high in arsenic.

However, groundwater pollution from human activity is widespread. Industrial, military, and urban wastes—from heavy metals to petroleum products to industrial solvents—can leach through soil and seep into aquifers. Pathogens and other pollutants can enter groundwater through improperly designed wells and from the pumping of hazardous waste below ground (▶ pp. 367–368). Nitrate from agricultural fertilizers has leached into aquifers in Canada and 49 U.S. states. Pesticides were detected in over half of the shallow aquifer sites tested in the United States in the mid-1990s, although generally below the standards set by the EPA for drinking water. Agriculture can also contribute pathogens; in 2000, the groundwater supply of Walkerton, Ontario, became contaminated with the bacterium *Escherichia coli*, or *E. coli*. Two thousand people became ill, and seven died.

Leakage from underground septic tanks, tanks of industrial chemicals, and tanks of oil and gas also create pollution. Across the United States, the Environmental Protection Agency (EPA) has embarked on a nationwide cleanup program to unearth and repair leaky tanks before they further degrade soil and groundwater quality. After more than a decade of work, the EPA in June 2006 had confirmed leaks from 460,000 tanks, had initiated cleanups on 430,000 of them, and had completed cleanups of 343,000.

Legislative and regulatory efforts have helped reduce pollution

As numerous as our freshwater pollution problems may seem, it is important to remember that many of them were worse a few decades ago, when the Cuyahoga River would catch fire (▶ pp. 37–38). Citizen activism and government response during the 1960s and 1970s in the United States resulted in laws such as the Clean Water Act of 1977. This act made it illegal to discharge pollution from a point source without a permit, set standards for industrial wastewater, set standards for contaminant levels in surface waters, and funded construction of sewage treatment plants. Thanks to such legislation, point-source pollution in the United States has been reduced, and rivers and lakes are cleaner than they have been in decades. In many other developed nations, citizens and governments have followed suit and enacted legislation to reduce pollution.

The Great Lakes of Canada and the United States represent a success story in fighting water pollution. In the 1970s these lakes, which hold 18% of the world's surface freshwater, were badly polluted with wastewater, fertilizers, and toxic chemicals. Algal blooms lined beaches, and Lake Erie was pronounced dead. Today, efforts of the Canadian and U.S. governments have paid off. According to Environment Canada, releases of seven toxic chemicals have been reduced by 71%, municipal phosphorus has been decreased by 80%, and chlorinated pollutants from paper mills are down by 82%. Levels of PCBs and DDE are down by 78% and 91%. Bird populations are rebounding, and Lake Erie is now home to the world's largest walleye fishery. The Great Lakes' troubles are by no means over—sediment pollution remains heavy, mercury and PCBs still fall from the air, and fish are not always safe to eat. However, the progress so far shows how conditions can improve when citizens push their governments to take action.

We treat our drinking water

Technological advances have also improved our ability to control pollution. The treatment of drinking water is a widespread and successful practice in developed nations today. The U.S. EPA sets standards for over 80 drinking water contaminants, which local governments and private water suppliers are obligated to meet. Before being sent to your tap, water from a reservoir or aquifer is treated with chemicals to remove particulate matter; is passed through filters of sand, gravel, and charcoal; and/or is disinfected with chlorine or other agents.

We treat our wastewater

Wastewater treatment (Figure 11.13) is also now a mainstream practice. **Wastewater** refers to water used by people, including water carrying sewage; water from showers, sinks, washing machines, and dishwashers; water used in manufacturing or cleaning processes by businesses and industries; and stormwater runoff. Although natural systems can process moderate amounts of wastewater, the

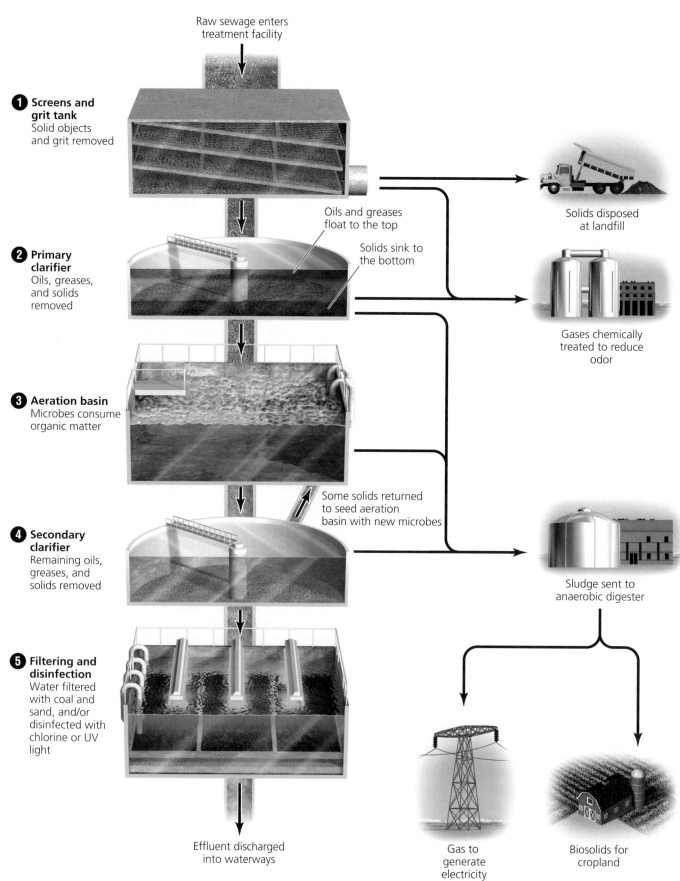

Raw sewage enters
treatment facility

**1 Screens and
grit tank**
Solid objects
and grit removed

Oils and greases
float to the top

Solids sink to
the bottom

**2 Primary
clarifier**
Oils, greases,
and solids
removed

Solids disposed
at landfill

Gases chemically
treated to reduce
odor

3 Aeration basin
Microbes consume
organic matter

Some solids returned
to seed aeration
basin with new microbes

**4 Secondary
clarifier**
Remaining oils,
greases, and
solids removed

Sludge sent to
anaerobic digester

**5 Filtering and
disinfection**
Water filtered
with coal and
sand, and/or
disinfected with
chlorine or UV
light

Effluent discharged
into waterways

Gas to
generate
electricity

Biosolids for
cropland

FIGURE 11.13   Shown here is a generalized process from a modern, environmentally sensitive wastewater treatment facility.

large amounts generated by our densely populated areas can harm ecosystems and pose threats to human health. Thus, attempts are now widely made to treat wastewater before releasing it into the environment.

In rural areas, **septic systems** are the most popular method of wastewater disposal. In a septic system, wastewater runs from the house to an underground septic tank, inside which solids and oils separate from water. The clarified water proceeds downhill to a drain field of perforated pipes laid horizontally in gravel-filled trenches underground. Microbes decompose the wastewater these pipes emit. Periodically, solid waste needs to be pumped from the septic tank and taken to a landfill.

In more densely populated areas, municipal sewer systems carry wastewater from homes and businesses to centralized treatment locations. There, pollutants are removed by physical, chemical, and biological means (see Figure 11.13).

At a treatment facility, **primary treatment**, the physical removal of contaminants in settling tanks or clarifiers, generally removes about 60% of suspended solids. Wastewater then proceeds to **secondary treatment**, in which water is stirred and aerated so that aerobic bacteria degrade organic pollutants. Roughly 90% of suspended solids may be removed after secondary treatment. Finally, the clarified water is treated with chlorine, and sometimes ultraviolet light, to kill bacteria. Most often, the treated water, called effluent, is piped into rivers or the ocean following primary and secondary treatment. Sometimes, however, "reclaimed" water is used for lawns and golf courses, for irrigation, or for industrial purposes such as cooling power plants.

The solid material removed as water is purified in the treatment process is termed *sludge*. Sludge is sent to digesting vats, where microorganisms decompose much of the matter. The result, a wet solution of "biosolids," is then dried and either disposed of in a landfill, incinerated, or used as fertilizer on cropland. Methane-rich gas created by the decomposition process is sometimes burned to generate electricity, helping to offset the cost of the treatment facility.

Artificial wetlands can aid treatment

Natural wetlands already perform the ecosystem service of water purification, and wastewater treatment engineers are now manipulating wetlands and even constructing wetlands *de novo* to employ them as tools to cleanse wastewater. Generally in this approach, wastewater that has gone through primary treatment at a conventional facility is pumped into the wetland, where microbes living amid the algae and aquatic plants decompose the remaining pollutants. Water cleansed in the wetland can then be released into waterways or allowed to percolate underground. Constructed wetlands also serve as havens for wildlife and areas for human recreation. A project in Arcata, California, was one of the first such attempts, and today over 500 artificially constructed or restored wetlands in the United States are performing this service.

Marine Systems

An adequate supply of clean freshwater is vital to our survival and to Earth's ecosystems, but recall that freshwater makes up only 2.5% of our planet's water. Fully 97.5% is salt water, comprising the oceans that cover 71% of our planet's surface. The oceans influence global climate, teem with biodiversity, facilitate transportation and commerce, and provide us many resources. Even landlocked areas far from salt water are affected by the oceans, which provide fish for people to eat in Iowa, supply oil to power cars in Ontario, and influence the weather in Nebraska.

The physical makeup of the ocean is complex

The world's five major oceans—Pacific, Atlantic, Indian, Arctic, and Antarctic—are all connected, comprising a single vast body of water. Ocean water contains roughly 96.5% H_2O by mass; most of the remainder consists of ions from dissolved salts. Ocean water is salty primarily because rivers and winds carry sediment and salts from the continents into the ocean. Evaporation then removes pure water, leaving a higher concentration of salts. If we were able to evaporate all the water from the oceans, the world's ocean basins would be covered with a layer of dried salt 63 m (207 ft) thick.

Surface waters of the oceans are warmer than subsurface waters because the sun heats them and because warmer water is less dense. Deep below the surface, water is dense and sluggish, unaffected by winds and storms, sunlight, and daily temperature fluctuations. Ocean water moves in **currents**, vast riverlike flows that move in the upper 400 m (1,300 ft) of water, horizontally and for great distances (Figure 11.14). Wind, solar heating and cooling, gravity, density differences, and the Coriolis effect (▶ p. 274) drive the global system of ocean currents. These long-lasting patterns influence global climate, El Niño and La Niña events (▶ pp. 290–291), and navigation and human history. Currents helped carry Polynesians to Easter Island, Darwin to the Galapagos, and Europeans to the New World.

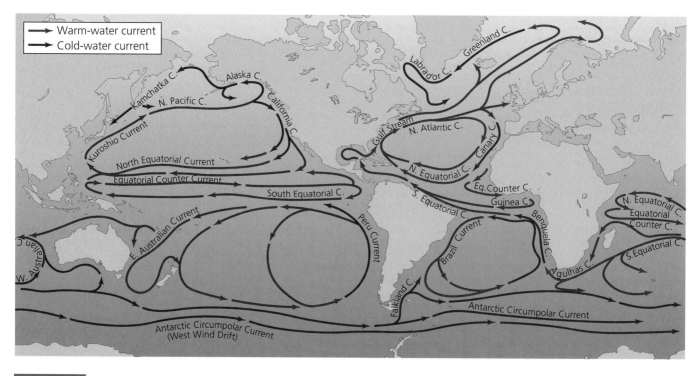

FIGURE 11.14 The upper waters of the oceans move in currents, which are long-lasting global patterns of water movement. Warm-water and cold-water currents interact with the planet's climate system (Chapter 12) and have been used by people for centuries to navigate the oceans. *Source:* Garrison, T.S. 1999. *Oceanography*, 3rd ed. Belmont, CA: Wadsworth.

Surface winds and heating also create vertical currents in seawater. **Upwelling** is the flow of cold, deep water toward the surface. Because upwelled water is rich in nutrients from the bottom, upwellings often support high primary productivity (▶ pp. 61–62) and lucrative fisheries. At **downwellings**, warm surface water rich in dissolved gases is displaced downward, providing an influx of oxygen for deep-water life.

Although oceans are depicted on most maps and globes as smooth, blue swaths, parts of the ocean floor are rugged and complex (Figure 11.15). Underwater volcanoes shoot forth enough magma to build islands above sea level, such as the Hawaiian Islands. Steep canyons similar in scale to Arizona's Grand Canyon lie just offshore of some continents. The deepest spot in the oceans is deeper than Mount Everest is high, by over a mile. Our planet's longest mountain range is under water—the Mid-Atlantic Ridge (▶ p. 71) runs the length of the Atlantic Ocean.

Some ocean regions support more life than others. The uppermost 10 m (33 ft) of ocean water absorbs 80% of the solar energy that reaches its surface. For this reason, nearly all of the oceans' primary productivity occurs in the well-lit top layer, or *photic zone*. Generally, the warm, shallow waters of continental shelves are most biologically productive and support the greatest species diversity.

Habitats and ecosystems occurring between the ocean's surface and floor are classified as **pelagic**, whereas those that occur on the ocean floor are classified as **benthic**.

Marine ecosystems are many and varied

With their variation in topography, temperature, salinity, nutrients, and sunlight, marine and coastal environments feature a variety of ecosystems.

Near-surface pelagic ecosystems In pelagic areas of the open ocean, photosynthetic productivity and animal life near the surface are concentrated in regions of nutrient-rich upwelling. Microscopic phytoplankton constitute the base of the marine food chain in the pelagic zone. These photosynthetic organisms feed zooplankton, which in turn become food for fish, jellyfish, whales, and other free-swimming animals (Figure 11.16a). Predators at higher trophic levels include larger fish, sea turtles, sharks, and seabirds.

Deep ocean In the tantalizing and little-known deep-water ecosystems, animals have adapted to deal with extreme water pressures and to live in the dark without food from photosynthesizers. Some of these often bizarre-looking

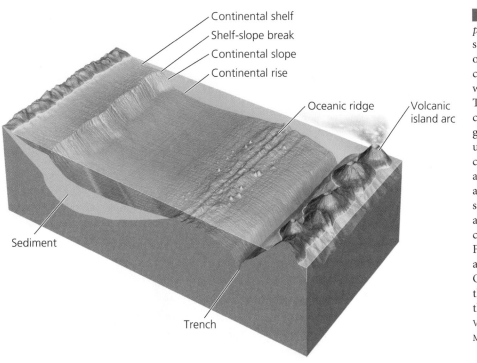

Continental shelf
Shelf-slope break
Continental slope
Continental rise
Oceanic ridge
Volcanic island arc
Sediment
Trench

FIGURE 11.15 A stylized *bathymetric profile* shows key geologic features of the submarine environment. Shallow regions of water exist around the edges of continents over the continental shelf, which drops off at the shelf-slope break. The relatively steep drop-off called the continental slope gives way to the more gradual continental rise, all of which are underlain by sediments from the continents. Vast areas of seafloor are flat abyssal plain. Seafloor spreading occurs at oceanic ridges, and oceanic crust is subducted in trenches. Volcanic activity along trenches often gives rise to island chains such as the Aleutian Islands. Features on the left side of this diagram are more characteristic of the Atlantic Ocean, and features on the right side of the diagram are more characteristic of the Pacific Ocean. Adapted from Thurman, H. V. 1990. *Essentials of oceanography*, 4th ed. New York: Macmillan.

creatures scavenge carcasses or detritus (organic particles) that fall from above. Others are predators, and still others attain food from symbiotic mutualistic (▸ p. 101) bacteria (Figure 11.16b). Some ecosystems cluster around **hydrothermal vents**, where heated water spurts from the seafloor, carrying minerals that precipitate to form large rocky structures. Tubeworms, shrimp, and other creatures in these recently discovered systems use symbiotic bacteria to derive their energy from chemicals in the heated water rather than from sunlight.

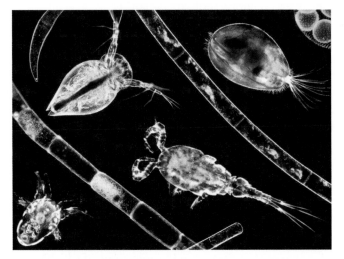

(a) Plankton in near-surface pelagic waters

(b) Angler fish in deep ocean

FIGURE 11.16 The uppermost reaches of ocean water contain billions upon billions of phytoplankton and zooplankton (**a**). Life is scarce in the dark depths of the deep ocean, but the creatures that do live there often appear bizarre to us. The anglerfish (**b**) lures prey toward its mouth with a bioluminescent (glowing) organ that protrudes from the front of its head.

Kelp forests Large brown algae, or **kelp** (often nick-named *seaweed*), grow from the floor of continental shelves, reaching toward the sunlit surface for up to 60 m (200 ft) in height. Dense stands of kelp form underwater forests in many temperate waters (Figure 11.17a). Kelp forests supply shelter and food for invertebrates and fish, which in turn provide food for predators such as seals, sharks, and sea otters (see Figure 5.6, ▶ p. 106). Kelp forests also absorb wave energy and protect shorelines from erosion. People derive alginates from kelp, which serve as thickeners in a wide range of consumer products, from cosmetics to foods to paints to paper to soaps.

Coral reefs A **coral reef** is an underwater outcrop of calcium carbonate composed of the skeletons of tiny marine animals known as *corals*. Corals attach to rock or existing reef and capture passing food with stinging tentacles. They also derive nourishment from symbiotic algae, known as *zooxanthellae*, that inhabit their bodies and produce food through photosynthesis. Most corals are colonial, and the colorful surface of a coral reef consists of thousands or millions of densely packed individuals. As corals die, their skeletons remain part of the reef while new corals grow atop them, increasing the reef's size.

Coral reefs absorb wave energy, protect shorelines, and host as much biodiversity as any other type of ecosystem (Figure 11.17b). The likely reason is that coral reefs provide complex physical structure (and thus many habitats) in shallow nearshore waters, which are regions of high primary productivity. Besides the staggering diversity of anemones, sponges, hydroids, tubeworms, and other sessile invertebrates, innumerable molluscs, crustaceans, flatworms, seastars, and urchins patrol the reefs, and thousands of fish species find food and shelter in reef nooks and crannies. Larger predators, such as grouper and moray eels, feed on the smaller fish.

Intertidal zones Where the ocean meets the land, **intertidal**, or *littoral*, ecosystems extend between the farthest reaches of the high and low tides. **Tides** are the periodic rising and falling of the ocean's height at a given location, caused by the gravitational pull of the moon and sun. Intertidal organisms spend part of each day submerged in water, part of the day exposed to the air and sun, and part of the day being lashed by waves. Life abounds in the crevices of rocky shorelines, which provide shelter and pools of water (tide pools) during low tides (Figure 11.18a). Sessile animals such as anemones, mussels, and barnacles live attached to rocks, filter-feeding on plankton in the water that washes over them. Urchins, sea slugs, chitons, and limpets eat algae or scrape

(a) Kelp forest

(b) Coral reef

FIGURE 11.17 Tall brown algae known as *kelp* grow from the floor of the continental shelf, creating kelp forests (**a**) in which many marine creatures forage or find refuge. Coral reefs (**b**) likewise provide structure and habitat for diverse communities of organisms.

food from the rocks. Seastars creep along, preying on the filter-feeders and herbivores. Crabs clamber around the rocks, scavenging detritus. The rocky intertidal zone is so diverse because environmental conditions change dramatically from the highest to the lowest reaches.

Mangrove forests In tropical and subtropical latitudes, mangrove forests occur along sandy and silty coasts. **Mangroves** are trees with odd roots, some of which curve upward like snorkels to attain oxygen lacking in the mud, and some of which curve downward, serving as stilts to support the tree in changing water levels (Figure 11.18b). Fish, shellfish, crabs, snakes, and other organisms thrive among the root networks, and birds feed and nest in the

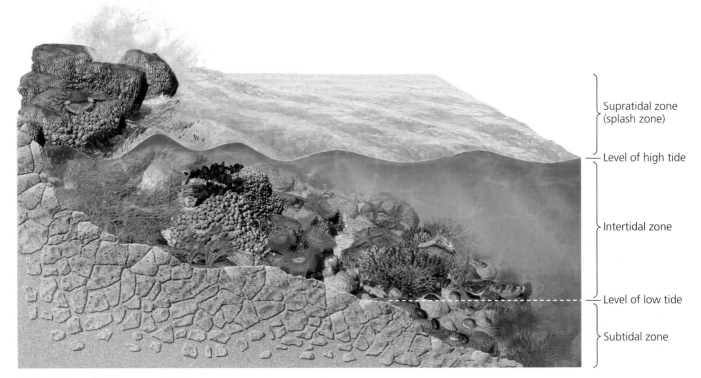

Supratidal zone
(splash zone)

Level of high tide

Intertidal zone

Level of low tide

Subtidal zone

(a) Rocky intertidal

(b) Mangroves

FIGURE 11.18 The rocky intertidal zone (**a**) stretches along rocky shorelines between the lowest and highest reaches of the tides, providing niches for a diversity of organisms including seastars (starfish), barnacles, crabs, sea anemones, mussels, and many others. Mangrove forests (**b**) line tropical and subtropical coastlines throughout the world. Mangrove trees, with their unique roots, are adapted for growing in salt water and provide habitat for many types of fish, birds, crabs, and other animals.

foliage of these coastal forests. Mangroves provide materials that people use for food, medicine, tools, and construction. They also protect coastlines; studies after the 2004 South Asian tsunami indicated that coasts with intact mangrove forests suffered less damage than deforested coasts.

Salt marshes Along many of the world's coastlines at temperate latitudes, **salt marshes** occur where the tides wash over gently sloping sandy or silty substrates. Rising and falling tides flow into and out of channels, and at highest tide water spills over onto elevated marsh flats. Marsh flats grow thick with rushes, shrubs, and grasses. Salt marshes boast very high primary productivity and provide critical habitat for shorebirds, waterfowl, and the adults and young of many commercially important fish and shellfish species.

Estuaries Many salt marshes and mangrove forests occur in or near **estuaries**, water bodies where rivers flow into the ocean, mixing freshwater with salt water. Biologically productive ecosystems, estuaries experience significant fluctuations in salinity as tidal currents and freshwater runoff vary daily and seasonally. For shorebirds and for many commercially important shellfish species, estuaries provide critical habitat. Mudflats, although not much to look at, are of key importance for these organisms. For fishes such as salmon, which spawn in streams and mature in the ocean, estuaries provide a transitional zone where young fish make the passage from freshwater to salt water.

All these coastal and nearshore ecosystems have borne the brunt of human impact because two-thirds of Earth's people live within 160 km (100 mi) of the ocean. Estuaries, salt marshes, mangroves, and coral reefs around the world have suffered from urban and coastal development, water pollution, habitat alteration, and overfishing. Huge stretches of mangrove forest have been cleared, often to make room for shrimp farms. Coral reefs are experiencing alarming declines as many reefs have undergone "coral bleaching," which occurs when zooxanthellae leave the coral, depriving it of nutrition. Coral bleaching is thought to result from increased sea surface temperatures associated with global climate change, from pollution, from unknown natural causes, or from some combination of these factors. Nutrient pollution in coastal waters also promotes the growth of algae, which are blanketing reefs in many regions.

Human Impact on the Oceans

Our species has a long history of interacting with the oceans. We have traveled across their waters, clustered our settlements along their coastlines, and been fascinated by their beauty, power, and vastness. We have also left our mark upon them by exploiting oceans for their resources and polluting them with our waste.

Oceans provide us transportation, energy, and minerals

The oceans have provided transportation routes for thousands of years and continue to provide affordable means of moving people and products over vast distances. Ocean transport has accelerated the global reach of some cultures and has promoted interaction among long-isolated peoples.

The oceans also contain sources of commercially valuable energy. Since the 1980s, about 25% of our crude oil and natural gas has come from exploitation of seafloor deposits (▸ p. 309). Ocean sediments also contain a novel potential source of fossil fuel energy, *methane hydrates* (▸ p. 312). The oceans also hold potential for providing renewable energy sources that do not emit greenhouse gases (▸ pp. 345–346).

We extract minerals from the ocean floor, as well. Using large hydraulic dredges, miners vacuum up sand, gravel, sulfur, and phosphorite. Other valuable minerals found on or beneath the seafloor include calcium carbonate (used in making cement), silica (used as fire-resistant insulation and in manufacturing glass), and deposits of copper, zinc, silver, and gold ore. Many minerals are found concentrated in manganese nodules, small ball-shaped accretions that are scattered across parts of the ocean floor. It is estimated that over 1.5 trillion tons of manganese nodules exist in the Pacific Ocean alone and that their reserves of metal exceed all terrestrial reserves. The logistical difficulty of mining them, however, has kept their extraction uneconomical so far.

Marine pollution threatens resources

People have long made the oceans a sink for waste and pollution. Oil, plastic, industrial chemicals, sewage, excess nutrients, and abandoned fishing gear all eventually make their way into the oceans.

Nets and plastic debris Plastic bags and bottles, fishing nets, gloves, fishing line, buckets, floats, abandoned cargo, and much else that people transport on the sea or deposit into it can harm marine organisms. Because most plastic is not biodegradable, it can drift for decades before washing up on beaches. Marine mammals, seabirds, fish, and sea turtles may mistake floating plastic debris for food and can die as a result of ingesting material they cannot digest or expel. Fishing nets that are lost or intentionally discarded may continue snaring animals for decades.

Of 115 marine mammal species, 49 are known to have eaten or become entangled in marine debris, and 111 of 312 species of seabirds are known to ingest plastic. We can all help minimize this type of harm by reducing our use of plastics, cutting the rings of six-pack beverage holders before throwing them away, and picking up trash from beaches.

Oil pollution Major oil spills—such as the one that occurred in 1989 when the *Exxon Valdez* struck a reef in Prince William Sound, Alaska, and spilled 42 million L (11 million gal) of crude oil—make headlines and cause serious environmental problems. Yet the majority of oil pollution in the oceans accumulates from innumerable widely spread small sources, including leakage from small boats and runoff from human activities on land. Moreover, the amount of petroleum spilled into the oceans in recent years is equaled by the amount that seeps up from naturally occurring seafloor deposits.

Nonetheless, minimizing the amount of oil we release into coastal waters is important, because petroleum pollution harms marine environments and human economies. Petroleum can physically coat and kill marine organisms, and ingested chemical components in petroleum can poison marine life. In response to headline-grabbing oil spills, governments around the world have begun to implement stricter safety standards for tankers, such as requiring industry to pay for tugboat escorts in sensitive or hazardous coastal waters, and to develop prevention and response plans for major oil spills. The U.S. Oil Pollution Act of 1990 created a $1 billion prevention and cleanup fund and required that by 2015 all oil tankers in U.S. waters be equipped with double hulls. Over the past three decades, the amount of oil spilled in U.S. and global waters has decreased (Figure 11.19), in part because of an increased emphasis on spill prevention and response.

Excess nutrients Pollution from fertilizer runoff or other nutrient inputs can create dead zones in coastal marine ecosystems, as we saw with the Gulf of Mexico (Chapter 3). Excessive nutrient concentrations also may give rise to population explosions among several species of marine algae

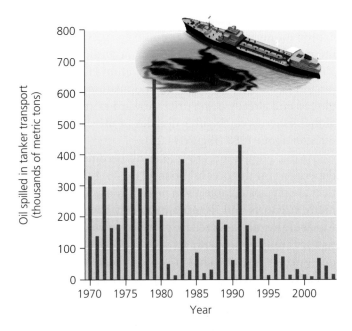

FIGURE 11.19 Of the 670 million metric tons of petroleum that people spill into the world's oceans each year, 72% comes from numerous diffuse non-point sources, especially runoff from rivers and coastal communities and leakage from two-stroke engines. Spills during petroleum transport account for 22%, and leakage during petroleum extraction accounts for 6%. Less oil is being spilled today in large tanker spills, thanks in part to regulations on the shipping industry and improved spill response techniques. The figure shows cumulative quantities of oil spilled worldwide from nonmilitary spills over 7 metric tons. Data from: National Research Council. 2003. *Oil in the sea III. Inputs, fates, and effects.* Washington, DC: National Academies Press; and International Tanker Owners Pollution Federation Ltd. 2005. *Oil tanker spill statistics: 2004.* London: ITOPF.

that produce powerful toxins that attack the nervous systems of vertebrates. Blooms of these algae are known as **harmful algal blooms.** Some algal species produce reddish pigments that discolor surface waters, and blooms of these species are nicknamed **red tides.** Harmful algal blooms can cause illness and death among zooplankton, birds, fish, marine mammals, and people as their toxins are passed up the food chain. They also cause economic loss for communities dependent on fishing or beach tourism.

As severe as the impacts of all these types of marine pollution can be, however, most marine scientists concur that the more worrisome dilemma is overharvesting.

Overfishing and Marine Conservation

The oceans and their biological resources have met human needs for thousands of years, but today we are placing unprecedented pressure on marine resources.

Half the world's marine fish populations are fully exploited, meaning that we cannot harvest them more intensively without depleting them, according to a 2004 U.N. Food and Agriculture Organization (FAO) report. An additional 25% of marine fish populations are overexploited and already being driven toward extinction. Total global fisheries catch, after decades of increases, leveled off after about 1988 (Figure 11.20).

As our population grows, we will become even more dependent on the oceans' bounty. Existing fishing practices are not sustainable given present consumption rates, many scientists and fisheries managers have concluded. Aquaculture may help relieve pressure on wild stocks, but fish farming comes with its own set of environmental dilemmas (▸ p. 164). Thus it is vital, these scientists and managers say, that we modify our priorities and improve our use of science in fisheries management.

Many fisheries are collapsing today

Modern industrialized fishing fleets can deplete marine populations with astonishing quickness. In a 2003 study, fisheries biologists Ransom Myers and Boris Worm analyzed fisheries data from FAO archives and concluded that the oceans today contain only one-tenth of the large-bodied animals they once did. The declines happened so fast in most regions that scientists never knew the original abundance of these animals.

Many fisheries have collapsed in recent years. These collapses are ecologically devastating and also take a severe economic toll on human communities that depend on fishing. A prime example took place in the 1990s in the North Atlantic off the Canadian and U.S. coasts. Groundfish (species that live in benthic habitats, such as Atlantic cod, haddock, halibut, and flounder) are major food sources that powered fishing economies in Newfoundland, Labrador, the Maritime Provinces, and the New England states for close to 400 years. Yet fishing pressure became so intense that most stocks collapsed, bringing fishing economies down with them. With Canada's cod stocks down by 99% and showing no sign of recovery, the Canadian government in 1992 ordered a complete ban on cod fishing in the Grand Banks region off Newfoundland and Labrador. The moratorium was partially lifted in 1998, after which catches declined, so the government reimposed it in 2003. To soften the economic blow, the government offered $50 million to affected fishers of the region.

On the U.S. side of the border, bans are helping to restore depleted fisheries. When the groundfish fisheries of Georges Bank in the Gulf of Maine collapsed in the mid-1990s, three areas totaling 17,000 km^2 (6,600 mi^2) were closed to fishing. The closures worked; five years later, haddock, flounder, and yellowtail were recovering, and scallops rebounded strongly, attaining sizes 9–14 times larger than before the closures. Fishers began having better luck, especially just outside the closed regions. Unfortunately, however, cod have not recovered.

Fisheries declines are masked by several factors

Although industrialized fishing has depleted fish stocks in region after region, the overall global catch has remained roughly stable for two decades (see Figure 11.20). This seeming stability can be explained by several factors that mask population declines. One is that fishing fleets have been traveling longer distances and fishing in deeper waters to reach less-exploited portions of the ocean. Moreover, fishing fleets have been spending more time fishing and have been setting out more nets and lines—expending increasing effort just to catch the same number of fish. Improved technology also helps explain high catches despite declining stocks. Today's

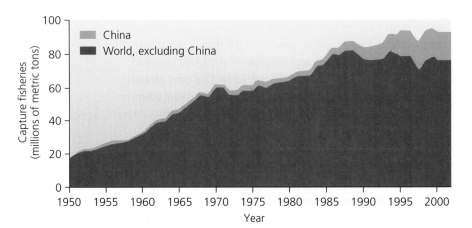

FIGURE 11.20 The total global fisheries catch has increased over the past half century, but in recent years growth has stalled. Many industry observers fear that a global catch decline is imminent if conservation measures are not taken immediately. The figure shows trends with and without China's data (which many scientists consider to be greatly inflated). With China's data, global catch has leveled off since the mid-1990s. Without China's data, catch has decreased slightly since 1988. Go to **GRAPHit!** at www.aw-bc.com/withgott. Data from U.N. Food and Agricultural Organization (FAO). 2004. *World review of fisheries and aquaculture.* Rome: FAO.

Japanese, European, Canadian, and U.S. fleets can reach almost any spot on the globe with boats that attain speeds of 80 kph (50 mph). They boast an array of technologies that militaries have developed for spying and for chasing enemy submarines, including advanced sonar mapping equipment, satellite navigation, and thermal sensing systems.

Numbers of fish do not tell the whole story. Analyses of fisheries data have revealed in case after case that as fishing increases, the size and age of fish caught decline. In addition, as particular species become too rare to fish profitably, fleets begin targeting other species that are more abundant. Generally this means shifting from large, desirable species to smaller, less desirable ones. Fleets have time and again depleted popular food fish such as cod and snapper and shifted their emphasis to species of lower value. Because this often entails catching species at lower trophic levels, this phenomenon has been termed "fishing down the food chain."

Some fishing practices kill nontarget animals

Fishing practices often catch more than just the species they target—a phenomenon called **by-catch**. Boats that drag huge *driftnets* through the water (Figure 11.21a) capture everything in their path, including substantial numbers of dolphins, seals, sea turtles, and nontarget fish. *Longline fishing* (Figure 11.21b), which involves dragging extremely long lines with baited hooks, kills turtles, sharks, and an estimated 300,000 seabirds each year. *Bottom-trawling* (Figure 11.21c) directly damages entire ecosystems. It involves dragging weighted nets over the seafloor to catch such benthic organisms as scallops and groundfish. Trawling crushes many organisms in its path and leaves long swaths of damaged sea bottom.

By exercising careful choice when we buy seafood, we as consumers can encourage fishing practices that are less environmentally damaging. Purchasing ecolabeled seafood products such as dolphin-safe tuna is one way to exercise choice. Several nonprofit organizations have devised concise guides to help consumers differentiate fish and shellfish that are overfished or whose capture is ecologically damaging from those that are harvested more sustainably.

Fisheries management has been based on maximum sustainable yield

For decades, fisheries managers have sought to use scientific assessments to ensure sustainable harvests. They have studied fish population biology and used that

(a) Driftnetting

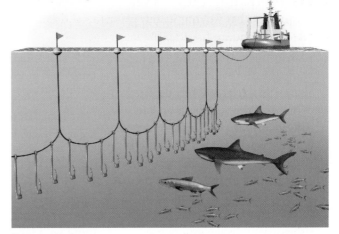

(b) Longlining

(c) Bottom-trawling

FIGURE 11.21 Commercial fishing fleets use several methods of capture. In driftnetting (**a**), huge nets are dragged through the open water to capture schools of fish. In longlining (**b**), lines with numerous baited hooks are pulled through the open water. In bottom-trawling (**c**), weighted nets are dragged along the floor of the continental shelf. All methods result in large amounts of by-catch, or capture of nontarget animals. Bottom-trawling can also result in severe structural damage to reefs and benthic habitats.

Do Marine Reserves Work?

In November 2001, a team of fisheries scientists published a paper in the journal *Science*, providing some of the first clear evidence that marine reserves can benefit nearby fisheries. The team, led by York University researcher Callum Roberts, focused on reserves off the coasts of Florida and the Caribbean island of St. Lucia.

Following the establishment in 1995 of the Soufrière Marine Management Area (SMMA), a network of reserves intended to help restore St. Lucia's severely depleted coral reef fishery, Roberts and his colleague, Julie Hawkins, conducted annual visual surveys of fish abundance in the reserves and nearby areas. Within 3 years, they found that the biomass of five commercially important families of fish—surgeonfishes, parrot fishes, groupers, grunts, and snappers—had tripled inside the reserves and doubled outside them (see the first figure).

Roberts and Hawkins also interviewed local fishers and found that

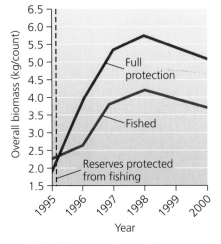

Established in 1995, the Soufrière Marine Management Area (SMMA), along the coast of St. Lucia, had a rapid impact. By 1998, fish biomass within the five reserves tripled, and in adjacent fished areas, it doubled. Data from Roberts, C., et al. 2001. Effects of marine reserves on adjacent fisheries. *Science* 294: 1920–1923.

those with large traps were catching 46% more fish per trip in 2000–2001 than they had in 1995–1996, and that fishers with small traps were catching 90% more (see the second figure).

Roberts and his colleagues concluded that "in 5 years, reserves have led to improvement in the SMMA fishery, despite the 35% decrease in area of fishing grounds."

Roberts and his coworkers also studied the oldest fully protected marine reserve in the United States, the Merritt Island National Wildlife Refuge (MINWR), established in 1962 as a buffer around what is today the Kennedy Space Center on Cape Canaveral, Florida. In a previous study, Darlene Johnson and James Bohnsack of the National Oceanic and Atmospheric Administration and Nicholas Funicelli of the United States Geological Survey had found that the reserve contained more and larger fish than did nearby unprotected areas. This team also found that some of the reserve's fish appeared to be migrating to nearby fishing areas.

Bohnsack, Roberts, and their colleagues corroborated the evidence for migration by analyzing trophy records from the

knowledge to regulate the timing, scale, and manner of harvests. The goal has been to allow maximal harvests of particular populations while keeping fish available for the future—the concept of maximum sustainable yield (▶ pp. 205–206).

Because a number of fish stocks have plummeted despite such efforts, many marine scientists and managers now feel it is time to rethink fisheries management. They suggest shifting the focus away from individual fish species and toward larger ecological systems. This means considering the impacts of fishing practices on habitat quality, species interactions, and other factors that may have indirect or long-term effects on populations. One key aspect of such ecosystem-based management (▶ p. 206) is to set aside areas of ocean where systems can function without human interference.

We can protect areas in the ocean

Marine protected areas (MPAs) have been established, most of them along the coastlines of developed countries. The United States now contains nearly 300 federally managed MPAs. However, despite their name, nearly all MPAs allow fishing or other extractive activities. As a recent report from an environmental advocacy group put it, they "are dredged, trawled, mowed for kelp, crisscrossed with

International Game Fish Association. They found that the proportion of Florida's record-sized fish caught near Merritt Island increased significantly after 1962. Nine years after the refuge was established, for instance, the number of spotted sea trout records from the Merritt Island area jumped dramatically. Bohnsack, Roberts, and their colleagues hypothesized that the reserve was providing a protected zone in which fish could grow to trophy size before migrating to nearby areas, where they were caught by recreational fishers.

Not everyone saw the St. Lucia and Merritt Island cases as proof that marine reserves could rescue depleted fisheries. In February 2002, several alternative interpretations were published as letters in *Science*. Mark Tupper, a fisheries scientist at the University of Guam, suggested that the St. Lucia results were relevant only to coral reef fisheries in developing nations, whereas Florida's boost in fish populations

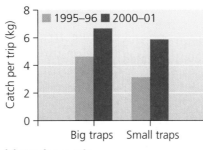

(a) Catch per trip

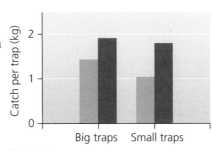

(b) Catch per trap

Callum Roberts and his colleagues studied biomass of fish caught at the SMMA over two 5-month periods in 1995–1996 and 2000–2001. For fishers with big traps, catch increased by 46%, and for those with small traps, it increased by 90%. Per trap, catch increased by 36% for big traps and by 80% for small traps. Data from Roberts, C., et al. 2001. Effects of marine reserves on adjacent fisheries. *Science* 294: 1920–1923.

was due primarily to limits on recreational fishing. Karl Wickstrom, editor-in-chief of *Florida Sportsman* magazine, suggested that the increase in trophy fish near MINWR was caused by commercial fishing regulations and changes in how trophies were recorded and promoted. And Ray Hilborn, a fisheries scientist at the University of Washington, challenged the study's scientific methods. In the St. Lucia case, he pointed out, there had been no control condition.

In response, Roberts and his colleagues reaffirmed the validity of their results while acknowledging some limitations. They agreed with Tupper that marine reserves are not always effective and often need to be complemented by other management tools, such as size limits. "We agree that inadequately protected reserves are useless," they wrote, "but our study shows that well-enforced reserves can be extremely effective and can play a critical role in achieving sustainable fisheries."

oil pipelines and fiber-optic cables, and swept through with fishing nets."

As a result, many scientists favor establishing areas where fishing is prohibited. Such "no-take" areas have come to be called **marine reserves**. Designed to preserve ecosystems intact, marine reserves are also intended to improve fisheries. Scientists argue that marine reserves can act as production factories for fish for surrounding areas, because fish larvae produced inside reserves will disperse outside and stock other parts of the ocean (see "The Science behind the Story," above). By serving both purposes, proponents argue, marine reserves are a win-win proposition for environmentalists and fishers alike.

Weighing the Issues:
Preservation on Land and at Sea

Almost 4% of U.S. land area is designated as wilderness, yet far less than 1% of coastal waters are protected in reserves. Why do you think it is taking so long for the preservation ethic to make the leap to the oceans?

Many fishers don't like the idea of no-take reserves, however. Nearly every marine reserve proposed has met with opposition from people and businesses who use the area for fishing or recreation. Commercial and sport fishers are concerned that marine reserves will simply put more areas off-limits to fishing.

Reserves can work for both fish and fishers

In the past decade, data from marine reserves around the world have been indicating that reserves *can* work as win-win solutions that benefit ecosystems, fish populations, and fishing economies. A comprehensive review of data from marine reserves as of 2001 revealed that within just 1–2 years after their establishment, marine reserves

▶ Increased densities of organisms on average by 91%
▶ Increased biomass of organisms on average by 192%
▶ Increased average size of organisms by 31%
▶ Increased species diversity by 23%

If marine reserves work in principle, the question becomes how best to design reserves and arrange them into networks. If reserve systems are designed to take advantage of ocean currents so that larvae disperse effectively, many scientists say, then they may well help "seed the seas" with life and lead us toward solving one of our most pressing environmental problems.

Conclusion

In the oceans, along coastlines, and on land, we face considerable challenges in maintaining saltwater and freshwater resources. The use of freshwater for drinking, irrigation, and other uses by a growing population threatens to deplete many of our water sources. Overfishing in the oceans imperils our fish stocks. Coastal development jeopardizes the natural systems that keep our planet functioning and provide us so many services. And pollution degrades water quality in our lakes, rivers, streams, and oceans.

There is plenty of reason for optimism, however. Water quality in many freshwater bodies has improved in recent decades, thanks to legislative action from policymakers and the efforts of millions of concerned citizens. In the oceans, marine reserves represent a new hope that we can restore marine ecosystems and fisheries at the same time. We must strive to attain sustainability in our water resources; potential solutions are numerous, and the issue is too important to ignore.

TESTING YOUR COMPREHENSION

1. Compare and contrast the main types of freshwater ecosystems. Name and describe the major zones of a typical pond or lake.
2. Describe three benefits of damming rivers, and three costs. What particular environmental, health, and social concerns has China's Three Gorges Dam and its reservoir raised?
3. Name three types of freshwater pollutants, and provide an example of each. What ecological effects do each of these produce? List three properties of water that scientists use to determine water quality.
4. Define *groundwater*. Why do many scientists consider groundwater pollution a greater problem than surface water pollution?
5. Describe and explain the major steps in the process of wastewater treatment. How can artificially constructed wetlands aid such treatment?

6. What proportion of Earth's surface do oceans cover? About how much salt does ocean water contain? What factors drive the system of ocean currents?
7. Describe three kinds of ecosystems found near coastal areas and the kinds of life they support. How are these systems being affected by human impact?
8. Describe three major forms of pollution in the oceans and the consequences of each.
9. Name three industrial fishing practices that create bycatch and harm marine life, and explain how they do so.
10. How does a marine protected area differ from a marine reserve? Why do many fishers oppose marine reserves? Explain why many scientists say no-take reserves will be good for fishers.

SEEKING SOLUTIONS

1. How can we decrease agricultural demand for water? Describe some ways that we can reduce household water use. How can we lessen industrial use of water?
2. Your state's governor has put you in charge of water policy. The aquifer beneath your state has been over-

pumped, and many wells have gone dry. Agricultural production last year decreased for the first time in a generation, and farmers are clamoring for you to do something. Meanwhile, the state's largest city is growing so fast that it needs more water for the burgeoning urban population. What policies would you consider to restore your state's water supply? Would you try to take steps to increase supply, to decrease demand, or both? Explain why you would choose such policies.

3. Having solved the water depletion problem in your state, your next task is to deal with pollution of the groundwater that provides your state's drinking water supply. Recent studies have shown that one-third of the state's groundwater has levels of pollutants that violate EPA standards for human health. The federal government is threatening enforcement, and citizens are fearful for their safety. What steps would you consider taking to safeguard the quality of your state's groundwater supply, and why?

4. What factors account for the trends in global fish capture over the past 20 years? What steps could be taken by (a) commercial fishers, (b) governments, and (c) consumers to help depleted fisheries rebound?

5. You are mayor of a coastal town where some residents are employed as commercial fishers and others make a living serving ecotourists who come to snorkel and scuba-dive at the nearby coral reef. In recent years, several fish stocks have crashed, and ecotourism is dropping off as fish disappear from the increasingly degraded reef. Scientists are urging you to help establish a marine reserve around portions of the reef, but most commercial and recreational fishers are opposed to this idea. What steps would you take to restore your community's economy and environment?

CALCULATING ECOLOGICAL FOOTPRINTS

In the United States, the EPA estimates that household water use averages 750 liters per person per day. One of the single greatest personal uses of water is for showering. Standard showerheads dispense 15 L of water per minute, but low-flow showerheads dispense only 9 L per minute. Given an average daily shower time of 10 minutes, calculate the amounts of water used and saved over the course of a year with standard versus low-flow showerheads, and record your results in the table below.

	Annual water use with standard showerheads (liters)	Annual water use with low-flow showerheads (liters)	Annual water savings with low-flow showerheads (liters)
You	54,750	32,850	21,900
Your class			
Your state			
United States			

Data from U.S. EPA, 1995. *Cleaner water through conservation: Chapter 1—How we use water in the United States.* EPA 841-B-95-002.

1. What percentage of personal water consumption would you calculate is used for showering?
2. How much additional water would you be able to save by shortening your average shower time from 10 minutes to 8 minutes? To 5 minutes?

3. Can you think of any factors that are not being considered in this scenario of water savings? Explain.

Take It Further

Go to www.aw-bc.com/withgott, where you'll find:

▶ Suggested answers to end-of-chapter questions
▶ Quizzes, animations, and flashcards to help you study
▶ *Research Navigator*™ database of credible and reliable sources to assist you with your research projects

▶ **GRAPHit!** tutorials to help you interpret graphs
▶ **INVESTIGATEit!** current news articles that link the topics that you study to case studies from your region to around the world

12 Atmospheric Science, Air Pollution, and Global Climate Change

Fishermen of the Maldives

Upon successfully completing this chapter, you will be able to:

▶ Describe the composition, structure, and function of Earth's atmosphere

▶ Outline the scope of outdoor air pollution and assess potential solutions

▶ Explain stratospheric ozone depletion and identify steps taken to address it

▶ Define acidic deposition and illustrate its consequences

▶ Characterize the scope of indoor air pollution and assess potential solutions

▶ Describe Earth's climate system and explain the variety of factors influencing global climate

▶ Characterize human influences on the atmosphere and global climate

▶ Delineate modern methods of climate research

▶ Summarize current consequences and potential future impacts of global climate change

▶ Evaluate the scientific, political, and economic debates concerning climate change

▶ Suggest potential responses to climate change

Central Case: Rising Temperatures and Seas May Take the Maldives Under

A nation of low-lying coral islands, or atolls, in the Indian Ocean, the Maldives is known for its spectacular tropical setting, colorful coral reefs, and sun-drenched beaches. For visiting tourists it is a paradise, and for 320,000 Maldives residents it is home. But residents and tourists alike now fear that the Maldives could soon be submerged by the rising seas that are accompanying global climate change.

Nearly 80% of the Maldives' land area of 300 km^2 (116 mi^2) lies less than 1 m (39 in.) above sea level. In a nation of 1,190 islands whose highest point is just 2.4 m (8 ft) above sea level, rising seas could be a matter of life or death. The world's oceans rose 10–20 cm (4–8 in.) during the 20th century as warming temperatures expanded ocean water and as melting icecaps discharged water into the ocean. Current projections are that sea level will rise another 9–88 cm (3.5–35 in.) by the year 2100.

Higher seas are expected to flood large areas of land in the Maldives and to cause salt water to intrude into drinking water supplies. Moreover, if climate change produces larger and more powerful storms, these could worsen flooding and damage the coral reefs that are so crucial to the nation's tourism- and fishing-driven economy. In recent years, the Maldives government has evacuated residents from several of the lowest-lying islands.

On December 26, 2004, the nation got a taste of what could be in store in the future. The massive *tsunami*, or tidal wave, that devastated coastal areas throughout the Indian Ocean hit the Maldives particularly hard. One hundred people were killed and 20,000 lost their homes, while schools, boats, tourist resorts, hospitals, and transportation and communication infrastructure were destroyed or badly damaged. On a per capita basis, the Maldives suffered the greatest economic shock from the tsunami of any nation. The

World Bank estimates that direct damage in the Maldives totaled $470 million, an astounding 62% of the nation's gross domestic product (GDP). Soil erosion, saltwater contamination of aquifers, and other environmental damage will result in still greater long-term economic losses.

The tsunami was caused *not* by climate change, but by an earthquake. Yet as sea level rises, the damage that such natural events—or ordinary storm waves—can inflict increases considerably. Maldives islanders are not alone in their predicament. Other island nations, from the Galapagos to Fiji to the Seychelles, are also fearing a future in which they may be constantly battling encroaching seawater. Mainland coastal areas of the world, such as the hurricane-battered coasts of Florida and Louisiana, will face similar issues. In one way or another, global climate change seems certain to affect each and every one of us for the remainder of our lifetimes.

The Atmosphere

Every breath we take reaffirms our connection to the **atmosphere**, the thin layer of gases that surrounds Earth. The atmosphere provides us oxygen, absorbs hazardous solar radiation, burns up incoming meteors, transports and recycles water and nutrients, and moderates climate.

Earth's atmosphere consists of roughly 78% nitrogen gas (N$_2$) and 21% oxygen gas (O$_2$). The remaining 1% is composed of argon gas (Ar) and minute concentrations of several other gases. Over our planet's long history, the atmosphere's chemical composition has changed. Oxygen gas began to build up about 2.7 billion years ago, with the emergence of microbes that released oxygen by photosynthesis. Today our own species is altering the quantities of some atmospheric gases, such as carbon dioxide (CO$_2$), methane (CH$_4$), and *ozone* (O$_3$).

The atmosphere consists of several layers

The atmosphere that seems to stretch so high above us is actually just a thin coating about 1/100th of Earth's diameter, like the fuzzy skin of a peach. This coating consists of four layers that atmospheric scientists recognize by measuring differences in temperature, density, and composition (Figure 12.1).

The bottommost layer, the **troposphere**, blankets Earth's surface and provides us the air we need to live. The movement of air within the troposphere is largely responsible for the planet's weather. Although it is thin (averaging 11 km [7 mi] high) relative to the atmosphere's other layers, the troposphere contains three-quarters of the atmosphere's mass, because air is denser near Earth's sur-face. On average, tropospheric air temperature declines by about 6 °C for each kilometer in altitude (or 3.5 °F per 1,000 ft), dropping to roughly –52 °C (–62 °F) at its highest point.

The **stratosphere** extends 11–50 km (7–31 mi) above sea level. Similar in composition to the troposphere, the stratosphere is 1,000 times drier and less dense. Its gases experience little vertical mixing, so once substances (including pollutants) enter it, they tend to remain for a long time. The stratosphere warms with altitude because its ozone and oxygen absorb and scatter much of the sun's ultraviolet (UV) radiation (► p. 59). Most of the atmosphere's minute amount of ozone is concentrated in a portion of the stratosphere roughly 17–30 km (10–19 mi) above sea level, a region that has

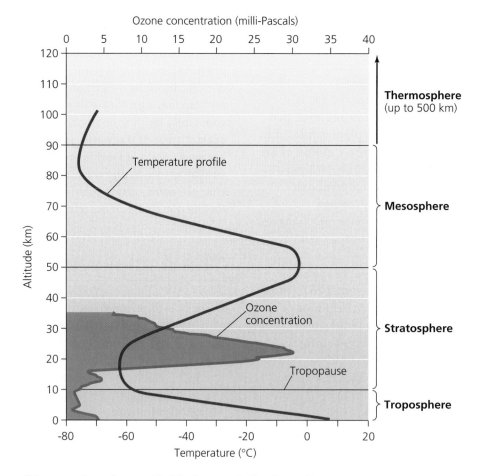

FIGURE 12.1 Aspects of the atmosphere change with altitude across its four layers. Temperature drops with altitude in the troposphere, rises with altitude in the stratosphere, drops in the mesosphere, and then rises again in the thermosphere. The shift in the trend in temperature at the *tropopause* acts like a cap, limiting mixing between the troposphere and the stratosphere. Ozone reaches a peak in a portion of the stratosphere, giving rise to the term *ozone layer*. Adapted from Jacobson, M. Z. 2002. *Atmospheric pollution: History, science, and regulation.* Cambridge: Cambridge University Press; and Parson, E. A. 2003. *Protecting the ozone layer: Science and strategy.* Oxford: Oxford University Press.

come to be called Earth's **ozone layer**. Because UV light can damage living tissue and induce genetic mutations, life on Earth depends on the ozone layer's protection.

The sun and the atmosphere drive weather and climate

An enormous amount of radiation from the sun—over 1,000 watts/m² —continuously bombards the upper atmosphere. Of that solar energy, about 70% is absorbed by the atmosphere and planetary surface, while the rest is reflected back into space (Figure 12.2). Land and surface water absorb solar energy, radiating some heat and causing some water to evaporate. Air

near Earth's surface therefore tends to be warmer and moister than air at higher altitudes. These differences set into motion a process of **convective circulation**. Warm air, being less dense, rises and creates vertical currents. As air rises into regions of lower atmospheric pressure, it expands and cools. Once the air cools, it descends and becomes denser, replacing warm air that is rising. The air picks up heat and moisture near ground level and prepares to rise again, continuing the process. Convective circulation helps guide both weather and climate.

In everyday speech, we often use the terms *weather* and *climate* interchangeably. However, these words have very distinct meanings. Both concepts involve the physical properties of the troposphere, such as temperature, pressure, humidity, cloudiness, and wind. However, **weather**

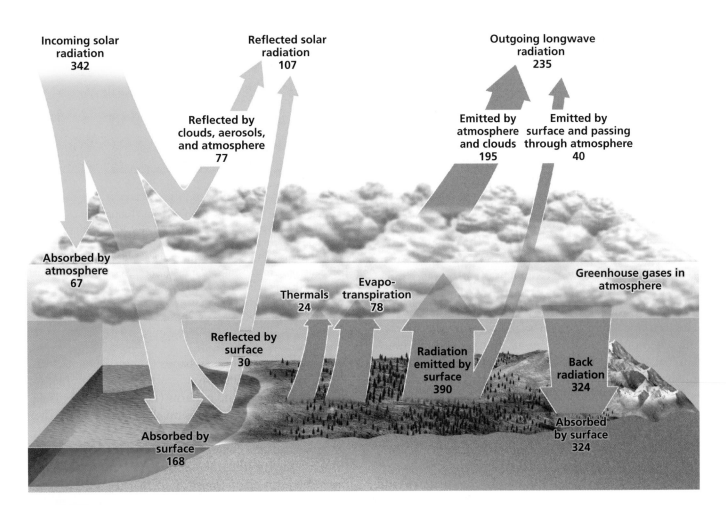

FIGURE 12.2 Earth's climate system is in rough equilibrium; our planet emits about the same amount of energy that it receives from the sun. As greenhouse gases accumulate in the atmosphere, however, they increase the amount of radiation that is emitted from the atmosphere back toward the surface. This illustration shows major pathways of energy flow in watts per square meter. Data from Kiehl, J. T., and K. E. Trenberth. 1997. Earth's annual global mean energy budget. *Bull. Amer. Meteorol. Soc.* 78: 197–208.

specifies atmospheric conditions over short time periods, typically hours or days, and within relatively small geographic areas. **Climate**, in contrast, describes the pattern of atmospheric conditions found across large geographic regions over long periods of time, typically seasons, years, or millennia. Mark Twain once noted the distinction between climate and weather by saying, "Climate is what we expect, weather is what we get."

Under most conditions, air in the troposphere decreases in temperature as altitude increases. Because warm air rises, this causes vertical mixing. Occasionally, however, a layer of relatively cool air occurs beneath a layer of warmer air. This departure from the normal temperature profile is known as a **temperature inversion**, or **thermal inversion** (Figure 12.3). The band of air in which temperature rises with altitude is called an **inversion layer** (because the normal direction of temperature change is inverted). The cooler air at the bottom of the inversion layer is denser than the warmer air at the top of the inversion layer, so it resists vertical mixing and remains stable. Whereas vertical mixing normally allows ground-level air pollution to be diluted upward, thermal inversions trap pollutants near the ground.

Global climate patterns result from large-scale circulation systems

At large geographic scales, convective air currents contribute to climatic patterns that are maintained over long periods of time (Figure 12.4a). Near the equator, solar radiation sets in motion a pair of convective cells known as *Hadley cells*. Here, where sunlight is most intense, surface air warms, rises, and expands. As it does so, it releases moisture, producing the heavy rainfall that gives rise to tropical rainforests near the equator. After releasing much of its moisture, this air diverges and moves in currents heading northward and southward. The air in these currents cools and descends back to Earth at about 30 degrees latitude north and south. Because the descending air has low relative humidity, the regions around 30 degrees latitude are quite arid, giving rise to deserts. Two pairs of similar but less intense convective cells, called *Ferrel cells* and *polar cells*, lift air and create precipitation around 60 degrees latitude north and south and cause air to descend at around 30 degrees latitude and in the polar regions.

These three pairs of cells account for the latitudinal distribution of moisture across Earth's surface and, along with temperature variation, help explain why biomes tend

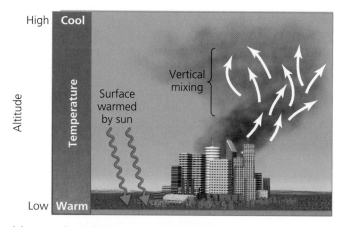

(a) Normal conditions

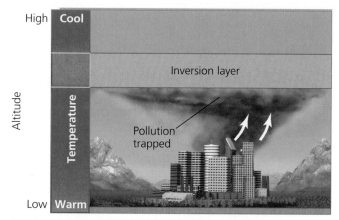

(b) Thermal inversion

FIGURE 12.3 A thermal inversion is a natural atmospheric occurrence that can exacerbate air pollution locally. Under normal conditions (**a**), tropospheric temperature decreases with altitude, and air of different altitudes mixes somewhat freely, dispersing most pollutants upward and outward from their sources. During a thermal inversion (**b**), cool air remains near the ground underneath an "inversion layer" of warmer air. No mixing occurs, and pollutants are trapped within the cool layer near the surface.

to be arrayed in latitudinal bands (Figure 5.10, ▶ p. 112). These convective cells also interact with Earth's rotation to produce global wind patterns (Figure 12.4b). As Earth rotates on its axis, locations on the equator spin faster than locations near the poles. As a result, the north-south air currents of the convective cells appear to be deflected from a straight path, because some portions of the globe move beneath them more quickly than others. This apparent deflection is called the *Coriolis effect*, and it results in the curving global wind patterns evident in Figure 12.4b. People used these global circulation patterns for centuries to facilitate ocean travel by wind-powered sailing ships. Today, understanding how the

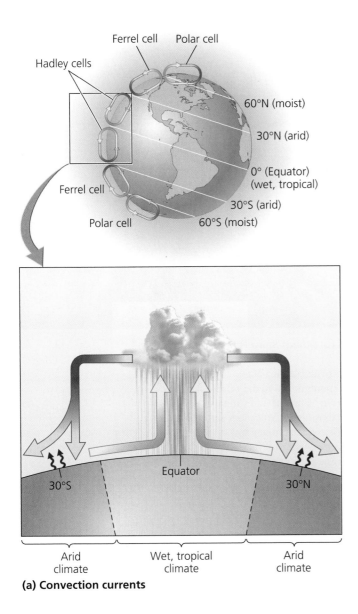

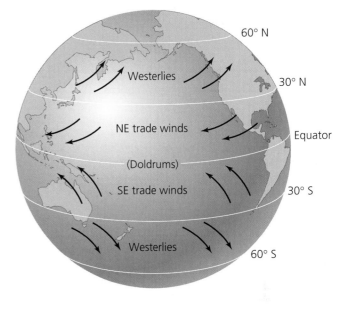

FIGURE 12.4 A series of large-scale convective cells (**a**) helps determine global patterns of humidity and aridity. Warm air near the equator rises, expands, and cools, and moisture condenses, giving rise to a warm, wet climate in tropical regions. Air travels toward the poles and descends around 30 degrees latitude. This air, which lost its moisture in the tropics, causes regions around 30 degrees latitude to be arid. This convective circulation, a Hadley cell, occurs on both sides of the equator. Between roughly 30 and 60 degrees latitude north and south, Ferrel cells occur; and between 60 and 90 degrees latitude, polar cells occur. As a result, air rises around 60 degrees latitude, creating a moist climate, and falls around 90 degrees, creating a dry climate. Global wind currents (**b**) show latitudinal patterns as well. Trade winds near the equator blow westward, while westerlies between 30 and 60 degrees latitude blow eastward.

(a) Convection currents

(b) Global wind patterns

atmosphere functions can help us comprehend how our pollution of the atmosphere can affect ecological systems, economies, and human health.

Outdoor Air Pollution

Throughout human history, we have made the atmosphere a dumping ground for our airborne wastes. Whether from primitive wood fires or modern coal-burning power plants, people have generated significant quantities of **air pollutants**, gases and particulate material added to the atmosphere that can affect climate or harm people or other organisms. **Air pollution** refers to the release of air pollutants. In recent decades, govern-

ment policy and improved technologies have helped us diminish *outdoor air pollution* (often called *ambient air pollution*) in industrialized nations. However, outdoor air pollution remains a problem, particularly in developing nations and in urban areas.

Natural sources can pollute

When we think of outdoor air pollution, we tend to envision smokestacks belching black smoke from industrial plants. However, natural processes produce a great deal of the world's air pollution (Figure 12.5). Winds sweeping over arid terrain can send huge amounts of dust aloft—sometimes across oceans from one continent to another. Volcanic eruptions release large quantities of particulate matter and sulfur dioxide into the troposphere, and major

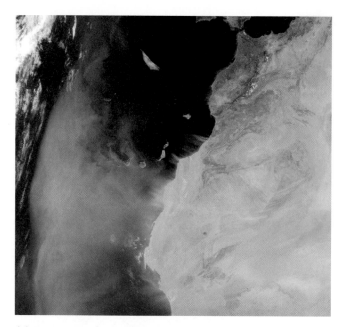

(a) Dust storm from Africa to the Americas

(b) Mount Saint Helens eruption, 1980

FIGURE 12.5 Massive dust storms are one type of natural air pollution. Storms blowing across the Atlantic Ocean from Africa to the Americas (**a**) carry fungal and bacterial spores linked to die-offs in Caribbean coral reef systems. Strong westerlies have lifted soil from deserts in Mongolia and China and carried it thousands of miles eastward across the Pacific Ocean to North America. Volcanoes are another source of natural air pollution, as shown by Mount Saint Helens (**b**), which erupted in Washington State in 1980. This eruption produced 1.1 billion m^3 (1.4 billion yd^3) of dust that circled the planet in 15 days. A third natural pollution source is fires in forests and grasslands. In 1997, a severe drought brought on by the 20th century's strongest El Niño event (▶ pp. 290–291) caused fires in Indonesia to rage out of control. Their smoke sickened 20 million people, caused cargo ships to collide, and even brought about an airplane crash.

eruptions may blow matter into the stratosphere, where it can remain for months or years. Fires also generate soot and gases, and over 60 million ha (150 million acres) of forest and grassland burn in a typical year.

Human activity and land-use policies can exacerbate some of these natural impacts. For instance, farming and grazing practices that strip vegetation from the soil and promote wind erosion can lead to intercontinental dust storms. And many fires in the tropics are set by people clearing and burning forest to make way for agriculture.

We create outdoor air pollution

Human activity also introduces new sources of air pollution. As with water pollution, air pollution can emanate from *point sources* or *non-point sources* (▶ pp. 252–253). Recall that a point source describes a specific spot—such as a factory's smokestacks—where large quantities of pollutants are discharged. In contrast, *non-point sources* are more diffuse, consisting of many small sources (such as millions of automobiles).

Primary pollutants, such as soot and carbon monoxide, are pollutants emitted into the troposphere in a form that can be directly harmful or that can react to form harmful substances. Harmful substances that are produced when primary pollutants interact, or when they react with constituents of the atmosphere, are called **secondary pollutants**.

Arguably the greatest air pollution problem today is our emission of greenhouse gases from fossil fuel combustion, which contribute to global climate change—a phenomenon we will explore in the second half of this chapter.

Clean Air Act legislation has addressed pollution in the United States

To address air pollution in the United States, Congress has passed a series of laws, including the **Clean Air Act**, first enacted in 1963 and amended multiple times since, particularly in 1970 and 1990. This body of legislation funds research into pollution control, sets standards for air quality, imposes limits on emissions from new stationary and mobile sources, and enables citizens to sue parties that violate the standards. It also introduced an emissions trading program (▸ p. 45) for sulfur dioxide. Beginning in 1995, businesses and utilities were allocated permits for emitting this pollutant and could buy, sell, or trade these allowances with one another. Each year the overall amount of allowed pollution was decreased. This successful market-based incentive program has spawned similar programs at state and regional levels and for other pollutants.

Under the Clean Air Act, the U.S. Environmental Protection Agency (EPA) sets nationwide standards for emissions of pollutants and for concentrations of pollutants in ambient air. However, it is largely up to the states to monitor air quality and to develop, implement, and enforce regulations within their boundaries. States submit implementation plans to the EPA for approval, and if a state's plans are not adequate, the EPA may take over enforcement in that state.

The EPA sets standards for "criteria pollutants"

The EPA and the states focus on six **criteria pollutants**, pollutants judged to pose especially great threats to human health—carbon monoxide (CO), sulfur dioxide (SO_2), nitrogen dioxide (NO_2), tropospheric ozone (O_3), particulate matter, and lead (Pb). For these, the EPA has established maximum concentrations allowable in ambient outdoor air.

Carbon monoxide *Carbon monoxide* is a colorless, odorless gas produced primarily by the incomplete combustion of fuel. In the United States in 2004, 87.2 million tons of CO were released, making it the most abundant air pollutant by mass. Vehicles account for 62% of these emissions, but other sources include lawn and garden equipment (10%), forest fires (6%), open burning of industrial waste (3%), and residential wood burning (2%). Carbon monoxide can bind irreversibly to hemoglobin in our red blood cells, preventing the hemoglobin from binding with oxygen. U.S. emissions of CO have decreased in recent decades largely because of cleaner-burning vehicle engines.

Sulfur dioxide Like CO, *sulfur dioxide* is a colorless gas. Of the 15.2 million metric tons of SO_2 released in the United States in 2004, about 70% resulted from the combustion of coal for electricity generation and industry. During combustion, elemental sulfur (S) in coal reacts with oxygen gas (O_2) to form SO_2. Once in the atmosphere, SO_2 may react to form sulfur trioxide (SO_3) and sulfuric acid (H_2SO_4), which may return to Earth in acidic deposition (▸ pp. 282–285).

Nitrogen dioxide *Nitrogen dioxide* is a foul-smelling reddish brown gas that contributes to smog and acidic deposition. Along with nitric oxide (NO), NO_2 belongs to a family of compounds called nitrogen oxides (NO_x). Nitrogen oxides result when atmospheric nitrogen and oxygen react at the high temperatures created by combustion engines. Of the 18.8 million tons of nitrogen oxides released in the United States in 2004, over half resulted from combustion in vehicle engines. Electrical utility and industrial combustion accounted for most of the rest.

Tropospheric ozone Although ozone in the stratosphere protects organisms from UV radiation, O_3 from human activity accumulates low in the troposphere. Here, this colorless gas is a secondary pollutant, created by the interaction of sunlight, heat, nitrogen oxides, and volatile carbon-containing chemicals. A major component of smog, O_3 poses health risks as a result of its instability as a molecule; this triplet of oxygen atoms will readily release one of its threesome, leaving a molecule of oxygen gas and a free oxygen atom. The free oxygen atom may then participate in reactions that can injure living tissues and cause respiratory problems. Although concentrations have fallen in the United States since 1982, *tropospheric ozone* is the pollutant that most frequently exceeds the EPA standard.

Particulate matter *Particulate matter* is composed of solid or liquid particles small enough to be suspended in the atmosphere and able to damage respiratory tissues when inhaled. Particulate matter includes primary pollutants such as dust and soot as well as secondary pollutants such as sulfates and nitrates. Most particulate matter is wind-blown dust (60%), but 2.5 million tons were released in the United States in 2004 as a result of human activities.

Lead *Lead* is a heavy metal that enters the atmosphere as a particulate pollutant. The lead-containing compounds tetraethyl lead and tetramethyl lead, when added to gasoline, improve engine performance. However, exhaust from leaded gasoline emits lead into the atmosphere, from which it can be inhaled or can be deposited on land and water. Lead can enter the food chain, accumulate within body tissues, and cause various ailments, including central nervous system malfunction and mental retardation in children. Since the 1980s, leaded gasoline has been phased out in many industrialized nations. U.S. lead emissions plummeted 93% from 1980 to 1990, and since then, emissions have remained low, coming primarily from industrial metal smelting. However, many developing nations continue to add lead to gasoline and experience significant lead pollution.

State and local agencies also monitor, calculate, and report to the EPA emissions of pollutants that affect ambient concentrations of the six criteria pollutants. These include the four criteria pollutants that are primary pollutants (carbon monoxide, sulfur dioxide, particulate matter, and lead), as well as all nitrogen oxides (because NO reacts to form NO_2, which is both a primary and secondary pollutant). Tropospheric ozone is a secondary pollutant only, so there are no emissions to monitor. Instead we monitor emissions of *volatile organic compounds (VOCs)*, carbon-containing chemicals (such as hydrocarbons; ▸ p. 57) that can react to produce ozone and other secondary pollutants. The largest sources of anthropogenic VOC emissions include industrial use of solvents (28%) and vehicle emissions (27%).

Air pollution has decreased markedly since 1970

Since the Clean Air Act of 1970, emissions of each of these six monitored pollutants have decreased, and total emissions of the six together have declined by 54% (Figure 12.6a). This decrease has occurred despite substantial increases in the nation's population, energy consumption, miles traveled by vehicle, and gross domestic product (Figure 12.6b). Because of this success in reducing emissions, air quality in the United States has improved markedly. EPA data show that the percentage of days U.S. citizens were exposed to unhealthy air dropped from 10% in 1988 to 3% in 2001.

Other chemicals known to cause serious health or environmental problems are classified as **toxic air pollutants**. These include substances known to cause cancer, reproductive defects, or neurological, developmental, immune system, or respiratory problems. The 1990 Clean Air Act identifies 188 toxic air pollutants, ranging from the heavy metal mercury (from coal-burning power plant emissions and other sources) to VOCs such as benzene (a component of gasoline). State and federal agencies do not monitor toxic air pollutants as extensively as they do the six criteria pollutants, but so far 300

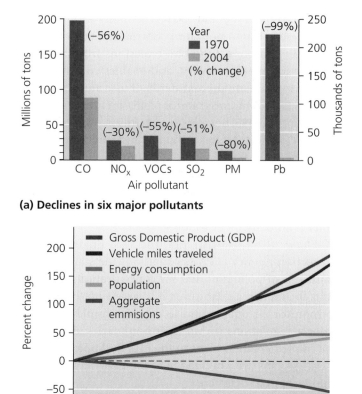

(a) Declines in six major pollutants

(b) Trends in major indicators

FIGURE 12.6 The EPA tracks emissions of several major pollutants into ambient air. Shown are emissions of four of the six "criteria pollutants," along with nitrogen oxides and volatile organic compounds. Each of these pollutants has shown substantial declines since 1970, and emissions from all six together have declined by 54% **(a)**. This decrease in emissions has occurred despite increases in U.S. population, energy consumption, vehicle miles traveled, and gross domestic product **(b)**. Go to **GRAPH**it! at www.aw-bc.com/withgott. Data from U.S. EPA, 2004.

monitoring sites are operating, and coverage is improving. The EPA estimates that because of Clean Air Act regulations, from 1990 to 1999 emissions of toxic air pollutants decreased by 30%.

Burning fossil fuels produces industrial smog

In response to an increasing incidence of fogs polluted by the smoke of Britain's industrial revolution, a British scientist coined the term *smog*. The world's worst smog event took place in London in 1952, when a thermal inversion trapped pollutants from factories and coal-burning home stoves over the city for several days, killing perhaps 12,000 people. Today the term *smog* is used worldwide to describe unhealthy mixtures of air pollutants that often form over urban areas. The smog that enveloped London in 1952 was what we now call **industrial smog**, or gray-air smog. When coal or oil is burned, some portion is completely combusted, forming CO_2; some is partially combusted, producing CO; and some remains unburned and is released as soot, or particles of carbon. Moreover, coal may contain contaminants including mercury and sulfur. Sulfur reacts with oxygen to form sulfur dioxide, which can undergo a series of reactions to form sulfuric acid and ammonium sulfate (Figure 12.7a). These chemicals and others produced by further reactions, along with soot, are the main components of industrial smog and produce its characteristic gray color.

Industrial smog is less common today in developed nations than it was 50–100 years ago. In the wake of the 1952 London episode and others, governments of most developed nations began regulating industrial emissions. However, in industrializing regions such as China, India, and Eastern Europe, coal burning (by industry and by citizens heating and cooking in their homes), combined with lax air pollution controls, produces industrial smog that poses significant health risks in many urban areas.

Weather plays a role in smog formation. In Donora, Pennsylvania, in 1948, a thermal inversion trapped smog containing particulate matter from a steel and wire factory. Twenty-one people were killed, and over 6,000 people—nearly half the town—became ill (Figure 12.7b). In Donora's "killer smog," air near the ground cooled during the night. Normally, morning sunlight warms the land and air, causing air to rise. However, because Donora is located in hilly terrain, too little sun reached the valley floor to warm and disperse the cold air. The resulting thermal inversion kept a pall of smog over the town. Hilly topography facilitates smog in many other cities where surrounding mountains trap air and create inversions—including the Los Angeles basin. Modern-day L.A., however, suffers from a different type of smog, one called photochemical smog.

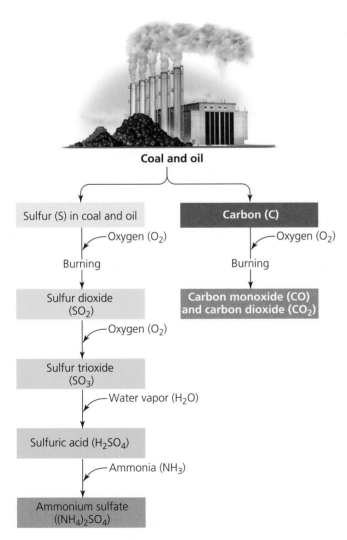

Coal and oil

Sulfur (S) in coal and oil — Oxygen (O_2) — Burning → Sulfur dioxide (SO_2) — Oxygen (O_2) → Sulfur trioxide (SO_3) — Water vapor (H_2O) → Sulfuric acid (H_2SO_4) — Ammonia (NH_3) → Ammonium sulfate (($NH_4)_2SO_4$)

Carbon (C) — Oxygen (O_2) — Burning → Carbon monoxide (CO) and carbon dioxide (CO_2)

(a) Burning sulfur-rich oil or coal without adequate pollution control technologies

FIGURE 12.7 Emissions from the combustion of coal and oil in manufacturing plants and from utilities without pollution control technologies can create industrial smog. Industrial smog consists primarily of sulfur dioxide and particulate matter, as well as carbon monoxide and carbon dioxide from the carbon component of fossil fuels. Sulfur contaminants in fossil fuels when combusted create the sulfur dioxide, which in the presence of other chemicals in the atmosphere can produce several other sulfur compounds **(a)**. Under certain weather conditions, industrial smog can blanket whole towns or regions, as it did in Donora, Pennsylvania, shown here in the daytime during its deadly 1948 smog episode **(b)**.

(b) Donora, Pennsylvania, at midday in the 1948 smog event

Photochemical smog is produced by a series of reactions

Photochemical smog, or brown-air smog, is formed by light-driven chemical reactions of primary pollutants and normal atmospheric compounds that produce a mix of over 100 different chemicals, tropospheric ozone often being the most abundant among them (Figure 12.8a). High levels of NO_2 cause photochemical smog to form a brownish haze (Figure 12.8b). Hot, sunny, windless days in urban areas provide perfect conditions for the formation of photochemical smog. Exhaust from morning traffic releases large amounts of NO and VOCs. Sunlight then promotes the production of ozone and other constituents of photochemical smog. Levels of photochemical pollutants in urban areas typically peak in mid-afternoon and, at sufficient levels, can irritate people's eyes, noses, and throats. *Peroxyacyl nitrates*, created by the reaction of NO_2 with hydro-carbons, can induce further reactions that damage living tissues in animals and plants.

Photochemical smog afflicts many major cities, especially those with topography and weather conditions that promote it. In Athens, Greece, site of the 2004 Olympics, the problem had been bad enough that the city government provided incentives to replace aging automobiles. It also mandated that cars with odd-numbered license plates be driven only on odd-numbered days, and those with even-numbered plates only on even-numbered days. According to Greek officials, smog has been reduced by 30% since 1990 as a result.

Synthetic chemicals deplete stratospheric ozone

As we have seen, ozone in the troposphere is a pollutant, but ozone in the stratosphere absorbs incoming ultraviolet radiation from the sun, shielding life on Earth's

(b) Photochemical smog over Mexico City

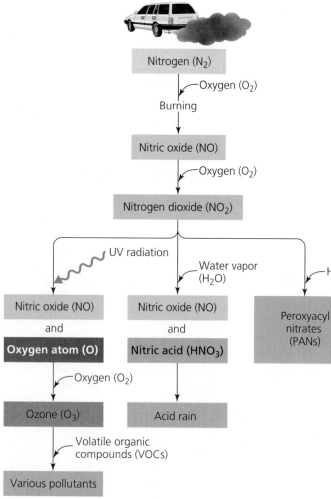

(a) Formation of photochemical smog

FIGURE 12.8 Nitric oxide, a key element of photochemical smog, can start a chemical chain reaction (**a**) that results in the production of other compounds, including nitrogen dioxide, peroxyacyl nitrates, nitric acid, and ozone. Nitric acid can contribute to acidic deposition as well as photochemical smog. Photochemical smog is common today over many urban areas, especially those with hilly topography or frequent thermal inversions. Mexico City (**b**) is one city that frequently experiences photochemical smog.

surface. In the 1960s, atmospheric scientists began wondering why their measurements of stratospheric ozone were lower than theoretical models predicted. researchers speculating that natural or artificial chemicals were depleting ozone finally pinpointed a group of human-made compounds derived from simple hydrocarbons, such as ethane and methane, in which hydrogen atoms are replaced by chlorine, bromine, or fluorine. One class of such compounds, **chlorofluorocarbons (CFCs)**, was being industrially produced at a rate of a million metric tons per year in the early 1970s.

In 1974, atmospheric scientists Sherwood Rowland and Mario Molina showed that CFCs could deplete stratospheric ozone by releasing chlorine atoms that split ozone molecules, creating from each of them an O_2 molecule and a ClO molecule. Three years before Rowland and Molina's study, researcher J. E. McDonald had predicted that ozone loss, by allowing more UV radiation to reach the surface, would result in thousands more skin cancer cases each year. This caught the attention of policymakers, environmentalists, and industry alike.

Then in 1985, scientists from the British Antarctic Survey announced that stratospheric ozone levels over Antarctica each autumn had declined by 40–60% in the previous decade, leaving a thinned ozone concentration that was soon dubbed the *ozone hole* (Figure 12.9). Over the next few years, researchers confirmed the link between CFCs and ozone loss in the Antarctic. They also indicated that depletion was occurring in the Arctic as well, and perhaps globally. Already concerned about skin cancer, scientists were becoming anxious over other possible effects of increased UV radiation, including harm to crops and to the productivity of ocean phytoplankton, the base of the marine food chain.

The Montreal Protocol addressed ozone depletion

International efforts to restrict CFC production finally bore fruit in 1987 with the **Montreal Protocol**. In this treaty, signatory nations (eventually numbering 180) agreed to cut CFC production in half. Five follow-up agreements deepened the cuts, advanced timetables for compliance, and addressed additional ozone-depleting chemicals. Today the production and use of ozone-depleting compounds has fallen by 95% since the late 1980s, and scientists can discern the beginnings of

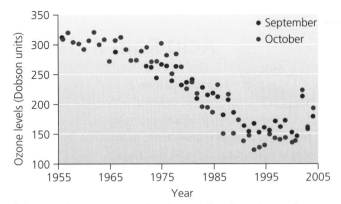

(a) Monthly mean ozone levels at Halley Bay, Antarctica

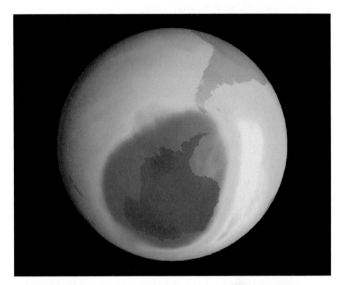

(b) The "ozone hole" over Antarctica, September 2000

FIGURE 12.9 The "ozone hole" consists of a region of thinned ozone density in the stratosphere over Antarctica and the southernmost ocean regions. It has reappeared seasonally each September in recent decades. Data from Halley Bay, Antarctica (**a**), show a steady decrease in stratospheric ozone concentrations from the 1960s to 1990. Ozone-depleting CFCs began to be regulated under the Montreal Protocol in 1987, and ozone concentrations stopped declining. Colorized satellite imagery from September 6, 2000 (**b**), shows the "ozone hole" (blue) at its maximal recorded extent to date. Data from British Antarctic Survey.

long-term recovery of the ozone layer (although much of the 5 billion kg of CFCs emitted into the troposphere has yet to diffuse up into the stratosphere, and CFCs are slow to dissipate or break down). Industry was able to shift to alternative, environmentally safer chemicals, which have largely turned out to be cheaper and more efficient.

Acid Rain at Hubbard Brook Research Forest

Acidic deposition involves incremental changes in pH levels that take place over long periods of time, so no single experiment can give us a complete picture of acidic deposition's effects. Nonetheless, one long-term study conducted in the Hubbard Brook Experimental Forest in New Hampshire's White Mountains has been critically important to our understanding of acidic deposition in the United States.

Established by the U.S. Forest Service in 1955, Hubbard Brook was initially devoted to research on hydrology, the study of water flow through forests and streams. In 1963, in collaboration with scientists at Dartmouth University, Hubbard Brook researchers broadened their focus to include a long-term study of nutrient cycling in forest ecosystems. Since then, they have collected and analyzed weekly samples of precipitation. The measurements make up the longest-running North American record of acid precipitation and have helped shape U.S. policy on sulfur and nitrogen emissions.

Throughout Hubbard Brook's 3,160 ha (7,800 acres), small plastic funnels channel precipitation into clean bottles, which researchers retrieve and replace each week.

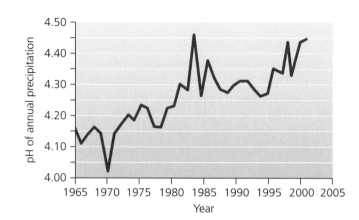

Over the past 40 years, precipitation at the Hubbard Brook Experimental Forest has become slightly less acidic. However, it is still far more acidic than natural precipitation. Data from Likens, G. E. 2004. *Ecology* 85: 2355–2362.

Hubbard Brook's laboratory measures acidity and conductivity, which indicates the amount of salts and other electrolytic contaminants dissolved in the water. Concentrations of sulfuric acid, nitrates, ammonia, and other compounds are measured elsewhere.

By the late 1960s, ecologists Gene Likens, F. Herbert Bormann, and others had found that precipitation at Hubbard Brook was several hundred times more acidic than natural rainwater. By the early 1970s, other studies indicated that precipitation from Pennsylvania to Maine had pH values averaging around 4, and that individual rain-

storms showed values as low as 2.1—almost 10,000 times more acidic than ordinary rainwater.

In 1978, the National Atmospheric Deposition Program was launched to monitor precipitation and dry deposition across the United States. Initially consisting of 22 sites, including Hubbard Brook, the program now comprises more than 200, each of which gathers weekly data on acidic deposition and deposition of other substances. By the late 1980s, this program had produced a nationwide map of pH values (see Figure 12.11, ▶ p. 285). The most severe problems were found to be in the

Because of its success in addressing ozone depletion, the Montreal Protocol is widely seen as a model for international cooperation on other pressing global problems, such as persistent organic pollutants (▶ p. 237), climate change (▶ pp. 288–298), and biodiversity loss (Chapter 8).

Acidic deposition presents another transboundary pollution issue

Just as stratospheric ozone depletion crosses political boundaries, so does **acidic deposition**, the deposition of acidic or acid-forming pollutants from the atmosphere

The effects of acidic deposition on trees can be seen in this forest on Mount Mitchell in western North Carolina.

Some long-term consequences of acidic deposition are now becoming clear. In 1996, researchers reported that approximately 50% of the calcium and magnesium in Hubbard Brook's soils had leached out. Meanwhile, acidic deposition had increased the concentration of aluminum in the soil, which can prevent tree roots from absorbing nutrients. The resulting nutrient deficiency slows forest growth and weakens trees, making them more vulnerable to drought and insects (see the second figure). It also reduces the ability of soil and water to neutralize acidity, making the ecosystem increasingly vulnerable to further acidic inputs.

In October 1999, researchers used a helicopter to distribute 50 tons of a calcium-containing mineral called wollastonite over one of Hubbard Brook's watersheds. Their objective was to raise the concentration of base cations to estimated historical levels. Over the next 50 years, scientists plan to evaluate the impact of calcium addition on the watershed's soil, water, and life. By providing a comparison to watersheds in which calcium remains depleted, the results should provide new insights into the consequences of acid rain and the possibilities for reversing its negative effects.

Northeast, where prevailing winds were blowing emissions from fossil-fuel-burning power plants in the Midwest. Scientists hypothesized that when sulfur dioxide, nitrogen oxides, and other pollutants arrived in the Northeast, they were absorbed by water droplets in clouds, converted to acidic compounds such as sulfuric acid, and deposited on farms, forests, and cities in rain or snow.

The Clean Air Act of 1970 helped reduce acidic deposition in the Northeast. At an area of Hubbard Brook known as Watershed 6, average pH increased slightly between 1965 and 1995, from about 4.15 to about 4.35 (see the first figure). In 1990, as a consequence of the Hubbard Brook study and the nationwide research that followed, the Clean Air Act was amended to further restrict emissions of sulfur dioxide and other acid-forming compounds. Nonetheless, acidic deposition continues to be a serious concern in the Northeast.

onto Earth's surface (see "The Science behind the Story," above). Acidic deposition can take place by precipitation (commonly referred to as *acid rain*, but also including acid snow, sleet, and hail), by fog, by gases, or by the deposition of dry particles. Acidic deposition is one type of **atmospheric deposition**, which refers more broadly to the wet or dry deposition on land of a wide variety of pollutants, including mercury, nitrates, organochlorines, and others.

Acidic deposition originates primarily with sulfur dioxide and nitrogen oxides, pollutants produced largely

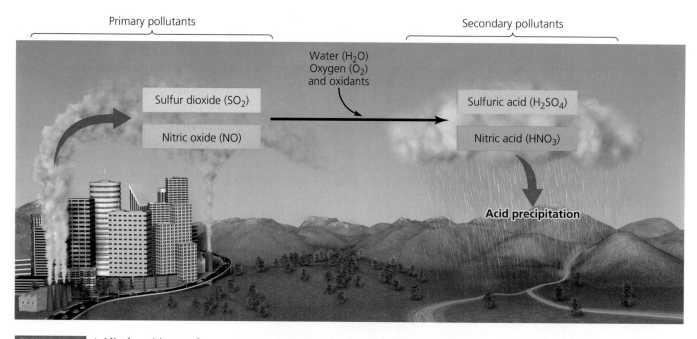

Primary pollutants

Secondary pollutants

Water (H₂O)
Oxygen (O₂)
and oxidants

Sulfur dioxide (SO₂)

Nitric oxide (NO)

Sulfuric acid (H₂SO₄)

Nitric acid (HNO₃)

Acid precipitation

FIGURE 12.10 Acidic deposition can have consequences many miles downwind from its source. Emissions containing sulfur dioxide and nitric oxide from industries and utilities begin the process. Sulfur dioxide and nitric oxide can be transformed into sulfuric acid and nitric acid through chemical reactions in the atmosphere, and these acidic compounds descend to Earth's surface in rain, snow, fog, and dry deposition.

through fossil fuel combustion by automobiles, electric utilities, and industrial facilities. Once emitted into the troposphere, these pollutants can react with water, oxygen, and oxidants to produce compounds of low pH (▸ pp. 56–57), primarily sulfuric acid and nitric acid. Suspended in the troposphere, droplets of these acids may travel for days or weeks, sometimes covering hundreds or thousands of kilometers (Figure 12.10).

Acidic deposition can have wide-ranging detrimental effects on ecosystems and on our built environment. Acids can leach basic minerals such as calcium and magnesium from soil, harming plants and soil organisms. Elevated aluminum in the soil hinders water and nutrient uptake by plants, and it may damage crops. Runoff may significantly acidify streams, rivers, and lakes—and thousands of northern lakes now contain water acidic enough to kill fish. In some regions, acid fog with a pH of 2.3 (equivalent to vinegar) can envelop forests for extended periods; large stands of trees have died in some eastern North American forests from these conditions. Moreover, acidic precipitation can erode stone buildings, eat away at cars, and erase the writing from tombstones. Ancient cathedrals in Europe, monuments in Washington, D.C., temples in Asia, and stone statues throughout the world

are experiencing irreparable damage as their features wear away.

Because the pollutants leading to acid deposition can travel long distances, their effects may be felt far from their sources—a situation that has led to political bickering among the leaders of states and nations (Figure 12.11).

New technology has helped reduce acidic deposition; "scrubbers" that filter pollutants in smokestacks have allowed factories to decrease emissions (▸ pp. 313, 359). As a result of declining emissions of SO₂, average sulfate precipitation in 1996–2000 was 10% lower than in 1990–1994 across the United States and 15% lower in the eastern states. However, because of increasing NOₓ emissions, average nitrate precipitation increased nationally by 3% between these time periods.

A recent report by scientists at New Hampshire's Hubbard Brook research forest, where acidic deposition's effects were first demonstrated in the United States, disputed the notion that the problem of acid deposition is being solved (see "The Science behind the Story," ▸ pp. 282–283). Instead, the report said, the effects are worse than first predicted, and the mandates of the 1990 Clean Air Act are not adequate to restore ecosystems in the northeastern United States.

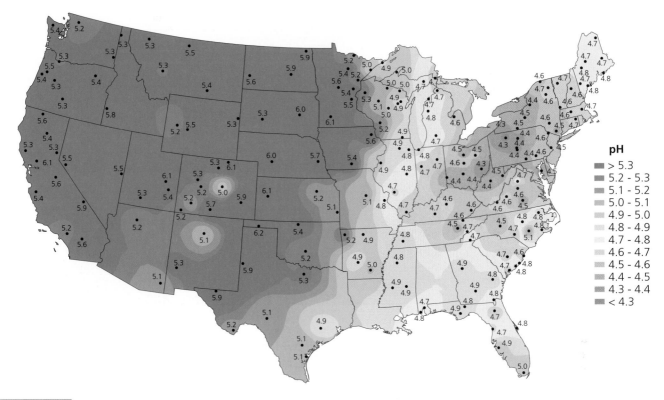

FIGURE 12.11 This map shows pH values for precipitation throughout the United States. Precipitation is most acidic in parts of the Northeast and Midwest, generally downwind from (roughly east of) areas of heavy industrial development. For instance, much of the pollution from power plants and factories in Pennsylvania, Ohio, and Illinois falls out in New York and the New England states, as well as in Ontario, Quebec, and Canada's Maritime Provinces. Data from National Atmospheric Deposition Program. 2005. Hydrogen ion concentration as pH from measurements made at the Central Analytical Laboratory, 2003.

Indoor Air Pollution

Indoor air generally contains higher concentrations of pollutants than does outdoor air. As a result, the health effects from *indoor air pollution* in workplaces, schools, and homes outweigh those from outdoor air pollution. One estimate, from the United Nations Development Programme in 1998, attributed 2.2 million deaths worldwide to indoor air pollution and 500,000 deaths to outdoor air pollution. Indoor air pollution alone, then, takes roughly 6,000 lives each day.

If this seems surprising, consider that the average U.S. citizen spends at least 90% of his or her time indoors. Then consider that in the past half century we have manufactured and sold a dizzying array of consumer products, many of which we keep in our homes and offices and have come to play major roles in our daily lives. Products such as insecticides and cleaning fluids exude volatile chemicals into the air, as can solid materials such as plastics and chemically treated wood.

Ironically, some attempts to be environmentally prudent during the "energy crisis" of 1973–1974 (▶ pp. 314, 316) worsened indoor air pollution in developed countries. To reduce heat loss and improve energy efficiency, building managers sealed off most ventilation in existing buildings, and building designers constructed new buildings with limited ventilation and with windows that did not open. These steps may have saved energy, but they worsened indoor air pollution by trapping stable, unmixed air—and its pollutants—inside.

Risks differ in developing and developed nations

Indoor air pollution has the greatest impact in the developing world, where poverty forces millions of people to burn wood, charcoal, animal dung, or crop waste inside their homes for cooking and heating with little or no ventilation (Figure 12.12). In the process, they inhale dangerous amounts of soot and carbon monoxide. In the air of such homes, concentrations of particulate matter are

FIGURE 12.12 In the developing world, many people build fires inside their homes for cooking and heating, as seen here in a South African kitchen. The fires expose family members to particulate matter and carbon monoxide. In most regions of the developing world, indoor air pollution is estimated to cause upwards of 3% of all health risks.

commonly 20 times above U.S. EPA standards, the World Health Organization (WHO) has found. Indoor air pollution from fuelwood burning, the WHO estimates, kills 1.6 million people each year, comprising over 5% of all deaths in some developing nations and 2.7% of the entire global disease burden.

Whereas particulate matter and chemicals from wood and charcoal smoke are the primary indoor air health risks in the developing world, the top risks in developed nations are cigarette smoke and radon. The health effects of smoking cigarettes are well known, and inhaling secondhand smoke causes many of the same problems, ranging from irritation of the eyes, nose, and throat, to exacerbation of asthma and other respiratory ailments, to lung cancer. Tobacco smoke is a brew of over 4,000 chemical compounds, many of which are known or suspected to be toxic or carcinogenic.

Radon is the second-leading cause of lung cancer in the United States, responsible for an estimated 20,000 deaths per year. As we saw in Chapter 10 (▶ p. 221), radon is a radioactive gas resulting from the natural decay of uranium in soil, rock, or water, which seeps up from the ground and can infiltrate buildings. Radon is colorless and odorless, and it can be impossible to predict where it will occur without knowing details of an area's underlying geology. As a result, the only way to determine whether radon is entering a building is to measure radon with a test kit. Testing in 1991 led the EPA to estimate that 6% of U.S. homes exceeded the EPA's maximum recommended level for radon. Since the mid-1980s, 18 million U.S. homes have been tested for radon, and 700,000 have undergone radon mitiga-

tion. Many new homes are now being built with radon-resistant features.

Many substances pollute indoor air

In our daily lives at home, we are exposed to many indoor air pollutants (Figure 12.13). The most diverse indoor pollutants are VOCs. These airborne carbon-containing compounds are released by everything from plastics to oils to perfumes to paints to cleaning fluids to adhesives to pesticides. VOCs evaporate from furnishings, building materials, color film, carpets, laser printers, fax machines, and sheets of paper. Some products, such as chemically treated furniture, release large amounts of VOCs when new and progressively less as they age. Other items, such as photocopying machines, emit VOCs each time they are used. The VOC formaldehyde—widely used in pressed wood, insulation, and other products—irritates mucous membranes, induces skin allergies, and causes other ailments.

VOCs are often thought responsible for *sick-building syndrome*, an illness produced by indoor pollution in which the specific cause is not identifiable. Besides VOCs, microorganisms such as bacteria, fungi, and mold can induce allergic responses and cause building-related illness. Heating and cooling systems in buildings make ideal breeding grounds for microbes, providing moisture, dust, and foam insulation as substrates, as well as air currents to carry the organisms aloft. The U.S. Occupational Safety and Health Administration (OSHA) has estimated that 30–70 million Americans have suffered ailments due to the environment of the building in which they live.

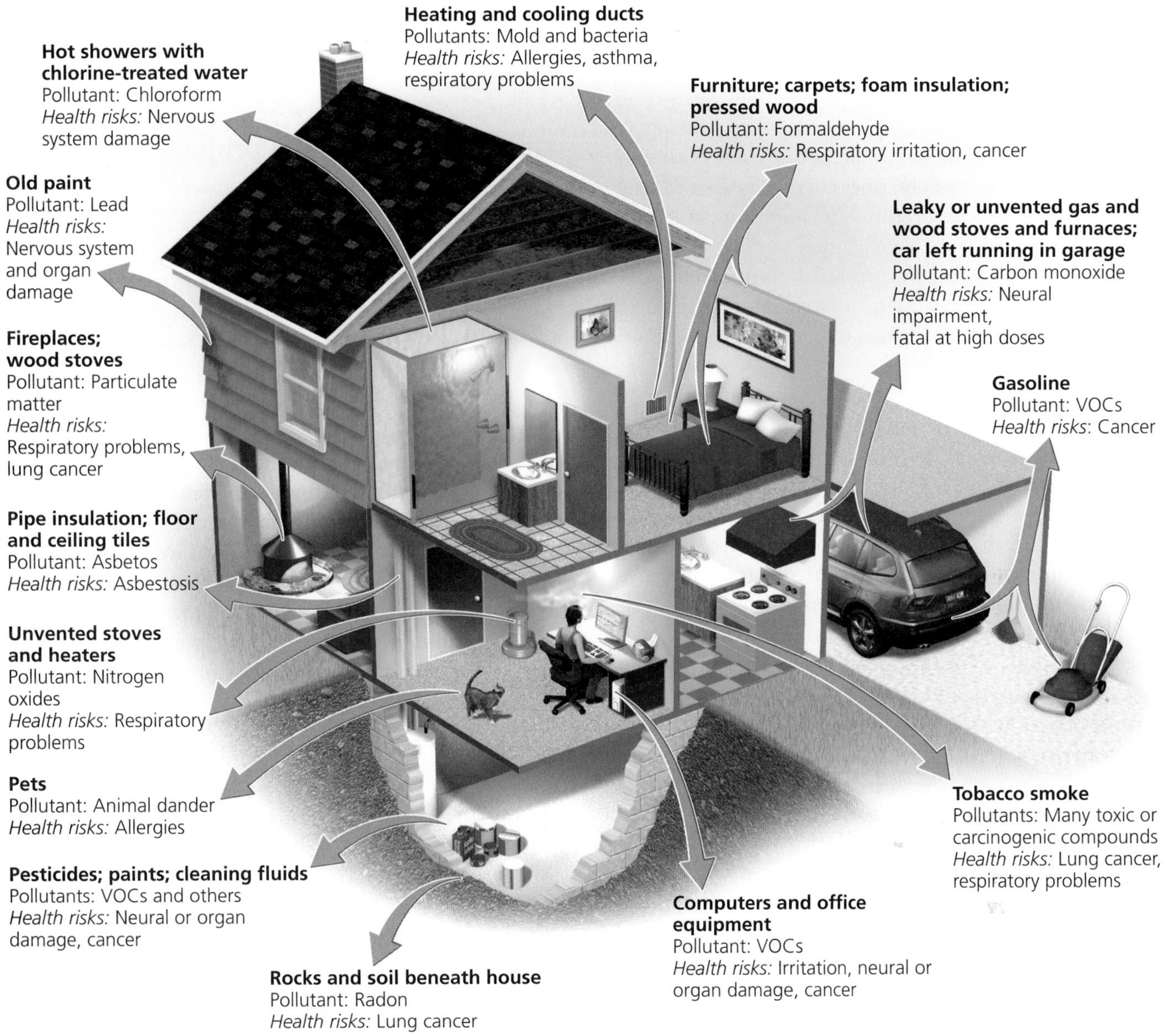

Hot showers with chlorine-treated water
Pollutant: Chloroform
Health risks: Nervous system damage

Old paint
Pollutant: Lead
Health risks: Nervous system and organ damage

Fireplaces; wood stoves
Pollutant: Particulate matter
Health risks: Respiratory problems, lung cancer

Pipe insulation; floor and ceiling tiles
Pollutant: Asbetos
Health risks: Asbestosis

Unvented stoves and heaters
Pollutant: Nitrogen oxides
Health risks: Respiratory problems

Pets
Pollutant: Animal dander
Health risks: Allergies

Pesticides; paints; cleaning fluids
Pollutants: VOCs and others
Health risks: Neural or organ damage, cancer

Heating and cooling ducts
Pollutants: Mold and bacteria
Health risks: Allergies, asthma, respiratory problems

Furniture; carpets; foam insulation; pressed wood
Pollutant: Formaldehyde
Health risks: Respiratory irritation, cancer

Leaky or unvented gas and wood stoves and furnaces; car left running in garage
Pollutant: Carbon monoxide
Health risks: Neural impairment, fatal at high doses

Gasoline
Pollutant: VOCs
Health risks: Cancer

Tobacco smoke
Pollutants: Many toxic or carcinogenic compounds
Health risks: Lung cancer, respiratory problems

Computers and office equipment
Pollutant: VOCs
Health risks: Irritation, neural or organ damage, cancer

Rocks and soil beneath house
Pollutant: Radon
Health risks: Lung cancer

FIGURE 12.13 The typical U.S. home contains a variety of potential sources of indoor air pollution. Shown are some of the most common sources, the major pollutants they emit, and some of the health risks they pose.

Weighing the Issues:
How Safe Is Your Indoor Environment?

Think about the amount of time you spend indoors. Name the potential indoor air quality hazards in your home, work, or school environment. Are these spaces well ventilated? What could you do to improve the safety of the indoor spaces you use?

Using low-toxicity materials, monitoring air quality, keeping rooms clean, and providing adequate ventilation are the keys to alleviating indoor air pollution in most situations. In the developed world, we can limit our use of plastics and treated wood where possible and limit our exposure to pesticides, cleaning fluids, and other known toxicants by keeping them in a garage or outdoor shed rather than in the house. The EPA recommends that we test our homes and offices for radon and install detectors for carbon monoxide. In addition, keeping rooms and air ducts clean and free of mildew and other biological pollutants will reduce potential irritants and allergens. Finally, it is important to keep our indoor spaces well ventilated to minimize concentrations of pollutants.

Global Climate Change

Air pollution has many effects on human health, but some types of air pollution appear to be causing far greater impacts because they affect our planet's climate. **Global climate change** describes changes in Earth's long-term patterns of temperature, precipitation, and storm frequency and intensity. People often use the term *global warming* synonymously in casual conversation, but *global warming* refers specifically to an increase in Earth's average surface temperature and is only one aspect of global climate change.

Our planet's climate has always changed naturally, but the climatic changes taking place today are unfolding at an exceedingly rapid rate. Moreover, most scientists agree that human activities, notably fossil fuel combustion and deforestation, are largely responsible for the current modification of Earth's atmosphere and climate. Climatic changes will likely have adverse consequences for ecosystems and for millions of people, from the residents of the Maldives to those of Florida, Louisiana, and other parts of the United States. For this reason, more and more scientists, policymakers, and ordinary citizens are seeking to take action to minimize and mitigate our impacts on the climate system.

"Greenhouse gases" warm the lower atmosphere

Three factors exert more influence on Earth's climate than all others combined. The first is the sun. Without it, Earth would be dark and frozen. The second is the atmosphere. Without it, Earth would be as much as 33 °C (59 °F) colder on average, and temperatures would vary far more between night and day. The third is the oceans, which shape climate by storing and transporting heat and moisture. As we saw in Figure 12.2 (▸ p. 273), Earth's atmosphere, clouds, land, ice, and water together reflect 30% of incoming solar radiation back into space, and absorb the remaining 70%.

As Earth's surface absorbs solar radiation, the surface increases in temperature and emits infrared radiation (▸ p. 59). Some atmospheric gases absorb this infrared radiation very effectively. These include water vapor, ozone, carbon dioxide, nitrous oxide (N_2O), methane, and halocarbons (a diverse group of gases that include chlorofluorocarbons [CFCs; ▸ p. 281]). Such gases are known as **greenhouse gases**. These gases subsequently re-emit infrared energy of slightly different wavelengths, warming the troposphere and the planet's surface in a phenomenon known as the **greenhouse effect**. Human activities have increased the concentrations of many greenhouse gases in the past 250–300 years, and we have thereby enhanced the greenhouse effect.

Carbon dioxide is the primary greenhouse gas

The atmospheric concentration of carbon dioxide has increased from around 280 parts per million (ppm) as recently as the late 1700s to 316 ppm in 1959 to 382 ppm in 2006 (Figure 12.14). The atmospheric CO_2 concentration is now at its highest level in at least 400,000 years, and likely the highest in the last 20 million years. Moreover, it is increasing faster today than at any time in at least 20,000 years.

What has changed since the 1700s to cause the atmospheric concentration of this greenhouse gas to increase so rapidly? As you may recall from our discussion of the carbon cycle in Chapter 3 (▸ pp. 61–64), and as we will see further in Chapter 13 (▸ pp. 304–305), a great deal of carbon is stored for long periods in sediments underground. The deposition, partial decay, and compression of organic matter (mostly plants) that grew in wetland or marine

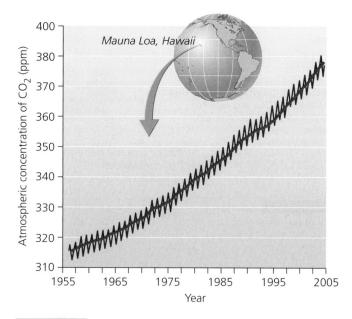

FIGURE 12.14 Atmospheric concentrations of carbon dioxide have risen steeply since 1958, when Charles Keeling of the Scripps Institution of Oceanography began collecting these data at the Mauna Loa Observatory in Hawaii. The jaggedness apparent within the overall upward trend of the so-called *Keeling curve* represents seasonal variation caused by the Northern Hemisphere's vegetation, which absorbs more carbon dioxide during the northern summer, when it is more photosynthetically active. Go to **GRAPHit!** at www.aw-bc.com/withgott. Data from National Oceanic and Atmospheric Administration, Climate Prediction Center, 2005.

areas millions of years ago has led to the formation of coal, oil, and natural gas in ancient sediments. In the absence of human activity, these carbon reservoirs would be practically permanent. However, over the past two centuries we have been burning increasing amounts of fossil fuels in our homes, factories, and automobiles. In so doing, we have transferred immense amounts of carbon from one reservoir (the underground deposits that stored the carbon for millions of years) to another (the atmosphere). This flux of carbon is the main reason atmospheric CO_2 concentrations have increased so dramatically.

At the same time, people have cleared and burned forests to make room for crops, pastures, villages, and cities. Deforestation reduces the biosphere's ability to absorb carbon dioxide from the atmosphere, and it has also thereby contributed to increasing atmospheric CO_2 concentrations.

Other greenhouse gases add to warming

Carbon dioxide is not the only greenhouse gas increasing in the atmosphere. Methane is also on the rise, and molecule-for-molecule, methane is 23 times as potent a greenhouse gas as is CO_2. We release methane into the atmosphere by tapping into fossil fuel deposits, raising livestock that release methane as a metabolic waste product, disposing of organic matter in landfills, and growing certain types of crops, especially rice. Since 1750, atmospheric methane concentrations have increased by 151%, and the current concentration is the highest in at least 400,000 years.

Nitrous oxide concentrations have also risen—by 17% since 1750. This greenhouse gas is a by-product of feedlots, chemical manufacturing plants, auto emissions, and use of nitrogen fertilizers in agriculture. Ozone concentrations in the troposphere have increased by 36% since 1750 as a result of smog formation (▶ pp. 277, 280). Halocarbon gases add greatly to the atmosphere's heat-absorbing ability on a per-molecule basis (although rare, each molecule packs up to 12,000 times the punch of a CO_2 molecule). Halocarbons' contribution to global warming, however, has begun to slow as a result of the Montreal Protocol and subsequent controls (▶ p. 281).

Water vapor is the most abundant greenhouse gas in the atmosphere. Its concentration varies, but if tropospheric temperatures continue to increase, the oceans and other water bodies should transfer increasingly more water vapor into the atmosphere. Such a positive feedback mechanism (▶ pp. 50–51) could amplify the greenhouse effect. Alternatively, higher water vapor concentrations could give rise to increased cloudiness, which might, in a negative feedback loop, slow global warming by reflecting more incoming solar radiation back into space.

Whereas greenhouse gases exert a warming effect, *aerosols*, microscopic droplets and particles, can have either a warming or cooling effect. Generally speaking, soot aerosols, also known as black carbon aerosols, can cause warming, but most tropospheric aerosols lead to climate cooling. Sulfate aerosols produced by fossil fuel combustion may slow global warming, at least in the short term.

The atmosphere is not the only factor that influences climate

Changes in Earth's rotation and orbit also affect climate. During the 1920s, Serbian mathematician Milutin Milankovitch described three kinds of changes in Earth's rotation and orbit around the sun. These variations, now known as **Milankovitch cycles**, result in slight changes in the relative amount of solar radiation reaching Earth's surface at different latitudes (Figure 12.15). As these cycles proceed, they change the way solar radiation is distributed over Earth's surface. This, in turn, contributes to changes in atmospheric heating and circulation that have triggered climate variation, such as periodic glaciation episodes.

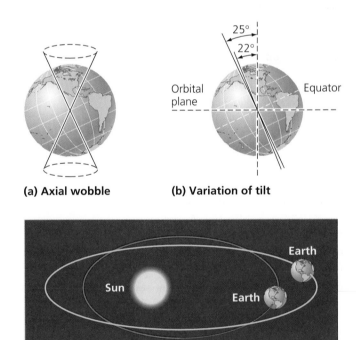

(a) Axial wobble   **(b) Variation of tilt**

(c) Variation of orbit

FIGURE 12.15   There are three types of Milankovitch cycles. The first is an axial wobble (**a**) that occurs on a 19,000- to 23,000-year cycle. The second is a 3-degree shift in the tilt of Earth's axis (**b**) that occurs on a 41,000-year cycle. The third is a variation in Earth's orbit from almost circular to more elliptical (**c**), which repeats itself every 100,000 years. These variations affect the intensity of solar radiation that reaches portions of Earth at different times, creating long-term changes in global climate.

The oceans also shape climate, by exchanging heat with the atmosphere and moving energy from place to place. In equatorial regions, such as around the Maldives, the oceans receive more heat from the sun and atmosphere than they emit. Near the poles, the oceans emit more heat than they receive. Because cooler water is denser than warmer water, the cooling water at the poles tends to sink, and the warmer surface water from the equator moves to take its place. This is one principle explaining global oceanic circulation (▶ pp. 257–258).

The best-known interaction among ocean circulation, atmospheric circulation, and global climate involves the phenomena named *El Niño* and *La Niña*. These terms refer to shifts in the temperature of surface waters in the middle latitudes of the Pacific Ocean that occur periodically every 2–7 years. Under normal conditions, prevailing winds blow from east to west along the equator in the Pacific Ocean, and form part of a large-scale convective loop, or atmospheric circulation pattern (Figure 12.16a). These winds push surface waters westward and cause

FIGURE 12.16 In these diagrams, red and orange colors indicate warmer water, and blue and green colors indicate cooler water. During normal conditions (**a**), prevailing winds push warm surface waters of the Pacific Ocean westward toward Indonesia. El Niño conditions (**b**) arise when the prevailing winds weaken and no longer hold the warm surface waters in the western Pacific. The warmer water "sloshes" back across the Pacific toward South America, altering weather patterns regionally and globally and leading to fishery collapses and other economic consequences. Data from National Oceanic and Atmospheric Administration, Tropical Atmospheric Ocean Project, 2001.

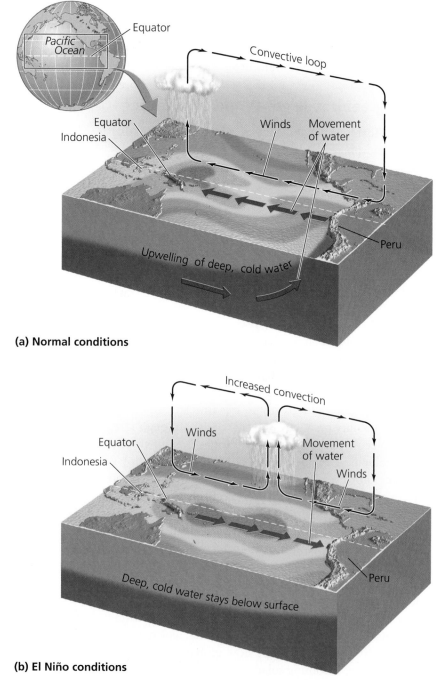

(a) Normal conditions

(b) El Niño conditions

water to "pile up" in the western Pacific. As a result, water near Indonesia is sometimes 50 cm (20 in.) higher and 8 °C warmer than water along the South American coast. The prevailing winds that move surface waters westward enable deep, cold, nutrient-rich water to form a nutrient-rich upwelling (▸ p. 258) along the coast of Peru.

El Niño conditions are triggered when prevailing winds weaken and the warm water that has collected in the western Pacific flows eastward toward South America (Figure 12.16b). This influx of water suppresses the upwelling along the coast of Peru, shutting down the delivery of nutrients that support the region's marine life and fisheries. **La Niña** is the opposite of El Niño; equatorial water sloshes to the west, bringing colder than normal surface waters far westward in the Pacific. Both these events alter weather patterns around the world in complex ways. Because warm surface water gives rise to heavy precipitation, its movement across the Pacific can create unusually intense rainstorms and floods in areas that are generally dry, and it can cause drought and fire in regions that typically experience wet weather. Scientists are exploring whether globally warming air and sea temperatures may be increasing the frequency and strength of these events.

Another growing concern involves the northern Atlantic Ocean, where currents carry warm surface water from the equator northward. As the surface water of the North Atlantic releases heat energy and cools, it becomes denser and sinks. Recently, scientists have realized that interrupting this circulation pattern could trigger rapid climate change. As global warming melts the Greenland Ice Sheet, freshwater runoff into the North Atlantic is increasing, making surface waters less dense. Further runoff could potentially shut down the northward flow of warm equatorial water, and much of Europe could cool rapidly as a result.

Scientists use several approaches to infer and measure climate change

To understand and predict how climate is changing, environmental scientists have developed a number of approaches and techniques.

Direct sampling One approach is to study present-day climate directly, by measuring atmospheric conditions. The data collected by the late Charles Keeling and shown in Figure 12.14 (▸ p. 288) are a prime example. Today Keeling's colleagues are continuing these carbon dioxide measurements, building on the single best long-term dataset we have of direct atmospheric sampling of any greenhouse gas.

Proxy indicators To predict future changes in climate, scientists must understand what climate was like in the past and how and why it has changed naturally over time. To this end, scientists have developed a number of creative methods to decipher clues from thousands or millions of years ago.

Earth's ice caps and glaciers hold clues about past climate. Over the ages, these huge expanses of snow and ice have accumulated to great depths, preserving within them tiny bubbles of the ancient atmosphere. Scientists can examine these trapped air bubbles by drilling into the ice and extracting long columns, or cores (Figure 12.17). From these ice cores, scientists can determine atmospheric composition, greenhouse gas concentrations, temperature trends, snowfall, solar activity, and even (from trapped soot particles) frequency of forest fires.

Scientists also drill cores into beds of sediment beneath bodies of water. Sediments often preserve pollen grains and

(a) Ice core

(b) Micrograph of ice core

FIGURE 12.17 In Greenland and Antarctica, scientists have drilled deep into ancient ice sheets and removed cores of ice like this one **(a)**, held by Dr. Gerald Holdsworth of the University of Calgary, to extract information about past climates. Bubbles (black shapes) trapped in the ice **(b)** contain small samples of the ancient atmosphere.

other remnants from plants that grew in the past, and as we saw with the study of Easter Island (▸ pp. 8–9), knowing what plants occurred in a location can tell us much about the climate at that time. Sources of data such as pollen from sediment cores and air bubbles from ice cores are known as **proxy indicators**, types of indirect evidence that serve as proxies for direct measurement and that indicate the nature of past climate. Other types of proxy indicators include data culled from coral reefs and the tree rings of long-lived trees.

Modeling To understand how climate systems function, and to predict future climate change, scientists attempt to simulate climate processes with sophisticated computer programs. *Coupled general circulation models* are computer programs that combine what is known about weather patterns, atmospheric circulation, atmosphere-ocean interactions, and feedback mechanisms to simulate climate processes. They couple, or combine, climate influences of the atmosphere and oceans in a single simulation. This requires manipulating vast amounts of data and complex mathematical equations—a task not possible until the advent of the supercomputer. As computing power increases and our ability to glean data from proxy indicators of past climate improves, these models are becoming more reliable and informative.

Climate Change Estimates and Predictions

The most thoroughly reviewed and widely accepted collection of scientific information concerning global climate change is a series of reports issued by the **Intergovernmental Panel on Climate Change (IPCC)**. This international panel of scientists and government officials was established in 1988 by the U.N. Environment Programme and the World Meteorological Organization, and its reports represent the consensus of climate experts around the world.

In 2001, the IPCC released its *Third Assessment Report*, summarizing how climate change has already influenced the weather, Earth's physical characteristics and processes, the habits of organisms, and our economies. The report concluded that average surface temperatures had increased by 0.6 °C (1.0 °F) during the 20th century (Figure 12.18). It inferred that glaciers, snow cover, and ice caps were melting worldwide. It found that hundreds of species of plants and animals were being

forced to shift their geographic ranges and the timing of their life cycles. These findings and many more were based on the assessment of thousands of scientific studies conducted worldwide over many years, judged by the world's top experts in these fields. Some of the report's major findings are shown in Table 12.1.

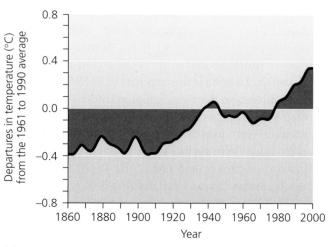

(a) Global temperature measured over the past 140 years

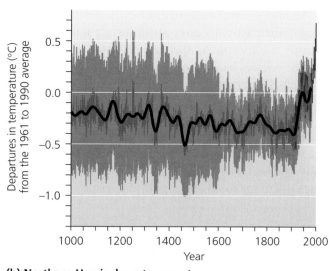

(b) Northern Hemisphere temperature over the past 1,000 years

FIGURE 12.18 Data from thermometers (**a**) show changes in Earth's average surface temperature since 1860. In (**b**), proxy indicators (blue data) and thermometer readings (red data) together show average temperature changes in the Northern Hemisphere over the past 1,000 years. The gray-shaded zone represents the 95% confidence range. This record shows that 20th-century warming has eclipsed the magnitude of change during both the "Medieval Warm Period" (10th–14th centuries) and the "Little Ice Age" (15th–19th centuries). Go to **GRAPHit!** at www.aw-bc.com/withgott. Data from the Intergovernmental Panel on Climate Change, 2001. *Third assessment report.*

Table 12.1   Major Findings of the IPCC Third Assessment Report, 2001

Weather indicators

▶ Earth's average surface temperature increased 0.6 °C (1.0 °F) during the 20th century

▶ Cold and frost days decreased for nearly all land areas in the 20th century

▶ Continental precipitation increased by 5–10% during the 20th century in the Northern Hemisphere, but it decreased in some regions

▶ The 1990s were the warmest decade of the past 1,000 years

▶ The 20th-century Northern Hemisphere temperature increase was the greatest in 1,000 years

▶ Droughts increased in frequency and severity

▶ Nighttime temperatures increased twice as fast as daytime ones

▶ Hot days and heat index increased

▶ Heavy precipitation events increased at northern latitudes

Physical indicators

▶ Average sea level increased 10–20 cm (4–8 in.) during the 20th century

▶ Rivers and lakes in the Northern Hemisphere were covered by ice 2 weeks less from the beginning to the end of the 20th century

▶ Arctic sea ice thinned by 10–40% in recent decades, depending on the season

▶ Mountaintop glaciers retreated widely during the century

▶ Global snow cover decreased by 10% since satellite observations began in the 1960s

▶ Permafrost thawed in many regions

▶ El Niño events became more frequent, persistent, and intense in the past 40 years in the Northern Hemisphere

▶ Growing seasons lengthened 1–4 days per decade in the last 40 years in northern latitudes

Biological indicators

▶ Geographic ranges of many plants, insects, birds, and fish shifted toward the poles and upward in elevation

▶ Plants are flowering earlier, migrating birds are arriving earlier, animals are breeding earlier, and insects are emerging earlier in the Northern Hemisphere

▶ Coral reefs are experiencing bleaching more frequently

Economic indicators

▶ Global economic losses due to weather events rose 10-fold over the past 40 years, partly because of climate factors

Data from Intergovernmental Panel on Climate Change. 2001. *Third assessment report.*

Climate change will cause many impacts

Because the consequences of climate change could be severe, the IPCC and other groups have attempted to predict future changes and their potential impacts. The U.S. Global Change Research Program (USGCRP), created by Congress in 1990 to coordinate federal climate research, issued a report in 2000–2001 that used coupled general circulation models to predict impacts of climate change in the United States (Table 12.2). These models, such as the Hadley model designed by British researchers and the Canadian model designed by British Columbian researchers, have allowed scientists to present graphical depictions summarizing predicted impacts of climate change (Figure 12.19).

Agriculture and ecosystems Drought and temperature extremes will likely threaten farms and natural systems worldwide. Agricultural productivity may increase in some high-latitude regions and decrease in many other regions. For this reason, some people argue that global warming could benefit their nations, but others point out that it may increase inequities between developed nations of the north and developing nations of the tropics.

Climate change could transform forests and other natural communities. Forests could become more productive, because additional atmospheric CO_2 will enhance photosynthesis. However, as productivity increases, the frequency and intensity of forest fires could rise by 10% or more. Forest communities should in general move poleward and upward in elevation. Alpine and subalpine plant

Table 12.2 Some Predicted Impacts of Climate Change in the United States

▶ Average U.S. temperatures will increase 3–5 °C (5–9 °F) in the next 100 years

▶ Droughts and flooding will worsen, and snowpack will be reduced

▶ Drought and other factors could decrease crop yields, but longer growing seasons and enhanced CO_2 could increase yields

▶ Water shortages will worsen

▶ Greater temperature extremes will increase health problems and human mortality. Some tropical diseases will spread north into temperate latitudes

▶ Forest growth may increase in the short term, but in the long term, drought, pests, and fire may alter forest ecosystems

▶ Alpine ecosystems and barrier islands will begin disappearing

▶ Southeastern U.S. forests will break up into savanna/grassland/forest mosaics

▶ Northeastern U.S. forests will lose sugar maples

▶ Loss of coastal wetlands and real estate due to sea level rise will continue

▶ Melting permafrost will undermine Alaskan buildings and roads

Adapted from National Assessment Synthesis Team, U.S. Global Change Research Program. 2000. *Climate change impacts on the United States: The potential consequences of climate variability and change: Overview.* Cambridge, U.K.: Cambridge University Press and USGCRP.

communities should become less common as the climate warms, because these mountaintop communities cannot move higher. Other forest types may expand, such as oak-hickory and oak-pine forests in the eastern United States.

Effects on plant communities comprise an important component of climate change, because by drawing in CO_2 for photosynthesis, plants act as sinks for carbon. The widespread regrowth of forests in eastern North America (▶ p. 209) has offset an estimated 25% of U.S. carbon emissions during the past four decades. If climate change increases overall vegetative growth, this could partially mitigate carbon emissions, in a process of negative feedback. However, if climate change decreases overall growth (as through drought or fire), then a positive feedback cycle could increase carbon flux to the atmosphere.

In regions where climate change increases precipitation and stream flow, erosion and flooding could pollute aquatic systems. Where precipitation decreases, lakes, ponds, wetlands, and streams would shrink, affecting the organisms that live in those habitats, as well as human health and well-being. By 2025, according to the IPCC, three-fifths of the world's people may live in areas with limited water supplies.

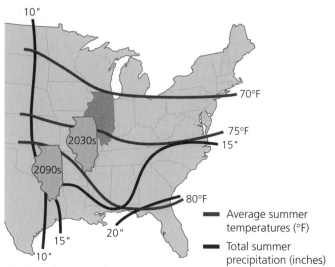

(a) Canadian model

—— Average summer temperatures (°F)

—— Total summer precipitation (inches)

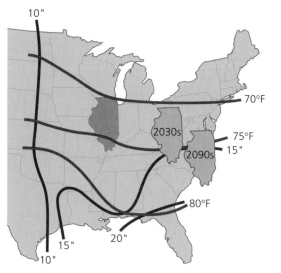

(b) Hadley model

FIGURE 12.19 Using parameters of the Canadian and Hadley models, researchers were able to portray the predicted future climate of the state of Illinois. The Canadian model **(a)** predicted Illinois to become warmer and drier, attaining in the 2030s a climate similar to Missouri's, and in the 2090s one similar to eastern Oklahoma's. The Hadley model **(b)** predicted Illinois to become warmer and moister, attaining in the 2030s a climate similar to West Virginia's, and in the 2090s one similar to North Carolina's. Data from National Assessment Synthesis Team, U.S. Global Change Research Program. 2001. *Climate change impacts on the United States: The potential consequences of climate variability and change: Foundation.* Cambridge, U.K.: Cambridge University Press and USGCRP.

Sea level rise Warming temperatures are causing glaciers to shrink and disappear worldwide. For the same reason, polar ice shelves that have been intact for millennia are breaking away and melting into the ocean. In fact, recent evidence indicates that Arctic and Antarctic ice sheets are melting at accelerating rates that have taken scientists by surprise. In addition, ocean temperatures are rising, and water expands as it warms. These factors account for the rise in sea level already being felt by people of the Maldives and many other island nations. Higher sea levels lead to beach erosion, coastal flooding, intrusion of saltwater into aquifers, destruction of coral reefs that protect coastlines, and other impacts. "With a mere 1-meter rise [in sea level]," Maldives' President Maumoon Abdul Gayoom warned in 2001, "a storm surge would be catastrophic, and possibly fatal to the nation."

The United States is not immune from coastal impacts—and 53% of U.S. residents live in coastal areas. The vulnerability to coastal flooding from storm surges became tragically apparent in 2005 when Hurricane Katrina struck New Orleans and the Gulf Coast, followed shortly thereafter by Hurricane Rita (Figure 12.20). Just outside New Orleans, marshes of the Mississippi River delta are being lost rapidly as dams upriver hold back the silt that used to maintain the delta, as land subsides due to petroleum extraction, and as rising seas eat away at coastal vegetation. Continued loss of these wetlands will mean that New Orleans will have less protection against future storm surges. Moreover, the record number of hurricanes and tropical storms in 2005 left many people wondering whether global warming was spawning more hurricanes, or hurricanes that are more powerful or long lasting. Scientists are not yet sure, but evidence from several recent studies indicates that warmer sea surface temperatures are likely increasing the destructive power of storms, and possibly their duration.

Human health As a result of climate change, people could face increased exposure to an array of health problems. These include heat stress, increased air pollution, expansion of tropical diseases such as malaria and dengue fever, sanitation problems when floods overcome sewage treatment systems, injuries and drowning if storms become more frequent or intense, and hunger-related ailments as human population growth puts greater demands on agricultural systems. However, a warmer world may also present fewer diseases and injuries that result from cold weather.

Despite scientific consensus, social debate over climate change continues

Virtually all environmental scientists agree that Earth's atmosphere and climate are changing. The great majority of them have concluded that human activity, particularly our emission of greenhouse gases, is the primary reason for this change. Despite this unusually strong scientific consensus, you have no doubt heard a great deal of debate over climate change. One reason is that some people, primarily nonscientists, continue to contest the consensus findings and interpretations of the scientific community. Some of these so-called greenhouse skeptics have vested interests in perpetuating our reliance on fossil fuels, and some of them have significant sway over policymakers, particularly in the United States. Moreover, the American news media by tradition try to portray both sides of any issue, even when one side of an issue is supported by stronger evidence than another side. By giving greenhouse skeptics equal time with mainstream scientists, many have argued, the media have amplified the views of greenhouse skeptics out of proportion to their actual prevalence in the scientific community.

Many details and aspects of climate change science remain uncertain, because climate systems are so complex. Nonetheless, the scientific community now feels that evidence for our role in influencing climate is strong enough

FIGURE 12.20 Coastal areas around the United States and the world are at risk of erosion, flooding, aquifer contamination, and other problems due to a rise in sea level. Rescues like this one in Mississippi from the floodwaters of Hurricane Katrina's storm surge in 2005 could become more frequent in a future world of higher sea levels.

that governments should take action to address greenhouse gas emissions. In June 2005, as the leaders of the "G8" industrialized nations met, the national academies of science from 11 nations (Brazil, Canada, China, France, Germany, India, Italy, Japan, Russia, the United Kingdom, and the United States) issued a joint statement urging these political leaders to take action. Such a broad consensus statement from the world's scientists was virtually unprecedented, on any issue. The statement read, in part:

> The scientific understanding of climate change is now sufficiently clear to justify nations taking prompt action. It is vital that all nations identify cost-effective steps that they can take now, to contribute to substantial and long-term reduction in net global greenhouse gas emissions. . . . A lack of full scientific certainty about some aspects of climate change is not a reason for delaying an immediate response that will, at a reasonable cost, prevent dangerous anthropogenic interference with the climate system.

Reducing Emissions

Once one accepts that climate change is real and poses significant ecological and economic threats, there remains plenty of room for disagreement over how our societies should respond. Political and economic debate over how we should respond stems from questions such as:

▶ Would the economic and political costs of reducing greenhouse gas emissions outweigh the costs of unabated emissions and resulting climate change?

▶ Should industrialized nations, developing nations, or both take responsibility for reducing greenhouse gas emissions?

▶ Should steps to reduce emissions occur voluntarily or as a result of government regulation?

▶ How should we allocate funds and human resources for reducing emissions and coping with climate change?

Electricity generation and transportation are the largest sources of emissions

Since 1990, the generation of electricity, largely through coal combustion, has produced the largest portion (34%, as of 2000) of U.S. greenhouse gas emissions (Figure 12.21). Transportation ranks second at 27%, industry produces 19%, agriculture and residential sources produce 8% each, and commercial sources account for 5%.

There are two ways to reduce the amount of fossil fuels we use in the generation of electricity: (1) encouraging conservation and efficiency (▶ pp. 317–319) and

FIGURE 12.21 Coal-fired electricity-generating power plants, such as this one in Maryland, are the largest contributors to U.S. greenhouse emissions.

(2) switching to renewable energy sources (Chapter 14). We can promote conservation and efficiency by using new high-efficiency technologies or by making ethical choices to reduce consumption.

Although most of us can barely imagine life without a car, reducing the fossil fuels burned in our vehicles would greatly lower emissions. If consumers demand better fuel efficiency of policymakers and manufacturers, efficiency could be enhanced through design improvements. Advancing technology is also making possible a number of alternatives to the traditional automobile, including hybrid vehicles (▶ pp. 318–319), electric vehicles, alternative fuels (▶ pp. 332–333), and hydrogen fuel cells (▶ pp. 347–348, 351). Moreover, many people today are making lifestyle choices that reduce their reliance on cars—choosing to live near one's place of employment, biking, walking, or using public transportation (▶ pp. 201–203).

Some international treaties address climate change

Because climate change is a global phenomenon, nations have tried to address it with international agreements. The U.N. *Framework Convention on Climate Change* (which came into force in 1994) outlined a plan for reducing greenhouse gas emissions to 1990 levels by the year 2000 through a voluntary, nation-by-nation approach. By the late 1990s, it was apparent that a voluntary approach was not likely to succeed. Between 1990 and 2003, for example, U.S. greenhouse emissions (in CO_2 equivalents) increased by 13.3%.

Thus, many nations embarked on an effort to create a binding international treaty that would *require* all signatory nations to reduce their emissions. Drafted in 1997 in Kyoto, Japan, the **Kyoto Protocol** mandates signatory nations, by the period 2008–2012, to reduce emissions of

Global Climate Change

VIEWPOINTS

What is scientific research telling us about global climate change and its potential consequences for our society and the environment? How should humanity respond to climate change?

The Science Is Settled Enough to Justify Action

Many things are abundantly clear in climate change research. Human activities have undoubtedly led to the highest atmospheric levels of CO_2 and CH_4 seen in almost 1 million years, and both are powerful greenhouse gases that warm the surface of the planet. Temperatures have risen around 0.7 °C since around 1900, of which a large proportion is attributable to this enhanced greenhouse effect. The significant increases of heat in the oceans match climate model calculations of the increasing energy imbalance of the planet, and the almost global retreat of mountain glaciers is a graphic symptom of the increasing warmth of the atmosphere. Some further climate change is inevitable because the climate has yet to come into equilibrium with current greenhouse gas levels.

There are also many remaining uncertainties in climate science, particularly related to the role of aerosols and clouds. However, despite occasional claims to the contrary, the science is settled enough to justify action to reduce greenhouse gas emissions.

Because of the inertia of energy infrastructure in today's societies, CO_2 emissions will continue to rise for decades to come, and given the long atmospheric lifetime of CO_2, atmospheric concentrations will continue to increase for even longer. Model projections indicate that this could have significant and costly impacts on the environment and sea level if this continues unabated. Given the long time scales involved (in society and in climate itself), decisions made now will only start to have impacts many decades hence. Therefore, by the time serious effects are obvious, it may be too late to avoid the worst consequences.

Sensible policy approaches should combine investment in an increased resilience to climate changes, along with long-term efforts to reduce emissions. Cuts in other warming factors such as CH_4, black carbon, and tropospheric ozone would be positive steps for tackling both climate change and air pollution.

Gavin Schmidt is a climate modeler at the NASA Goddard Institute for Space Studies in New York. He is an associate editor for the *Journal of Climate* and was recently cited by *Scientific American* as one of the 50 Research Leaders of 2004.

Climate Changes Are Mostly Natural

Climate is never constant but varies, sometimes dramatically, on timescales ranging from years to eons. On the human scale, decades to centuries, the major cause seems to be cyclical variations of solar radiation. Since the end of the most recent ice age, some 10,000 years ago, there have been many such cycles. In recent history we have seen the Medieval Warming Period, when England produced wines and Vikings colonized Greenland. Then followed the Little Ice Age, which disappeared only around A.D. 1850, about the time when thermometers first became available in much of the world. The global climate then warmed strongly until about 1940, followed by a cooling until 1975 that provoked great fears of a return to an ice age. All these changes are believed to be of natural origin, even though during this time there had been a steady increase in levels of atmospheric greenhouse gases from human activities.

Since 1979, weather satellites have reported a slight warming trend, which could be partially anthropogenic, as theoretical climate models suggest. Teasing out the small human contribution from the natural "noise" is difficult and the focus of ongoing scientific debate.

Nonetheless, we can draw certain conclusions. Anthropogenic global warming is not a significant problem, amounting to less than 1 °C by 2100. On the whole, it will be beneficial, with higher levels of carbon dioxide speeding growth of crops and forests. And realistically, there is little that can be done to stem the rise of emissions, especially from nations such as China and India. The best policy is one of "no regrets"—energy conservation because it pays, and strengthening our ability to adapt by fighting poverty around the world.

S. Fred Singer is professor emeritus of environmental science at the University of Virginia, and president of the Science and Environmental Policy Project. He was the first director of the U.S. Weather Satellite Service and served 5 years as vice-chair of the National Advisory Committee on Oceans and Atmospheres. With Dennis T. Avery, he is the co-author of *Unstoppable Global Warming—Every 1500 Years* (Rowman & Littlefield Publishers, 2006).

Explore this issue further by accessing **Viewpoints** at www.aw-bc.com/withgott.

six greenhouse gases to levels equal to or lower than those of 1990. In 2005, the Kyoto Protocol came into force after Russia became the 127th nation to ratify it.

The United States, the world's largest emitter of greenhouse gases, has refused to ratify the Kyoto Protocol. U.S. leaders have called the treaty unfair because it requires industrialized nations to reduce emissions but does not require the same of developing nations, even rapidly industrializing ones such as China and India. Proponents of the Kyoto Protocol justify the differential requirements by pointing out that the industrialized world created the current problem and therefore should make the sacrifices necessary to solve it.

Despite the U.S. government's refusal to ratify Kyoto, many state and local governments across the nation have adopted policies to reduce their greenhouse gas emissions. In fall 2006, California legislators passed, and Governor Arnold Schwarzenegger pledged to sign, a landmark law committing the state to reduce its emissions by 25% by the year 2020. The law directed the state's Air Resources Board to develop limits and regulations for industries and encouraged it to set up an emissions trading program (▶ p. 45). Many business leaders opposed to the law predicted that emissions limits would increase energy prices and cause businesses to flee the state. Supporters of the law countered that it would boost the state's economy by making California a leader in developing environmentally cleaner technology that world markets are demanding.

With global climate change, as with many other environmental issues, we may never be entirely certain of the precise outcomes of our actions until after they have occurred. Advocates of the precautionary principle (▶ p. 160) assert that if a threat is reasonably suspected, we should take precautionary action without waiting for full scientific certainty regarding cause and effect.

Weighing the Issues:
The Precautionary Principle

Critics of the precautionary approach say that it will impede economic growth and innovation. Advocates of the precautionary principle say the stakes are too high to gamble with climate. What do you think? Is the precautionary approach an appropriate guide for dealing with climate change? What role should economics play in the discussion?

Conclusion

We have seen that many factors, including human activities, can shape atmospheric composition and global climate. We have also seen that scientists and policymakers are beginning to understand anthropogenic climate change and its environmental, economic, and social consequences more fully. As time passes, fewer experts are arguing that the changes will be minor. Sea level rise and other consequences of global climate change will affect far-flung places such as the Maldives, but they will also influence populated mainland areas such as Louisiana and Florida. The challenges inherent in slowing and reversing climate change are immense, but our success so far in addressing many sources of outdoor air pollution should provide hope that success is feasible.

TESTING YOUR COMPREHENSION

1. How and why is stratospheric ozone beneficial for people, and tropospheric ozone harmful? How do chlorofluorocarbons (CFCs) deplete stratospheric ozone, and what was done to address this problem?

2. Name three natural sources of outdoor air pollution and three sources stemming from human activity. Give one example of how human activity can worsen a natural pollution source.

3. What is smog? How is smog formation influenced by the weather? By topography? How does photochemical, or brown-air, smog differ from industrial, or gray-air, smog?

4. Why are the effects of acidic deposition often felt in areas far from where the primary pollutants are produced?

5. Name three common sources of indoor pollution. For each, describe one way to reduce one's exposure.

6. What happens to solar radiation after it reaches Earth? How do greenhouse gases warm the lower atmosphere?

7. Compare and contrast three major approaches scientists use to study climate so that they can better predict future climate change.

8. Why are sea levels rising? In what ways can rising sea levels create problems for people?

9. Describe at least one likely impact of climate change on (1) agriculture, (2) natural systems, and (3) human health.

10. What are the largest two sources of greenhouse gas emissions in the United States, and why? In what ways can we try to reduce these emissions? What roles have international treaties played in addressing climate change? Give two specific examples.

SEEKING SOLUTIONS

1. Describe how and why emissions of major pollutants have been reduced by over 50% in the United States since 1970, despite increases in population and economic activity.

2. Consider volatile chemicals, such as formaldehyde and other VOCs, that may be emitted at very low levels over long periods of time from manufactured products. What do you think are the best ways to lessen the health impacts of such indoor pollutants?

3. You have just become the head of your county health department, and the EPA has informed you that your county has failed to meet the national ambient air quality standards for ozone, sulfur dioxide, and nitrogen dioxide. Your county is partly rural but is home to a city of 200,000 people and 10 sprawling suburbs. There are several large and aging coal-fired power plants, a number of factories with advanced pollution control technology, and no public transportation system. What steps would you urge the county government to take to meet the air quality standards? Explain how you would prioritize these steps.

4. Today, many people argue that we need "more proof," or "better science" before we commit to substantial changes in the way we live our lives. How much "science," or certainty, do you think we need before we have enough to guide our decisions regarding climate change? How much certainty do you need in your own life before you make a change? Should nations and elected officials follow a different standard? Is the precautionary principle is an appropriate standard in the case of global climate change? Why or why not?

5. Describe several ways that greenhouse gas emissions from transportation can be reduced. Which approach do you think is most realistic, which approach do you think is least realistic, and why? Do you think the U.S. should ratify the Kyoto Protocol? Why or why not?

CALCULATING ECOLOGICAL FOOTPRINTS

U.S. energy consumption from fossil fuels currently totals 306 gigajoules per person per year. Carbon emitted from fossil fuel combustion is "sequestered" when plants take up the emitted carbon dioxide through photosynthesis and store it as organic matter. Researchers have estimated that for each 100 gigajoules of fossil fuel burned, 1 hectare of ecologically productive land is required for carbon sequestration. Considering these data, calculate the component of the ecological footprint required to sequester carbon from fossil fuel emissions.

1. The land area of the United States is about 916 million hectares. What percentage of that land would need to be set aside to sequester all carbon from the fossil fuel consumption of 300 million Americans?

	Hectares of land to sequester carbon
You	3.06
Your class	
Your state	
United States	

Data from Wackernagel, M., and W. Rees. 1996. *Our ecological footprint: Reducing human impact on the earth.* Gabriola Island, British Columbia, Canada: New Society Publishers.

2. What is the environmental fate of carbon dioxide that is released from the combustion of fossil fuels and is not sequestered by plants? Why is this a concern?

3. Name four things you could do to lessen the modification of the global environment caused by your own personal energy consumption.

Take It Further

Go to www.aw-bc.com/withgott, where you'll find:

▶ Suggested answers to end-of-chapter questions
▶ Quizzes, animations, and flashcards to help you study
▶ *Research Navigator*™ database of credible and reliable sources to assist you with your research projects

▶ **GRAPHit!** tutorials to help you interpret graphs

▶ **INVESTIGATEit!** current news articles that link the topics that you study to case studies from your region to around the world

13 Nonrenewable Energy Sources, Their Impacts, and Energy Conservation

Caribou crossing haul road at Prudhoe Bay

Upon successfully completing this chapter, you will be able to:

▶ Survey the energy sources that we use

▶ Describe the nature and origin of coal, natural gas, and petroleum, and evaluate their extraction and use

▶ Outline the future depletion of global oil supplies and assess the concerns over "peak oil"

▶ Review and assess environmental, political, social, and economic impacts of fossil fuel use

▶ Specify strategies for conserving energy

▶ Describe nuclear energy and how it is harnessed

▶ Assess the benefits and drawbacks of nuclear power and outline the societal debate over this energy source

Central Case: Oil or Wilderness on Alaska's North Slope?

"The roar alone of road building, drilling, trucks, and generators would pollute the wild music of the Arctic and be as out of place there as it would be in the heart of Yellowstone or the Grand Canyon."
—FORMER PRESIDENT JIMMY CARTER, 2000

"There is absolutely no indication that environmentally responsible exploration will harm the 129,000-member porcupine caribou herd."
—ALASKA SENATOR FRANK MURKOWSKI, 2002

Above the Arctic Circle, at the top of the North American continent, the land drops steeply down from the jagged mountains of Alaska's spectacular Brooks Range and stretches north in a vast, flat expanse of tundra until it meets the icy waters of the Arctic Ocean. Few Americans have been to this remote region, yet it has come to symbolize a struggle between two values in our modern life.

For some U.S. citizens, Alaska's North Slope is one of the last great expanses of wilderness in their sprawling industrialized country—one of the last places humans have left untouched. For these millions of Americans, simply knowing that this wilderness still exists is of tremendous value. For millions of others, this land represents something else entirely—a source of petroleum, the natural resource that, more than any other, fuels our society and shapes our way of life. To these people, it seems wrong to leave such an important resource sitting unused in the ground. Those who advocate drilling for oil here accuse wilderness preservationists of neglecting the country's economic interests, whereas advocates for wilderness argue that drilling will sacrifice the nation's natural heritage for little gain.

Ever since oil was found seeping from the ground in this area a century ago, these two visions for Alaska's North Slope have competed. Now they exist side by side across three regions of this vast swath of land (Figure 13.1). The westernmost portion of the North Slope was set aside in 1923 by the U.S. government as an emergency reserve for petroleum. This parcel of land, the size of Indiana, is today called the National Petroleum Reserve–Alaska and was intended to remain untapped for oil unless the nation faced an emergency. So far, most of this region's 9.5 million ha (23.5 million acres) remains undeveloped.

East of the National Petroleum Reserve are state lands that experienced widespread development and extraction after oil was discovered at Prudhoe Bay in 1968. Since drilling began in 1977, over 14 billion barrels (1 barrel = 159 L or 42 gal) of crude oil have been extracted from 19 oil fields spread over 160,000 ha (395,000 acres) of this region. The oil is transported across the state of Alaska by the 1,300-km (800-mi) trans-Alaska pipeline south to the port of Valdez, where it is loaded onto tankers.

East of the Prudhoe Bay region lies the Arctic National Wildlife Refuge (ANWR), an area the size of South Carolina consisting of federal lands set aside in 1960 and 1980 mainly to protect wildlife and preserve pristine ecosystems of tundra, mountains, and seacoast. This scenic region is home to 160 nesting bird species, numerous fish and marine mammals, grizzly bears, polar bears, Arctic foxes, timber wolves, and musk oxen. In most years, thousands of caribou arrive from the south to spend the summer, giving birth to and raising their calves. Because of the vast caribou herd and the other large mammals, ANWR has been called "the Serengeti of North America."

ANWR has been the focus of debate for decades. Advocates of oil drilling have tried to open its lands for development, and proponents of wilderness preservation have fought for its preservation. Scientists, oil industry experts, politicians, environmental groups, citizens, and Alaska residents have all been part of the debate. So have the two Native groups in the area, the Gwich'in and the Inupiat, who disagree over whether the refuge should be opened to oil development. The Gwich'in depend on hunting caribou and fear that oil industry activity will reduce caribou herds, whereas the Inupiat, who rely less on caribou, see oil extraction as one of the few opportunities for economic development in the area.

In a compromise in 1980, the U.S. Congress put most of the refuge off limits to oil but reserved for future decision making a 600,000-ha (1.5-million-acre) area of coastal plain. This region, called the 1002 Area (after Section 1002 of the bill that established it), remains undeveloped for oil but can be opened for development

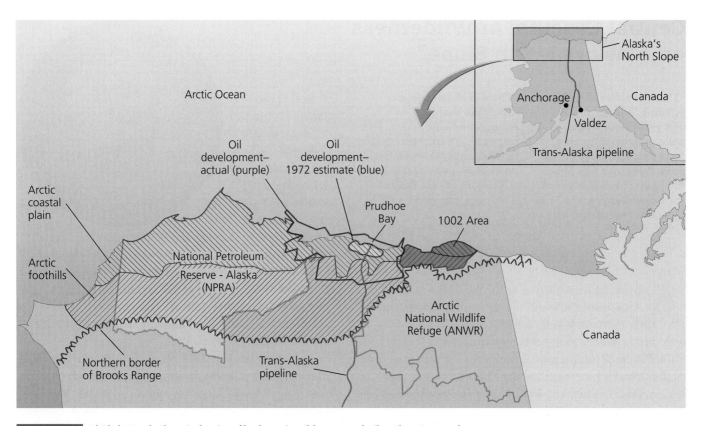

FIGURE 13.1 Alaska's North Slope is the site of both arctic wilderness and oil exploration. In the western portion of this region, the U.S. government established the National Petroleum Reserve–Alaska as an area in which to drill for oil if it is needed in an emergency. It is well explored but not widely developed. To the east of this area, the Prudhoe Bay region is the site of widespread oil extraction, and since 1977 it has produced over 14 billion barrels of oil. Oil development has expanded much farther than experts estimated it would in 1972. Farther east lies the Arctic National Wildlife Refuge, home to untrammeled Arctic wilderness and the focus of debate for years. Proponents of oil extraction and proponents of wilderness preservation have been battling over whether the 1002 Area of the coastal plain north of the Brooks Range should be opened to oil development.

by a vote of both houses of Congress. Its unsettled status has made it the center of the oil-versus-wilderness debate, and Congress has been caught between passionate feelings on both sides for a quarter of a century.

In 2005 and 2006, the Republican-controlled Congress made several efforts to open the refuge to drilling—more than once with provisions slipped into unrelated budget legislation. As of this writing, each attempt had been narrowly blocked.

Behind the noisy policy debate over ANWR, scientists have attempted to inform the dialogue through research. Geologists have tried to ascertain how much oil lies underneath the refuge. Biologists have tried to predict the impacts of oil drilling on Arctic ecosystems. Many environmental scientists have concluded that a small amount of conservation would free up more oil than drilling in ANWR would produce.

Moreover, scientists and nonscientists alike are debating the relevance of the oil beneath the refuge for the security and prosperity of the nation. We will examine these questions by revisiting Alaska's North Slope as we survey the fossil fuel energy we use to heat and light our homes, power our machinery, and provide the comforts, conveniences, and mobility to which technology has accustomed us.

Sources of Energy

The debate over drilling for oil in the Arctic National Wildlife Refuge is a thoroughly modern debate, pitting the culturally new concept of wilderness preservation against the desire to exploit a resource that has come to guide the world's economy only in the past 150 years.

However, people have used—and fought over—energy in one way or another for all of our history.

We use a variety of energy sources

Our planet receives energy from several sources, and people have developed many ways to harness renewable and nonrenewable forms of energy (Table 13.1). Most of Earth's energy comes from the sun. We can harness energy from the sun's radiation directly, but solar radiation also makes possible several other energy sources. Sunlight drives the growth of plants, from which we take wood as a fuel source. After their death, plants may impart their stored chemical energy to **fossil fuels** (such as oil, coal, and natural gas; ▸ p. 5), which are highly combustible substances formed from the remains of organisms from past geological ages. Solar radiation also helps drive wind patterns and the hydrologic cycle, making possible other forms of energy, such as wind power and hydroelectric power.

A great deal of energy also emanates from Earth's core, enabling us to harness geothermal power. A much smaller amount of energy results from the gravitational pull of the moon and sun, and we are just beginning to harness power from the ocean tides that these forces generate. An immense amount of energy resides within the bonds among protons and neutrons in atoms, and this energy provides us with nuclear power.

As we first noted in Chapter 1 (▸ pp. 3–4), energy sources such as sunlight, geothermal energy, and tidal energy are considered *renewable* because they will not be depleted by our use of them. In contrast, energy sources such as oil, coal, and natural gas are considered *nonrenewable*, because at our current rates of consumption we will use up Earth's accessible store of them in a matter of decades to centuries. Nuclear power as currently harnessed through fission of uranium (▸ pp. 319–320) is considered nonrenewable to the extent that uranium ore is in limited supply. Although these nonrenewable fuels result from ongoing natural processes, the timescales on which they are created are so long that, once the fuels are depleted, they cannot be replaced in any time span useful to our civilization. It takes a thousand years for the biosphere to generate the amount of organic matter that must be buried to produce a single day's worth of fossil fuels for our society. For this reason, and because fossil fuels exert severe environmental impacts, renewable energy sources increasingly are being developed as alternatives to fossil fuels, as we will see in Chapter 14.

Citizens of developed nations generally consume far more energy than do those of developing nations. (Figure 13.2). Developing nations devote a greater proportion of energy to subsistence activities, such as growing and preparing food and heating homes, whereas industrialized countries use a greater proportion for transportation and industry. Because industrialized nations rely more on mechanized equipment and technology, they use more fossil fuels. In the United States, fossil fuels supply 89% of energy needs.

Fossil fuels provide most of the energy we consume because their high energy content makes them efficient to burn, ship, and store. Besides providing for transportation, heating, and cooking, we use these fuels to generate **electricity**, a secondary form of energy that is easier to transfer over long distances and apply to a variety of uses. Global

Table 13.1 Energy Sources We Use Today		
Energy source	**Description**	**Type of energy**
Crude oil	Fossil fuel extracted from ground	Nonrenewable
Natural gas	Fossil fuel extracted from ground	Nonrenewable
Coal	Fossil fuel extracted from ground	Nonrenewable
Nuclear energy	Energy from atomic nuclei of uranium mined from ground and processed	Nonrenewable
Biomass energy	Chemical energy from photosynthesis stored in plant matter	Renewable
Hydropower	Energy from running water	Renewable
Solar energy	Energy from sunlight directly	Renewable
Wind energy	Energy from wind	Renewable
Geothermal energy	Earth's internal heat rising from core	Renewable
Tidal and wave energy	Energy from tidal forces and ocean waves	Renewable

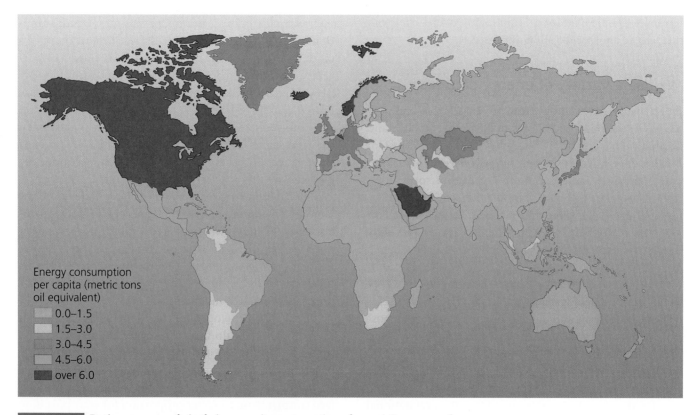

FIGURE 13.2 Regions vary greatly in their per capita consumption of energy. Per person, the most-industrialized nations use up to 100 times more energy than do the least-industrialized nations. This map combines all types of energy, standardized to metric tons of "oil equivalent," that is, the amount of fuel needed to produce the energy gained from combusting one metric ton of crude oil. Data from British Petroleum. 2005. *Statistical review of world energy 2005.*

consumption of the three main fossil fuels has risen steadily for years and is now at its highest level ever (Figure 13.3).

Fossil fuels are indeed fuels created from "fossils"

The fossil fuels we burn today in our vehicles, homes, industries, and electrical power plants were formed from the tissues of organisms that lived 100–500 million years ago. The energy these fuels contain came originally from the sun and was converted to chemical-bond energy as a result of photosynthesis (▶ pp. 59–60). The chemical energy in these organisms' tissues was then concentrated as these tissues decomposed and their hydrocarbon compounds were altered and compressed (Figure 13.4).

Most organisms, after death, do not end up as part of a coal, gas, or oil deposit. A tree that falls and decays as a rotting log undergoes mostly **aerobic** decomposition; in the presence of air, bacteria and other organisms that use oxygen break down plant and animal remains into simpler molecules that are recycled through the ecosystem.

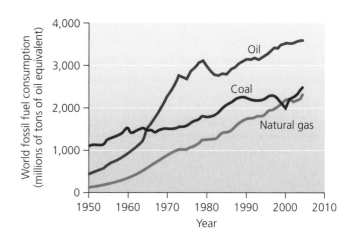

FIGURE 13.3 Global consumption of fossil fuels has risen greatly over the past half century. Oil use rose steeply during the 1960s to overtake coal, and today it remains our leading energy source. Data from Worldwatch Institute. 2005. *Vital signs 2005.*

Fossil fuels are produced only when organic material is broken down in an **anaerobic** environment, one with little or no oxygen. Such environments include the bottoms of lakes, swamps, and shallow seas.

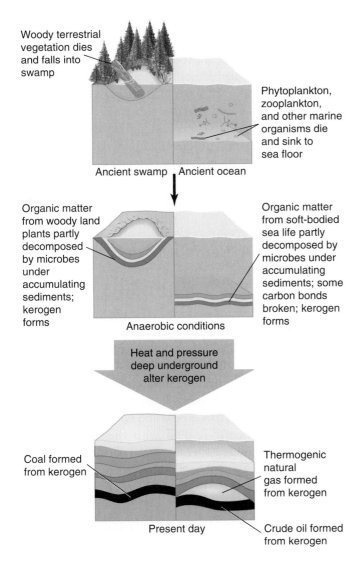

Table 13.2	Nations with Largest Proven Reserves of Fossil Fuels	
Oil (% world reserves)	**Natural gas** (% world reserves)	**Coal** (% world reserves)
Saudi Arabia, 22.0	Russia, 26.6	United States, 27.1
Iran, 11.5	Iran, 14.9	Russia, 17.3
Iraq, 9.6	Qatar, 14.3	China, 12.6
Kuwait, 8.5	Saudi Arabia, 3.8	India, 10.2
United Arab Emirates, 8.1	United Arab Emirates, 3.4	Australia, 8.6
Venezuela, 6.6	United States, 3.0	South Africa, 5.4
Russia, 6.2	Nigeria, 2.9	Ukraine, 3.8
Kazakhstan, 3.3	Algeria, 2.5	Kazakhstan, 3.4
Libya, 3.3	Venezuela, 2.4	Poland, 1.5
Nigeria, 3.0	Iraq, 1.8	Brazil, 1.1

Data from British Petroleum. 2006. *Statistical review of world energy 2006.*

FIGURE 13.4 The fossil fuels we use for energy today consist of the remains of organic material from plants (and to a lesser extent, animals) that died millions of years ago. Their formation begins when organisms die and end up in oxygen-poor conditions, such as when trees fall into lakes and are buried by sediment, or when phytoplankton and zooplankton drift to the seafloor and are buried. Organic matter that undergoes slow anaerobic decomposition deep under sediments forms kerogen. Geothermal heating acts on kerogen to create crude oil and natural gas. Natural gas can also be produced nearer the surface by anaerobic bacterial decomposition of organic matter. Oil and gas come to reside in porous rock layers beneath dense, impervious layers. Coal is formed when plant matter is compacted so tightly that there is little decomposition.

Fossil fuel reserves are unevenly distributed

Fossil fuel deposits are localized and unevenly distributed over Earth's surface, so some regions have substantial reserves of fossil fuels whereas others have very few. How long each nation's fossil fuel reserves will last depends on how much the nation extracts, how much it consumes, and how much it imports from and exports to other nations. Nearly two-thirds of the world's proven reserves of crude oil lie in the Middle East. The Middle East is also rich in natural gas, but Russia contains more than twice as much natural gas as any other country. Russia is also rich in coal, as is China, but the United States possesses more coal than any other nation (Table 13.2).

Coal

Coal is organic matter (generally woody plant material) that was compressed under very high pressure to form dense, solid carbon structures. Coal typically results when little decomposition takes place because the material cannot be digested or appropriate decomposers are not present. The proliferation 300–400 million years ago of swampy environments where organic material was buried has resulted in substantial coal deposits throughout the world. In fact, coal is the world's most abundant fossil fuel.

Coal varies in its qualities

Coal varies from deposit to deposit in its water content and in the amount of potential energy it contains. Organic material that is broken down anaerobically but remains wet, near the surface, and poorly compressed is called *peat.* As peat decomposes further, as it is buried

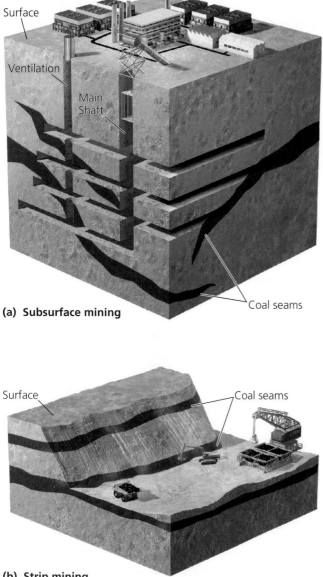

(a) Subsurface mining

(b) Strip mining

 Coal is mined in two major ways. In subsurface mining (**a**), miners work below ground in shafts and tunnels blasted through the rock; these passageways provide access to underground seams of coal. This type of mining poses dangers and long-term health risks to miners. In strip mining (**b**), soil is removed from the surface, exposing coal seams from which coal is mined. This type of mining can cause substantial environmental impact.

more deeply under sediments, as pressure and heat increase, and as time passes, water is squeezed from the material, and carbon compounds are packed more tightly together, forming coal. The greater the compression, the greater is the energy content per unit volume.

Coal deposits also vary in the amount of impurities they contain, including sulfur, mercury, arsenic, and other trace metals. Coal in the eastern United States tends to be high in sulfur because it was formed in marine sediments, where sulfur from seawater was present. When high-sulfur coal is burned, it produces sulfate air pollutants, which contribute to industrial smog and acidic deposition (▶ pp. 278–279, 282–285). Combustion of mercury-rich coal emits mercury that can bioaccumulate in organisms' tissues, poisoning animals as it moves up food chains. Such pollution problems commonly occur downwind of coal-fired power plants. Scientists and engineers are seeking ways to cleanse coal of its impurities so that it can continue to be used as an energy source while minimizing impact on the environment (▶ p. 313).

Coal is mined from the surface and from below ground

We extract coal using two major methods (Figure 13.5). We reach underground deposits with **subsurface mining**. Shafts are dug deep into the ground, and networks of tunnels are dug or blasted out to follow coal seams. Coal is removed systematically and shipped to the surface. When coal deposits are at or near the surface, strip-mining methods are used. In **strip mining**, heavy machinery removes huge amounts of earth to expose and extract the coal. The pits are subsequently refilled with the soil that was removed. Strip-mining operations can occur on immense scales; in some cases entire mountaintops are lopped off (▶ p. 313).

We generate electricity with coal

In the early days of the industrial revolution, we used coal for direct heating and for running steam engines. Today we burn coal largely to generate electricity. In coal-fired power plants, coal combustion converts water to steam, which turns a turbine to create electricity (Figure 13.6). Today coal provides over half the electrical generating capacity of the United States. China and the United States are the primary producers and consumers of coal (Table 13.3).

Table 13.3 Top Producers and Consumers of Coal	
Production (% world production)	**Consumption** (% world consumption)
China, 38.4	China, 36.9
United States, 20.0	United States, 19.6
Australia, 7.0	India, 7.3
India, 6.9	Japan, 4.1
South Africa, 4.8	Russia, 3.8

Data from British Petroleum. 2006. *Statistical review of world energy 2006.*

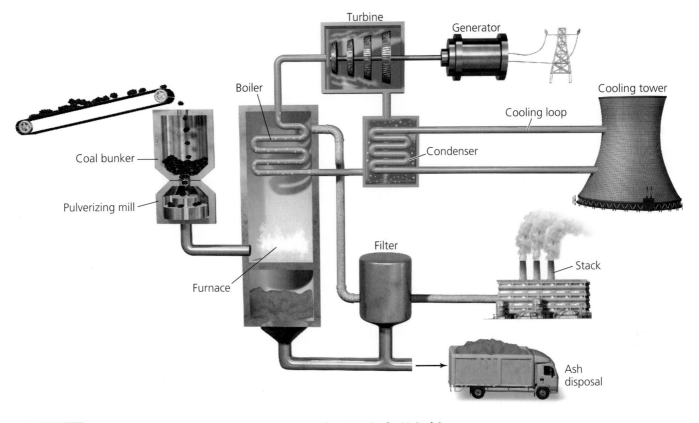

FIGURE 13.6 Coal is the primary fuel source used to generate electricity in the United States. Pieces of coal are pulverized and blown into a high-temperature furnace. Heat from the combustion boils water, and the resulting steam turns a turbine, generating electricity by passing magnets past copper coils. The steam is then cooled and condensed in a cooling loop and returned to the furnace. "Clean coal" technologies (▶ p. 313) help filter out pollutants from the combustion process, and toxic ash residue is disposed of in hazardous waste disposal sites.

Natural Gas

Natural gas is the fastest-growing fossil fuel in use today and provides for one-quarter of global commercial energy consumption. World supplies of natural gas are projected to last perhaps 60 more years.

Natural gas is formed in two ways

Natural gas can arise from either of two processes. *Biogenic* gas is created at shallow depths by the anaerobic decomposition of organic matter by bacteria. An example is the "swamp gas" you can sometimes smell when stepping into the muck of a swamp. In contrast, *thermogenic* gas results from compression and heat deep underground. Thermogenic gas may be formed directly, along with crude oil, or it may be formed from crude oil that is altered by heating. Biogenic gas is nearly pure methane, (CH_4), but thermogenic gas contains small amounts of other hydrocarbon gases as well. Most gas extracted commercially is thermo-

genic and found above deposits of crude oil or seams of coal, so its extraction often accompanies the extraction of those fossil fuels. Natural gas deposits are greatest in Russia and the United States, and these two nations lead the world in both gas production and gas consumption (Table 13.4).

Natural gas extraction becomes more challenging with time

To access some natural gas deposits, prospectors need only drill an opening, because pressure and low molecular weight drive the gas upward naturally. The first gas fields to be tapped were of this type. Most fields remaining today, however, require that gas be pumped to Earth's surface. As with oil and coal, many of the most accessible natural gas reserves have already been exhausted. Thus, much extraction today makes use of sophisticated techniques to break into rock formations and pump gas to the surface. One such "fracturing technique" is to pump salt water under high pressure into rocks to crack them. Sand or small glass beads are inserted to hold the cracks open once the water is withdrawn.

Table 13.4	Top Producers and Consumers of Natural Gas

Production (% world production)	Consumption (% world consumption)
Russia, 21.6	United States, 23.0
United States, 19.0	Russia, 14.7
Canada, 6.7	United Kingdom, 3.4
United Kingdom, 3.2	Canada, 3.3
Algeria, 3.2	Iran, 3.2

Data from British Petroleum. 2006. *Statistical review of world energy 2006.*

Oil

The world's most-used fuel is oil, which today accounts for 37% of global commercial energy consumption. Its use worldwide over the past decade has risen roughly 16%, and today our global society produces and consumes nearly 750 L (200 gal) of oil each year for every man, woman, and child. Table 13.5 shows the top oil-producing and oil-consuming nations.

Heat and pressure underground form petroleum

The sludgelike liquid we know as **crude oil**, or **petroleum**, tends to form under temperature and pressure conditions often found 1.5–3 km (1–2 mi) below the surface. The crude oil of Alaska's North Slope was formed when dead plant material (and small amounts of animal material) drifted down through coastal marine waters millions of years ago and was buried in ocean sediments. These organic remains were then transformed by time, heat, and pressure into the petroleum of today.

Table 13.5	Top Producers and Consumers of Oil

Production (% world production)	Consumption (% world consumption)
Saudi Arabia, 13.5	United States, 24.6
Russia, 12.1	China, 8.5
United States, 8.0	Japan, 6.4
Iran, 5.1	Russia, 3.4
Mexico, 4.8	Germany, 3.2

Data from British Petroleum. 2006. *Statistical review of world energy 2006.*

Petroleum geologists infer the location and size of deposits

Because petroleum forms only under certain conditions, it occurs in isolated deposits, tending to collect in porous layers beneath dense, impermeable layers. Oil deposits are not large black underground pools, but instead consist of small droplets within holes in porous rock, like a hard sponge full of oil. Geologists searching for oil (or other fossil fuels) drill rock cores and conduct ground, air, and seismic surveys to map underground rock formations, understand geological history, and predict where fossil fuel deposits might lie. Using such techniques, geologists from the U.S. Geological Survey (USGS) in 1998 estimated, with 95% certainty, the total amount of oil underneath ANWR's 1002 Area to be between 11.6 and 31.5 billion barrels. The geologists' average estimate of 20.7 billion represents their best guess as to the number of barrels of oil the 1002 Area holds.

Some portion of oil will be impossible to extract using current technology. Thus, estimates are generally made of *technically recoverable* amounts of fuels. In its 1998 estimates, the USGS calculated technically recoverable amounts of oil under the 1002 Area to be between 4.3 and 11.8 billion barrels, with an average estimate of 7.7 billion barrels (an amount equal to roughly one year's supply for the United States at current consumption rates). However, oil companies will not be willing to extract these entire amounts. Some oil will be so difficult to extract that the expense of doing so would exceed the income the company would receive from the oil's sale. USGS scientists calculated that at a price of $30 per barrel, 3.0–10.4 billion barrels would be economically worthwhile to recover from the 1002 Area. The USGS did not present estimates for today's much higher prices, but as prices climb, *economically recoverable* amounts approach technically recoverable amounts.

Thus, technology sets a limit on the amount that *can* be extracted, whereas economics determines how much *will* be extracted. The amount of oil, or any other fossil fuel, in a deposit that is technologically and economically feasible to remove under current conditions is termed the **proven recoverable reserve** of that fuel.

We drill to extract oil

Once geologists have identified an oil deposit, an oil company will typically conduct *exploratory drilling*, drilling small holes that descend to great depths. If enough oil is encountered, extraction begins. Just as you would squeeze a sponge to remove its liquid, pressure is required to extract oil from porous rock. Oil is typically already under pressure—from above by rock or trapped gas, from below by groundwater, or internally from natural gas dissolved in the oil. All these forces are held in place by surrounding rock until drilling reaches the deposit, whereupon oil will often rise to the surface of its own accord. Once pressure is relieved, however, oil becomes more difficult to extract, and may need to be pumped out.

We drill for oil and natural gas not just on land but also in the seafloor on the continental shelves. Offshore drilling has required us to develop technology that can withstand wind, waves, and ocean currents. Some drilling platforms are fixed standing platforms built with unusual strength. Others are resilient floating platforms anchored in place above the drilling site. Over 25% of the oil and gas extracted in the United States comes from offshore sites, primarily in the Gulf of Mexico and off the southern California coast. This is why Hurricanes Katrina and Rita

in 2005 disrupted U.S. oil supplies. By battering offshore oil platforms in the Gulf, and by damaging refineries onshore, the storms interrupted a substantial portion of the nation's oil supply, and prices rose accordingly.

Petroleum products have many uses

Once we extract crude oil, we refine it. Crude oil is a mixture of hundreds of different types of hydrocarbon molecules characterized by carbon chains of different lengths (▶ p. 57). A chain's length affects its chemical properties, which has consequences for human use, such as whether a given fuel burns cleanly in a car engine. Oil refineries sort the various hydrocarbons of crude oil, separating those intended for use in gasoline engines from those, such as tar and asphalt, used for other purposes (Figure 13.7).

(a) Distillation columns

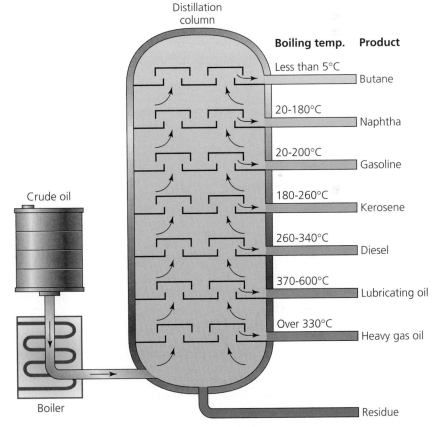

(b) Distillation process

FIGURE 13.7 Crude oil is shipped to petroleum refineries, where distillation columns **(a)** are used to refine it into a number of different types of fuel. In the distillation process **(b)**, crude oil is boiled, causing its many hydrocarbon constituents to volatilize and proceed upward through a distillation column. Constituents that boil only at the highest temperatures and condense readily once the temperature drops will condense at low levels in the column. Constituents that volatilize readily at lower temperatures will continue rising through the column and condense at higher levels, where temperatures are lower. In this way, heavy oils (generally consisting of long hydrocarbon molecules) are separated from lighter oils (generally those with short hydrocarbon molecules).

Separating crude oil's components helps us create many types of petroleum products. Since the 1920s, refining techniques and chemical manufacturing have greatly expanded our uses of petroleum to include a wide array of products and applications, from lubricants to plastics to fabrics to pharmaceuticals. Today, petroleum-based products are all around us in our everyday lives (Figure 13.8).

Because petroleum products have become so central to our lives, it should concern us that oil production will soon decline as we continue to deplete the world's oil reserves.

We may already have depleted half our oil reserves

Many scientists and oil industry analysts calculate that we have already extracted about half of the world's oil reserves. So far we have used up about 1 trillion barrels of oil, and most estimates hold that an additional 1 trillion barrels, or somewhat more, remain. To estimate how long this remaining oil will last, analysts calculate the **reserves-to-production ratio**, or **R/P ratio**, by dividing the amount of total remaining reserves by the annual rate of production (i.e., extraction and processing). At current levels of production (30 billion barrels globally per year), most analysts estimate that world oil supplies will last about 40 more years.

Unfortunately, this does not mean that we have 40 years to figure out what to do once the oil runs out. A growing number of scientists and analysts insist that we will face a crisis not when the last drop of oil is pumped, but when the rate of production first begins to decline. They point out that when production declines as demand continues to increase (because of rising global population and consumption), we will experience an oil shortage immediately. Because production tends to decline once reserves are depleted halfway, most of these experts calculate that this crisis will likely begin within the next several years.

To understand the basis of these concerns, we need to turn back the clock to 1956. In that year, Shell Oil geologist M. King Hubbert calculated that U.S. oil production would peak around 1970. His prediction was ridiculed at the time, but it proved to be accurate; U.S. production peaked in that very year and has continued to fall since

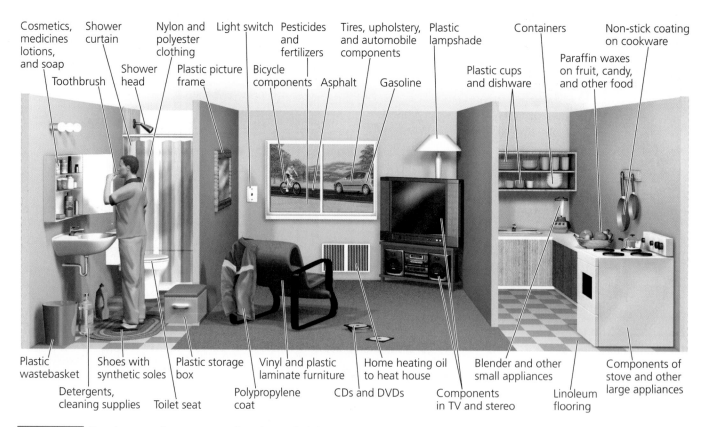

FIGURE 13.8 Petroleum products are everywhere in our daily lives. The gasoline and other fuels we use for transportation and heating are just a few of the many products we derive from petroleum. These products include many of the fabrics that we wear and most of the plastics that help make up countless items we use every day.

then (Figure 13.9a). The peak in production came to be known as **Hubbert's peak**. In 1974, Hubbert analyzed data on technology, economics, and geology to predict that global oil production would peak in 1995. It grew past 1995, but many scientists using better data today predict that at some point in the coming decade, production will begin to decline (Figure 13.9b). Discoveries of new oilfields peaked 30 years ago, and since then we have been extracting and consuming more oil than we have been discovering.

Predicting an exact date for the coming decline in production is difficult, and we will not know for certain when we have reached the peak until a few years after we have passed it. But the divergence of supply and demand will likely have momentous economic, social, and political consequences that will profoundly affect the lives of each and every one of us. One prophet of "peak oil," writer James Howard Kunstler, has sketched a frightening scenario of our post-peak world during what he calls "the long emergency": Lacking cheap oil with which to transport goods long distances, today's globalized economy would collapse and our economies would become intensely localized. Large cities could no longer be supported without urban agriculture, and even by expanding agricultural land we could only feed a fraction of the world's 6.5 billion people without petroleum-based fertilizers and pesticides. The American suburbs would be hit particularly hard because of their utter dependence on the automobile; Kunstler argues that they will become the slums of the future, a bleak landscape littered with the hulls of rusted-out SUVs.

FIGURE 13.9 Because fossil fuels are nonrenewable resources, supplies at some point pass the midway point of their depletion, and annual production begins to decline. U.S. oil production peaked in 1970, just as geologist M. King Hubbert had predicted decades previously; this high point is referred to as "Hubbert's peak" **(a)**. Today many analysts believe global oil production is about to peak. Shown **(b)** is the latest projection, from a 2004 analysis by scientists at the Association for the Study of Peak Oil. Go to **GRAPHit!** at www.aw-bc.com/withgott. Data from (a) Deffeyes, K. S. 2001. *Hubbert's peak: The impending world oil shortage.* (b) Colin J. Campbell and Association for the Study of Peak Oil, 2004.

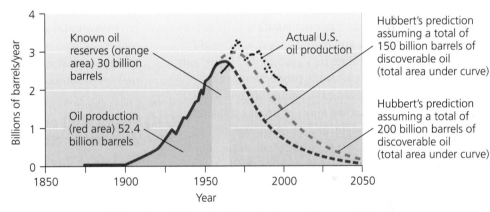

(a) Hubbert's prediction of peak in U.S. oil production, with actual data

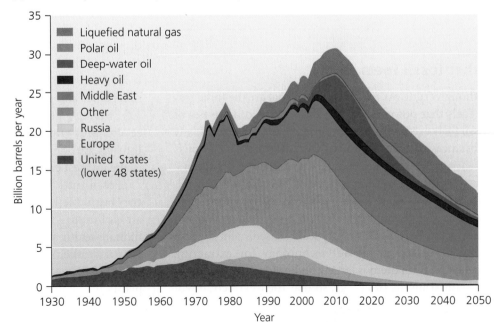

(b) Modern prediction of peak in global oil production

More optimistic observers argue that as oil supplies dwindle, rising prices will create powerful incentives for businesses, governments, and individuals to conserve energy and develop renewable energy sources (Chapter 14)—and that these developments will save us from major disruptions caused by the coming oil peak. Indeed, to achieve a sustainable society, we will need to switch to renewable energy sources, and energy conservation (▸ pp. 317–319) can extend the time we have in which to make this transition. However, the research needed to develop new technologies and the work needed to construct their infrastructure both depend on cheap oil, and the time we will have to make this enormous transition will be quite limited.

Weighing the **Issues:**
The End of Oil

Physicist David Goodstein has calculated that the gap between rising demand and falling supply after world oil production begins to decline may amount to 5% per year. As a result, just 10 years after the production peak we will have only half the oil availability we had at the peak. He worries that we may not be able to modify our infrastructure and institutions fast enough to accommodate other energy sources before the economic impacts of an oil shortage undermine our ability to do so.

Do you think our society could adapt to a 50% decrease in oil availability over 10 years? How do you think we would most likely respond? How do you think we should respond?

Other fossil fuels exist

Although crude oil, natural gas, and coal are the three fossil fuels that power our civilization today, other types of fossil fuels exist. *Oil sands* or *tar sands* are dense, hard, oily substances that can be mined from the ground. *Shale oil* is essentially kerogen, sedimentary rock filled with organic matter that was not buried deeply enough to form oil. *Methane hydrates* are ice-like solids consisting of methane molecules embedded in a crystal lattice of water molecules, and are found in many seafloor sediments. These sources are abundant, but technology for extracting usable fuel from them is largely undeveloped and requires substantial inputs of energy, so that their extraction will likely remain inefficient and expensive.

Many advocates of sustainability believe it would be a mistake to try to switch to these sources once our conven-tional fossil fuels become scarce. Such sources will require extensive mining and will emit at least as much carbon dioxide, methane, and other air pollutants as do coal, oil, and gas. As such, they will exacerbate the severe environmental impacts that fossil fuels are already causing.

Impacts of Fossil Fuel Use

Our society's love affair with fossil fuels and the many petrochemical products we have developed from them has boosted our material standard of living beyond what our ancestors could have dreamed, has eased constraints on travel, and has helped lengthen our life spans. However, it has also harmed the environment and human health, and it serves as a source of political and economic instability. Concern over environmental, social, political, and economic impacts is a prime reason many scientists, environmental advocates, businesspeople, and policy-makers are increasingly looking toward renewable sources of energy that exert less impact on natural systems and may be more socially sustainable.

Fossil fuel emissions cause pollution and drive climate change

When we burn fossil fuels, we alter certain flux rates in Earth's carbon cycle (▸ pp. 61–64). We essentially take carbon that has been effectively retired into a long-term reservoir underground and release it into the air. This occurs as carbon from within the hydrocarbon molecules of fossil fuels unites with oxygen from the atmosphere during combustion, producing carbon dioxide (CO_2). Carbon dioxide is a greenhouse gas (▸ p. 288), and CO_2 from fossil fuel combustion has been inferred to warm our planet and drive changes in global climate (Chapter 12). Because global climate change may have diverse, severe, and widespread ecological and socioeconomic impacts, carbon dioxide pollution is becoming recognized as the greatest environmental impact of fossil fuel use.

Fossil fuels release more than carbon dioxide when they burn. Methane is a potent greenhouse gas, and other air pollutants from fossil fuel combustion can affect human health and the environment. Deposition of mercury and other pollutants from coal-fired power plants is increasingly recognized as a substantial health risk. Power plants and vehicles release sulfur dioxide (SO_2) and nitrogen oxides (NO_X), which contribute to industrial and photochemical smog and acidic deposition (▸ pp. 280, 282–285). Gasoline combustion in automobiles releases pollutants that irritate the nose, throat, and lungs, as well as hydrocarbons (such as benzene and toluene) and impurities (such as lead and arsenic) known to cause cancer or other serious health risks.

To try to make emissions from coal combustion less toxic, U.S. policymakers have appropriated billions of dollars in recent years toward developing "clean coal" technologies. Such technologies include *scrubbers*, which are materials based on minerals such as calcium or sodium that absorb and remove sulfur dioxide from smokestack emissions. They also include chemical reactions that strip away nitrogen oxides, breaking them down into elemental nitrogen and water. In addition, multilayered filtering devices are used to capture tiny ash particles. The most successful of these efforts have reduced emissions by up to 95% at some power plants. However, some energy analysts and environmental advocates question a policy emphasis on clean coal. Coal, they maintain, is an inherently dirty means of generating power and should be replaced outright with cleaner energy sources.

Coal mining affects health and the environment

Surface strip mining for coal can destroy large swaths of habitat, cause extensive soil erosion, and lead to water pollution. Regulations in the United States require mining companies to restore strip-mined land following min-

ing, but impacts are severe and long-lasting just the same. Most other nations exercise less oversight.

Mountaintop removal (Figure 13.10) has even greater impacts than conventional strip mining. When tons of rock and soil are removed from atop a mountain, it is difficult to keep material from sliding downhill, where immense areas of habitat can be degraded or destroyed and creek beds can be polluted and clogged. Loosening of U.S. government restrictions in 2002 enabled mining companies to legally dump mountaintop rock and soil into valleys and rivers below, regardless of the consequences for ecosystems, wildlife, and local residents.

Subsurface mining raises the greatest health concerns for miners and is one of our society's most dangerous occupations. Besides risking injury or death from dynamite blasts and collapsing tunnels, miners constantly inhale coal dust, which can lead to respiratory diseases, including fatal black lung disease.

Fossil fuels pollute water

Our extraction and use of coal and oil can also pollute rivers, oceans, and groundwater. Water pollution from oil results from large tanker spills but mostly from a variety of non-point sources, as we discussed in Chapter 11 (▶ p. 263). Strip mining for coal often causes chemical runoff into waterways through the process of **acid drainage**, which occurs when sulfide minerals in newly

FIGURE 13.10 Strip mining in some areas is taking place on massive scales, such that entire mountain peaks are leveled, as at this site in West Virginia. Such "mountaintop removal" can cause enormous amounts of erosion into streams that flow from near the mine into surrounding valleys, affecting ecosystems over large areas, as well as the people who live there.

exposed rock surfaces react with oxygen and rainwater to produce sulfuric acid. As the sulfuric acid runs off, it leaches metals from the rocks, many of which can be toxic to organisms.

With both air and water pollution, the costs of alleviating all these health and environmental impacts are high, and the public eventually pays them in an inefficient manner. The reason is that the costs are generally not internalized (▶ pp. 28–29) in the relatively cheap, subsidized prices of fossil fuels.

Oil and gas extraction can alter the environment

Drilling for oil or gas in itself has fairly minimal environmental impact, but much more than drilling is involved in the development of an oil or gas field. Road networks must be constructed, and many sites may be explored in the course of prospecting. These activities can fragment habitats and can be noisy and disruptive enough to affect wildlife. The extensive infrastructure that must be erected to support a full-scale drilling operation typically includes housing for workers, access roads, transport pipelines, and waste piles for removed soil. Ponds may be constructed for collecting sludge, the toxic leftovers that remain after the useful components of oil have been removed.

To predict the possible ecological effects of drilling in ANWR's 1002 Area, scientists have examined the effects of development on arctic vegetation, air quality, water quality, and wildlife (including caribou, grizzly bears, and a variety of bird species) in Prudhoe Bay and other Alaska locales with environments similar to that of ANWR's coastal plain (Figure 13.11). Scientists have compared different areas and have contrasted single areas or populations before and after drilling. In addition, scientists have run small-scale manipulative experiments when possible, for example, to study the effects of ice roads and secondary extraction methods such as seawater flushing. In one way or another they have examined the effects of road building, oil pad construction, worker presence, oil spills, accidental fires, trash buildup, permafrost melting, off-road vehicle trails, and dust from roads.

Based on these studies, many scientists anticipate damage to vegetation and wildlife if drilling takes place in ANWR. Vegetation can be killed when saltwater pumped in for flushing deposits is spilled or when plants are buried under gravel pits or roads. Plants grow slowly in the Arctic, so even minor changes can have long-lasting repercussions. For example, tundra vegetation at Prudhoe Bay still has not fully recovered from temporary roads last

FIGURE 13.11 Alaska's North Slope is home to a variety of large mammals, including grizzly bears, polar bears, wolves, arctic foxes, and large herds of caribou. Whether and how oil development may negatively affect these animals are highly controversial issues, and scientific studies are ongoing. The caribou herd near Prudhoe Bay has increased since oil extraction began there, but not by as much as have herds in other parts of Alaska. Grizzly bears such as the ones shown here can sometimes be found near, or even walking atop, the trans-Alaska pipeline.

used 30 years ago during the exploratory phase of development. In addition, air and water quality can be degraded by fumes from equipment and drilling operations, burning of natural gas associated with oil extraction, sludge ponds, waste pits, and oil spills.

Other scientists contend that drilling operations in ANWR would have little environmental impact. Roads would be built of ice that will melt in the summer, they point out, and most drilling activity would be confined to the winter, when caribou are elsewhere. Moreover, drilling proponents maintain, much of the technology used at Prudhoe Bay is now outdated, and ANWR would be developed with more environmentally sensitive technology and approaches.

Oil supply and prices affect the economies of nations

Hurricanes Katrina and Rita and the increased gasoline prices they caused in 2005 served to remind us how much we rely on a cheap and ever-increasing supply of petroleum. The hurricanes' economic impact should have come as no surprise, for we have experienced "oil shocks" before. In 1973, with domestic sources in decline, the United States was importing more and more oil, depending on a constant flow from abroad to

Drilling in the Arctic National Wildlife Refuge

Should we drill for oil in the Arctic National Wildlife Refuge?

ANWR Oil Means Better Living, Not Eco-Apocalypse

Oil development has taken place in the Arctic for over 30 years. Originally there were fears for the environment, just as we are hearing today. The Inupiat feared harm to the caribou and to their lifestyles while environmentalists claimed that oil development meant environmental apocalypse and that, moreover, it would yield only 2 years' supply of oil.

Well, "2 years" of oil turned into over 28; 3,000 caribou turned into over 32,000; and the Inupiat have turned into the number-one supporters of development. Technology has allowed a 140-acre drill pad to be reduced to 5 acres in size. Improvements in directional drilling now allow us to extract oil 8 miles from drill-point. The EPA, the North Slope Borough, and Alaska Department of Fish and Game monitor activities daily, so one can hardly claim an eco-apocalypse. Misinformation is rife in the ANWR debate.

Furthermore, the NIMBY argument doesn't work with ANWR; the people who live off the 1002 Area want this badly. To them, drilling in the 1002 Area will provide jobs, schools, clinics—a future. They have seen firsthand how their fears of 30 years ago were unfounded. They have seen their communities prosper.

The energy future of the United States should be multi-sourced. Wind turbines, solar panels, and hydrogen cells *all* need oil to build their working parts. None of these sources can produce plastic, bitumen, tires, paints, medicines, or a thousand other oil-based products we use daily. One can scream against oil, yet every one of us uses it every day in every aspect of our lives. We don't know any alternative. We should invest heavily in finding alternatives, but we should not think for a second that we can cut off our oil supply, domestic or foreign, and live as happily as we do now.

Adrian Herrera is a consultant for the lobby group Arctic Power, lobbying Congress to open the 1002 Area of ANWR to responsible oil development. Formerly, Herrera worked as an engineering assistant at Prudhoe Bay and for an independent ecological research company tasked with monitoring the wildlife and environment in and around the oil fields. He is an Alaskan resident of over 30 years and has worked and lived in the Arctic extensively.

Drilling the Arctic Refuge Is Not a Solution to Our Energy Problems—It's a Distraction

The Arctic National Wildlife Refuge is one of the last unspoiled wild areas in the United States. Its 1.5-million-acre coastal plain is rich in biodiversity, home to nearly 200 species, including polar bears, musk oxen, caribou, and millions of migratory birds.

There is no way to drill in the refuge without permanently harming this unique ecosystem or destroying the culture of the native Gwich'in people, who have depended on caribou for thousands of years. The little oil beneath the refuge is scattered in more than 30 small deposits. To extract it, roads, pipelines, airstrips, and other industrial infrastructure would be built across the entire area.

Drilling the Arctic Refuge would do nothing to lower gas prices or lessen our nation's dependence on imported oil. According to the U.S. Geological Survey, the refuge holds less economically recoverable oil than what Americans consume in a year, and it would take 8–10 years for that oil to reach the market. A recent U.S. Energy Department report found that oil from the Arctic Refuge would have little impact on the price of gasoline, lowering gas prices by less than a penny and a half per gallon—in 2025.

If we boosted the fuel economy performance of our cars and trucks by just 1 mile per gallon annually over the next 15 years, we would save more than 10 times the oil that could be recovered from the refuge. We have the technology today to accomplish that goal.

The United States has 3% of the world's oil reserves but consumes 25% of all oil produced each year. We cannot drill our way to lower gas prices. By focusing on efficiency and alternative fuels, we can improve our energy security and preserve the Arctic Refuge for future generations.

Karen Wayland is the Natural Resources Defense Council's legislative director and an adjunct professor at Georgetown University. Dr. Wayland, who holds a dual Ph.D. in geology and resource development, was a legislative fellow for Sen. Harry Reid (D-Nev.) on nuclear waste, water, energy, and Native American issues before joining NRDC's staff.

Explore this issue further by accessing **Viewpoints** at www.aw-bc.com/withgott.

keep cars on the road and industries running. Then the predominantly Arab nations of the *Organization of Petroleum Exporting Countries (OPEC)* resolved to stop selling oil to the United States. OPEC wished to raise prices by restricting supply, and opposed U.S. support of Israel in the Arab-Israeli Yom Kippur War. The embargo created panic in the West and caused oil prices to skyrocket, spurring inflation. Short-term oil shortages drove American consumers to wait in long lines at gas pumps.

To counter dependence on a few major supplier nations, the United States has diversified its sources of petroleum and now receives most of it from non–Middle Eastern nations, including Venezuela, Canada, Mexico, and Nigeria. Major trade relations among nations and regions of the world are depicted in Figure 13.12.

The fact that so many nations' economies are utterly tied to fossil fuels means that those economies are tremendously vulnerable to supplies' becoming suddenly unavailable or extremely costly. In the United States, concern over reliance on foreign oil sources has repeatedly driven the proposal to open ANWR to drilling, despite critics' charges that such drilling would do little to decrease the nation's dependence. The United States currently imports 65% of its crude oil. With the majority of world oil reserves located in the politically unstable Middle East, crises such as the 1973 embargo are a constant concern for U.S. policymakers. The United States has cultivated a close relationship with Saudi Arabia, the owner of 22% of world oil reserves, even though that country's political system allows for little of the democracy that U.S. leaders claim to cherish and promote. The world's third-largest holder of oil reserves, at almost 10%, is Iraq, which is why many people around the world believe the U.S.-led invasion of Iraq in 2003 was motivated primarily to secure access to oil.

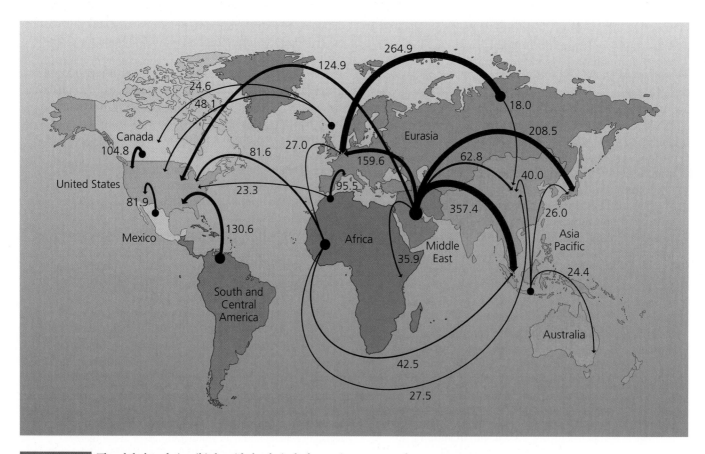

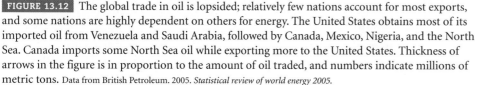

FIGURE 13.12 The global trade in oil is lopsided; relatively few nations account for most exports, and some nations are highly dependent on others for energy. The United States obtains most of its imported oil from Venezuela and Saudi Arabia, followed by Canada, Mexico, Nigeria, and the North Sea. Canada imports some North Sea oil while exporting more to the United States. Thickness of arrows in the figure is in proportion to the amount of oil traded, and numbers indicate millions of metric tons. Data from British Petroleum. 2005. *Statistical review of world energy 2005.*

Residents may or may not benefit from their fossil fuel reserves

People who live in areas where fossil fuels are extracted may experience benefits or drawbacks. In Alaska, citizens have seen economic gain from Prudhoe Bay's development. Alaska's constitution requires that one-quarter of state oil revenues be placed in the Permanent Fund, which pays yearly dividends to all Alaska residents. Since 1982, each Alaska resident has received annual payouts ranging from $331 to $1,964. Because development of ANWR would add to this fund and create jobs, most Alaska residents support oil drilling in ANWR. The Native people who live on the North Slope are split over the proposed drilling. The Inupiat want income for health care, police and fire protection, and other services that are scarce in this remote region. The Gwich'in oppose drilling because they fear it will threaten the caribou herds and Arctic ecosystems they depend on.

Alaska's distribution of revenue among its citizenry is unusual. In most parts of the world where fossil fuels are extracted, local residents have not seen benefits. In Nigeria, oil was discovered in 1958 in the territory of the native Ogoni people, and the Shell Oil Company moved in to develop oil fields. Although Shell extracted $30 billion of oil from Ogoni land over the years, the Ogoni still live in poverty, with no running water or electricity. The profits from oil extraction on Ogoni land have gone to Shell and to the military dictatorships of Nigeria. The development resulted in oil spills, noise, and constantly burning gas flares, all of which caused illness among people living nearby. From 1962 until his death in 1995, Ogoni activist and leader Ken Saro-Wiwa worked for fair compensation to the Ogoni for oil extraction and environmental degradation on their land. After years of persecution by the Nigerian government, Saro-Wiwa was arrested in 1994, given a trial widely regarded in the international human rights community as a sham, and put to death by military tribunal.

Energy Conservation

Given that fossil fuel supplies are limited and that their use has health, environmental, political, and socioeconomic consequences, a sustainable future requires that we move toward replacing them with clean and renewable energy sources. As we make this historic transition, it will benefit us to minimize our energy use so as to prolong the availability of fossil fuels. **Energy conservation** is the practice of reducing energy use to extend the lifetimes of our nonrenewable energy supplies, be less wasteful, and reduce our environmental impact.

Energy conservation has often been a function of economic need

In the United States, many people first saw the value of conserving energy following the OPEC embargo of 1973–1974. In response to that event, the U.S. government enacted policies such as a mandated increase in the miles-per-gallon (mpg) fuel efficiency of automobiles and a reduction in the national speed limit to 55 miles per hour.

Three decades later, many of these conservation initiatives have been abandoned. Government funding for research into alternative energy sources has decreased, speed limits have increased, and recent bills to raise the mandated average fuel efficiency of vehicles have failed in Congress. The average fuel efficiency of new vehicles has fallen from a high of 22.1 mpg in 1988 to 21.0 mpg in 2005 (Figure 13.13). This decrease is due to increased sales of light trucks (averaging 18.2 mpg), including sport-utility vehicles, relative to cars (averaging 24.7 mpg). Transportation accounts for two-thirds of U.S. oil use, and passenger vehicles consume over half this energy. Thus, the failure to improve vehicular fuel economy over the past 20 years, despite the existence of technology to do so, has added greatly to U.S. oil consumption.

Moreover, because the government heavily subsidizes fossil fuel production, market prices do not tell us the true costs of fossil fuels. This disparity decreases our economic incentives to conserve.

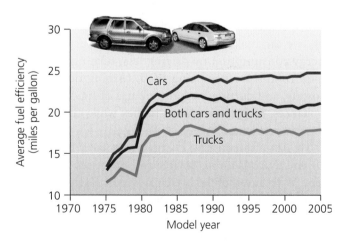

FIGURE 13.13 Fuel efficiency of automobiles in the United States rose dramatically in the late 1970s as a result of legislative mandates, but it has declined slightly since 1988. The decline is due to a lack of further legislation for improved fuel economy and to the increased popularity in recent years of sport-utility vehicles. Data from U.S. Environmental Protection Agency. 2005. *Light-duty automotive technology and fuel economy trends: 1975 through 2005.*

More Miles, Less Gas

If you drive an automobile, what gas mileage does it get? How does it compare to the vehicle averages in Figure 13.13? If your vehicle's fuel efficiency were 10 mpg greater, and you drove the same amount, how many gallons of gasoline would you no longer need to purchase each year? How much money would you save? If all U.S. vehicles were mandated to increase fuel efficiency by 10 mpg, how much gasoline do you think the over 200 million Americans who drive could conserve? What other strategies can you think of to conserve fossil fuels, and how might they compare in effectiveness to a rise in fuel efficiency standards?

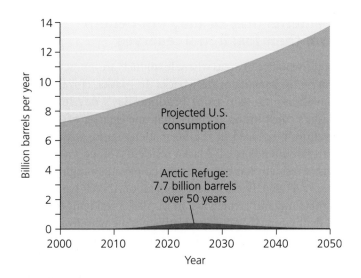

FIGURE 13.14 Opponents of oil drilling in the Arctic National Wildlife Refuge contend that the amount of oil estimated to be recoverable would make only a small contribution toward overall U.S. oil demand. In this graph, the best USGS estimate of oil from ANWR's 1002 Area is shown in red, in the context of total U.S. oil consumption, assuming that current consumption trends are extrapolated into the future and oil production takes place over many years. The actual ANWR contribution, if it comes to pass, would depend greatly on the amount of oil actually present under ANWR, the time it would take to extract it, and future trends in consumption. Adapted from Natural Resources Defense Council. 2002. *Oil and the Arctic National Wildlife Refuge*; and U.S. Geological Survey. 2001. *Arctic National Wildlife Refuge, 1002 Area, petroleum assessment, 1998, including economic analysis.*

Many critics of oil drilling in the Arctic National Wildlife Refuge point out the vast amounts of oil wasted by our fuel-inefficient automobiles. They argue that a small amount of conservation would save the nation far more oil than it would ever obtain from ANWR. As we noted, the USGS's average estimate for recoverable oil in the 1002 Area, 7.7 billion barrels, represents just one year's supply for the United States at current consumption rates. Spread over a period of extraction of many years, the proportion of U.S. oil needs that ANWR would fulfill appears strikingly small (Figure 13.14).

Personal choice and increased efficiency are two routes to conservation

We can conserve energy in two primary ways. As individuals, we can make conscious choices to reduce our own energy consumption by driving less, turning off lights when rooms are not being used, turning down thermostats, and cutting back on the use of machines and appliances. Such steps save us money while also helping conserve resources.

We can also conserve energy as a society by making our energy-consuming devices and processes more efficient. With automobiles, we already possess the technology to increase fuel efficiency far above the current average of 21 mpg. We could accomplish such improvement with more efficient gasoline engines or with alternative technology vehicles such as electric/gasoline hybrids (Figure 13.15).

The efficiency of power plants can be improved through **cogeneration**, in which excess heat produced during electricity generation is captured and used to heat workplaces and homes and to produce other kinds of power. Cogeneration can nearly double the efficiency of a power plant.

In homes and public buildings, inadequate insulation causes a significant amount of heat loss in winter and heat retention in summer. Improvements in the design of homes and offices can reduce energy required to heat and cool them. Such design changes may involve the building's location, insulation, or even the color of its roof (light colors keep buildings cooler by reflecting the sun's rays).

Among consumer products, scores of appliances, from refrigerators to light bulbs, have been reengineered through the years to increase energy efficiency. Even so, there remains room for improvement. Energy-efficient lighting, for example, can reduce energy use by 80%, and federal standards for energy-efficient appliances have already reduced per-person home electricity use below what it was in the 1970s. While manufacturers can improve the energy efficiency of appliances, consumers need to "vote with their wallets" by purchasing these energy-efficient appliances, so that they are kept commercially available. The U.S. Environmental Protection Agency (EPA) estimates that if all U.S. households purchased energy-efficient appliances, the national annual

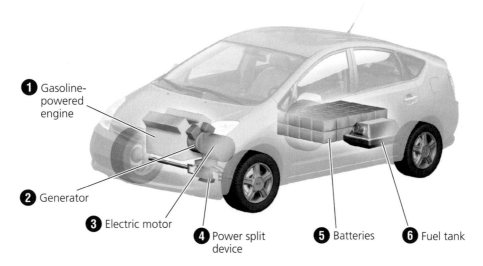

1 Gasoline-powered engine

2 Generator

3 Electric motor

4 Power split device

5 Batteries

6 Fuel tank

FIGURE 13.15 A hybrid car, such as the Toyota Prius diagrammed here, uses a small, clean, and efficient gasoline-powered engine (1) to produce power that the generator (2) can convert to electricity to drive the electric motor (3). The power split device (4) integrates the engine, generator, and motor, serving as a continuously variable transmission. The car automatically switches between all-electrical power, all-gas power, and a mix of the two, depending on the demands being placed on the engine. Typically, the motor provides power for low-speed city driving and adds extra power on hills. The motor and generator charge a pack of nickel-metal-hydride batteries (5), which can in turn supply power to the motor. Energy for the engine comes from gasoline carried in a typical fuel tank (6).

energy expenditure would be reduced by $200 billion. For the individual consumer, most studies show that the savings on utility bills rapidly offset the higher costs of energy-efficient appliances.

In the energy bill endorsed by the Bush administration and passed by the U.S. Congress in 2005, some conservation measures were adopted, including tax credits for consumers who buy hybrid cars or improve energy efficiency in their homes. However, these measures were a relatively minor component of the legislation, and most of its funding went toward subsidies for production in various energy sectors. Among the beneficiaries was the nuclear power industry. Increasing numbers of people across the political spectrum recognize that we may see a resurgence of nuclear power as one alternative to fossil fuels.

Nuclear Power

Nuclear power occupies an odd and conflicted position in our modern debate over energy. Free of the air pollution produced by fossil fuel combustion, it has long been put forth as an environmentally friendly alternative to fossil fuels. At the same time, nuclear power's great promise has been clouded by nuclear weaponry, the dilemma of radioactive waste disposal, and the long shadow of Chernobyl and other power plant accidents.

First developed commercially in the 1950s, nuclear power has expanded 15-fold worldwide since 1970, experiencing most of its growth during the 1970s and 1980s. The United States generates the most electricity from nuclear energy—nearly a third of the world's production—yet only 20% of U.S. electricity comes from nuclear power. A number of other nations rely more heavily on nuclear power (Table 13.6).

Fission releases nuclear energy in reactors to generate electricity

Strictly defined, **nuclear energy** is the energy that holds together protons and neutrons within the nucleus of an atom. We harness this energy by converting it to thermal energy, which can then be used to generate electricity. The reaction that drives the release of nuclear energy in power plants is **nuclear fission**, the splitting apart of atomic nuclei (Figure 13.16). In fission, the nuclei of large, heavy atoms, such as uranium or plutonium, are bombarded with neutrons (▸ p. 53). Ordinarily neutrons move too quickly to split nuclei when they collide with them, but if neutrons are slowed down they can break apart nuclei. Each split nucleus emits multiple neutrons, together with substantial heat and radiation. These neutrons (two to three in the case of fissile uranium-235 isotopes) can in turn bombard other uranium-235 (^{235}U) atoms in the vicinity, resulting in a self-sustaining chain reaction.

Table 13.6 Top Consumers of Nuclear Power

Nation	Nuclear power consumed*	Number of plants†	Percentage of electricity generation from nuclear power plants‡
United States	185.9	104	19
France	102.4	59	78
Japan	66.3	53	23
Germany	36.9	18	28
Russia	33.9	30	16
South Korea	33.2	19	37
Canada	20.8	16	13
Ukraine	20.1	13	45
United Kingdom	18.5	27	22
Sweden	16.3	11	50

Data from International Atomic Energy Agency, British Petroleum, and International Energy Agency.

*In million metric tons of oil equivalent, 2005 data.

† 2003 data.

‡ 2003 data.

If not controlled, this chain reaction becomes a runaway process of positive feedback—the process that creates the explosive power of a nuclear bomb. Inside a nuclear power plant, however, fission is controlled so that only one of the two or three neutrons emitted with each fission event goes on to induce another fission event. In this way, the chain reaction maintains a constant output of energy at a controlled rate.

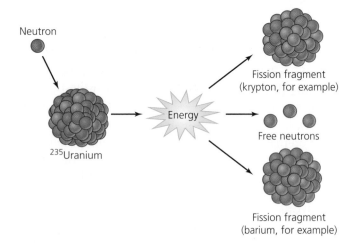

Neutron

Energy

Fission fragment (krypton, for example)

Free neutrons

^{235}Uranium

Fission fragment (barium, for example)

FIGURE 13.16 In nuclear fission, atoms of uranium-235 are bombarded with neutrons. Each collision splits uranium atoms into smaller atoms and releases two or three neutrons, along with energy and radiation. Because the neutrons can continue to split uranium atoms and set in motion a runaway chain reaction, engineers at nuclear plants must absorb excess neutrons with control rods to regulate the rate of the reaction.

For fission to begin in a nuclear reactor, the neutrons bombarding uranium are slowed down with a substance called a *moderator*, most often water or graphite. As fission proceeds, it becomes necessary to soak up the excess neutrons produced when uranium nuclei divide, so that on average only a single uranium atom from each nucleus goes on to split another nucleus. For this purpose, *control rods*, made of a metallic alloy that absorbs neutrons, are placed into the reactor among the water-bathed *fuel rods* of uranium. Engineers move these control rods into and out of the water to maintain the fission reaction at the desired rate. All this takes place within the reactor core and is the first step in the electricity-generating process of a nuclear power plant (Figure 13.17).

Nuclear energy comes from processed and enriched uranium

Uranium is used for nuclear power because it is radioactive, emitting subatomic particles and high-energy radiation as it decays into a series of daughter isotopes (▶ pp. 54, 56). Over 99% of the uranium in nature occurs as the isotope uranium-238. Uranium-235 (with three fewer neutrons) makes up less than 1% of the total. Because ^{238}U does not emit enough neutrons to maintain a chain reaction when fissioned, we use ^{235}U for commercial nuclear power. So, mined uranium ore must be processed to enrich the concentration of ^{235}U to at least 3%. The enriched uranium is formed into small pellets of uranium dioxide (UO_2), which are incorporated into the fuel rods used in reactors. After several years in a reactor, enough uranium has decayed so that the fuel loses its ability to generate adequate energy, and it must be replaced. Some spent fuel is reprocessed to recover what usable energy may be left, but most is disposed of as radioactive waste.

Uranium is an uncommon mineral, and uranium ore is in finite supply, so nuclear power is generally considered a nonrenewable energy source.

Nuclear power delivers energy more cleanly than fossil fuels

Using conventional fission, nuclear power plants generate electricity without creating the air pollution from stack emissions that fossil fuels do. After considering all the steps involved in building plants and generating power, researchers from the International Atomic Energy Agency (IAEA) have calculated that nuclear power lowers emissions 4–150 times below fossil fuel combustion. IAEA scientists estimate that nuclear power helps us avoid emitting 600 million metric tons of carbon each year, equivalent to 8% of global greenhouse gas emissions. Moreover, for residents

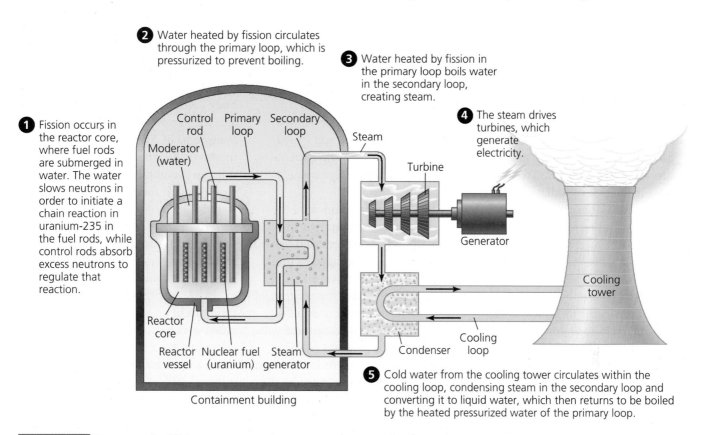

2 Water heated by fission circulates through the primary loop, which is pressurized to prevent boiling.

3 Water heated by fission in the primary loop boils water in the secondary loop, creating steam.

4 The steam drives turbines, which generate electricity.

1 Fission occurs in the reactor core, where fuel rods are submerged in water. The water slows neutrons in order to initiate a chain reaction in uranium-235 in the fuel rods, while control rods absorb excess neutrons to regulate that reaction.

Control rod   Primary loop   Secondary loop

Moderator (water)

Steam

Turbine

Generator

Reactor core

Cooling tower

Reactor vessel   Nuclear fuel (uranium)   Steam generator

Cooling loop

Condenser

Containment building

5 Cold water from the cooling tower circulates within the cooling loop, condensing steam in the secondary loop and converting it to liquid water, which then returns to be boiled by the heated pressurized water of the primary loop.

FIGURE 13.17 In a pressurized light water reactor, the most common type of nuclear reactor, uranium fuel rods are placed in water, which slows neutrons so that fission can occur (1). Control rods that can be moved into and out of the reactor core absorb excess neutrons to regulate the chain reaction. Water heated by fission circulates through the primary loop (2) and warms water in the secondary loop, which turns to steam (3). Steam drives turbines, which generate electricity (4). The steam is then cooled by water from the cooling tower and returns to the containment building (5), to be heated again by heat from the primary loop. Containment buildings, with their meter-thick concrete and steel walls, are constructed to prevent leaks of radioactivity due to accidents or natural catastrophes such as earthquakes.

living downwind from power plants, scientists calculate that nuclear power poses far fewer chronic health risks from pollution than does fossil fuel combustion.

Nuclear power has additional environmental advantages over fossil fuels, coal in particular. Because uranium generates far more power than coal by weight or volume, less of it needs to be mined, so uranium mining causes less damage to landscapes and generates less solid waste than coal mining. Moreover, in the course of normal operation, nuclear power plants are safer for workers than coal-fired plants.

Nuclear power poses small risks of large accidents

Nuclear power also has drawbacks. One is that the waste it produces is radioactive and must be disposed of safely. The second main drawback is that if an accident occurs at a power plant, or if a plant is sabotaged, the consequences can potentially be catastrophic.

The nuclear industry's first major accident took place at the **Three Mile Island** plant in Pennsylvania in 1979. Through a combination of mechanical failure and human error, coolant water drained from the reactor vessel, temperatures rose inside the reactor core, and metal surrounding the uranium fuel rods began to melt, releasing radiation. This process is termed a *meltdown*, and at Three Mile Island it proceeded through half of one reactor core. Area residents stood ready to be evacuated as the nation held its breath, but fortunately most radiation remained trapped inside the containment building. The accident was brought under control, the damaged reactor was shut down, and multibillion-dollar cleanup efforts stretched on for years. Three Mile Island is best regarded as a near-miss; the emergency could have been far worse had the meltdown proceeded through the entire stock of uranium fuel, or had the containment building not trapped the radiation.

Health Impacts of Chernobyl

In the wake of the nuclear power plant accident at Chernobyl in 1986, medical scientists from around the world rushed to study how the release of radiation might affect human health.

Determining long-term health impacts of a discrete event is difficult, so it is not surprising that the hundreds of researchers trying to pin down Chernobyl's impacts sometimes came up with very different conclusions. In an effort to reach some consensus, researchers at the Nuclear Energy Agency (NEA) of the Organization for Economic Cooperation and Development (OECD) reviewed studies through 2002 and issued a report summarizing what scientists had learned in the 16 years since the accident.

Doctors had documented the most severe effects among plant workers and firefighters who battled to contain the incident in its initial hours and days. Medical staff treated and recorded the progress of 237 patients who had been admitted to area hospitals diagnosed with acute radiation sickness (ARS). Radiation destroys cells in the body, and if the destruction outpaces the body's abilities to repair the damage, the person will soon die. Symptoms of ARS include vomiting, fever, diarrhea, thermal burns, mucous membrane damage, and weakening of the immune system. In total, 28 (11.8%) of these people died from acute effects soon after the accident, and those who died had had the greatest estimated exposure to radiation.

IAEA scientists in 1990 studied residents of areas highly contaminated with radioactive cesium and compared their health with people of the same ages living in uncontaminated settlements nearby.

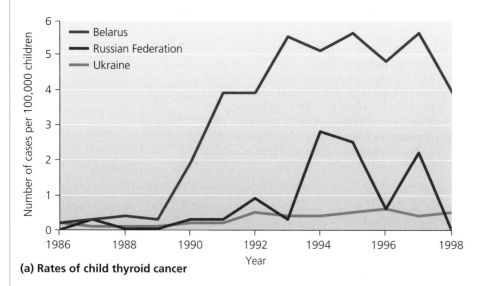

(a) Rates of child thyroid cancer

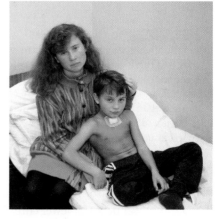

(b) Cancer patient with mother after surgery

The incidence of thyroid cancer **(a)** jumped in Belarus, Ukraine, and southwestern Russia starting 4 years after the Chernobyl accident released high levels of radioactive iodine isotopes. Many babies and young children **(b)** at the time of the accident developed thyroid cancer in later years. Most have undergone treatment and survived. Data from Nuclear Energy Agency, OECD, 2002.

Chernobyl saw the worst accident yet

In 1986 an explosion at the **Chernobyl** plant in Ukraine (part of the Soviet Union at the time) caused the most severe nuclear power plant accident the world has seen. Engineers had turned off safety systems to conduct tests, and human error, combined with unsafe reactor design, led to explosions that destroyed the reactor and sent clouds of radioactive debris billowing into the atmosphere. For 10 days radiation escaped from the plant while emergency crews risked their lives putting out fires. Most residents of the surrounding countryside remained at home for these 10

Medical exams of 1,356 people showed no significant differences between the two groups or health abnormalities attributable to radiation exposure. However, the study was criticized for the quality of its data, its small sample size, and potential conflict of interest (the IAEA is charged with promoting the nuclear industry). Moreover, the study was conducted only 4 years after the accident, before many cancers would be expected to appear.

Nonetheless, studies by the World Health Organization and others have come to similar conclusions. Overall, the NEA summary concluded, there is little evidence for long-term physical health effects resulting from Chernobyl (although psychological and social effects among residents displaced from their homes have been substantial). If cancer rates have risen among exposed populations, they have risen so little as to be statistically indistinguishable from normal variation in background levels of cancer.

The one exception is thyroid cancer, for which numerous studies have documented a real and perceptible increase among Chernobyl-area residents, particularly children (who have large and active thyroid glands). The thyroid gland (located in the neck) is where the human body concentrates iodine, and one of the most common radioactive isotopes released early in the disaster was iodine-131 (^{131}I).

Realizing that thyroid cancer induced by radioisotopes of iodine might be a problem, medical workers took measurements of iodine activity from the thyroid glands of hundreds of thousands of people— 60,000 in Russia, 150,000 in Ukraine, and several hundred thousand in Belarus—in the months immediately following the accident. They also measured food contamination and had people fill out questionnaires on their food consumption. These data showed that drinking cows' milk was the main route of exposure to ^{131}I for most people, although fresh vegetables also contributed.

As doctors had feared, in the years following the accident rates of thyroid cancer began rising among children in regions of highest exposure (see the figure). The yearly number of thyroid cancer cases in the 1990s, particularly in Belarus, far exceeded numbers from years before Chernobyl. Multiple studies found linear dose-response relationships (▶ p. 229) in data from Ukraine and Belarus. Fortunately, treatment of thyroid cancer has a high success rate, and as of 2002 only 3 of the 1,036 children cited in our figure had died of thyroid cancer. By comparing the Chernobyl-region data to background rates elsewhere, researchers calculated that Ukraine would eventually suffer 300 thyroid cancer cases more than normal and that the nearby region of Russia (with a population of 4.3 million) would suffer 349 extra cases, a 3–6% increase above the normal rate.

Critics pointed out that any targeted search tends to turn up more of whatever medical problem is being looked for. But the magnitude of the increase in childhood thyroid cancer was large enough that most experts judge it to be real. The rise in thyroid cancer, the NEA concluded, "should be attributed to the Chernobyl accident until proven otherwise."

Thyroid cancer also appears to have risen in adults. Adult cases in Belarus in the 12 years before the accident totaled 1,392, but in the 12 years after Chernobyl totaled 5,449. In the most contaminated regions of Russia, thyroid cancer incidence rose to 11 per 100,000 women and 1.7 per 100,000 men, compared to normal rates of 4 and 1.1 for Russia as a whole. And although rates of childhood cancer may now be falling, rates for adults are still rising. As new cancer cases accumulate in the future, continued research will be needed to measure the full scope of health effects from Chernobyl.

days, exposed to radiation, before the Soviet government belatedly began evacuating more than 100,000 people. In the months and years afterwards, workers erected a gigantic concrete sarcophagus around the demolished reactor, scrubbed buildings and roads, and removed irradiated materials (Figure 13.18). However, the landscape for at least 30 km (19 mi) around the plant remains contaminated today, and an international team plans to build a larger sarcophagus around the original one, which is deteriorating.

The accident at Chernobyl killed 31 people directly and sickened or caused cancer in many more (see "The Science behind the Story," above). Atmospheric currents

(a) The Chernobyl sarcophagus

(b) Technicians measuring radiation

FIGURE 13.18 The world's worst nuclear power plant accident unfolded in 1986 at Chernobyl, in present-day Ukraine (then part of the Soviet Union). As part of the extensive cleanup operation, the destroyed reactor was encased in a massive concrete sarcophagus **(a)** to contain further radiation leakage. Technicians scoured the landscape surrounding the plant **(b)**, measuring radiation levels, removing soil, and scrubbing roads and buildings.

carried radioactive fallout across much of the Northern Hemisphere, particularly Ukraine, Belarus, and parts of Russia and Europe.

Nuclear waste disposal remains a problem

Even if nuclear power generation could be made completely safe, we would still be left with the conundrum of what to do with spent fuel rods and other radioactive waste, which will continue emitting radiation for thousands of years. Currently, such waste is held in temporary storage at nuclear power plants across the world. Spent fuel rods are sunken in pools of cooling water or encased in thick casks of steel, lead, and concrete to minimize escape of radiation.

In total, U.S. power plants are storing over 49,000 metric tons of radioactive waste, enough to fill a football field to the depth of 3.3 m (10 ft). This waste is held at 125 sites spread across 39 states (Figure 13.19). The U.S. Department of Energy (DOE) estimates that over 161 million U.S. citizens live within 125 km (75 mi) of temporarily stored waste, and a 2005 National Academy of Sciences report judged that most of these sites were vulnerable to terrorist attacks.

Because storing waste at many dispersed sites creates a large number of potential hazards, nuclear waste managers have long wanted to send all waste to a central repository that can be heavily guarded. In the United States, the search homed in on Yucca Mountain, a remote site in the desert of southern Nevada, 160 km (100 mi) from Las Vegas. If given final approval, Yucca Mountain is expected to begin receiv-

ing waste from nuclear plants and military installations in 2010. Waste would be stored in a network of tunnels 300 m (1,000 ft) underground, yet 300 m (1,000 ft) above the water table (Figure 13.20). Scientists and policymakers chose Yucca Mountain because they determined that it is remote and unpopulated, has minimal chance of earthquakes, receives little rain that could contaminate groundwater with radioactivity, has a deep water table atop an isolated aquifer, and is on federal land that can be protected from sabotage. However, some scientists, antinuclear activists, and concerned Nevadans have challenged these conclusions.

Waste would need to be transported to Yucca Mountain from the 125 current storage areas and on a regular basis from all current and future nuclear plants and military installations. Because this would involve many thousands of shipments by rail and truck across hundreds of public highways through almost every state of the union, many people worry that the risk of accident or sabotage is unacceptably high.

Weighing the Issues:
How to Store Waste?

Which do you think is a better option—to transport nuclear waste cross-country to a single repository or to store it permanently at numerous power plants and military bases scattered across the nation? Would your opinion be affected if you lived near the repository site? Near a power plant? On a highway route along which waste was transported?

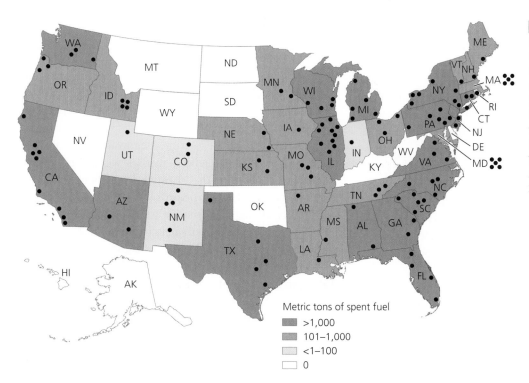

FIGURE 13.19 Nuclear waste from civilian reactors is currently stored at 125 sites in 39 states across the United States. In this map, dots indicate each storage site, and the four shades of color indicate the total amount of waste stored in each state. Data from the Office of Civilian Radioactive Waste Management, U.S. Department of Energy; and from the Nuclear Energy Institute, Washington, D.C.

Metric tons of spent fuel

- >1,000
- 101–1,000
- <1–100
- 0

Multiple dilemmas have slowed nuclear power's growth

Dogged by concerns over waste disposal, safety, and expensive cost overruns, nuclear power's growth has slowed. Since the late 1980s, nuclear power worldwide has grown by 2.5% per year, about the same rate as electricity generation overall. Public anxiety in the wake of Chernobyl made utilities less willing to invest in new plants. So did the enormous expense of building, maintaining, operating, and ensuring the safety of nuclear facilities. Almost every nuclear plant has turned out to be more expensive than expected. In addition, plants have aged more quickly than expected because of problems that were underestimated, such as corrosion in coolant pipes. The plants that have been shut down—well over 100 around the world to date—have served on average less than half their expected lifetimes. Moreover,

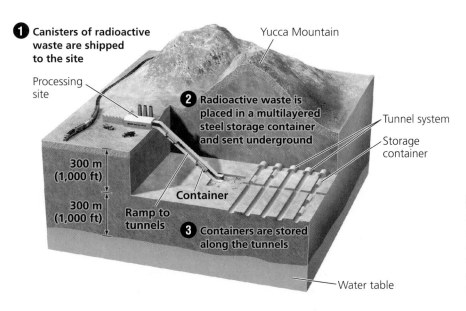

1 Canisters of radioactive waste are shipped to the site

Processing site

2 Radioactive waste is placed in a multilayered steel storage container and sent underground

Yucca Mountain

Tunnel system

Storage container

300 m (1,000 ft)

300 m (1,000 ft)

Container

Ramp to tunnels

3 Containers are stored along the tunnels

Water table

FIGURE 13.20 Yucca Mountain, in a remote part of Nevada, awaits final approval as the central repository site for all the nuclear waste in the United States. Waste would be buried in a network of tunnels deep underground yet still high above the water table.

shutting down, or decommissioning, a plant can sometimes be more expensive than the original construction. As a result of these economic issues, electricity from nuclear power today remains more expensive than electricity from coal and other sources. Governments are still subsidizing nuclear power to keep consumer costs down, but many private investors lost interest long ago.

Nonetheless, nuclear power remains one of the few currently viable alternatives to fossil fuels with which we can generate large amounts of electricity in short order. This is why both the Bush administration and an increasing number of environmental advocates propose expanding U.S. nuclear capacity using a new generation of reactors designed to be safer and less expensive.

With slow growth predicted for nuclear power, fossil fuels in limited supply, an oil production peak looming, and climate change worsening, where will our growing human population turn for clean and sustainable energy? As we will see in our next chapter, people increasingly are turning to renewable sources of energy: energy sources that cannot be depleted by our use.

Conclusion

Over the past 200 years, fossil fuels have helped us build complex industrialized societies capable of exploring (and exploiting) all parts of the world, and even venturing beyond our planet. Today, however, we are approaching a turning point in history: The availability of fossil fuels will begin to decline, just as we are becoming increasingly aware of the negative impacts of their use. Nuclear power showed promise to be a pollution-free and highly efficient alternative form of energy, but high costs and public fears over safety in the wake of accidents at Chernobyl and Three Mile Island stalled its growth.

We can respond to this new challenge in creative ways, particularly by encouraging energy conservation and developing renewable energy sources. Or we can continue our current dependence on fossil fuels and wait until they near depletion before we try to develop new technologies and ways of life. The path we choose will have far-reaching consequences for our environment and our civilization.

TESTING YOUR COMPREHENSION

1. Why are fossil fuels our most prevalent source of energy today? Why are they considered nonrenewable sources of energy?

2. How are fossil fuels formed? Why are they often concentrated in localized deposits?

3. Describe how coal is used to generate electricity.

4. How do we create petroleum-based products? Look around you, and at what you are wearing and carrying with you, and provide examples of several such products you use in your everyday life.

5. Explain why many scientists and oil experts are predicting that global oil production will soon begin to decline. What could be the social and economic consequences of such a decline?

6. Describe at least two major impacts of fossil fuel emissions. Describe at least one major impact of fossil fuel reliance for national economies or governments. Compare some of the contrasting views regarding the environmental impacts of drilling for oil in ANWR.

7. Describe two main approaches to energy conservation, and give a specific example of each. Name one barrier or disincentive to energy conservation.

8. Describe how nuclear fission works. How do nuclear plant engineers control fission and prevent a runaway chain reaction?

9. In terms of greenhouse gas emissions, how does nuclear power compare to coal, oil, and natural gas?

10. In what ways did the incident at Three Mile Island differ from that at Chernobyl? What consequences resulted from each of these incidents?

SEEKING SOLUTIONS

1. Imagine we were living in the 1950s and the United States were facing an oil shortage. Do you think the nation would be debating whether to drill in ANWR? Why do you think so many people today are concerned about wildlife and about wilderness preservation? Now fast forward to the 2020s. Do you think the nation will still be debating whether to drill in ANWR? Why or why not?

2. In February 2006, noted petroleum geologist Kenneth Deffeyes analyzed oil production data and calculated that the world had reached its peak of oil production on December 16, 2005. Let's suppose he is correct. What do you think our policymakers, businesses, and consumers should do in response? Now let's suppose that instead, the oil peak will not come until the year 2020. How do you think our society should react in response to this information?

3. If the United States and other developed countries relinquished dependence on foreign oil and on fossil fuels in general, do you think that their economies would benefit or suffer? Imagine that oil became 50% less available than it is today and that prices rose four times as high as today's. What sorts of incentives would these events create for the development of alternative energy sources and for investments in conservation and efficiency?

4. Nuclear power has by now been widely used for over three decades, and the world has experienced only one major accident (Chernobyl) responsible for any significant number of deaths. Would you call this a good safety record? Do you think we should maintain, decrease, or increase our reliance on nuclear power? Explain the reasons for your answer.

5. Imagine that you are the head of the national department of energy in a country that has just experienced a minor accident at one of its nuclear plants. A partial meltdown released radiation, but the radiation was fully contained inside the containment building, and there were no health impacts on area residents. However, citizens are terrified, and the media is playing up the dangers of nuclear power. Your country relies on its 10 nuclear plants for 25% of its energy and 50% of its electricity needs. It has no fossil fuel deposits and recently began a promising but still-young program to develop renewable energy options. What will you tell the public at your next press conference, and what policy steps will you recommend taking to assure a safe and reliable national energy supply?

CALCULATING ECOLOGICAL FOOTPRINTS

 Wackernagel and Rees calculated the energy component of our ecological footprint by estimating the amount of ecologically productive land required to absorb the carbon released from fossil fuel combustion (see Chapter 12, "Calculating Ecological Footprints," ▶ p. 299). For the average American, this translates into 2.9 ha of their 5.1-ha ecological footprint. Another way to think about our footprint, however, is to estimate how much land would be needed to grow biomass with an energy content equal to that of the fossil fuel we burn. Assume that you are an average American who burns 287 gigajoules of fossil fuels per year and that average terrestrial net primary productivity (▶ pp. 61–62) can be expressed as 160 megajoules/ha/year. Calculate how many hectares of land it would take to supply our fuel use by present-day photosynthetic production. A gigajoule is 10^9 joules; a megajoule is 10^6 joules.

	Hectares of land for fuel production
You	1,794
Your class	
Your state	
United States	

Data from Wackernagel, M., and W. Rees. 1996. *Our ecological footprint: Reducing human impact on the Earth.* Gabriola Island, British Columbia: New Society Publishers.

1. Compare the energy component of your ecological footprint calculated in this way with the 2.9 ha calculated using the method of Wackernagel and Rees. Explain how and why results from the two methods differ.

2. Earth's total land area is approximately 1.5×10^{10} ha. Compare this to the hectares of land for fuel production from the table.

3. How large a human population could Earth support at the level of consumption of the average American, if all of Earth's land were devoted to fuel production? Do you consider this realistic? Provide two reasons why or why not.

Take It Further

Go to www.aw-bc.com/withgott, where you'll find:

▶ Suggested answers to end-of-chapter questions
▶ Quizzes, animations, and flashcards to help you study

▶ *Research Navigator*™ database of credible and reliable sources to assist you with your research projects
▶ **GRAPHit!** tutorials to help you interpret graphs
▶ **INVESTIGATEit!** current news articles that link the topics that you study to case studies from your region to around the world

Renewable Energy Alternatives

Iceland's Blue Lagoon and Svartsengi geothermal power plant

Upon successfully completing this chapter, you will be able to:

▶ Discuss the reasons for seeking alternatives to fossil fuels

▶ Outline the major sources of renewable energy and assess their potential for growth

▶ Describe the major sources of biomass energy and discuss their benefits and drawbacks

▶ Describe the scale, methods, and impacts of hydroelectric power

▶ Describe solar energy and evaluate its advantages and disadvantages

▶ Describe wind energy and evaluate its advantages and disadvantages

▶ Describe geothermal energy and evaluate its advantages and disadvantages

▶ Describe ocean energy sources and the potential ways they can be harnessed

▶ Explain hydrogen fuel cells and assess future options for energy storage and transportation

Central Case: Iceland Moves toward a Hydrogen Economy

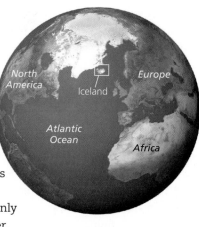

"I believe that water will one day be employed as fuel, that hydrogen and oxygen which constitute it, used singly or together, will furnish an inexhaustible source of heat and light. . . . Water will be the coal of the future."
—JULES VERNE, IN THE MYSTERIOUS ISLAND, 1874

"Our long-term vision is of a hydrogen economy."
ROBERT PURCELL, JR., EXECUTIVE DIRECTOR, GENERAL MOTORS, 2000

The Viking explorers who first set foot centuries ago on the remote island of Iceland were trailblazers. Today the citizens of the nation of Iceland are blazing a bold new path. Iceland aims to become the first nation to leave fossil fuels behind and convert to an economy based completely on renewable energy.

Iceland is essentially a hunk of lava the size of Kentucky that has risen out of the North Atlantic from the rift between tectonic plates known as the Mid-Atlantic Ridge (▶ pp. 71–72). Most of Iceland's 290,000 people live in the capital city of Reykjavik, leaving much of the island an unpeopled and starkly beautiful landscape of volcanoes, hot springs, and glaciers.

The magma that gave birth to the island heats its groundwater in many places, giving Iceland some of the world's best sources of geothermal energy. Iceland has also dammed some of its many rivers for hydropower. Together these two renewable energy sources provide 73% of the country's energy supply and virtually all its electricity generation.

Yet the nation, like most others, also depends on fossil fuels. Oil powers its automobiles, some of its factories, and its economically vital fishing fleet—and together these have given Iceland one of the highest per capita rates of greenhouse gas emission in the world. Because it possesses no fossil fuel reserves, all fossil fuels must be imported—a weak link in an otherwise robust economy that has given its citizens one of the highest per capita incomes in the world.

Enter Bragi Árnason, a University of Iceland professor popularly known as "Dr. Hydrogen." Árnason began arguing in the 1970s that Iceland could achieve independence from fossil fuel imports, boost its economy, and serve as a model to the world by converting from fossil fuels to a renewable energy economy based on hydrogen. He suggested zapping water with Iceland's cheap and renewable electricity in a chemical reaction called electrolysis, splitting the hydrogen from the oxygen and then using the hydrogen to power fuel cells that would produce and store electricity. The process is clean; nothing is combusted, and the only waste product is water.

In the late 1990s Árnason's countrymen began to listen. The nation's leaders decided to embark on a grand experiment to test the efficacy of switching to a "hydrogen economy." By setting an example for the rest of the world to follow, and by getting a head start at producing and exporting hydrogen fuel, these leaders hoped Iceland could become a "Kuwait of the North" and get rich by exporting hydrogen to an energy-hungry world, as Kuwait has done by exporting oil.

The leaders planning the shift to a hydrogen economy sketched a stepped transition in which fossil fuels would be phased out over 30–50 years. Conversion of the Reykjavik bus fleet to run on hydrogen fuel is the first step. After that, Iceland's 180,000 private cars would be powered by fuel cells, and then the fishing fleet would be converted to hydrogen. The final stage would be the export of hydrogen fuel to mainland Europe.

To make this happen, Icelanders in 1999 teamed up with corporate partners looking to develop technology for the future. Auto company Daimler-Chrysler is producing hydrogen buses, oil company Royal Dutch Shell is running hydrogen filling stations to fuel the buses, and Norsk Hydro is providing electrolysis technology.

In 2003, the world's first commercial hydrogen filling station opened in Reykjavik, and three hydrogen-fueled buses began operation. The public-private consortium, *Icelandic New Energy (INE)*, is monitoring the technology's effectiveness and the costs of developing infrastructure. Iceland's citizens are behind the effort; a recent poll showed 93% support among Icelanders.

Meanwhile, Daimler-Chrysler has introduced trios of hydrogen-fueled buses to nine other European cities. Hydrogen buses are also being developed by other companies and run in cities in Europe and throughout the world, from Tokyo to Chicago to Perth to Winnipeg. Hydrogen refueling stations are being demonstrated in Japan, Singapore, and the United States, and fuel-cell passenger automobiles are being tested in Japan and California. A global hydrogen economy could be closer than we suspect.

Renewable Energy Sources

Iceland's bold drive toward a hydrogen economy is one facet of a global move toward renewable energy. Across the world, nations are searching for ways to move away from fossil fuels while ensuring a continued supply of energy for their economies.

Fossil fuels helped to drive the industrial revolution and increase our material prosperity. However, these nonrenewable energy sources will not last forever. As we saw in Chapter 13, oil production is thought to be peaking, and easily extractable supplies of oil and natural gas may not last half a century more. Moreover, our use of coal, oil, and natural gas entails substantial environmental impacts (Chapters 12 and 13). For these reasons, most scientists and energy experts, as well as many economists and policymakers, accept that the world's economies will need to shift from fossil fuels to energy sources that are less easily depleted and gentler on our environment.

People have developed a range of alternatives to fossil fuels. The main nonrenewable alternative is nuclear energy (▶ pp. 319–326). Renewable alternatives include biomass, hydropower, solar, wind, geothermal, and ocean energy sources. Biomass and hydropower are well-established and widely used sources. The remainder are often termed "new renewables" because they are less used and are harnessed using technologies that are still in a rapid phase of development.

Renewable energy sources promise several benefits over fossil fuels. As they replace fossil fuels, they help alleviate air pollution and the greenhouse gas emissions that drive global climate change (Chapter 12). Unlike fossil fuels, many renewable sources are inexhaustible on time scales relevant to human societies. Developing renewables can also help diversify an economy's mix of energy, lowering price volatility and protecting against supply restrictions such as those caused by the 1973 oil embargo or by Hurricane Katrina in 2005 (▶ pp. 314, 316).

Renewable sources provide little of our energy, but are growing fast

Today's economies are powered largely by fossil fuels; 80% of all primary energy comes from oil, coal, and natural gas (Figure 14.1a). These three fuels also power two-thirds of the world's electricity generation (Figure 14.1b).

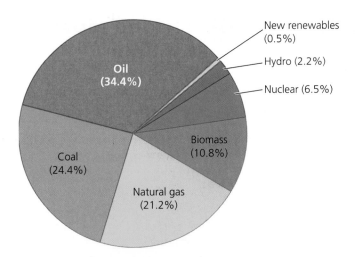

(a) World total primary energy supply

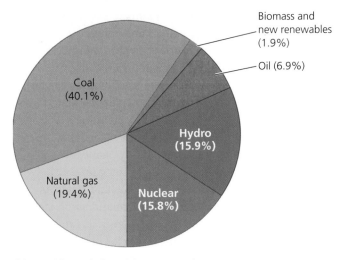

(b) World total electricity generation

FIGURE 14.1 Fossil fuels account for 80% of the world's total supply of primary energy. Fuelwood and other biomass sources provide 10.8%, hydropower provides 2.2%, and the "new renewable" energy sources account for only 0.5% (**a**). Fossil fuels provide two-thirds of global electricity. Less than 18% of our electricity comes from renewable sources, and of this amount, hydropower accounts for 90% (**b**). Data from International Energy Agency. 2005. *Key world energy statistics 2005.* Paris: IEA.

Nations and regions vary in the renewable sources they use. Developing nations account for most use of combustible renewables, or biomass, such as fuelwood. In the United States, most energy supplied by renewable sources comes from hydropower and biomass, in nearly equal proportions (Figure 14.2a). Of electricity generated in the United States from renewables, hydropower accounts for nearly 75% (Figure 14.2b).

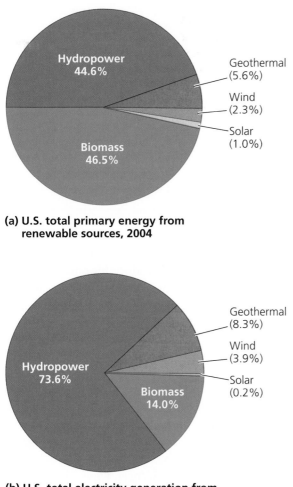

(a) U.S. total primary energy from renewable sources, 2004

(b) U.S. total electricity generation from renewable sources, 2004

FIGURE 14.2 Only 6% of the total primary energy consumed in the United States each year comes from renewable sources. Of this amount, most derives from hydropower and biomass energy. Geothermal energy accounts for 5.6% of this amount, wind for 2.3%, and solar for 1.0% (**a**). Similarly, only 9% of electricity generated in the United States comes from renewable energy sources, predominantly hydropower and biomass. Geothermal energy accounts for 8.3%, wind for 3.9%, and solar for only 0.2% (**b**). Data from Energy Information Administration, U.S. Department of Energy. 2005. *Annual Energy Review, 2004.*

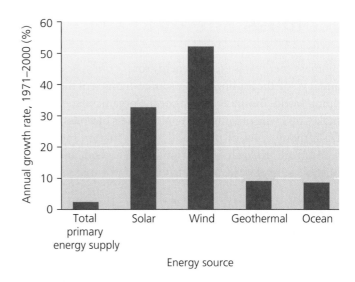

FIGURE 14.3 The "new renewable" energy sources are growing substantially faster than the total primary energy supply. Solar power has grown by 32% each year since 1971, and wind power has grown by 52% each year. Because these sources began from such low starting levels, however, their overall contribution to our energy supply is still small. Go to **GRAPHit!** at www.aw-bc.com/ withgott. Data from International Energy Agency Statistics, 2002.

Although they comprise only a minuscule proportion of our energy budget, the new renewable energy sources are growing at much faster rates than are conventional energy sources. Over the past three decades, solar, wind, geothermal, and ocean energy sources have grown far faster than has the overall primary energy supply (Figure 14.3). Among renewable sources, the leader in growth is wind power, which has expanded by about 50% *each year* over the past three decades. Because these sources started from such low levels of use, however, it will take them some time to catch up to conventional sources. For instance, the absolute amount of energy added by a 50% increase in wind power is still less than the amount added by just a 1% increase in oil, coal, or natural gas.

Rapid growth in renewable energy sectors seems likely to continue as population and consumption grow, global energy demand expands, fossil fuel supplies decline, technology advances, and citizens demand cleaner environments. Although most renewables currently lack adequate technological development and lack infrastructure to transfer power on the required scale, engineers have made dramatic improvements in technology and infrastructure in recent years. Most remaining barriers are political; renewable energy sources have received far less in subsidies, tax breaks, and other incentives from governments than have conventional sources.

If our civilization is to persist in the long term, we will need to shift to renewable energy sources. A key question is whether we will be able to shift soon enough and smoothly enough to avoid widespread war, social unrest, and further damage to the environment. The answer to this question will largely determine the quality of life for all of us in the coming decades.

Biomass Energy

Biomass energy was the first energy source our species used, and it is still the leading energy source in much of the developing world. **Biomass** consists of the organic material that makes up living organisms. People harness **biomass energy** from many types of plant and animal matter, including wood from trees, charcoal from burned wood, and combustible animal waste products such as cattle manure.

Biomass energy comes from diverse sources

Over 1 billion people still use wood from trees as their principal energy source. In developing nations, especially in rural areas, families gather fuelwood to burn in or near their homes for heating, cooking, and lighting (Figure 14.4). In these nations, fuelwood, charcoal, and manure account for fully 35% of energy use—in the poorest nations, up to 90%.

Considering the global problem of deforestation (▸ pp. 206–208), it is clear that biomass is renewable only if it is not overharvested. At moderate rates of use, trees and other plants can replenish themselves over months to decades. However, when forests are cut too quickly, or when overharvesting leads to soil erosion and forests fail to grow back, then biomass is not effectively replenished.

Besides the fuelwood, charcoal, and manure traditionally used, biomass energy sources in today's world include a number of sources for which innovative uses have

Table 14.1 Major Sources of Biomass Energy
▸ Wood cut from trees (fuelwood)
▸ Charcoal
▸ Manure from domestic animals
▸ Crops grown specifically for biomass energy production
▸ Crop residues (such as corn stalks)
▸ Forestry residues (such as wood waste from logging)
▸ Processing wastes (such as solid or liquid waste from sawmills, pulp mills, and paper mills)
▸ Components of municipal solid waste

recently been developed (Table 14.1). Some of these sources can be burned to produce **biopower**, generating heat or electricity. Other new biomass sources can be converted into fuels used primarily to power automobiles; these are termed **biofuels**.

Biofuels can power automobiles

Liquid fuels from biomass sources are helping to power millions of vehicles on today's roads, and some vehicles can run entirely on biofuels. The two primary types of such fuels developed so far are ethanol (for gasoline engines) and biodiesel (for diesel engines).

Ethanol is the alcohol in beer, wine, and liquor. It is produced as a biofuel by fermenting biomass, generally from carbohydrate-rich crops, such as corn, in a process similar to brewing beer. In fermentation, carbohydrates

FIGURE 14.4 Hundreds of millions of people in the developing world rely on fuelwood for heating and cooking. Wood cut from trees remains the major source of biomass energy used today. In theory biomass is renewable, but in practice it may not be renewable if forests are overharvested.

are converted to sugars and then to ethanol. Spurred by the 1990 Clean Air Act amendments and generous government subsidies, ethanol is widely added to gasoline in the United States to reduce automotive emissions. In 2005 in the United States, nearly 15 billion L (3.9 billion gal) of ethanol were produced from corn at nearly 100 facilities. Each of the U.S. "big three" automakers is now producing *flexible fuel vehicles* that run on E-85, a mix of 85% ethanol and 15% gasoline. In Brazil, half of all new cars are flexible-fuel cars, and ethanol from sugarcane accounts for 44% of all automotive fuel used.

Biodiesel is produced from vegetable oil, used cooking grease, or animal fat. The oil or fat is mixed with small amounts of ethanol or methanol (wood alcohol) in the presence of a chemical catalyst. Vehicles with diesel engines can run on 100% biodiesel, or biodiesel can be mixed with conventional petrodiesel; a 20% biodiesel mix (called B20) is common today.

Biodiesel cuts down on emissions compared with petrodiesel. Its fuel economy is almost as good, and it costs just 10–20% more. It is also nontoxic and biodegradable. Increasing numbers of environmentally conscious individuals in North America and Europe are fueling their cars with biodiesel from waste oils, and some buses and recycling trucks are now running on biodiesel. Governments are encouraging its use, too; Minnesota, for instance, mandates that all diesel sold must include a 2% biodiesel component, and already over 40 state and federal fleets are using biodiesel blends.

Some enthusiasts eliminate the processing step that biodiesel requires and instead use straight vegetable oil in their diesel engines. One notable effort is the "B.I.O. Bus," a bus fueled entirely on waste oil from restaurants that a group of students, environmentalists, and artists drives across North America (Figure 14.5). Each summer the group goes on tour with the bus, hosting festive events that combine music and dancing with seminars on environmental sustainability, and spreading the word about nonpetroleum fuels. In the summers of 2003, 2004, and 2005, the bus traveled 25,000 miles.

Biopower generates electricity from biomass

We can harness biopower in various ways. The most common strategy is to burn biomass in the presence of large amounts of air. This can be done on small scales, with furnaces and stoves to produce heat for domestic needs, or on large industrial scales to produce district heating and to generate electricity. Power plants built to combust biomass operate in a similar way to those fired by fossil fuels (▶ Figure 13.6, p. 307); the combustion heats water, creating steam to turn turbines and generators, thereby generating electricity. Many of these plants generate both electricity and heating through cogeneration (▶ p. 318) and are often located where they can take advantage of waste material from the forest products industry.

Biomass is also increasingly being combined with coal in existing coal-fired power plants in a process called *co-firing*. Up to 15% of the coal can be substituted with biomass, with only minor equipment modification and no appreciable loss of efficiency. Co-firing can be a relatively

FIGURE 14.5 Each summer a group of alternative-fuel advocates goes on tour in the B.I.O. Bus, a bus fueled entirely on used vegetable oil from restaurants. Their "Bio Tours" sponsor events across North America that include music, dancing, and seminars on environmental sustainability and alternative fuels. Their motto: "Solar-powered sound, veggie-powered bus."

easy and inexpensive way for fossil-fuel-based utilities to expand their use of renewable energy.

The anaerobic bacterial breakdown of waste in landfills produces methane and other components, and this "landfill gas" is now being captured at many solid waste landfills and sold as fuel (▸ p. 360). Methane and other gases can also be produced in a more controlled way in anaerobic digestion facilities. This "biogas" can then be burned in a power plant's boiler to generate electricity.

The process of *gasification* can also provide biopower, as well as biofuels. In this process, biomass is vaporized at extremely high temperatures in the absence of oxygen, creating a gaseous mixture including hydrogen, carbon monoxide, and methane. This mixture can generate electricity when used in power plants to turn a gas turbine to propel a generator. Gas from gasification can also be treated in various ways to produce methanol, synthesize a type of diesel fuel, or isolate hydrogen for use in hydrogen fuel cells (▸ pp. 347–348, 351). An alternative method of heating biomass in the absence of oxygen results in *pyrolysis*, which produces a mix of solids, gases, and liquids. This includes a liquid fuel called pyrolysis oil, which can be burned to generate electricity.

Biomass energy brings a mix of benefits and drawbacks

Biomass energy has one overarching environmental benefit: It is essentially carbon-neutral, releasing no net carbon into the atmosphere. Although burning biomass emits plenty of carbon, the carbon released is simply the carbon that was pulled from the atmosphere by photosynthesis to create the biomass being burned. That is, because biomass is the product of recent photosynthesis, the carbon released in its combustion is balanced by the carbon taken up in the photosynthesis that created it. Therefore, when biomass sources take the place of fossil fuels, net carbon flux to the atmosphere is reduced. However, this holds only if biomass sources are not overharvested. Deforestation will increase carbon flux to the atmosphere because less vegetation means less carbon uptake by plants for photosynthesis.

Shifting from fossil fuels to biomass energy also can have economic benefits. As a resource, biomass tends to be well spread geographically, so using it should help support rural economies and reduce many nations' dependence on imported fuels. Biomass also tends to be inexpensive for burning in power plants, and improved energy efficiency brings lower prices for consumers. By increasing energy efficiency and recycling waste products,

use of biomass energy helps move our industrial systems toward greater sustainability. The U.S. forest products industry now obtains over half its energy by combusting the biomass waste it recycles, including woody waste and liquor from pulp mill processing.

Relative to fossil fuels, biomass also benefits human health. By replacing coal in co-firing and direct combustion, biomass reduces emissions of nitrogen oxides and particularly sulfur dioxide. By replacing gasoline and petrodiesel and burning more cleanly, biofuels reduce emissions of various air pollutants.

However, burning fuelwood and other biomass in traditional ways for cooking and heating leads to health hazards from indoor air pollution (▸ pp. 285–287). In addition, harvesting fuelwood at an unsustainably rapid rate leads to deforestation, soil erosion, and desertification, damaging landscapes, diminishing biodiversity, and impoverishing human societies. In arid regions that are heavily populated and support meager woodlands, fuelwood harvesting can have enormous impacts.

Another drawback of biomass energy is that growing crops specifically to produce biofuels establishes monoculture agriculture, with all its impacts (▸ pp. 152–153, 160), on precious land that might otherwise be used to grow food, be developed for other purposes, or be left as wildlife habitat. Fully 10% of the U.S. corn crop is used to make ethanol, and most U.S. biodiesel is produced from oil from soybeans grown specifically for this purpose.

Growing bioenergy crops also requires substantial inputs of energy. We currently operate farm equipment using oil, and apply petroleum-based pesticides and fertilizers to increase yields. Moreover, growing corn for ethanol yields only small amounts of ethanol per acre of crop grown; 1 bushel of corn creates only 9.4 L (2.5 gal) of ethanol. It is not efficient to grow high-input, high-energy crops merely to reduce them to a few liters of fuel. Indeed, even using ethanol as a gasoline additive to reduce emissions has a downside; it lowers fuel economy very slightly, so that more gasoline needs to be used.

On the positive side, crops grown for energy—fast-growing grasses such as bamboo, fescue, and switchgrass; grain and oil-producing crops such as corn and soybeans; and fast-growing trees such as poplar, willow, and cottonwood—typically receive lower inputs of pesticides and fertilizers than those grown for food. Researchers are refining techniques to produce ethanol from the cellulose that gives structure to plant material, and not just from starchy crops like corn. If these techniques can be made widely feasible, then ethanol could be produced primarily from low-quality waste and crop residues, rather than from high-quality crops.

- -

Weighing the ISSUES:
Ethanol

Do you think producing and using ethanol from corn or other crops is a good idea? Do the benefits outweigh the drawbacks? Can you suggest ways of using biofuels that would minimize environmental impacts?

- -

Hydroelectric Power

Next to biomass, we draw more energy from the motion of water than from any other renewable resource. In **hydroelectric power**, or **hydropower**, we use the kinetic energy of moving water to turn turbines and generate electricity. We examined hydropower and its environmental impacts in our discussion of freshwater resources in Chapter 11 (▸ pp. 245–248), but we will now take a closer look at hydropower as an energy source.

Modern hydropower uses two approaches

Most of our hydroelectric power today comes from impounding water in reservoirs behind concrete dams that block the flow of river water, and then letting that water pass through the dam. Because immense amounts of water are stored behind dams, this is called the *storage* technique. As reservoir water passes through a dam, it turns the blades of turbines, which cause a generator to generate electricity (Figure 14.6). Electricity generated in the powerhouse of a dam is transmitted to the electric grid by transmission lines, while the water is allowed to flow into the riverbed below the dam to continue down-river. By storing water in reservoirs, dam operators can ensure a steady and predictable supply of electricity, even during times of naturally low river flow.

An alternative approach is the *run-of-river* approach, in which we generate electricity without greatly disrupting the flow of river water. This approach sacrifices the reliability of water flow across seasons that the storage approach guarantees, but it minimizes many of the environmental impacts of large dams. The run-of-river approach can be followed using several methods, one of which is to divert a portion of a river's flow through a pipe or channel, pass it

through a powerhouse, and then return it to the river. Run-of-river systems are particularly useful in areas remote from electrical grids and in regions without the economic resources to build and maintain large dams.

Hydropower is clean and renewable, but also has negative impacts

For producing electricity, hydropower has two clear advantages over fossil fuels. First, it is renewable; as long as precipitation falls from the sky and fills rivers and reservoirs, we can use water to turn turbines. When pressure from human population or consumption is intense enough, water supplies may be used faster than they are replenished, and hydropower may not be sustainable. However, in most cases hydropower continues today to provide the renewable energy that dams were built to provide.

The second advantage of hydropower over fossil fuels is its cleanliness. Because no carbon compounds are burned in the production of hydropower, no carbon dioxide or other pollutants are emitted into the atmosphere. Of course, fossil fuels are used in the construction and maintenance of dams. Moreover, recent evidence indicates that large reservoirs may release the greenhouse gas methane as a result of anaerobic decay in deep water. But overall, hydropower is thought to release only a fraction of the greenhouse gas emissions typical of fossil fuel combustion.

Although it is renewable and produces little air pollution, hydropower does create other environmental impacts. These include altering riverine habitat above and below dam sites, disrupting natural flooding cycles and sediment deposition, causing thermal pollution of downstream waters, and reducing fish populations and aquatic biodiversity. These ecological impacts generally translate into negative social and economic consequences for local communities. We discussed the environmental, economic, and social impacts of dams, and their advantages and disadvantages, more fully in Chapter 11 (▸ pp. 245–248).

Hydropower is widely used, but it may not expand much more

Hydropower accounts for 2.2% of the world's primary energy supply and 16.2% of the world's electricity production. For nations with large amounts of river water and the economic resources to build dams, hydroelectric power has been a keystone of their development and wealth. Canada, Brazil, Norway, Austria, Switzerland, and

(a) Ice Harbor Dam, Washington

(b) Turbine generator inside MacNary Dam, Columbia River

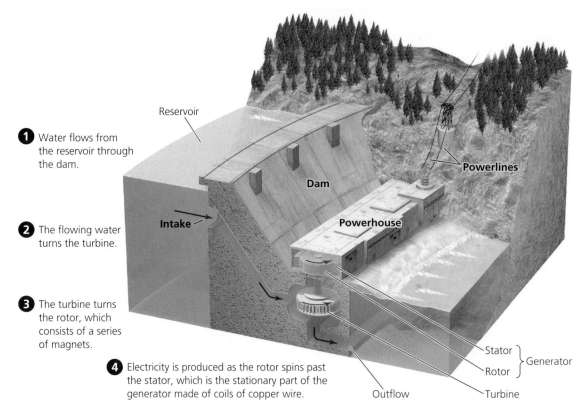

1 Water flows from the reservoir through the dam.

2 The flowing water turns the turbine.

3 The turbine turns the rotor, which consists of a series of magnets.

4 Electricity is produced as the rotor spins past the stator, which is the stationary part of the generator made of coils of copper wire.

Reservoir

Powerlines

Dam

Intake

Powerhouse

Stator
Rotor } Generator

Outflow

Turbine

(c) Hydroelectric power

FIGURE 14.6 Large dams, such as the Ice Harbor Dam on the Snake River in Washington (**a**), generate substantial amounts of hydroelectric power. Inside these dams, flowing water is used to turn turbines (**b**) and generate electricity. Water is funneled from the reservoir through a portion of the dam (**c**) to rotate turbines, which turn rotors containing magnets. The spinning rotors generate electricity as their magnets pass coils of copper wire. Electrical current is transmitted away through power lines, and the river's water flows out through the base of the dam.

many other nations today obtain large amounts of their energy from hydropower (Table 14.2). Iceland developed hydropower early on and is planning to use it as a primary source of electricity for powering its production of hydrogen.

Use of hydropower is growing slightly more quickly than overall energy use, and some gargantuan projects such as China's Three Gorges Dam (▸ pp. 246–247) are being planned and carried out. However, unlike other renewable energy sources, hydropower is not likely to

Table 14.2 Top Consumers of Hydropower

Nation	Hydropower consumed[1]	Percentage of electricity generation from hydropower[2]
China	90.8	14.9
Canada	81.7	57.5
Brazil	77.0	83.8
United States	60.6	7.5
Russia	39.6	17.2
Norway	30.9	98.9
India	21.7	11.9
Japan	19.8	9.9
Venezuela	17.6	66.0
Sweden	15.5	39.3

Data from British Petroleum and International Energy Agency.
[1]In million metric tons of oil equivalent, 2005 data.
[2]2003 data.

expand very much more. One reason is that most of the world's large rivers that offer excellent opportunities for hydropower have already been dammed. Another reason is that people's awareness of the ecological impacts of dams has grown, and in some nations residents are fighting dam construction and proposing to dismantle some dams and restore free-flowing rivers (▶ pp. 246–248). Iceland will need considerably more electricity than it currently generates to power its hydrogen economy, but many Icelanders are opposed to building new dams and flooding scenic areas. Although plans for new dams are mostly going ahead, some Icelanders have instead urged that the nation obtain additional electricity from new geothermal and wind sources. In the United States, 98% of rivers appropriate for dam construction already are dammed, and many of the remaining 2% are protected under the Wild and Scenic Rivers Act.

Solar Energy

The sun provides energy for almost all biological activity on Earth (▶ pp. 59, 273). Each square meter of our planet's surface receives about 1 kilowatt of solar energy—17 times the energy of a lightbulb. The amount of energy Earth receives from the sun each day, if it could be collected in full for our use, would be enough to power human consumption for 27 years.

Clearly, the potential for using sunlight to meet our energy needs is tremendous. However, all this "free" energy from the sun cannot be harnessed just yet. We are still in the process of developing solar technologies and learning the most effective and cost-efficient ways to put the sun's energy to use.

The most commonly used way to harness solar energy is through **passive solar** energy collection. In this approach, buildings are designed and building materials are chosen to maximize direct absorption of sunlight in winter, even as they keep the interior cool in the heat of summer. In contrast, **active solar** energy collection makes use of technological devices to focus, move, or store solar energy.

One passive solar design technique involves installing low, south-facing windows to maximize sunlight capture in the winter (in the Northern Hemisphere; north-facing windows are used in the Southern Hemisphere). Overhangs block light from above, shading these windows in the summer, when the sun is high in the sky and when cooling, not heating, is desired. Passive solar techniques also include the use of construction materials (often called *thermal mass*) that absorb heat, store it, and release it later. Thermal mass (of straw, brick, concrete, or other materials) most often makes up floors, roofs, and walls, but also can comprise portable blocks. By heating buildings in cold weather and cooling them in warm weather, passive solar methods help conserve energy and reduce energy costs.

One active method for harnessing solar energy involves using *solar panels* or *flat-plate solar collectors*, most often installed on rooftops. These panels generally consist of dark-colored, heat-absorbing metal plates mounted in flat boxes covered with glass panes. Water, air, or antifreeze solutions run through tubes that pass through the collectors, transferring heat throughout a building. Heated water can be pumped to tanks to store the heat for later use and through pipes designed to release the heat into the building. Such systems have proven especially effective for heating water for homes.

Concentrating solar rays magnifies energy

We can magnify the strength of solar energy by gathering sunlight from a wide area and focusing it on a single point. This is the principle behind *solar cookers*, simple portable ovens that use reflectors to focus sunlight onto food and cook it (see Figure 2.11, ▶ p. 40). Such cookers are proving extremely useful in parts of the developing world.

The principle of concentrating the sun's rays has also been put to work by utilities in large-scale, high-tech approaches to generating electricity. In one approach, mirrors concentrate sunlight onto a receiver atop a tall

FIGURE 14.7 At the Solar Two facility operated by Southern California Edison in the desert of southern California, mirrors are spread across wide expanses of land to concentrate sunlight onto a receiver atop a "power-tower." Heat is then transported through fluid-filled pipes to a steam-driven generator that produces electricity.

"power-tower" (Figure 14.7). From the receiver, heat is transported by fluids (often molten salts) that are piped to a steam-driven generator to create electricity. These solar power plants can harness light from large mirrors spread across many hectares of land. The world's largest such plant so far—a collaboration among government, industry, and utility companies in the California desert—produces power for 10,000 households.

Photovoltaic cells generate electricity directly

A more direct approach to producing electricity from sunlight involves photovoltaic (PV) systems. **Photovoltaic (PV) cells** collect sunlight and convert it to electrical energy directly by making use of the *photovoltaic* or *photoelectric effect*. This effect occurs when light strikes one of a pair of metal plates in a PV cell, causing the release of electrons, which are attracted by electrostatic forces to the opposing plate. The flow of electrons from one plate to the other creates an electrical current, which can be converted into alternating current (AC) and used for residential and commercial electrical power (Figure 14.8). Photovoltaic cells can be connected to batteries that store the accumulated charge until it is needed.

You may be familiar with small PV cells that power your watch or your calculator. Atop the roofs of homes and other buildings, multiple PV cells are arranged in modules. These modules can comprise panels, which can be gathered together in flat arrays. Increasingly, people are using PV roofing tiles, which look like normal roofing shingles but generate electricity by the photovoltaic effect.

In some remote areas, such as Xcalak, Mexico, PV systems are being used in combination with wind turbines (▸ pp. 340–341) and a diesel generator to power entire villages.

Solar power is little used but fast growing

Although active solar technology dates from the 18th century, it was pushed to the sidelines as fossil fuels came to dominate our energy economy. Largely because of a lack of investment, solar power currently contributes only a minuscule portion of our energy production. In 2004, solar accounted for only 0.06%—less than 6 parts in 10,000—of the U.S. primary energy supply, and only 0.02% of U.S. electricity generation. However, use of solar energy has grown by nearly one-third annually worldwide in the past 35 years, a growth rate second only to that of wind power. Solar power is proving especially attractive in developing countries, many of which are rich in sun but poor in power infrastructure, and where hundreds of millions of people are still without electricity. Some multinational corporations that built themselves on fossil fuels are now investing in alternative energy as well. BP Solar, British Petroleum's solar energy wing, recently completed $30 million projects in the Philippines and Indonesia and is working on a $48 million project to supply electricity to 400,000 people in 150 villages.

Sales of PV cells are growing fast—by 25% per year in the United States and by 63% annually in Japan, where PV roofing tiles are widely used. Use of solar technology is expected to continue increasing as prices fall, technologies improve, and economic incentives are enacted. However, the very small amount of energy currently produced by

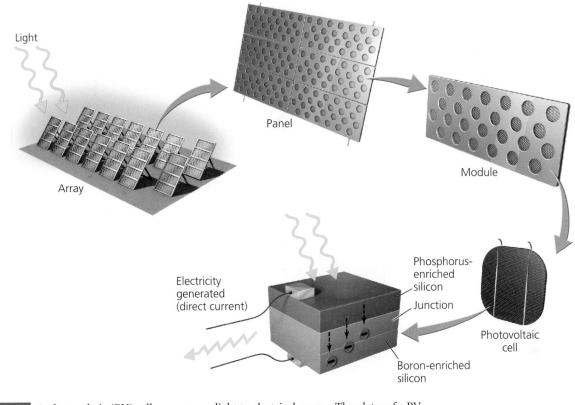

FIGURE 14.8 A photovoltaic (PV) cell converts sunlight to electrical energy. The plates of a PV cell are made primarily of silicon, enriched on one side with phosphorus and on the other with boron. Silicon is a semiconductor, so it conducts and controls the flow of electricity. Because of the chemical properties of boron and phosphorus, the phosphorus-enriched side has excess electrons, and the boron-enriched side has fewer electrons. When sunlight strikes the cell surface, it transfers energy and causes electrons to move. When wires connect the two sides, electricity is created as electrons flow from the phosphorus-enriched side to the boron-enriched side. PV cells are grouped in modules, which can comprise panels, which can be erected in arrays.

solar power means that its market share will likely remain small for years or decades to come—unless governments, businesses, and consumers become more motivated by the benefits that solar energy can provide.

Solar power offers many benefits

The fact that the sun will continue burning for another 4–5 billion years makes it practically inexhaustible as an energy source for human civilization. Moreover, the amount of solar energy reaching Earth should be enough to power our civilization once solar technology is adequately developed. These advantages of solar energy are clear, but the technologies themselves also provide benefits. PV cells and other solar technologies use no fuel, are quiet and safe, contain no moving parts, require little maintenance, and do not require a turbine or generator to create electricity. An average unit can produce energy for 20–30 years.

Another advantage of solar systems is that they allow for local, decentralized control over power. Homes, businesses, and isolated communities can use solar power to produce their own electricity and may not need to be near a power plant or connected to the grid of a city. This is especially helpful in rural areas of the developing world. In industrialized nations, most PV systems today are connected to the regional electric grid. As a result, owners of houses with PV systems can sell their excess solar energy to their local power utility through a process called *two-way metering*. The value of the power the consumer sells to the utility is subtracted from the consumer's monthly utility bill.

Finally, a major advantage of solar power over fossil fuels is that it does not pollute the air with greenhouse gas emissions and other air pollutants. The manufacture and maintenance of photovoltaic cells *does* currently require fossil fuel use, but once it is up and running, a PV system produces no emissions.

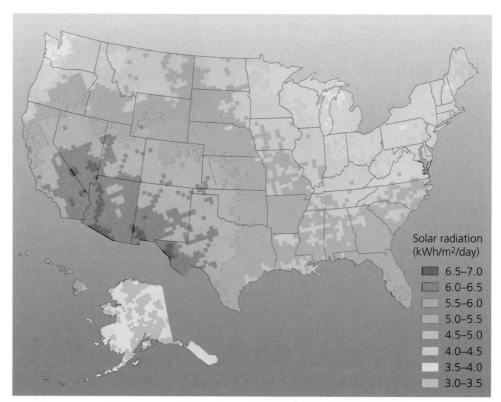

FIGURE 14.9 Because some locations receive more sunlight than others, harnessing solar energy is more profitable in some areas than in others. In the United States, many areas of Alaska and the Pacific Northwest receive only 3–4 kilowatt-hours per square meter per day, whereas most areas of the Southwest receive 6–7 kilowatt-hours per square meter per day. Data from National Renewable Energy Laboratory, U.S. Department of Energy, 2005.

Solar radiation
(kWh/m²/day)

- 6.5–7.0
- 6.0–6.5
- 5.5–6.0
- 5.0–5.5
- 4.5–5.0
- 4.0–4.5
- 3.5–4.0
- 3.0–3.5

Location and cost can be drawbacks

Solar power currently has two major disadvantages. One is that not all regions are sunny enough to provide adequate power, given current technology. Although Earth as a whole receives vast amounts of sunlight, not every location on Earth does (Figure 14.9). People in cities such as Seattle might find it difficult to harness enough sunlight most of the year to rely on solar power. Daily or seasonal variation in sunlight can also pose problems for stand-alone solar systems if storage capacity in batteries or fuel cells is inadequate or if backup power is not available from a municipal electricity grid.

The primary disadvantage of current solar technology— as with other renewable sources—is the up-front cost of investing in the equipment. The investment cost for solar is higher than that for fossil fuels, and indeed, solar power remains the most expensive way to produce electricity. However, decreases in price and improvements in energy efficiency of solar technologies so far are encouraging, even in the absence of significant financial commitment from government and industry. At their advent in the 1950s, solar technologies had efficiencies of around 6%, while costing $600 per watt. Recent single-crystal silicon PV cells are showing 15% efficiency commercially and 24% efficiency in lab research, suggesting that future solar technologies may be more efficient than any energy technologies we have today. Solar systems have become less expensive and now can often pay for themselves in 10–15 years. After that time, they provide energy virtually for free as long as the equipment lasts. With future technological advances, some experts believe that the time to recoup investment could fall to 1–3 years.

Wind Energy

Wind energy can be thought of as an indirect form of solar energy, because it is the sun's differential heating of air masses on Earth that causes wind to blow. We can harness power from wind by using devices called **wind turbines**, mechanical assemblies that convert wind's kinetic energy (▸ p. 58), or energy of motion, into electrical energy.

Wind turbines convert kinetic energy to electrical energy

Wind blowing into a turbine turns the blades of the rotor, which rotate the machinery inside a compartment called a *nacelle*, which sits atop a tall tower (Figure 14.10). Inside the nacelle are a gearbox and a generator, as well as equipment to monitor and control the turbine's activity. Most of today's towers range from 40 to 100 m (131–328 ft) in height, so the largest are taller than a football field is long.

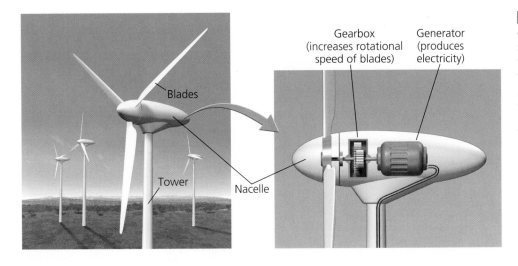

FIGURE 14.10 A wind turbine converts wind's energy of motion into electrical energy. Wind causes the blades of a wind turbine to spin, turning a shaft that extends into the nacelle that is perched atop the tower. Inside the nacelle, a gearbox converts the rotational speed of the blades, which can be up to 20 revolutions per minute (rpm) or more, into much higher rotational speeds (over 1,500 rpm). These high speeds provide adequate motion for a generator inside the nacelle to produce electricity.

Higher is generally better, to minimize turbulence (and potential damage) and to maximize wind speed. Most rotors consist of three blades and measure 42–80 m (138–262 ft) across. Engineers have designed turbines to begin turning at specific wind speeds to harness wind energy as efficiently as possible. Turbines are also designed to yaw, or rotate back and forth in response to changes in wind direction, ensuring that the motor faces into the wind at all times. Turbines can be erected singly, but they are most often erected in groups called wind parks, or *wind farms*. The world's largest wind farms contain several hundred or thousand turbines spread across the landscape.

Wind power is growing fast

Like solar energy, wind provides only a minuscule proportion of the world's power needs. However, wind power is the fastest growing energy sector today, having expanded by nearly 30% per year globally between 2000 and 2004. Wind provided 3.9% of U.S. renewable electricity generation in 2004—a small amount but nearly 20 times more than solar power. So far, wind energy production is geographically concentrated; only five nations (Germany, the United States, Spain, Denmark, and India) account for 82% of the world's wind energy output. California and Texas account for two-thirds of the wind power generated within the United States. Denmark is a leader in wind power; there, a series of wind farms supplies over 20% of the nation's electricity needs.

Experts agree that wind power's rapid growth will continue because only a very small portion of this resource is currently being tapped and because wind power at favorable locations already generates electricity nearly as cheaply as do conventional fossil fuels.

Meteorological evidence suggests that wind power could be expanded in the United States to meet the electrical needs of the entire country.

Offshore sites can be promising

Wind speeds on average are roughly 20% greater over water than over land. There is also less air turbulence over water than over land. For these reasons, offshore wind turbines are becoming popular (Figure 14.11). Although costs to erect and maintain turbines in water are higher, the stronger, less turbulent winds produce more power and make offshore wind potentially more profitable. Currently, offshore wind farms are limited to shallow water, where towers are sunk into sediments singly or with a tripod configuration to stabilize them. In the future, towers may also be placed on floating pads anchored to the seafloor in deep water. At great distances from land, it may be best to store the generated electricity as hydrogen and then ship or pipe this to land (instead of building submarine cables to carry electricity to shore), but further research is needed.

Denmark erected the first offshore wind farm in 1991. Over the next decade, nine more came into operation across northern Europe, where the North and Baltic Seas offer strong winds. The power output of these farms increased by 43% annually as larger turbines were erected. In Iceland, wind advocates are considering developing 240 offshore wind turbines in the nation's waters to meet future electricity demand for its hydrogen economy.

Wind power has many benefits

Like solar power, wind produces no emissions once the necessary equipment is manufactured and installed. As a replacement for fossil fuel combustion in the average U.S.

FIGURE 14.11 Many wind farms are being developed offshore, because offshore winds tend to be stronger yet less turbulent. Denmark is a world leader in wind power, and much of it comes from offshore turbines. This Danish wind farm is one of several that provide over 20% of the nation's electricity.

utility generator, the U.S. Environmental Protection Agency has calculated that running a 1-megawatt wind turbine for one year prevents the release of more than 1,500 tons of carbon dioxide, 6.5 tons of sulfur dioxide, 3.2 tons of nitrogen oxides, and 60 lb of mercury. The amount of carbon pollution that all U.S. wind turbines together prevent from entering the atmosphere is greater than the cargo of a 50-car freight train, with each car holding 100 tons of solid carbon, each and every day.

Wind power appears considerably more energy-efficient than conventional power sources. One recent study, which compared the amount of energy that various types of technology produce to the amount they consume, found that wind turbines produce 23 times as much as they consume. For nuclear energy, the ratio was 16:1; for coal it was 11:1; and for natural gas it was 5:1. Wind farms also use less water than do conventional power plants.

Wind turbine technology can be used on many scales, from a single tower for local use to fields of thousands that supply large regions. Small-scale turbine development can help make local areas more self-sufficient, just as solar energy can. For instance, the Rosebud Sioux Tribe in 2003 set up a single turbine on their reservation in South Dakota. The turbine is producing electricity for 220 homes and brings the tribe an estimated $15,000 per year in revenue. Wind resources are rich in this region, and the tribe plans to develop a wind farm nearby in coming years.

Another benefit of wind power is that landowners can lease their land for wind development, which provides them extra revenue while also increasing property tax income for rural communities. A single large turbine can bring in $2,000–4,500 in annual royalties while occupying just a quarter-acre of land. Because each turbine takes up only a small area, most of the land can still be used for farming, ranching, or other uses.

Wind energy has some downsides

Wind is an intermittent resource, and we have no control over when wind will occur. This poses little problem, however, if wind is only one of several sources contributing to a utility's power generation. Moreover, several technologies are available to address problems posed by relying on intermittent wind resources. For example, batteries or hydrogen fuel can store energy generated by wind and release it later when needed.

Just as wind varies from time to time, it varies from place to place; some areas are windier than others. Global wind patterns combine with local topography—mountains, hills, water bodies, forests, cities—to create local wind patterns, and companies study these patterns closely before investing in a wind farm. Meteorological research has given us information with which to judge prime areas for locating wind farms. A map of average wind speeds across the United States (Figure 14.12a) shows that mountainous regions are best, along with areas of the Great Plains. Based on such information, the young wind power industry has located much of its generating capacity in states with high wind speeds (Figure 14.12b).

Good wind resources, however, are not always near population centers that need the energy. Thus, transmission networks would need to be greatly expanded. Moreover, when wind farms *are* proposed near population centers, local residents often oppose them. Turbines

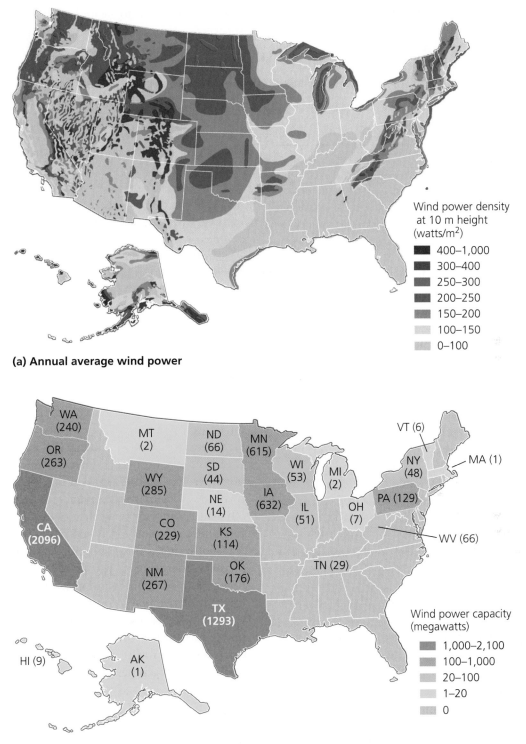

(a) Annual average wind power

Wind power density
at 10 m height
(watts/m²)

■ 400–1,000
■ 300–400
■ 250–300
■ 200–250
■ 150–200
■ 100–150
■ 0–100

(b) Wind generating capacity, 2004

WA (240)
OR (263)
MT (2)
ND (66)
MN (615)
VT (6)
SD (44)
WI (53)
MI (2)
NY (48)
MA (1)
WY (285)
IA (632)
PA (129)
NE (14)
IL (51)
OH (7)
CA (2096)
CO (229)
KS (114)
WV (66)
NM (267)
OK (176)
TN (29)
TX (1293)
HI (9)
AK (1)

Wind power capacity
(megawatts)

■ 1,000–2,100
■ 100–1,000
■ 20–100
■ 1–20
■ 0

FIGURE 14.12 Wind's capacity to generate power varies according to wind speed. Meteorologists have measured wind speed to calculate the potential generating capacity from wind in different areas. The map in (**a**) shows average wind power in watts per square meter at a height of 10 m (33 ft) above ground across the United States. Such maps are used to help guide placement of wind farms. The development of U.S. wind power so far is summarized in (**b**), which shows the megawatts of generating capacity developed in each state through the end of 2004. *Sources:* (a) Elliott, D. L., et al. 1987. *Wind energy resource atlas of the United States.* Golden, CO: Solar Energy Research Institute; (b) National Renewable Energy Laboratory, U.S. Department of Energy, 2005.

are generally located in exposed, conspicuous sites, and many people object to wind farms for aesthetic reasons, feeling that the structures clutter the landscape. Wind turbines are also known to pose a threat to birds and bats, which can be killed when they fly into the rotating blades.

Geothermal Energy

Geothermal energy is one form of renewable energy that does not originate from the sun. Instead, it is generated deep within Earth. The radioactive decay of elements amid the extremely high pressures deep in the interior of our planet generates heat that rises to the surface through magma (molten rock, ▸ pp. 70–71) and through fissures and cracks. Where this energy heats groundwater, natural spurts of heated water and steam are sent up from below and may erupt through the surface as *geysers*. Iceland is built from magma that extruded above the ocean's surface and cooled—magma from the Mid-Atlantic Ridge (▸ pp. 71–72), along the spreading boundary of two tectonic plates. Because of the geothermal heat in this region, volcanoes and geysers are numerous in Iceland.

Geothermal power plants use the energy of naturally heated water to generate power. Rising underground water and steam are harnessed to turn turbines and create electricity.

Geothermal energy is renewable in principle (its use does not affect the amount of heat produced in Earth's interior), but the power plants we build to use this energy may not all be capable of operating indefinitely. If a geothermal plant uses heated water at a rate faster than the rate at which groundwater is recharged, the plant will eventually run out of water. This is occurring at The Geysers in Napa Valley, California, where the first generator was built in 1960. In response, operators have begun injecting municipal wastewater into the ground to replenish the supply. A second reason geothermal energy may not always be renewable is that patterns of geothermal activity in Earth's crust shift naturally over time, so an area that produces hot groundwater now may not always do so.

Geothermal energy is harnessed for heating and electricity

Geothermal energy can be harnessed directly from geysers at the surface, but most often wells must be drilled down hundreds or thousands of meters toward heated groundwater. Generally, water at temperatures of 150–370 °C (300–700 °F) or more is brought to the sur-

face and converted to steam by lowering the pressure in specialized compartments. The steam is then employed in turning turbines to generate electricity (Figure 14.13).

Hot groundwater can also be used directly for heating homes, offices, and greenhouses; for driving industrial processes; and for drying crops. Iceland heats most of its homes through direct heating with piped hot water. Iceland began putting geothermal energy to use in the 1940s, and today 30 municipal district heating systems and 200 small private rural networks supply heat to 86% of the nation's residences. Such direct use of naturally heated water is cheap and efficient, but it is feasible only in areas where geothermal energy sources are available and near where the heat must be transported.

Geothermal *ground source heat pumps* (GSHPs) use thermal energy from near-surface sources of earth and water. Soil varies in temperature from season to season less than air does, so the pumps heat buildings in the winter by transferring heat from the ground into buildings, and they cool buildings in the summer by transferring heat from buildings into the ground. Both types of heat transfer are accomplished by a single network of underground plastic pipes that circulate water. Because heat is simply moved from place to place rather than being produced using outside energy inputs, heat pumps can be highly energy-efficient. Roughly half a million GSHPs are already used to heat U.S. residences. Compared to conventional electric heating and cooling systems, GSHPs heat spaces 50–70% more efficiently, cool them 20–40% more efficiently, can reduce electricity use by 25%–60%, and can reduce emissions by up to 72%.

Geothermal sources provide less than 0.5% of total primary energy used worldwide. They provide more power than solar and wind combined, but only a small fraction of the power from hydropower and biomass. Geothermal energy in the United States provides enough power to supply electricity to over 1.4 million homes. Currently Japan, China, and the United States lead the world in use of geothermal power.

Geothermal power has benefits and limitations

Like other renewable sources, geothermal power greatly reduces emissions relative to fossil fuel combustion. Geothermal sources can release variable amounts of gases dissolved in their water, including carbon dioxide, methane, ammonia, and hydrogen sulfide. However, these gases are generally in small quantities, and geothermal facilities on average release only one-sixth of the carbon dioxide produced by plants fueled by natural gas.

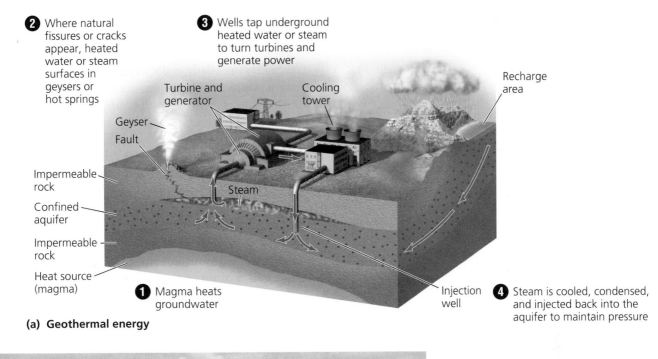

❷ Where natural fissures or cracks appear, heated water or steam surfaces in geysers or hot springs

❸ Wells tap underground heated water or steam to turn turbines and generate power

Recharge area

Turbine and generator

Cooling tower

Geyser

Fault

Impermeable rock

Confined aquifer

Impermeable rock

Heat source (magma)

Steam

❶ Magma heats groundwater

Injection well

❹ Steam is cooled, condensed, and injected back into the aquifer to maintain pressure

(a) Geothermal energy

(b) Nesjavellir geothermal power station, Iceland

FIGURE 14.13 With geothermal energy (**a**) magma heats groundwater deep within Earth (1), some of which is let off naturally through surface vents such as geysers (2). Geothermal facilities tap into heated water below ground and channel steam through turbines in buildings to generate electricity (3). After being used, the steam is often condensed and pumped back into the aquifer to maintain pressure (4). At Nesjavellir geothermal power station in Iceland (**b**), steam is piped from four wells to a condenser at the plant, where cold water pumped from distant lakeshore wells is heated. The heated water is sent through an insulated 270-km (170-mi) pipeline to Reykjavik and environs, where residents use it for washing and space heating.

Geothermal facilities using the latest filtering technologies produce even fewer emissions. By one estimate, each megawatt of geothermal power prevents the emission of 3.5 million kg (7.8 million lb) of carbon dioxide emissions and 860 kg (1,900 lb) of other pollutant emissions from gas-fired plants each year.

On the negative side of the ledger, geothermal sources, as we have seen, may not always be truly sustainable. In addition, the water of many hot springs is laced with salts and minerals that corrode equipment and pollute the air. These factors may shorten the lifetime of plants, increase maintenance costs, and add to pollution. Moreover, use of geothermal energy is limited to areas where the energy can be tapped. Unless technology is developed to penetrate far more deeply into the ground, geothermal energy use will remain more localized than solar, wind, biomass, or hydropower.

Ocean Energy Sources

The oceans are home to several underexploited energy sources. Each involves continuous natural processes that could potentially provide sustainable energy for our needs. Of the three approaches developed so far, two involve motion and one involves temperature.

We can harness energy from tides and waves

Just as dams on rivers use flowing freshwater to generate hydroelectric power, some scientists, engineers, businesses, and governments are developing ways to use kinetic energy from the motion of ocean water to generate electrical power.

The rising and falling of ocean tides (▸ p. 260) twice each day moves large amounts of water past any given point on the world's coastlines. Differences in height between low and high tides are especially great in long, narrow bays such as Alaska's Cook Inlet or the Bay of Fundy between New Brunswick and Nova Scotia. Such locations are best for harnessing tidal energy, which is accomplished by erecting dams across the outlets of tidal basins. The incoming tide flows through sluices past the dam, and as the outgoing tide passes through the dam, it turns turbines to generate electricity. Some designs allow for generating electricity from water moving in both directions. The world's largest tidal generating facility, the La Rance facility in France, has operated for over 30 years. Smaller facilities now operate in China, Russia, and Canada. Tidal stations release few or no pollutant emissions, but they can affect the ecology of estuaries and tidal basins.

People are also working to harness the motion of wind-driven waves at the ocean's surface and convert this mechanical energy into electricity. Many designs for machinery to harness wave energy have been invented, but few have been adequately tested. Some designs are for offshore facilities, involving floating devices that move up and down with the waves. Other designs are for coastal onshore facilities. Some of these designs funnel waves from large areas into narrow channels and elevated reservoirs, from which water is then allowed to flow out, generating electricity as hydroelectric dams do. Other coastal designs use rising and falling waves to push air into and out of chambers, turning turbines to generate electricity. No commercial wave energy facilities are operating yet, but some have been deployed as demonstration projects in several western European nations.

The ocean stores thermal energy

Other oceanic energy sources we have not yet effectively tapped include the motion of ocean currents, chemical gradients in salinity, and the immense thermal energy contained in the oceans. The concept of **ocean thermal energy conversion (OTEC)** has been most fully devel-oped. Each day the tropical oceans absorb an amount of solar radiation equivalent to the heat content of 250 billion barrels of oil—enough to provide 20,000 times the electricity used daily in the United States. The ocean's surface is warmer than its deep water (▸ p. 257), and OTEC approaches are based on this temperature gradient.

In one approach, warm surface water is piped into a facility to evaporate chemicals, such as ammonia, that boil at low temperatures. The evaporated gases spin turbines to generate electricity. Cold water piped in from ocean depths then condenses the gases so they can be reused. In another approach, warm surface water is evaporated in a vacuum, and its steam turns turbines and then is condensed by cold water. Research on OTEC systems has been conducted in Hawaii and other locations, but costs remain high, and as of yet no facility is commercially operational.

Weighing the issues:
Your Island's Energy?

Imagine you have been elected the president of an island nation the size of Iceland, and your nation's congress is calling on you to propose a national energy policy. Unlike Iceland, your country is located in equatorial waters. Your geologists do not yet know whether there are fossil fuel deposits or geothermal resources under your land, but your country gets a lot of sunlight and a fair amount of wind, and broad, shallow shelf regions surround its coasts. Your island's population is moderately wealthy but is growing fast, and importing fossil fuels from mainland nations is becoming increasingly expensive.

What approaches would you propose in your energy policy? What specific steps would you urge your congress to fund immediately? What trade relationships would you seek to establish with other countries? What questions would you ask of your economic advisors? What questions would you fund your country's scientists to research?

Hydrogen

All the renewable energy sources we have discussed can be used to generate electricity more cleanly than can fossil fuels. As useful as electricity is to us, however, it cannot be stored easily in large quantities for use when and where it is needed. This is why vehicles rely on fossil fuels for power. The development of fuel cells and hydrogen fuel shows promise to store energy conveniently and in considerable quantities, and to produce electricity at least as cleanly and efficiently as renewable energy sources.

In the "hydrogen economy" that Iceland's leaders and many energy experts worldwide envision, hydrogen fuel, together with electricity, would serve as the basis for a clean, safe, and efficient energy system. This system would use as a fuel the universe's simplest and most abundant element. Electricity generated from intermittent renewable sources, such as wind or solar energy, could be used to produce hydrogen. Fuel cells (Figure 14.14) could then employ hydrogen to produce electrical energy as needed to power vehicles, computers, cell phones, home heating, and countless other applications.

Fuel cell technology has been used since the 1960s in NASA's space flight programs. Basing an energy system on hydrogen could alleviate dependence on foreign fuels and help fight climate change. For these reasons, governments are funding research into hydrogen and fuel cell technology, and automobile companies are investing in research and development to produce vehicles that run on hydrogen (Figure 14.15).

Hydrogen fuel may be produced from water or from other matter

Hydrogen gas (H_2) does not tend to exist freely on Earth; rather, hydrogen atoms bind to other molecules, becoming incorporated in everything from water to organic mole-

cules. To obtain hydrogen gas for fuel, we must force these substances to release their hydrogen atoms, and this requires an input of energy. Several potential ways of producing hydrogen are being studied (see "The Science behind the Story," ▸ p. 349). In **electrolysis**, the process being pursued by Iceland, electricity is input to split hydrogen atoms from the oxygen atoms of water molecules:

$$2H_2O \rightarrow 2H_2 + O_2$$

Electrolysis produces pure hydrogen, and it does so without emitting the carbon- or nitrogen-based pollutants of fossil fuel combustion. However, whether this strategy for producing hydrogen will cause pollution over its entire life cycle depends on the source of the electricity used for the electrolysis. If coal is burned to create the electricity, then the entire process will not reduce emissions compared with reliance on fossil fuels. If, however, the electricity is produced by some less-polluting renewable source, then hydrogen production by electrolysis would create much less pollution and greenhouse warming than reliance on fossil fuels. The "cleanliness" of a future hydrogen economy in Iceland or anywhere else would, therefore, depend largely on the source of electricity used in electrolysis.

The environmental impact of hydrogen production will also depend on the source material for the hydrogen. Besides water, hydrogen can be obtained from biomass and fossil fuels. Obtaining hydrogen from these sources generally requires less energy input, but it results in

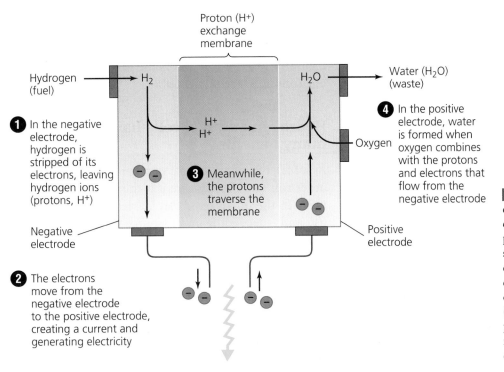

FIGURE 14.14 Hydrogen fuel drives electricity generation in a fuel cell, creating water as a waste product. Atoms of hydrogen are first stripped of their electrons (1). The electrons move from a negative electrode to a positive one, creating a current and generating electricity (2). Meanwhile, the hydrogen ions pass through a proton exchange membrane (3) and combine with oxygen to form water molecules (4).

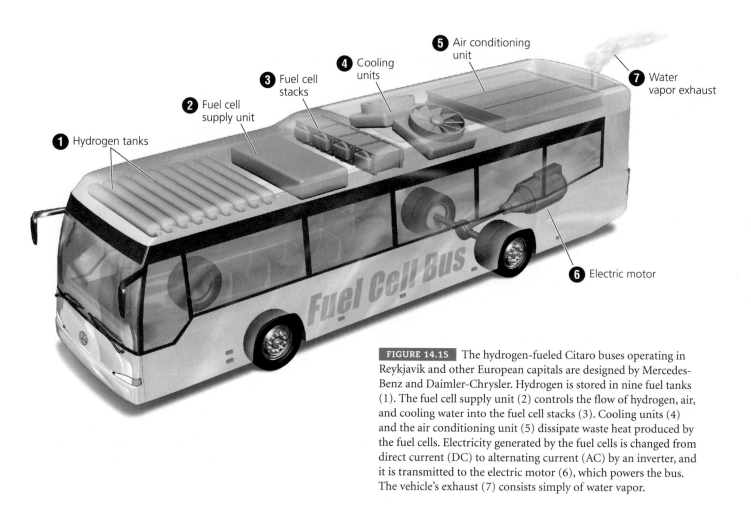

① Hydrogen tanks
② Fuel cell supply unit
③ Fuel cell stacks
④ Cooling units
⑤ Air conditioning unit
⑥ Electric motor
⑦ Water vapor exhaust

FIGURE 14.15 The hydrogen-fueled Citaro buses operating in Reykjavik and other European capitals are designed by Mercedes-Benz and Daimler-Chrysler. Hydrogen is stored in nine fuel tanks (1). The fuel cell supply unit (2) controls the flow of hydrogen, air, and cooling water into the fuel cell stacks (3). Cooling units (4) and the air conditioning unit (5) dissipate waste heat produced by the fuel cells. Electricity generated by the fuel cells is changed from direct current (DC) to alternating current (AC) by an inverter, and it is transmitted to the electric motor (6), which powers the bus. The vehicle's exhaust (7) consists simply of water vapor.

emissions of carbon-based pollutants. For instance, extracting hydrogen from the methane (CH_4) in natural gas entails producing one molecule of the greenhouse gas carbon dioxide for every four molecules of hydrogen gas:

$$CH_4 + 2H_2O \rightarrow 4H_2 + CO_2$$

Thus, whether a hydrogen-based energy system is environmentally cleaner than a fossil fuel system depends on how the hydrogen is extracted.

In addition, some new research suggests that leakage of hydrogen from its production, transport, and use could potentially deplete stratospheric ozone and lengthen the atmospheric lifetime of the greenhouse gas methane. Research into these questions is ongoing, because scientists do not want society to switch from fossil fuels to hydrogen without first knowing the possible risks from hydrogen.

Once hydrogen gas has been isolated, it can be used as a fuel to produce electricity within fuel cells. The chemical reaction involved in a fuel cell is simply the reverse of that shown for electrolysis; an oxygen molecule and two

hydrogen molecules each split so that their atoms can bind and form two water molecules:

$$2H_2 + O_2 \rightarrow 2H_2O$$

Figure 14.14 (▸ p. 347) shows the way this occurs within one common type of fuel cell.

Hydrogen and fuel cells have many benefits

As a fuel, hydrogen offers a number of benefits. We will never run out of hydrogen; it is the most abundant element in the universe. It can be clean and nontoxic to use, and—depending on the source of the hydrogen and the source of electricity for its extraction—it may produce few greenhouse gases and other pollutants. Pure water and heat may be the only waste products from a hydrogen fuel cell, along with negligible traces of other compounds. In terms of safety for transport and storage, hydrogen can catch fire, but if it is kept under pressure, it is probably no more dangerous than gasoline in tanks.

Algae as a Hydrogen Fuel Source

As scientists search for new ways to generate energy, some are looking to an unlikely power source—pond scum. Algae are being studied as an innovative way to generate large amounts of hydrogen to move society toward a more sustainable energy future.

Hydrogen's benefits hinge on how hydrogen fuel is produced. Some methods release substantial amounts of carbon dioxide, and other, nonpolluting, processes can be costly. These drawbacks have kept scientists searching for new hydrogen sources.

At the University of California at Berkeley, biologist Anastasios Melis thought one possible hydrogen source might be a single-celled aquatic plant. The alga *Chlamydomonas reinhardtii* was known to emit small amounts of hydrogen for brief periods of time when deprived of light. Melis set up an experiment with energy experts at the National Renewable Energy Laboratory in Colorado, aiming to develop ways to tweak the alga's basic biological functions so that the plant produced large quantities of hydrogen.

Green algae photosynthesize, drawing in carbon dioxide and water, absorbing energy from light that converts those nutrients into food, and then expelling oxygen as a waste product. Additional nutrients from soil or water, and catalysts called *enzymes* within the plant, keep this process running smoothly. To conduct photosynthesis effectively, *Chlamydomonas reinhardtii* needs sulfur as a nutrient. The alga

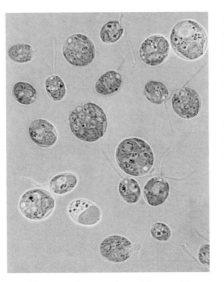

Could green algae such as this provide hydrogen for our energy needs?

also contains an enzyme called *hydrogenase*, which can trigger the alga to stop producing oxygen as a metabolic by-product and start releasing hydrogen instead.

Hydrogenase normally is active only after *Chlamydomonas reinhardtii* has been deprived of light. When deprived of light, the light-dependent reactions of photosynthesis ebb, little oxygen is produced, and hydrogenase is activated. When light returns and the alga begins producing oxygen again, hydrogenase is deactivated, and the hydrogen release stops.

Melis's team wanted to activate hydrogenase so that more hydrogen would be produced. But simply keeping the algae in the dark would not escalate hydrogen production because the alga's metabolic functions slowed without light. So the researchers decided to try limiting

the alga's oxygen output another way, by putting it on a sulfur-free, bright-light regimen. The lack of sulfur would hinder photosynthesis, limiting oxygen output enough to activate hydrogenase and trigger hydrogen production. The presence of light would keep the algae metabolically active and releasing large amounts of by-products.

The researchers cultured large quantities of the algae in bottles in labs. Then they deprived the cultures of sulfur but kept them exposed to light for long periods of time—up to 150 hours. After the sustained light exposure, gas and liquids were extracted from the bottles and analyzed.

The analysis supported the team's hypothesis. Without sulfur or photosynthesis, the algae were not producing oxygen. This low-oxygen, or anaerobic, environment had induced hydrogenase, which spurred the algae to begin splitting water molecules and releasing substantial amounts of hydrogen gas. Hydrogen also dominated the alga's emissions—in gas collection analysis, approximately 87% of the gas was hydrogen, 1% was carbon dioxide, and the remaining 12% was nitrogen with traces of oxygen. The research teams published their findings in the journal *Plant Physiology* in 2000.

Many questions remain about algae-derived hydrogen, particularly how much fuel can be harvested continuously using this *photobiological* process. Nevertheless, the research results so far are helping to fuel the momentum of a future hydrogen economy.

Hydrogen and Renewable Energy

Is establishing a "hydrogen economy," as Iceland is trying to do, the best way to reduce the use of fossil fuels?

The Role of Renewable Energy for the Hydrogen Economy

Abundant, reliable, and affordable energy is an essential component of a healthy economy. Because hydrogen can be produced from a wide variety of domestically available resources and can be used in heat, power, and fuel applications, it is uniquely positioned to contribute to our growing energy demands, particularly for resource-constrained communities. However, if we are to realize the true benefits of a hydrogen economy, other renewables must play a substantial role in the efficient and affordable production of the hydrogen.

Several renewable options could make a substantial impact in the production of hydrogen: electrolysis powered by wind, photovoltaic, solar-thermal electric, hydropower, and geothermal energy; use of microorganisms and semiconductors to split water; and the thermal and biological conversion of biomass and wastes. Researchers around the globe are working on improving these renewable technologies. As a result, costs continue to drop. Technologies for renewable hydrogen production, coupled with advances in hydrogen production equipment (e.g., electrolyzers) can supply cost-competitive hydrogen and will ultimately play a substantial role in our energy supply.

In addition to the potential supply of affordable hydrogen, these technologies also offer a wide variety of opportunities for developing new centers of economic growth. Most investments in renewable energy are spent on materials and workmanship to build and maintain the facilities, rather than on costly energy imports. Therefore, funds are usually spent regionally and even locally, leading to new jobs and investments in local economies. Because of this synergistic relationship, the shift toward a hydrogen economy will naturally facilitate the advancement of renewable energy. By diversifying our energy supply, we not only will reduce our dependence on imported fuels, but also will benefit from cleaner technologies and investment in our communities.

Susan Hock directs the Electric and Hydrogen Technologies and Systems Center of the National Renewable Energy Laboratory. The center conducts research activities in four areas: distributed power systems integration, hydrogen technologies and systems, geographic information system analysis, and solar measurements and instrumentation.

Is Hydrogen the Answer?

We'll never use the last drop of oil, the last chunk of coal, the last cubic foot of natural gas, or the last pound of uranium. Eventually though, these fossil and nuclear fuels will become too expensive to extract, or politics will make one or more of them unavailable, leaving us to ask how we'll satisfy our voracious appetite in the future.

We should immediately apply all practical energy conservation strategies. Mother Nature is out there making more fossil fuels as we speak, but we don't have time to wait the few million years that will take. The short list of renewables: solar, wind, hydro, biomass, geothermal, waves, tides, and ocean thermal energy conversion. These are all relatively benign and abundant.

An alternative: hydrogen. It can either be burned or electrochemically used in fuel cells to provide useful energy. The by-product, or "exhaust," is water. You start with water, get some energy, and end up with water, making it renewable. Another form of hydrogen energy is fusion, hydrogen atoms fusing to form helium plus a lot of energy, the way the sun does it. The catch? It takes about as much energy to extract hydrogen gas from water (by electrolysis) as you get back from your energy conversion device. Until it becomes cheaper (economically and in physical terms), fossil fuels will continue to rule the energy world. The breakthrough may involve using our renewable energy resources to separate hydrogen from other molecules.

Arguably, to reduce our dependence on fossil fuels, the priority list for this country should be:
1. Energy conservation
2. Wind
3. Passive solar
4. Biomass
5. Active solar
6. Hydrogen (chemical)
7. Hydroelectricity
8. Hydrogen (fusion)
9. Others (geothermal, tides, waves, ocean thermal)

Daryl Prigmore has studied energy and the environment since the late 1960s. After receiving bachelor and master of science degrees in mechanical engineering from Colorado State University, he spent 10 years in industry with a company developing solar, geothermal, and low-pollution automotive power systems. He has taught energy science classes for the past 23 years, 20 at the University of Colorado (Colorado Springs).

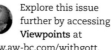

Explore this issue further by accessing **Viewpoints** at www.aw-bc.com/withgott.

Hydrogen fuel cells are energy-efficient. Depending on the type of fuel cell, 35% to 70% of the energy released in the reaction can be used. If the system is designed to capture heat as well as electricity, then the energy efficiency of fuel cells can rise to 90%. These rates are comparable or superior to most nonrenewable alternatives.

Fuel cells are also silent and nonpolluting. Unlike batteries (which also produce electricity through chemical reactions), fuel cells will generate electricity whenever hydrogen fuel is supplied, without ever needing recharging. For all these reasons, hydrogen fuel cells are being used to power vehicles, including the buses (see Figure 14.15) now operating on the streets of Reykjavik and many other European, American, and Asian cities.

Conclusion

The coming decline of fossil fuel supplies and the increasing concern over air pollution and global climate change have convinced many people that we will need to shift to renewable energy sources that will not run out and will pollute less. Biomass and hydropower already play important roles in our energy use and electricity production, but they are not always strictly renewable. Renewable sources with promise for sustaining our civilization without greatly degrading our environment include solar energy, wind energy, geothermal energy, and ocean energy sources. Moreover, by using electricity from renewable sources to produce hydrogen fuel, we may be able to use fuel cells to produce electricity when and where it is needed, helping convert our transportation sector to a nonpolluting, renewable basis.

Most renewable energy sources have been held back by inadequate funding for research and development, and by artificially cheap market prices for nonrenewable sources that do not include external costs. Despite these obstacles, renewable technologies have progressed far enough to offer hope that we can shift from fossil fuels to renewable energy with a minimum of economic and social disruption. Whether we can also limit environmental impact will depend on how soon and how quickly we make the transition, and to what extent we put efficiency and conservation measures into place.

TESTING YOUR COMPREHENSION

1. About how much of our energy now comes from renewable sources? What is the most prevalent form of renewable energy we use? What form of renewable energy is most used to generate electricity?

2. What factors and concerns are causing renewable energy sectors to expand? Which renewable source is experiencing the most rapid growth?

3. List five sources of biomass energy. How does biomass energy use differ between developed and developing nations?

4. Describe and contrast two major approaches to generating hydroelectric power.

5. Contrast passive and active solar heating. Describe how each works, and give examples. Now define the photoelectric effect, and explain how photovoltaic (PV) cells function and are used.

6. How do modern wind turbines generate electricity? What factors affect where wind turbines are placed?

7. Define geothermal energy, and explain how it is obtained and used. In what ways is it renewable, and in what way is it not renewable?

8. List and describe three approaches to obtaining energy from ocean water.

9. For each major type of renewable energy (biomass, hydropower, solar, wind, geothermal, and ocean), name and describe at least one advantage and one disadvantage of its use, relative to fossil fuels.

10. How is hydrogen fuel produced? Is this a clean process? What factors determine the amount of pollutants hydrogen production will emit?

SEEKING SOLUTIONS

1. Do you think we can develop and implement renewable energy resources to replace fossil fuels without great social, economic, and environmental disruption? What steps would we need to take? Will market forces alone suffice to bring about this transition? Do you think such a shift will be good for the economy?

2. There are many different sources of biomass and many ways of harnessing energy from biomass. Discuss one

that seems particularly beneficial to you, and one with which you see problems. What biomass energy sources and strategies do you think our society should focus on investing in?

3. Discuss the advantages and impacts of hydropower. If there is a hydroelectric dam near where you live, consider what economic benefits and environmental impacts it has had in your area. Given this mix of effects, would you favor the construction of more dams and greater reliance on hydropower?

4. Iceland is giving itself many years to phase in its planned hydrogen economy. Do you think large industrialized nations such as the United States and Canada could transition to a hydrogen economy more quickly,

less quickly, or not at all? Why? What steps could your nation take to accelerate such a transition? Do you think it should do so?

5. Imagine you are the CEO of a company that develops wind farms. Your staff is presenting you with three options, listed below, for sites for your next development. Describe at least one likely advantage and at least one likely disadvantage you would expect to encounter with each option. What further information would you like to know before deciding which to pursue?

▶ Option A: A remote rural site in North Dakota
▶ Option B: A ridge-top site in the suburbs of Philadelphia
▶ Option C: An offshore site off the Florida coast

CALCULATING ECOLOGICAL FOOTPRINTS

 Assume that average per capita residential consumption of electricity is 12 kilowatt-hours per day, that photovoltaic cells have an electrical output of 10% incident solar radiation, and that PV cells cost $800 per square meter. Now refer to Figure 14.9 on ▶ p. 340, and estimate the area and cost of the PV cells needed to provide all of the residential electricity used by each group in the table.

	Area of photovoltaic cells	Cost of photovoltaic cells
You	25	$20,000
Your class		
Your state		
United States		

1. What additional information do you need in order to increase the accuracy of your estimates for the areas in the table above?

2. Considering the distribution of solar radiation in the United States, where do you think it will be most feasible to greatly increase the percentage of electricity generated from photovoltaic solar cells?

3. The purchase price of a photovoltaic system is considerable. What other costs and benefits should you consider, in addition to the purchase price, when contemplating "going solar"?

Take It Further

Go to www.aw-bc.com/withgott, where you'll find:

▶ Suggested answers to end-of-chapter questions
▶ Quizzes, animations, and flashcards to help you study
▶ *Research Navigator*™ database of credible and reliable sources to assist you with your research projects

▶ **GRAPHit!** tutorials to help you interpret graphs
▶ **INVESTIGATEit!** current news articles that link the topics that you study to case studies from your region to around the world

Waste Management

Fresh Kills Landfill,
Staten Island, New York

Upon successfully completing this chapter, you will be able to:

▶ Summarize and compare the types of waste we generate

▶ List the major approaches to managing waste

▶ Delineate the scale of the waste dilemma

▶ Describe conventional waste disposal methods: landfills and incineration

▶ Evaluate approaches for reducing waste: source reduction, reuse, composting, and recycling

▶ Discuss industrial solid waste management and principles of industrial ecology

▶ Assess issues in managing hazardous waste

Central Case: Transforming New York's Fresh Kills Landfill

> "An extraterrestrial observer might conclude that conversion of raw materials to wastes is the real purpose of human economic activity."
> —GARY GARDNER AND PAYAL SAMPAT, WORLDWATCH INSTITUTE

> "Recycling is one of the best environmental success stories of the late 20th century."
> —U.S. ENVIRONMENTAL PROTECTION AGENCY

The closure of a landfill is not the kind of event that normally draws politicians and the press, but the Fresh Kills Landfill was no ordinary dump. The largest landfill in the world, Fresh Kills was the primary repository of New York City's garbage for half a century. On March 22, 2001, New York City Mayor Rudolph Giuliani and New York Governor George Pataki were on hand to celebrate as a barge arrived on the western shore of New York City's Staten Island and dumped the final load of 650 tons of trash at Fresh Kills.

The landfill's closure was a welcome event for Staten Island's 450,000 residents, who had long viewed the landfill as a bad-smelling eyesore, health threat, and civic blemish. The 890-ha (2,200-acre) landfill featured six gigantic mounds of trash and soil. The highest, at 69 m (225 ft), was higher than the nearby Statue of Liberty.

New York City had grandiose plans for the site. It planned to transform the old landfill into a world-class public park—a verdant landscape of rolling hills and wetlands teeming with wildlife, and a mecca for recreation for New York's residents. The site certainly had potential. It was two-and-a-half times bigger than Central Park. It was the region's largest remaining complex of saltwater tidal marshes and freshwater creeks and wetlands, which still attracted birds and wildlife. And the mounds offered panoramic views of the Manhattan skyline. The city sponsored an international competition to select a landscape architecture firm to design plans for the new park.

Meanwhile, with its only landfill closed, New York City began to develop an efficient network of stations to package and transfer the waste and ship it outward by barge and railroad. However, these plans soon fell apart amid neighborhood opposition, economic misjudgments, and accusations of political favoritism and mob influence. The city instead found itself paying contractors exorbitant prices to haul its garbage away one truckload at a time. In the years following the Fresh Kills closure, trucks full of trash rumbled through neighborhood streets, carrying 12,000 tons of waste each day bound for 26 different landfills and incinerators in New York, New Jersey, Virginia, Pennsylvania, and Ohio. The city sanitation department's budget nearly doubled, and budget woes caused the city to scale back its recycling program. Some New Yorkers suggested reopening Fresh Kills.

The landfill *was* reopened, but not for a reason anyone could have foreseen. After the September 11, 2001, terrorist attacks, the 1.8 million tons of rubble from the collapsed World Trade Center towers, including unrecoverable human remains, were taken by barge to Fresh Kills. A monument will be erected at the site as part of the new park.

Today, plans for the park are forging ahead. Field Operations, the design firm that won the competition, completed a preliminary master plan in 2005, incorporating suggestions from the public. The plan involves everything from ecological restoration of the wetlands to construction of roads, ball fields, sculptures, and roller-blading rinks. People will be able to bicycle on trails paralleling tidal creeks of the region's largest estuary and reach stunning vistas atop the hills. This undertaking will be one of the largest public works projects in the world. Designers, city officials, and Staten Island residents hope the first portions of the new park will open between 2008 and 2012.

Approaches to Waste Management

As the world's human population rises, and as we produce and consume more material goods, we generate more waste. **Waste** refers to any unwanted material or substance that results from a human activity or process. Waste can degrade water quality, soil quality, air quality,

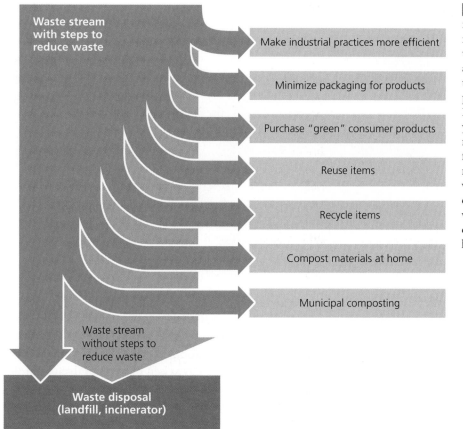

Waste stream with steps to reduce waste

Make industrial practices more efficient

Minimize packaging for products

Purchase "green" consumer products

Reuse items

Recycle items

Compost materials at home

Municipal composting

Waste stream without steps to reduce waste

Waste disposal (landfill, incinerator)

FIGURE 15.1 The most effective way to manage waste is to minimize the amount of material that enters the waste stream. To do this, manufacturers can increase efficiency, and consumers can buy "green" products that have minimal packaging or are produced in ways that minimize waste. Individuals can compost food scraps and yard waste at home and can reuse items rather than buying new ones. When we are finished using products, many of us can recycle some materials and compost yard waste through municipal recycling and composting programs. For all remaining waste, waste managers attempt to find disposal methods that minimize impact on human health and environmental quality.

and human health. Waste indicates inefficiency, so reducing waste can potentially save money and resources. Waste is also unpleasant aesthetically. For all these reasons, waste management has become a vital pursuit.

For management purposes, we divide waste into several main categories. **Municipal solid waste** is nonliquid waste that comes from homes, institutions, and small businesses. **Industrial solid waste** comes from production of consumer goods, mining, agriculture, and petroleum extraction and refining. **Hazardous waste** refers to solid or liquid waste that is toxic, chemically reactive, flammable, or corrosive. Another type of waste is *wastewater*, water we use in our households, businesses, industries, or public facilities and drain or flush down our pipes, as well as the polluted runoff from our streets and storm drains. We discussed wastewater in Chapter 11 (▸ pp. 255–257).

There are three main components of **waste management**: (1) minimizing the amount of waste we generate, (2) recovering waste materials and finding ways to recycle them, and (3) disposing of waste safely and effectively. Minimizing waste at its source—called *source reduction*—is the preferred approach. There are several ways to reduce the amount of waste that enters the **waste stream**, the

flow of waste as it moves from its sources toward disposal destinations (Figure 15.1). In this chapter we first examine how each of these approaches are used to manage municipal solid waste, and then we address approaches for managing industrial solid waste and hazardous waste.

Municipal Solid Waste

Municipal solid waste is waste produced by consumers, public facilities, and small businesses. It is what we commonly refer to as "trash" or "garbage." Everything from paper to food scraps to roadside litter to old appliances and furniture is considered municipal solid waste.

Waste generation is rising

In the United States since 1960, the generation of municipal solid waste has increased by 2.7 times, and waste generation per person has risen by 66%. Paper, yard debris, food scraps, and plastics are the principal components of municipal solid waste in the United States, together accounting for 70% of the waste stream. Paper products are the largest component (even after recycling), but plastics,

FIGURE 15.2 U.S. waste generation has increased by more than 2.7 times since 1960. Paper products comprise the largest component of the municipal solid waste stream, but plastic waste is the only type that has taken up a substantially greater percentage of the waste stream through time. Data from U.S. Environmental Protection Agency, 2005. *Municipal solid waste generation, recycling, and disposal in the United States: Facts and figures for 2003.* EPA530-F-05-003. Washington, D.C.: EPA.

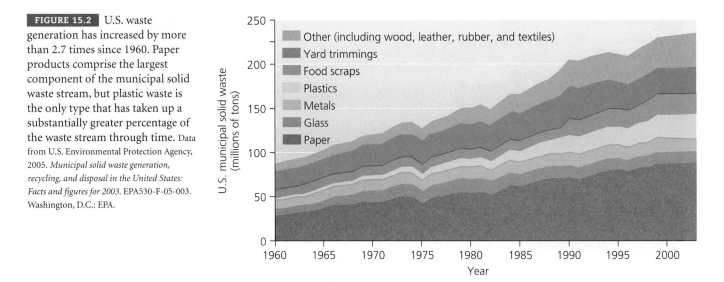

which came into wide consumer use only after 1970, have accounted for the greatest relative increase in the waste stream during the last several decades (Figure 15.2).

As we acquire more goods, we generate more waste. In 2003, U.S. citizens produced 236 million tons of municipal solid waste, almost 1 ton per person. This means that the average American generates about 2.0 kg (4.4 lb) of trash per day, more than the citizens of any other nation. The relative wastefulness of the U.S. lifestyle, with its excess packaging and reliance on nondurable goods (goods meant to be discarded after a short period of use), has caused critics to label the United States "the throwaway society."

People in developing nations, where consumption is lower, generate considerably less waste—but in nations from Mexico to Kenya to Indonesia, consumption and waste generation are now rising rapidly. This change reflects rising material standards of living, but an increase in packaging is also to blame. Over the past three decades, per capita waste generation rates have more than doubled in Latin America and have risen more than fivefold in the Middle East. Like U.S. consumers in the "throwaway society," wealthy consumers in developing nations often discard items that can still be used. In fact, at many dumps in the developing world, poor people support themselves by selling items they scavenge (Figure 15.3).

Wealthier nations can afford to invest more in waste collection and disposal, so they are often better able to manage their waste proliferation and minimize impacts on human health and the environment. Moreover, in many industrialized nations, recycling, composting, reduction, and reuse are taking care of an increasingly larger portion

FIGURE 15.3 Tens of thousands of people used to scavenge each day from the dump at Payatas, outside Manila in the Philippines, finding items for themselves and selling material to junk dealers for 100–200 pesos (U.S. $2–4) per day. That so many people could support themselves this way testifies to the immense amount of usable material needlessly discarded by wealthier portions of the population. The dump was closed in 2000 after an avalanche of trash killed hundreds of people.

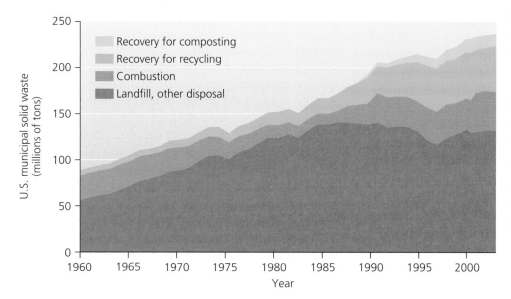

FIGURE 15.4 Since the 1980s, recycling and composting have grown in the United States, allowing a smaller proportion of waste to go to landfills. As of 2003, 55.4% of U.S. municipal solid waste was going to landfills and 14.0% to incinerators, while 30.6% was being recovered for composting and recycling. Go to **GRAPHit!** at www.aw-bc.com/ withgott. Data from U.S. Environmental Protection Agency, 2005. *Municipal solid waste generation, recycling, and disposal in the United States: Facts and figures for 2003.* EPA530-F-05-003. Washington, D.C.: EPA.

of waste (Figure 15.4). We will examine these approaches shortly, but let's first assess how we dispose of waste.

Sanitary landfills are regulated by health and environmental guidelines

In modern **sanitary landfills**, waste is buried in the ground or piled up in large, carefully engineered mounds designed to prevent waste from contaminating the environment and threatening public health (Figure 15.5). Most municipal landfills in the United States are regulated locally or by

the states, but they must meet national standards set by the U.S. Environmental Protection Agency (EPA), under the federal *Resource Conservation and Recovery Act (RCRA)*, enacted in 1976 and amended in 1984.

In a sanitary landfill, waste is layered along with soil, a method that speeds decomposition, reduces odor, and lessens infestation by pests. Limited infiltration of rainwater allows for biodegradation by aerobic and anaerobic bacteria. To protect against environmental contamination, U.S. regulations require that landfills be located away from

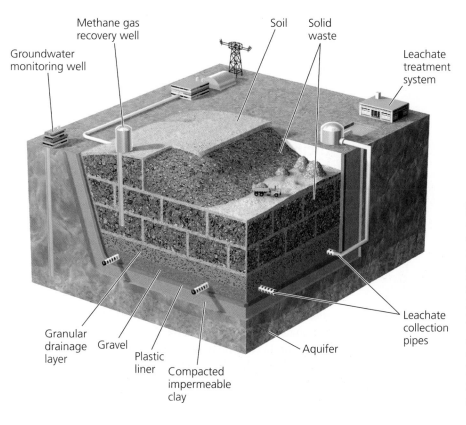

FIGURE 15.5 Sanitary landfills are engineered to prevent waste from contaminating soil and groundwater. Waste is laid in a large depression lined with plastic and impervious clay designed to prevent liquids from leaching out. Pipes of a leachate collection system draw out these liquids from the bottom of the landfill. Waste is layered along with soil until the depression is filled, and the waste continues to be built up until the landfill is capped. Landfill gas produced by anaerobic bacteria may be recovered, and waste managers monitor groundwater for contamination.

wetlands and earthquake-prone faults and be at least 6 m (20 ft) above the water table. The bottoms and sides of sanitary landfills must be lined with heavy-duty plastic and 60-120 cm (2-4 ft) of impermeable clay to help prevent contaminants from seeping into aquifers. Sanitary landfills also have systems of pipes, collection ponds, and treatment facilities to collect and treat **leachate**, liquid that results when substances from the trash dissolve in water as rainwater percolates downward. After a landfill is closed, it is capped with an engineered cover consisting of layers of plastic, gravel, and soil, and managers are required to maintain leachate collection systems for 30 years.

Although it was considered a model for advanced landfill technology at the time of its construction, the Fresh Kills Landfill predated most of the EPA guidelines. As a result, it caused some environmental contamination. However, engineers have retrofitted the landfill with clay liners and a sophisticated leachate collection system. Three of the six mounds have been capped with a "final cover," and the remaining mounds will soon be capped. Because these safeguards need to be maintained and monitored for 30 years after closure, designs for a public park at Fresh Kills have had to work around them.

In 1988 the United States had nearly 8,000 landfills, but today it has fewer than 1,800. One reason is that waste managers have closed many smaller landfills and consolidated the trash stream into larger landfills. A growing number of cities have been converting closed landfills into public parks (Figure 15.6). The Fresh Kills redevelopment endeavor will be the world's largest such landfill conversion project.

FIGURE 15.6 Old landfills, once properly capped, can serve other purposes. A number of them, such as Cesar Chavez Park in Berkeley, California, shown here, have been developed into areas for recreation.

Landfills have drawbacks

Despite improvements in liner technology and landfill siting, liners can be punctured, and leachate collection systems eventually cease to be maintained. Moreover, landfills are kept dry to reduce leachate, but the bacteria that break down material thrive in wet conditions. Dryness, therefore, slows waste decomposition.

Another problem is finding suitable areas to locate landfills, because most communities do not want them in their midst. This *not-in-my-backyard (NIMBY)* reaction is one reason why New York decided to export its waste and why residents of states receiving that waste are increasingly protesting. Landfills are rarely sited in neighborhoods that are home to wealthy and educated people with the political power to keep them out. Instead, they are disproportionately sited in poor and minority communities, as environmental justice advocates have frequently pointed out.

One famous case of long-distance waste transport illustrates the unwillingness of most communities to accept garbage. In Islip, New York, in 1987, the town's landfills were full, prompting town administrators to ship waste by barge to a methane production plant in North Carolina. Prior to the barge's arrival, it became known that the shipment was contaminated with 16 bags of medical waste, including syringes, hospital gowns, and diapers. Because of the medical waste, the methane plant rejected the entire load. The barge sat in a North Carolina harbor for 11 days before heading for Louisiana. Louisiana, however, would not permit the barge to dock. The barge traveled toward Mexico, but the Mexican navy prevented it from entering that nation's waters. In the end, the barge traveled 9,700 km (6,000 mi) before eventually returning to New York, where, after several court battles, the waste was finally incinerated at a facility in Queens.

Incinerating trash reduces pressure on landfills

Just as sanitary landfills are an improvement over open dumping, incineration in specially constructed facilities can be an improvement over open-air burning of trash. **Incineration**, or combustion, is a controlled process in which mixed garbage is burned at very high temperatures (Figure 15.7). At incineration facilities, waste is generally sorted and metals removed. Metal-free waste is chopped into small pieces to aid combustion and then is burned in a furnace. Incinerating waste reduces its weight by up to 75% and its volume by up to 90%.

However, the ash remaining after trash is incinerated contains toxic components and therefore must be disposed of in hazardous waste landfills (▸ p. 367). Moreover,

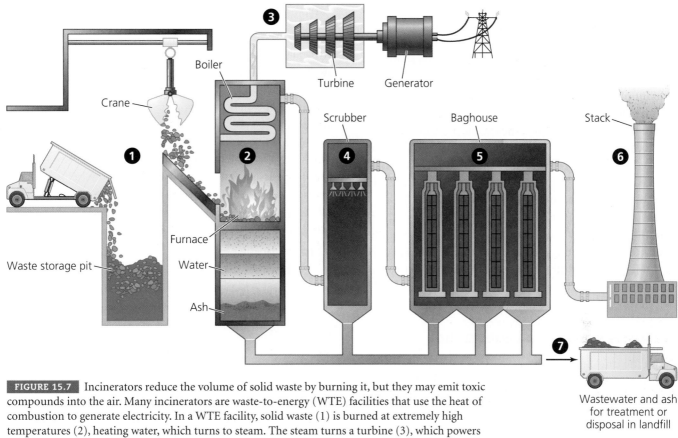

FIGURE 15.7 Incinerators reduce the volume of solid waste by burning it, but they may emit toxic compounds into the air. Many incinerators are waste-to-energy (WTE) facilities that use the heat of combustion to generate electricity. In a WTE facility, solid waste (1) is burned at extremely high temperatures (2), heating water, which turns to steam. The steam turns a turbine (3), which powers a generator to create electricity. In an incinerator outfitted with pollution control technology, a scrubber (4) chemically mitigates toxic gases produced by combustion , and airborne particulate matter is filtered physically in a baghouse (5) before air is emitted from the stack (6). Ash remaining from the combustion process is disposed of (7) in a landfill.

when trash is burned, hazardous chemicals—including dioxins, heavy metals, and PCBs (Chapter 10)—can be created and released into the atmosphere. Such releases caused a backlash against incineration from citizens concerned about public health.

Most developed nations now regulate incinerator emissions, some have banned incineration outright, and several technologies have been developed to mitigate emissions. *Scrubbers* chemically treat the gases produced in combustion to remove hazardous components and neutralize acidic gases, such as sulfur dioxide and hydrochloric acid, turning them into water and salt. Scrubbers generally do this either by spraying liquids formulated to neutralize the gases or by passing the gases through dry lime. Particulate matter is physically removed from incinerator emissions in a system of huge filters known as a *baghouse*. These tiny particles, called fly ash, often contain some of the worst dioxin and heavy metal pollutants. In addition, burning garbage at especially high temperatures can destroy certain pollutants, such as PCBs. Even all these measures, however, do not fully eliminate toxic emissions.

Weighing the Issues:
Environmental Justice?

Do you know where your trash goes? Where is your landfill or incinerator located? Who lives closest to the facility? Are the people in this neighborhood wealthy, poor, or middle-class? What race or ethnicity are they? Do you know whether the people of this neighborhood protested against the introduction of the landfill or incinerator?

We can gain energy from trash

Incineration was initially practiced simply to reduce the volume of waste, but it often serves to generate electricity as well. Most North American incinerators today are **waste-to-energy (WTE)** facilities that use the heat produced by waste combustion to boil water, creating steam that drives electricity generation or that fuels heating systems. When burned, waste generates approximately 35% of the energy generated by burning coal. Revenues from power generation, however, are usually not enough to offset the financial cost of building and running incinerators. Because it can take many years for a WTE facility to become profitable, many companies that build and operate these facilities require communities contracting with them to guarantee the facility a minimum amount of garbage. On occasion, such long-term commitments have interfered with communities' later efforts to reduce their waste through recycling and source-reduction.

Combustion in WTE plants is not the only way to gain energy from waste. Deep inside landfills, bacteria decompose waste in an oxygen-deficient environment. This anaerobic decomposition produces *landfill gas*, a mix of gases consisting of roughly half methane (▸ p. 56). Landfill gas can be collected, processed, and used in the same way as natural gas (▸ p. 307). At Fresh Kills, collection wells pull landfill gas upward through a network of pipes by vacuum pressure, and the gas should soon provide enough energy for 25,000 homes.

Reducing waste is a better option

Reducing the amount of material entering the waste stream avoids costs of disposal and recycling, helps conserve resources, minimizes pollution, and can often save consumers and businesses money. Preventing waste generation in this way is known as **source reduction**.

Much of our waste stream consists of materials used to package goods. Packaging serves worthwhile purposes—preserving freshness, preventing breakage, protecting against tampering, and providing information—but much packaging is extraneous. Consumers can help reduce packaging waste by choosing minimally packaged goods, buying unwrapped fruit and vegetables, and buying food in bulk. Consumer preference can give manufacturers incentive to reduce packaging. In addition, manufacturers can use packaging that is more recyclable. They can also reduce the size or weight of goods and materials, as they already have with many items, such as aluminum cans, plastic soft drink bottles, and personal computers. Finally, consumer choice can influence manufacturers to create goods that last longer.

Reuse is one main strategy for waste reduction

To reduce waste, you can save items to use again, or substitute disposable goods with durable ones. Habits as simple as bringing your own cup to coffee shops or bringing cloth bags to the grocery store can, over time, have substantial impact. You can also donate unwanted items and shop for used items yourself at yard sales and resale centers. Over 6,000 reuse centers exist in the United States, including stores run by organizations that resell donated

Table 15.1 Some Everyday Things You Can Do to Reduce and Reuse
▸ Donate used items to charity
▸ Reuse boxes, paper, plastic wrap, plastic containers, aluminum foil, bags, wrapping paper, fabric, packing material, and so on
▸ Rent or borrow items instead of buying them, when possible . . . and lend your items to friends
▸ Buy groceries in bulk
▸ Decline bags at stores when you don't need them
▸ Bring reusable cloth bags when you shop
▸ Make double-sided photocopies
▸ Bring your own cup to coffee shops
▸ Pay a bit extra for durable, long-lasting reusable goods
▸ Buy rechargeable batteries
▸ Select goods with less packaging
▸ Compost kitchen and yard wastes
▸ Buy clothing and other items at resale stores and garage sales
▸ Use cloth napkins and rags rather than paper napkins and towels
▸ Write to companies to tell them what you think about their packaging and products
▸ When solid waste policy is being debated, let your government representatives know your thoughts
▸ Support organizations that promote waste reduction

items, such as Goodwill Industries and the Salvation Army. Table 15.1 presents a sampling of actions that we all can take to reduce the waste we generate.

Composting recovers organic waste

Composting is the conversion of organic waste into mulch or humus through natural biological processes of decomposition. The compost can then be used to enrich soil. Householders can place waste in compost piles, underground pits, or specially constructed containers, where heat from microbial action builds in the interior and decomposition proceeds. Banana peels, coffee grounds, grass clippings, autumn leaves, and countless other organic items can be converted into rich, high-quality compost through the actions of earthworms, bacteria, soil mites, sow bugs, and other detritivores and decomposers. Home composting is a prime example of how we can live more sustainably by mimicking natural cycles and incorporating them into our daily lives.

Municipal composting programs—3,800 across the United States at last count—divert food and yard waste to central composting facilities, where they decompose into mulch that community residents can use for gardens and landscaping. Nearly half of U.S. states now ban yard waste from the municipal waste stream, helping accelerate the drive toward composting. Approximately one-fifth of the U.S. waste stream is made up of materials that can be easily composted. Composting reduces landfill waste, enriches soil and helps it resist erosion, encourages soil biodiversity, makes for healthier plants and more pleasing gardens, and reduces the need for chemical fertilizers.

Recycling consists of three steps

Recycling, too, offers many benefits. **Recycling** consists of collecting materials that can be broken down and reprocessed to manufacture new items. The recycling loop contains three basic steps (Figure 15.8). The first step is collecting and processing used recyclable goods and materials. Communities may designate locations where residents can drop off recyclables or receive money for them. Many of these have now been replaced by the more convenient option of curbside recycling, in which trucks pick up recyclable items in front of houses, usually in conjunction with municipal trash pickup. Curbside recycling has grown rapidly, and its convenience has helped boost recycling rates. More than 9,000 curbside recycling programs across all 50 U.S. states now serve nearly half of all U.S residents.

Items collected are taken to **materials recovery facilities (MRFs)**, where workers and machines sort items using magnetic pulleys, optical sensors, water currents, and air classifiers that separate items by weight and size. The facilities clean the materials, shred them, and prepare them for reprocessing.

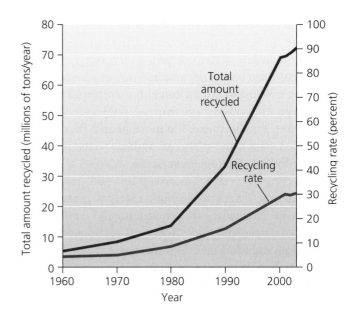

FIGURE 15.9 Recycling has risen sharply in the United States over the past 40 years. Today over 70 million tons of material is recycled, comprising more than 30% of the waste stream. Data from U.S. EPA, 2005.

Once readied, these materials are used in manufacturing new goods. Newspapers and many other paper products use recycled paper, many glass and metal containers are now made from recycled materials, and some plastic containers are of recycled origin. The pages in this textbook are made from recycled paper that is up to 20% post-consumer waste.

If the recycling loop is to function, consumers and businesses must complete the third step in the cycle by purchasing ecolabeled products (▶ pp. 45–46) made from recycled materials. Buying recycled goods provides economic incentive for industries to recycle materials, and for new recycling facilities to open or existing ones to expand.

Recycling has grown rapidly

The thousands of curbside recycling programs in place today have sprung up only in the last 20 years. Recycling in the United States has risen from 6.4% of the waste stream in 1960 to 23.5% in 2003 (and 30.6% if you include composting), according to EPA data (Figure 15.9).

Recycling rates vary greatly from one product or material type to another (e.g., from 22% for glass containers to 93% for auto batteries). Recycling rates among U.S. states also vary greatly, from less than 1% to nearly 50%.

Recycling's growth has been propelled in part by economic forces as established businesses see opportunities to save money and as entrepreneurs see opportunities to start new businesses. It has also been driven by the desire of municipalities to reduce waste and by the satisfaction

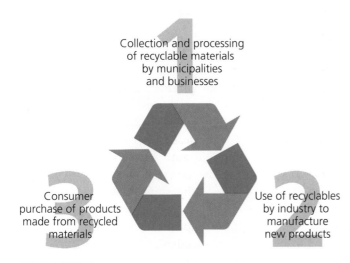

FIGURE 15.8 The familiar recycling symbol consists of three arrows to represent the three components of a sustainable recycling strategy: collection and processing of recyclable materials, use of the materials in making new products, and consumer purchase of these products.

people take in recycling. These two forces have driven the rise in recycling even though it has often not been financially profitable. In fact, many of the increasingly popular municipal recycling programs are run at an economic loss. The expense required to collect, sort, and process recycled goods is often more than recyclables are worth in the market. Furthermore, the more people recycle, the more glass, paper, and plastic is available to manufacturers for purchase, driving down prices.

Recycling advocates, however, point out that market prices do not take into account external costs (▸ pp. 28–29)—in particular, the environmental and health impacts of *not* recycling. For instance, it has been estimated that globally, recycling saves enough energy to power 6 million households per year. And recycling aluminum cans saves 95% of the energy required to make the same amount of aluminum from mined virgin bauxite, its source material.

As more manufacturers use recycled products, and as more technologies and methods are developed to use recycled materials in new ways, markets should continue to expand, and new business opportunities may arise. We are still at an early stage in the shift from an economy that moves linearly from raw materials to products to waste, to an economy that moves circularly, using waste products as raw materials for new manufacturing processes. The steps we have taken in recycling so far are central to this transition, which many analysts view as key to building a sustainable economy.

Weighing the Issues:
Costs of Recycling and Not Recycling

Should recycling programs be subsidized by governments even if they are run at an economic loss? What external costs—costs not reflected in market prices—do you think would be involved in not recycling, say, aluminum cans? Do you feel these costs justify sponsoring recycling programs even when they are not financially self-supporting? Why or why not?

Financial incentives can help address waste

Waste managers have employed economic incentives as tools to reduce the waste stream. The "pay-as-you-throw" approach to garbage collection uses a financial incentive to influence consumer behavior. In these programs, municipalities charge residents for home trash pickup according to the amount of trash they put out. The less waste the household generates, the less the resident has to pay. Over 4,000 of these programs now exist in the United States.

"Bottle bills" represent another approach that hinges on financial incentive. Eleven U.S. states have these laws, which allow consumers to return bottles and cans to stores after use and receive a refund—generally 5 cents per bottle or can. In states where they have been enacted, these laws have proved profoundly effective and resoundingly popular; bottle bills are recognized as among the most successful state legislation of recent decades.

One Canadian city showcases the shift from disposal to reduction and recycling

Edmonton, Alberta, has created one of the world's most advanced waste management programs. As recently as 1998, fully 85% of the city's waste was being landfilled, and space was running out. Today, just 35% goes to the new sanitary landfill, while 15% is recycled, and an impressive 50% is composted. Edmonton's citizens are proud of the program, and 81% of them participate in its curbside recycling program.

When Edmonton's residents put out their trash, city trucks take it to their new co-composting plant—at the size of eight football fields, the largest in North America. The bulk of the waste is mixed with dried sewage sludge for 1–2 days in five large rotating drums, each the length of six buses. The resulting mix travels on a conveyor to a screen that removes nonbiodegradable items. It is aerated for several weeks in the largest stainless steel building in North America (Figure 15.10). The mix is then passed through a finer screen and finally is left outside for 4–6 months. The resulting compost—80,000 tons annually—is made

FIGURE 15.10 Edmonton, Alberta, boasts one of North America's most successful waste management programs. Inside the aeration building, which is the size of 14 professional hockey rinks, mixtures of solid waste and sewage sludge are exposed to oxygen and composted for 14–21 days.

Recycling
VIEWPOINTS

Will we need to make recycling more economically profitable if it is to continue growing? What, if anything, should we do to encourage recycling? Or should we instead focus on other ways of managing waste?

How to Enhance Recycling?

Certainly recycling needs to be profitable if it is to survive and expand. The question is, how should it be made profitable?

Government subsidies are one common answer; many communities provide small subsidies to keep their recycling programs going. These subsidies will undoubtedly continue because people like municipal recycling programs. But subsidies are not likely to expand much beyond their current, modest level.

In some cases, new technologies are needed. Recycling plastics requires expensive sorting and separation of different plastics to create a high-quality product. With today's technology, recycling of unsorted, mixed plastics yields a low-value product with limited uses. New inventions that improve the sorting process or improve the quality of recycled mixed plastics could make plastics recycling much more profitable.

In other cases, we need new recycling programs and opportunities to keep up with changing lifestyles. Recycling rates are declining for beverage cans and bottles because so many beverages are consumed (and so many containers are discarded) away from home, at parks, beaches, and other public places. A system for recycling in public places could collect the growing quantities of beverage containers, newspapers, and other recyclable materials that are thrown out by people on the go.

Some products should be carefully recycled because it is hazardous to throw them out. Automobile batteries contain large amounts of lead, a toxic substance that should not be tossed in the trash. Some states have laws requiring a deposit on every battery that is sold—which makes it worthwhile to return a dead battery for a refund instead of discarding it. Similar approaches could and should be used with other potentially hazardous products.

Finally, it will never be possible to recycle everything. Along with continuing efforts to expand recycling, we must ensure that there are safe, clean opportunities for disposing of the remaining, nonrecyclable, waste.

Frank Ackerman is an economist at Tufts University's Global Development and Environ-ment Institute. He has advised the U.S. EPA and state and local agencies on waste management policies. His books include *Why Do We Recycle? Markets, Values, and Public Policies* (Island Press, 1997), and, with Lisa Heinzerling, *Priceless: On Knowing the Price of Everything and the Value of Nothing* (The New Press, 2004).

Recycling: A Mixed Bag

Recycling will continue as long as it is profitable. If it becomes more profitable, we will see more of it. Currently, 55% of all aluminum cans are recycled, a high rate compared to the 30% recycling rate that the Environmental Protection Agency says is average for solid waste. The reason: Aluminum companies can save money by using recycled materials because making cans from bauxite ore is expensive. Paper and cardboard are recycled to a great extent too, partly because cardboard can be made from many kinds of paper.

Plastics aren't recycled nearly as much (about 9%, according to the EPA). One reason is that different kinds of plastic resins can't be mixed. It is expensive for companies to separate the plastics before reprocessing them.

To increase recycling, governments would probably have to force people to recycle and require them to buy products made of recycled materials. To boost the city's recycling rate, Seattle has already made it illegal for residents to put recyclable materials in their regular trash.

This could make sense if recycling always saved resources. But does it?

Not always. The goal of recycling is to save resources, but mandatory recycling often wastes them. Additional trucks must go out into the community, using more energy and adding to air pollution. And reprocessing recyclables is just another form of manufacturing, which inevitably causes some pollution and waste.

Fortunately, there are additional ways to deal with trash. Modern landfills are scientifically engineered to keep garbage dry so that harmful leakage doesn't occur. And there is plenty of space. One widely cited estimate is that the United States could bury all its trash for the next century in a landfill 225 feet deep and 10 miles square.

We should recycle when it makes sense, but we shouldn't be afraid to use other means as well.

Jane S. Shaw is a senior fellow of the Property and Environment Research Center (PERC), a nonprofit institute in Bozeman, Montana, dedicated to improving environmental quality through property rights and markets. With Michael Sanera, she is coauthor of *Facts, Not Fear: Teaching Children about the Environment* (Regnery, 1999) and editor of the Greenhaven Press book series, *Critical Thinking about Environmental Issues*.

Explore this issue further by accessing **Viewpoints** at www.aw-bc.com/withgott.

available to area farmers and residents. Edmonton's program also includes a state-of-the-art MRF that handles 30,000–40,000 tons of waste annually, a leachate treatment plant, a research center, public education programs, and a wetland and landfill revegetation program. In addition, 100 pipes collect enough landfill gas to power 4,000 homes, bringing thousands of dollars to the city and helping power the new waste management center. Five area businesses reprocess the city's recycled items. Newsprint and magazines are turned into new newsprint and cellulose insulation, and cardboard and paper are converted into building paper and shingles. Household metal is made into rebar and blades for tractors and graders, and recycled glass is used for reflective paint and signs.

Industrial Solid Waste

Each year, U.S. industrial facilities generate about 7.6 billion tons of waste, according to the EPA, about 97% of which is wastewater. Thus, very roughly, 228 million or so tons of solid waste are generated by 60,000 facilities each year—an amount about equal to that of municipal solid waste. In the United States, industrial solid waste is defined as solid waste that is considered neither municipal solid waste nor hazardous waste under the Resource Conservation and Recovery Act. Whereas the federal government regulates municipal solid waste, state or local governments regulate industrial solid waste (with federal guidance). Industrial waste includes waste from factories, mining activities, agriculture, petroleum extraction, and more.

Regulation and economics both influence industrial waste generation

Most methods and strategies of waste disposal, reduction, and recycling by industry are similar to those for municipal solid waste. Businesses that dispose of their own waste on site must design and manage landfills in ways that meet state, local, or tribal guidelines. Other businesses pay to have their waste disposed of at municipal disposal sites. Regulation varies greatly from place to place, but in most cases, state and local regulation of industrial solid waste is less strict than federal regulation of municipal solid waste. In many areas, industries are not required to have permits, install landfill liners or leachate collection systems, or monitor groundwater for contamination.

The amount of waste generated by a manufacturing process is one measure of its efficiency; the less waste produced per unit or volume of product, the more effi-

cient that process is, from a physical standpoint. However, physical efficiency is not always equivalent to economic efficiency. Often it is cheaper for industry to manufacture its products or perform its services quickly but messily. That is, it can be cheaper to generate waste than to avoid generating waste. In such cases, economic efficiency is maximized, but physical efficiency is not. The frequent mismatch between these two types of efficiency is a major reason why the output of industrial waste is so great.

Rising costs of waste disposal, however, enhance the financial incentive to decrease waste and increase physical efficiency. Once either government or the market makes the physically efficient use of raw materials also economically efficient, businesses have financial incentives to reduce their waste.

Industrial ecology seeks to make industry more sustainable

To reduce waste, growing numbers of industries today are experimenting with industrial ecology. A holistic approach that integrates principles from engineering, chemistry, ecology, and economics, **industrial ecology** seeks to redesign industrial systems to reduce resource inputs and to maximize both physical and economic efficiency. Industrial ecologists would reshape industry so that nearly everything produced in a manufacturing process is used, either within that process or in a different one.

The larger idea behind industrial ecology is that industrial systems should function more like ecological systems, in which almost everything produced is used by some organism, with very little being wasted. This principle brings industry closer to the ideal of ecological economists, in which human economies attain sustainability by functioning in a circular fashion rather than a linear one (▶ p. 29).

Industrial ecologists pursue their goals in several ways. For one, they examine the entire life cycle of a given product—from its origins in raw materials, through its manufacturing, to its use, and finally its disposal—and look for ways to make the process more efficient. This strategy is called **life-cycle analysis**. Industrial ecologists also try to identify points at which waste products from one manufacturing process can be used as raw materials for a different process. In addition, industrial ecologists examine industrial processes with an eye toward eliminating environmentally harmful products and materials. Finally, they study the flow of materials through industrial systems to look for ways to create products that are more durable, recyclable, or reusable.

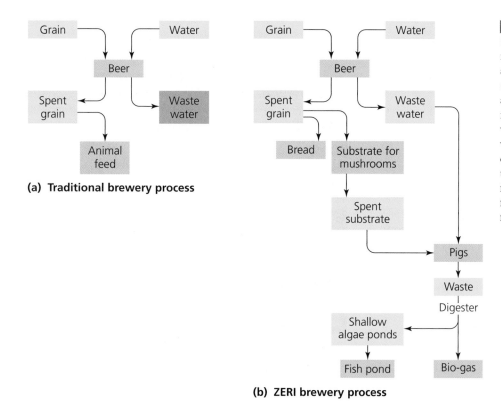

(a) Traditional brewery process

(b) ZERI brewery process

FIGURE 15.11 Traditional breweries (**a**) produce only beer while generating much waste, some of which goes toward animal feed. ZERI-sponsored breweries (**b**) use their waste grain to make bread and to farm mushrooms. Waste from the mushroom farming, along with brewery wastewater, goes to feed pigs. The pigs' waste is digested in containers that capture natural gas and collect nutrients used to nourish algae for growing fish in fish farms. The brewer derives income from bread, mushrooms, pigs, gas, and fish, as well as beer.

Attentive businesses are taking advantage of the insights of industrial ecology to save money while reducing waste. The Swiss Zero Emissions Research and Initiatives (ZERI) Foundation sponsors dozens of innovative projects worldwide that attempt to create goods and services in closed-loop systems, without generating waste. In so doing, they cut down on waste while increasing output and income, and often they generate new jobs as well. One example involves breweries, currently being pursued in Canada, Sweden, Japan, and Namibia (Figure 15.11).

Hazardous Waste

Hazardous wastes are diverse in their chemical composition and may be liquid, solid, or gaseous. By EPA definition, **hazardous waste** is waste that is one of the following:

▶ *Ignitable.* Substances that easily catch fire (for example, natural gas or alcohol).

▶ *Corrosive.* Substances that corrode metals in storage tanks or equipment.

▶ *Reactive.* Substances that are chemically unstable and readily react with other compounds, often explosively or by producing noxious fumes.

▶ *Toxic.* Substances that harm human health when they are inhaled, are ingested, or contact human skin.

Hazardous wastes have diverse sources

Industry, mining, households, small businesses, agriculture, utilities, and building demolition all create hazardous waste. Industry generates the largest amounts, but in most developed nations industrial waste disposal is highly regulated. This regulation has reduced the amount of hazardous waste entering the environment from industrial activities. As a result, households currently are the largest source of unregulated hazardous waste.

Household hazardous waste includes a wide range of items, such as paints, batteries, oils, solvents, cleaning agents, lubricants, and pesticides. U.S. citizens generate 1.6 million tons of household hazardous waste annually, and the average home contains close to 45 kg (100 lb) of it in sheds, basements, closets, and garages.

Many hazardous substances become less hazardous over time as they degrade, but some show especially persistent effects. Radioactive substances are an example, and we discussed radioactive waste and the dilemma of its disposal in Chapter 13 (▶ pp. 324–325). Other types of persistent hazardous substances include organic compounds and heavy metals.

Organic compounds and heavy metals can be hazardous

In our day-to-day lives, we rely on the capacity of synthetic organic compounds and petroleum-derived compounds to resist bacterial, fungal, and insect activity. Items such as plastic containers, rubber tires, pesticides, solvents, and wood preservatives are useful to us precisely because they resist decomposition. However, the resistance of these compounds to decay is a double-edged sword, for it also makes them persistent pollutants. Many synthetic organic compounds are toxic because they can be readily absorbed through the skin of humans and other animals and can act as mutagens, carcinogens, teratogens, and endocrine disruptors (▶ pp. 223–224).

Heavy metals such as lead, chromium, mercury, arsenic, cadmium, tin, and copper are used widely in industry for wiring, electronics, metal plating, metal fabrication, pigments, and dyes. Heavy metals enter the environment when paints, electronic devices, batteries, and other materials are disposed of improperly. Lead from fishing weights and from hunters' lead shot has accumulated in many rivers, lakes, and forests. In older homes, lead from pipes contaminates drinking water, and lead paint remains a problem, especially for infants. Heavy metals that are fat-soluble and break down slowly are prone to bioaccumulate (▶ p. 227).

Computers, televisions, VCRs, cell phones, and other electronic devices represent major new sources of potential heavy metal contamination. These products have short lifetimes before people judge them obsolete, and most are discarded after only a few years. The amount of this electronic waste—sometimes called *e-waste*—is growing. In the United States alone, there are well over 300 million television sets, and the National Safety Council has estimated that 500 million computers will have been retired between 1997 and 2007. Most e-waste is still disposed of in landfills as conventional solid waste, but recent research suggests that it should instead be treated as hazardous waste (see "The Science behind the Story," ▶ pp. 368–369).

Several steps precede the disposal of hazardous waste

Since the 1980s, many communities have designated sites or special collection days to gather household hazardous waste, or facilities for the exchange and reuse of substances (Figure 15.12). Once consolidated in such sites, the waste is transported for treatment and ultimate disposal.

FIGURE 15.12 Many communities designate collection sites or collection days for household hazardous waste. Here, workers handle waste from an Earth Day collection event near Los Angeles.

Under the Resource Conservation and Recovery Act, the EPA sets standards by which states are to manage hazardous waste. RCRA also requires large generators of hazardous waste to obtain permits, and it mandates that hazardous materials be tracked "from cradle to grave." As hazardous waste is generated, transported, and disposed of, the producer, carrier, and disposal facility must each report to the EPA the type and amount of material generated; its location, origin, and destination; and the way it is being handled.

Because current U.S. law makes disposing of hazardous waste quite costly, some irresponsible companies have been found guilty of illegally dumping waste, creating health risks for residents and financial headaches for local governments forced to deal with the mess (Figure 15.13). However, high costs have also encouraged responsible businesses to invest in reducing their hazardous waste. Many biologically hazardous materials can be broken down by incineration at high temperatures in cement kilns. Some hazardous materials can be treated by exposure to bacteria that break down harmful components and synthesize them into new compounds. Various plants have now been bred or engineered to take up specific contaminants from soil, then break down organic contaminants into safer compounds or concentrate heavy metals in their tissues. The plants are eventually harvested and disposed of.

FIGURE 15.13 Unscrupulous individuals or businesses sometimes dump hazardous waste illegally to avoid disposal costs.

We have three disposal methods for hazardous waste

We have developed three primary methods for disposing of hazardous waste: landfills, surface impoundments, and injection wells. These do nothing to lessen the hazardous nature of the waste, but they do help keep it isolated from people, wildlife, and ecosystems. Design and construction standards for hazardous waste landfills are stricter than those for ordinary sanitary landfills. Hazardous waste landfills must have several impervious liners and leachate removal systems, and they must be located far from aquifers. Dumping of hazardous waste in ordinary landfills has long been a problem. In New York City, Fresh Kills largely managed to keep hazardous waste out, but most of the city's older landfills were declared to be hazardous sites because of past toxic waste dumping.

A method for storing liquid hazardous waste or waste in dissolved form is in **surface impoundments**. A shallow depression is dug and lined with plastic and an impervious material, such as clay. Water containing dilute hazardous waste is placed in the pond and allowed to evaporate, leaving a residue of solid hazardous waste on the bottom. This process is repeated until the dry material is removed and transported elsewhere for permanent disposal. Impoundments are not ideal. The underlying layer can crack and leak waste. Some material may evaporate or be blown into surrounding areas. Rainstorms may cause waste to overflow and contaminate nearby areas. For these reasons, surface impoundments are used only for temporary storage.

The third method is intended for long-term disposal. In **deep-well injection** (Figure 15.14), a well is drilled

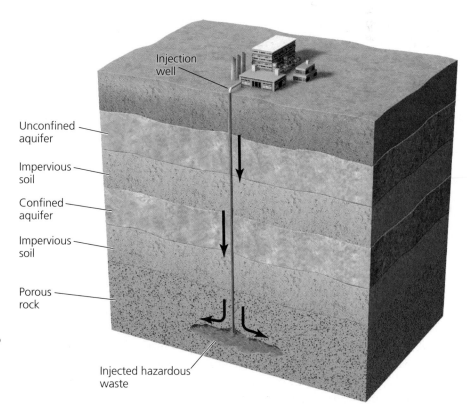

FIGURE 15.14 One way to dispose of liquid hazardous waste is to pump it deep underground, in deep-well injection. The well must be drilled below any aquifers, into porous rock separated by impervious clay. The technique is expensive, however, and leakage from the well shaft into groundwater may occur.

Injection well

Unconfined aquifer

Impervious soil

Confined aquifer

Impervious soil

Porous rock

Injected hazardous waste

Testing the Toxicity of "E-Waste"

As we began to conduct more of our business, learning, and communication with computers and other electronic devices, many people predicted that our waste, particularly paper waste, would decrease. Instead, the proliferation of computers, printers, VCRs, fax machines, cell phones, and other gadgets has created a substantial new source of waste.

Most of this electronic waste, or "e-waste," is landfilled as conventional solid waste. However, most electronic appliances contain heavy metals that can cause environmental health risks. For instance, over 6% of a typical computer is composed of lead.

At the University of Florida, Gainesville, the EPA funded Timothy Townsend's lab to determine whether e-waste is toxic enough to be classified as hazardous waste under the Resource Conservation and Recovery Act.

With students and colleagues, Townsend determined in 1999–2000 that cathode ray tubes (CRTs) from computer monitors and color televisions leach an average of 18.5 mg/L of lead, far above the regulatory threshold of 5 mg/L. Following this research, the EPA proposed classifying CRTs as hazardous waste, and several U.S. states banned these items from conventional landfills.

Discarded electronic waste can leach heavy metals and should be considered hazardous waste, researchers say.

Then in 2004, Townsend's lab group completed experiments on 12 other types of electronic devices. To measure their toxicity, Townsend's group used the EPA's standard test, the Toxicity Characteristic Leaching Procedure (TCLP), designed to mimic the process by which chemicals leach out of solid waste in landfills. In the TCLP, waste is ground up into fine pieces and 100 g (3.5 oz) is put in a container with 2 L (0.53 gal) of an acidic leaching fluid. The container is rotated for 18 hours, after which the leachate is analyzed for its chemical content. Researchers look

for eight heavy metals—arsenic, barium, cadmium, chromium, lead, mercury, selenium, and silver—and determine for each whether their concentration exceeds that allowed by EPA regulations. Of these elements, electronic devices contain notable amounts of cadmium, chromium, lead, and mercury.

To conduct the standard TCLP, Townsend's team ground up the central processing units (CPUs) of personal computers, creating a mix made up by weight of 15.8% circuit board, 7.5% plastic, 68.2% ferrous metal, 5.4% nonferrous metal, and 3.1% wire and cable. However, grinding up a computer into small bits is no easy task, and it is hard to obtain a sample that accurately represents all components and materials. So the researchers also designed a modified TCLP test in which they placed whole CPUs—with the parts disassembled but not ground up—in a rotating 55-gallon drum full of leaching liquid. Then they tested their 12 types of devices using a combination of the standard and modified TCLP methods.

The team's results are summarized in the accompanying bar graph. Lead was the only heavy metal found to exceed the EPA's regulatory threshold, but this threshold (5 mg/L) was exceeded in the majority of trials. Computer moni-

deep beneath the water table, into porous rock, and wastes are injected into it. The aim is that waste will remain deep underground, isolated from groundwater and human contact. However, wells can corrode and leak wastes into soil, allowing them to enter aquifers. Roughly 34 billion L (9 billion gal) of hazardous waste are placed in U.S. injection wells each year.

Contaminated sites are being cleaned up, slowly

In 1980 the U.S. Congress passed the Comprehensive Environmental Response Compensation and Liability Act (CERCLA). This legislation established a federal program to clean up U.S. sites polluted with hazardous waste from past activities. The EPA administers this cleanup program,

tors leached the most lead (47.7 mg/L on average), as expected, because monitors include the cathode ray tubes already known to be a problem. However, laptops, color TVs, smoke detectors, cell phones, and computer mice also leached high levels of lead. Next came remote controls, VCRs, keyboards, and printers, all of which leached more lead on average than the EPA threshold, and did so in 50% or more of the trials. Whole CPUs and flat panel monitors were the only devices to leach less than 5 mg/L of lead on average, but even these exceeded the threshold more than one-quarter of the time.

The researchers found that items containing more ferrous metals (such as iron) tended to leach less lead. Further experiments confirmed that ferrous metals were chemically reacting with lead and stopping it from leaching.

Townsend says the work suggests that many electronic devices have the potential to be classified as hazardous waste. However, EPA scientists must decide how to judge results from the modified TCLP methods, and must evaluate other research, before determining whether to alter regulatory standards.

Furthermore, lab tests may not accurately reflect what actually happens in landfills. So

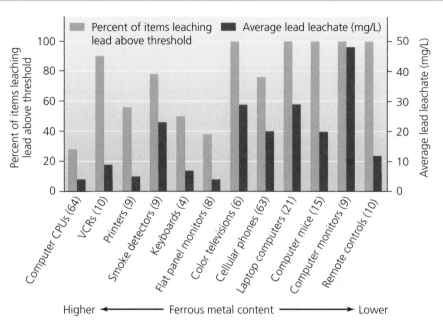

Some proportion of all 12 devices tested exceeded the EPA regulatory standard for lead leachate. Devices with higher ferrous metal content tended to leach less lead. Where both standard and modified TCLP methods were used, results are averaged. Data from Townsend, T. G., et al., 2004. RCRA toxicity characterization of computer CPUs and other discarded electronic devices. July 15, 2004, report to the U.S. EPA.

Townsend's team is filling columns measuring 24 cm (2 ft) wide by 4.9 m (16 ft) long with e-waste and municipal solid waste, burying them in a Florida landfill, and then testing the leachate that results.

As the EPA and more states move toward keeping e-waste out of conventional landfills, more computers and accessories are being recycled. Although there are serious concerns about the health risks this

may pose to workers doing the disassembly, recycling done responsibly seems likely to be the way of the future.

In many North American cities, businesses, nonprofit organizations, or municipal services now recycle used computers and related devices. So next time you upgrade to a new computer, TV, DVD player, VCR, or cell phone, check out what opportunities may exist in your area to recycle your old ones.

---

called the **Superfund.** Under EPA auspices, experts identify sites polluted with hazardous chemicals, take action to protect groundwater near these sites, and clean up the pollution. The objective of the act is to charge responsible parties for cleanup of sites (according to the *polluter-pays principle*). For many polluted sites, however, the responsible parties cannot be found or held liable, and in such

cases—roughly one of every four so far—Superfund activities are covered by taxpayers' funds. These funds come from the federal budget ($1.4 billion in 2005), and from a trust fund (which went bankrupt in 2004) established from a tax on chemical raw materials.

Once a Superfund site is identified, EPA scientists evaluate how close the site is to human habitation, whether wastes are currently confined or likely to spread, and whether the site threatens drinking water supplies. Sites that appear harmful are placed on EPA's National Priority List, ranked according to the level of risk to human health that they pose. Cleanup proceeds on a site-by-site basis as funds are available. Throughout the process, the EPA is required to hold public hearings to inform area residents of their findings and to receive feedback.

As of mid-2006, 975 of the 1,553 Superfund sites on the National Priority List had been cleaned up. The average cleanup has cost $25 million and has taken 12–15 years. Increasingly, emphasis in the United States and elsewhere is being placed on preventing hazardous waste contamination in the first place.

Conclusion

Our societies have made great strides in addressing our waste problems. Modern methods of waste manage-

ment are far safer for people and gentler on the environment than past practices of open dumping and open burning. In many countries, recycling and composting are making rapid strides. The United States has gone in a few decades from a country that virtually did not recycle to a nation in which 30% of all solid waste is diverted from disposal. The continuing growth of recycling, driven by market forces, government policy, and consumer behavior, shows potential to further alleviate our waste problems.

Despite these advances, our prodigious consumption habits have created more waste than ever before. Our waste management efforts are marked by a number of difficult dilemmas, including the cleanup of Superfund sites, safe disposal of hazardous and radioactive waste, and frequent local opposition to disposal sites. These dilemmas make clear that the best solution to our waste problem is to reduce our generation of waste. Finding ways to reduce, reuse, and efficiently recycle the materials and goods that we use stands as a key challenge for the new century.

TESTING YOUR COMPREHENSION

1. Describe five major methods of managing waste. Why do we practice waste management?
2. Why have some people labeled the United States "the throwaway society"? How much solid waste do Americans generate, and how does this amount compare to that of people from other countries?
3. Name several guidelines by which sanitary landfills are regulated. Describe at least three problems with landfills.
4. Describe the process of incineration or combustion. What happens to the resulting ash? What is one drawback of incineration?

5. Explain composting. How does it help reduce input to the waste stream?
6. What are the three elements of a sustainable process of recycling?
7. Describe the goals of industrial ecology.
8. What four criteria are used to define hazardous waste? Why are heavy metals and synthetic organic compounds particularly hazardous?
9. What are the largest sources of hazardous waste? Describe three ways to dispose of hazardous waste.
10. What is the Superfund program? How does it work?

SEEKING SOLUTIONS

1. How much waste do you generate? Look into your waste bin at the end of the day and categorize and measure the waste there. List all other waste you may have generated in other places throughout the day. How much of this waste could you have avoided generating? How much could have been reused or recycled?
2. Some people have criticized current waste management practices as merely moving waste from one medium to another. How might this criticism apply to the methods now in practice? What are some potential solutions?

3. Of the various waste management approaches covered in this chapter, which ones are your community or campus pursuing, and which are they not pursuing? Would you suggest that your community or campus start pursuing any new approaches? If so, which ones, and why?
4. Imagine you head a major corporation that produces containers for soft drinks and other consumer products. Your company's shareholders are asking that you improve the company's image—while not cutting into

profits—by taking steps to reduce waste. What steps would you consider taking?

5. Think of several industries or businesses in your community, as well as the ways these interact with facilities on your campus. Bearing in mind the principles of industrial ecology, can you think of any novel ways that these entities might mutually benefit from one another's services, products, or waste materials? Are there waste products from one business, industry, or campus facility that another might put to good use? Can you design an eco-industrial park that might work in your community or on your campus?

CALCULATING ECOLOGICAL FOOTPRINTS

 The 14th annual "State of Garbage in America" survey documents the prodigious ability of U.S. residents to generate municipal solid waste (MSW). According to the survey, on a per capita basis, South Dakotans generate the least MSW (3.72 lb/day), and Kansans the most (9.48 lb/day). The average for the entire country is 7.17 lb MSW per person per day. In comparison, a 1995 study estimated that people of high-income nations generate an average of 2.64 lb/day, while the world average was 1.47 lb/day. Calculate the amount of MSW generated in 1 day and in 1 year by each of the groups indicated at each of the rates shown in the accompanying table.

Groups generating municipal solid waste	Per-capita MSW generation rates							
	U.S. average (7.17 lb/day)		South Dakota (3.72 lb/day)		"High-income" countries (2.64 lb/day)*		World average (1.47 lb/day)*	
	Day	Year	Day	Year	Day	Year	Day	Year
You	7.17	2,617.05						
Your class								
Your town								
Your state								
United States								
World								

Data sources: Kaufman, S. M. et al. 2004. The state of garbage in America. *BioCycle* 45: 31–41.

*Beede, D. N. and D. E. Bloom. 1995. The economics of municipal solid waste. *World Bank Research Observer* 10: 113–150.

1. Suppose your town of 50,000 people has just approved construction of a landfill nearby. Estimates are that it will accommodate 1 million tons of MSW. Assuming the landfill is serving only your town, for how many years will it accept waste before filling up? How much longer would a landfill of the same capacity serve a town of the same size in another industrialized ("high-income") country?

2. Why do you think U.S. residents generate so much more MSW than people in other "high-income" countries, when standards of living in those countries are comparable?

Take It Further

Go to www.aw-bc.com/withgott, where you'll find:

▶ Suggested answers to end-of-chapter questions
▶ Quizzes, animations, and flashcards to help you study
▶ *Research Navigator*™ database of credible and reliable sources to assist you with your research projects
▶ **GRAPHit!** tutorials to help you interpret graphs
▶ **INVESTIGATEit!** current news articles that link the topics that you study to case studies from your region to around the world

Epilogue   Sustainable Solutions

The notion of sustainability has run throughout this book. Sustainability is an outlook and approach relevant to every environmental issue—and indeed, to every aspect of our daily lives. As more and more people come to appreciate Earth's limited capacity to accommodate our rising population and consumption, they are voicing concern that we will need to modify our behaviors, institutions, and technologies if we wish to sustain our civilization and the natural environment on which it depends.

When people speak of *sustainability*, what precisely do they mean to sustain? Generally they mean to sustain human institutions in a healthy and functional state—and also to sustain the natural environment, its biodiversity, and its ecological systems in a healthy and functional state. The short-term needs of human society and of environmental sustenance are often cast as being in opposition, but environmental scientists recognize that our civilization cannot exist without an intact and functioning natural environment. We wholly depend on the contributions of biodiversity (▸ pp. 179–181) and ecosystem goods and services (▸ pp. 25–27, 30–32), although we have long taken them for granted.

We can develop sustainably

In recent years, people have increasingly realized how our quality of life depends on environmental quality. Moreover, we now recognize that society's poorer people often suffer the most from environmental degradation. This realization has led advocates of environmental protection, economic development, and social justice to work together toward common goals. This cooperative approach has given rise to the modern drive for sustainable development (▸ p. 20, 40).

Not long ago, most people might have thought "sustainable development" to be an oxymoron, a phrase that contradicts itself. *Development*—making purposeful changes intended to improve the quality of human life—often so degrades the natural environment that it threatens the very improvements for human life that were intended. The question "Can we develop in a sustainable way?" may well be the single most important question in the world today. Increasingly, efforts being made by governments, businesses, industries, organizations, and individuals across the globe are giving people optimism that

developing in sustainable ways is possible. Students at colleges and universities are playing a crucial role. They are creating models for the wider world by leading sustainability initiatives on their campuses (Figure E.1).

Environmental protection can enhance economic well-being

What is good for the environment can be good for people, and protecting environmental quality can improve the economic bottom line. For individuals, businesses, and institutions, reducing resource consumption and waste can save money. For society, attention to environmental quality can enhance economic opportunity by providing new types of employment. In a transition to a sustainable economy, some industries will decline, while others will spring up to take their place. As jobs in logging, mining, and manufacturing decreased in developed nations over the past few decades, jobs greatly increased in many service occupations and high-technology sectors. If we decrease our dependence on fossil fuels, experts predict, jobs and investment opportunities will blossom in renewable energy sectors, such as wind power and fuel cell technology (Chapter 14).

Moreover, people desire to live in areas that have clean air and water, intact forests, and parks and open space. Environmental protection increases a region's attractiveness, drawing residents and increasing property values and the tax revenues that fund social services. As a result, regions that act to protect their environments are generally the ones that retain and increase their wealth and quality of life. Thus, environmental protection need not lead to economic stagnation, but instead is likely to enhance economic opportunity. Indeed, a recent U.S. government review concluded that the economic benefits of environmental regulations greatly exceed their economic costs. Both the U.S. economy and the global economy have expanded rapidly in the past 30 years, the very period during which environmental protection measures have proliferated.

We can consume less

Although environmental protection can enhance economic well-being, economic *growth* is another matter (▸ p. 29). Economic growth is driven largely by consumption,

(a) Lewis Center at Oberlin

(b) Gardening at Middlebury

(c) "Landfill on the lawn" at UNC-Greensboro

(d) Low-emission buses at UC Davis

FIGURE E.1 Students are helping to guide sustainability efforts on hundreds of college and university campuses, serving as models for the wider world. Oberlin College's Adam Joseph Lewis Center for Environmental Studies (**a**) was one of the first "green buildings" on a U.S. campus. A number of campuses now include gardens such as this one at Middlebury College in Vermont (**b**), where students grow organic vegetables used for meals in dining halls. In "landfill on the lawn" events such as this one at the University of North Carolina at Greensboro (**c**), students retrieve recyclable items from piles of rubbish dumped onto a campus open space, offering a dramatic demonstration of just how many recyclables are needlessly thrown away. Transportation is a frequent component of campus sustainability efforts; most buses at the University of California at Davis (**d**) are low-emission vehicles that run on compressed natural gas, and one runs on a novel mixture of natural gas and hydrogen.

the purchase of material goods and services (and thus the use of resources involved in their manufacture) by consumers (Figure E.2). Our tendency to believe that more, bigger, and faster are always better is reinforced by advertisers seeking to sell us more goods more quickly. The United States, with less than 5% of the world's population, consumes 30% of the world's energy resources and 40% of total global resources. U.S. houses are larger than ever, sports-utility vehicles are among the most popular automobiles, and many citizens have more material belongings than they know what to do with. Our lavishly consumptive lifestyles are a brand-new phenomenon on Earth. We are enjoying the greatest material prosperity in all of human history, but if we do not find ways to make our wealth sustainable, the

FIGURE E.2 Citizens of the United States consume more than the people of any other nation. Unless we find ways to increase the sustainability of our manufacturing processes, our rising rate of consumption cannot be sustained in the long run.

party may not last much longer. Many of Earth's natural resources are limited and nonrenewable, so if we do not shift to sustainable practices of resource use, per capita consumption will drop for rich and poor alike as resources dwindle.

Fortunately, material consumption is only one way to measure prosperity, and consumption alone does not reflect a person's quality of life. True prosperity consists of an increase in human happiness, not simply growth in material wealth. Every one of us can find ways to reduce our consumption while enhancing our quality of life. At the outset, choices to consume less may seem like sacrifices, but people who have slowed down the pace of their busy lives and freed themselves of an attachment to excess material possessions say it can feel tremendously liberating.

Population growth must eventually cease

Just as continued growth in consumption is not sustainable, neither is growth in the human population. We have seen (▶ pp. 88–89) that populations may grow exponentially for a time but eventually encounter limiting factors and level off at a carrying capacity. We have increased Earth's carrying capacity for our species with the help of technology, but our population growth cannot continue forever; sooner or later, the human population will stop growing. The question is how: through war, plagues, and famine, or through voluntary means as a result of wealth and education?

The demographic transition (▶ pp. 131–132) that many nations are undergoing provides reason to hope that population sizes will stabilize and begin to fall. This transition is already far along in many developed nations thanks to urbanization, wealth, education, and the empowerment of women. If the demographic transition occurs for today's developing nations, then there is hope that humanity may halt its population growth while creating more prosperous and equitable societies.

Technology can help us toward sustainability

It is largely technology—developed with the agricultural revolution, the industrial revolution, and advances in medicine and health—that has spurred our population increase. Technology has magnified our impacts on Earth's environmental systems. However, technology also can give us ways to reduce environmental impact. The shortsighted use of high-impact technologies may have gotten us into this mess, but wiser use of environmentally friendly, or "green," technologies can help get us out.

In recent years, technology has exacerbated environmental impact in developing countries as industrial technologies from the developed world have been exported to poorer nations eager to industrialize. In developed nations, meanwhile, green technologies have begun mitigating our environmental impact. Catalytic converters on cars have reduced emissions, as have scrubbers on industrial smokestacks. Recycling technology and advances in wastewater treatment are helping reduce our waste output. Solar, wind, and geothermal energy technologies are producing cleaner renewable energy. Countless technological advances such as these are one reason that people of the United States and western Europe today enjoy cleaner environments, although they consume far more, than people of eastern Europe and rapidly industrializing nations such as China.

We can follow several strategies for sustainability

Shifts in outlook regarding consumption, population, and the deployment of technology are already occurring, but they will likely need to accelerate if we are to find truly lasting win-win solutions for the human condition and for Earth's environmental systems. Sustainable solutions to environmental problems are numerous, and we have seen specific examples throughout this book. Let's now survey 10 broad strategies or approaches that can spawn sustainable solutions (Table E.1).

Table E.1 Some Major Approaches to Sustainability

▶ Refine our ideas about economic growth and quality of life

▶ Reduce unnecessary consumption

▶ Limit population growth

▶ Encourage green technologies

▶ Mimic natural systems by promoting closed-loop industrial processes

▶ Enhance local self-sufficiency

▶ Vote with our wallets

▶ Vote with our ballots

▶ Think in the long term

▶ Promote research and education

We have already touched on four of these approaches: redefining our priorities on economic growth and quality of life, reducing unnecessary consumption, limiting population growth, and encouraging green technologies. Other vital economic goals are to implement green taxes (▶ p. 45), phase out harmful subsidies (▶ pp. 44–45) and incorporate external costs (▶ pp. 28–29) into the market prices of goods and services. Currently, goods and services are priced as though pollution and resource extraction involved no costs to society. If we can make our accounting practices reflect indirect negative consequences and provide a clearer view of the full costs and benefits of any given action or product, then the free market can become a force for improving environmental quality and our quality of life.

Additionally, we can strive to make our industrial systems circular and recycling-oriented. We have an excellent model: nature itself. As we saw in Chapter 3 and throughout this book, environmental systems tend to operate in cycles consisting of feedback loops and the circular flow of materials. In natural systems, output is recycled into input. In contrast, human manufacturing processes have run on a linear model in which raw materials are input, go through processing, and create a product, while by-products and waste are generated and disposed of. Some forward-thinking industrialists are already taking steps toward making their industrial processes sustainable (▶ pp. 364–365).

In another approach, many proponents of sustainability believe that encouraging local self-sufficiency is an important element of building sustainable societies. When people are tied closely to the area in which they live, they tend to value the area and seek to sustain its environment and human communities. This line of argument is frequently made regarding the growing and distribution of food, specifically in encouraging locally based organic or sustainable agriculture.

However, even a global capitalist free-market system driven by consumerism holds one great asset for sustainability: Consumers can exercise influence through what they choose to buy. When products produced sustainably are ecolabeled (▶ pp. 45–46), consumers can "vote with their wallets" by preferentially purchasing these products. Consumer choice already drives sales of everything from recycled paper to organic produce to "dolphin-safe" tuna.

Although economic leverage is powerful, many changes needed to attain sustainable solutions require policymakers to usher them through. Policymakers respond to whoever exerts influence. Corporations and interest groups employ lobbyists to push politicians in one direction or another all the time. Citizens in a democratic republic have the same power, *if* they choose to exercise it. You can exercise this power at the ballot box, by attending public hearings, and by writing letters and making phone calls to policymakers. Today's environmental laws came about because citizens pressured their governmental representatives to do something about environmental problems. We owe it to our children and future generations to be engaged and to act responsibly now so that they have a better world in which to live.

Whatever solutions we pursue, we must base our decisions on long-term thinking, because to be sustainable, a solution must work in the long term (Figure E.3).

FIGURE E.3 Sustainable solutions require thinking in the long term. These Sri Lankan children are planting tree seedlings on deforested and eroded hillsides around their village. In so doing, they are investing in their own future.

Policymakers in democracies often act for short-term good because they compete to produce immediate, positive results so that they will be reelected. Yet many environmental dilemmas are cumulative, worsen gradually, and can be resolved only over a long period. Often the costs for addressing environmental problems are short-term, whereas the benefits are long-term. In such a situation, citizen pressure on policymakers becomes especially vital.

Finally, individuals can educate others with information and serve as role models through their actions. The discipline of environmental science plays a key role in providing information people can use to make wise decisions about environmental problems. By promoting scientific research and by educating the public about environmental science, we can all play an important role in the pursuit of sustainable solutions.

Time is precious

By shifting our conventional patterns of thinking and following the approaches outlined above, we can bring sustainable solutions within reach. However, time is getting short, and many human impacts on the natural systems we depend on—deforestation, overfishing, land clearing, wetland draining, resource extraction—are becoming more severe and widespread. Even if we can visualize ways to mitigate such impacts, how can we find the time to implement sustainable solutions before we have done irreparable damage to our environment and our future?

On May 25, 1961, U.S. President John F. Kennedy announced that within the decade the United States would be "landing a man on the moon and returning him safely to the Earth." It was a bold and astonishing statement; the technology to achieve this unprecedented, almost unimaginable, feat did not yet exist. Yet just eight years later, astronauts walked on the moon. The United States accomplished this milestone in human history by building public support for a goal and by giving its scientists and engineers the wherewithal to focus on developing technology and strategies to meet the goal.

Today humanity faces a challenge more important than any previous one—the challenge to achieve sustainability. Attaining sustainability is a larger and more complex process than traveling to the moon. However, it is one to which every single person on Earth can contribute; in which government, industry, and citizens can all coop-

erate; and toward which governments of all nations can work together. If America was able to reach the moon in a mere eight years, then certainly humanity can begin down the road to sustainability with comparable speed. Human ingenuity is capable of it; it is just a question of rallying public resolve and engaging our governments, institutions, and entrepreneurs in the race.

Fortunately, in our global society today we have many thousands of scientists who study Earth's processes and resources. For this reason, we have access to an accumulated knowledge and an ever-developing understanding of our dynamic planet, what it offers us, and what impacts it can bear. The challenge for our society is to support that science so that we can accurately judge false alarms from real problems, and distinguish legitimate concerns from thoughtless denial. This science, this study of Earth and of ourselves, offers us hope for our future.

We must think of Earth as an island

We began this book with the vision of Earth as an island. Islands can be paradise, as Easter Island (▸ pp. 8–9) likely was when the Polynesians first reached it. But if people do not live sustainably on islands, they can turn paradises into desolate graveyards. When Europeans arrived at Easter Island, they witnessed the scene of a civilization that had depleted its resources, degraded its environment, and collapsed as a result. People of the once-mighty culture had cut trees unsustainably, kicking the base out from beneath their elaborate and prosperous civilization.

As Easter Island's trees disappeared, some individuals must have spoken out for conservation and for finding ways to live sustainably amid dwindling resources. Others likely ignored those calls and went on extracting more than the land could bear, assuming that somehow things would turn out all right. Indeed, whoever cut down the last tree from atop the most remote mountaintop could have looked out across the island and seen that it was the last tree. And yet that person cut it down.

It would be tragic folly to let such a fate occur to our planet as a whole. By recognizing this, by deciding to shift our individual behavior and our cultural institutions in ways that encourage sustainable practices, and by employing science to help us achieve these ends, we may yet be able to live sustainably and happily on our wondrous island, Earth.

Appendix A *Some Basics on Graphs*

Presenting data in ways that are clear and that help make trends and patterns visually apparent is a vital part of the scientific endeavor. Scientists' primary tool for presenting data and expressing patterns is the graph. Thus, the ability to interpret graphs is a skill that you will want to cultivate early in your study of the sciences. This appendix provides basic information on four of the most common types of graphs—line plots, bar charts, scatter plots, and pie charts—and the rationale for the use of each.

Line Plot

A line plot is drawn when a data set involves a sequence of some kind, such as a sequence through time or across distance (Figure A.1; see ▶ p. 90 and ▶ p. 282). Using a line plot allows us to see increasing or decreasing trends in the data. Line plots are appropriate when the variable measured by the y axis (the vertical axis) represents continuous numerical data, and when the variable measured by the x axis (the horizontal axis) represents either continuous numerical data or sequential categories, such as years.

One useful technique is to plot two data sets together on the same graph (Figure A.2; see Figure 5.2, ▶ p. 99). This allows us to compare trends in the two data sets to see whether and how they may be related.

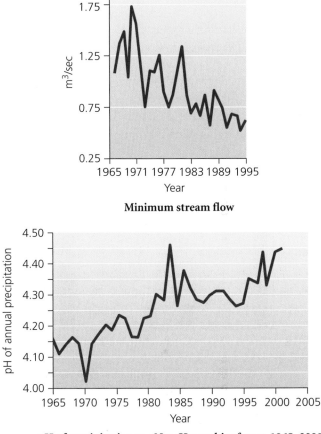

Minimum stream flow

pH of precipitation at a New Hampshire forest, 1965–2001

FIGURE A.1

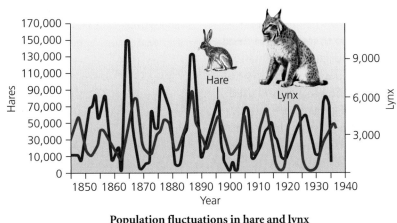

Population fluctuations in hare and lynx

FIGURE A.2

Bar Chart

A bar chart is most often used when one of the variables represents categories rather than numerical values (Figure A.3; see Figure 9.6a, ▶ p. 202). Bar charts allow us to visualize how a variable differs quantitatively among categories.

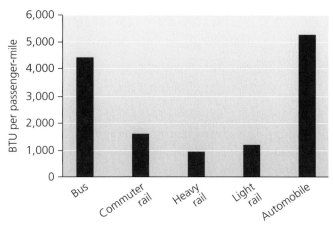

Energy consumption for different modes of transit

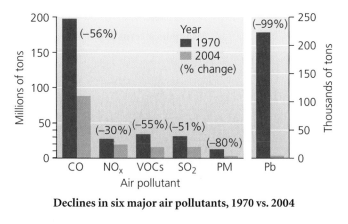

Declines in six major air pollutants, 1970 vs. 2004

It is often instructive to graph two or more variables together to reveal patterns and relationships (Figure A.4; see Figure 12.6a, ▶ p. 278). Many of the bar charts you will see in this book illustrate several types of information at once in this manner.

Bar charts are usually arrayed so that the bars extend vertically. Sometimes, however, a horizontal orientation may make for a clearer presentation. One special type of horizontally oriented bar chart is the age pyramid used by demographers (Figure A.5; see Figure 6.8a, ▶ p. 129). Age categories are displayed on the *y* axis, with bars representing the population size of each age group varying in width instead of height.

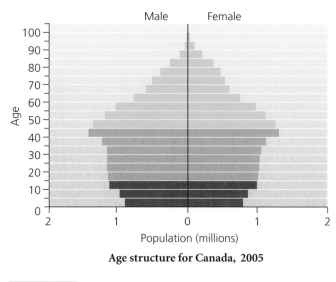

Age structure for Canada, 2005

Scatter Plot

A scatter plot is used most often when there is no sequential aspect to the data, and each data point is independent, having no particular connection to other data points (Figure A.6; see Figure 6.12, ▸ p. 133). Scatter plots allow you to visualize a broad positive or negative correlation between variables on the two axes.

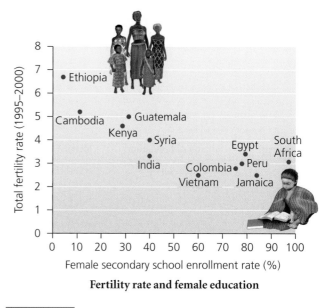

Fertility rate and female education

FIGURE A.6

Pie Chart

A pie chart is used when we wish to compare the proportions of some whole that are taken up by each of several categories (Figure A.7; see Figure 7.3, ▸ p. 145). A pie chart is appropriate when one variable is categorical and one is numerical. Each category is represented visually like a slice from a pie, with the size of the slice reflecting the percentage of the whole that is taken up by that category.

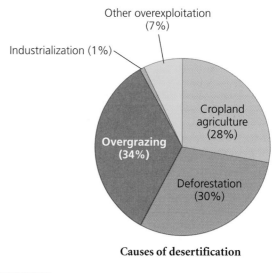

Causes of desertification

FIGURE A.7

But don't stop here. Take advantage of the **GRAPHit!** tutorials on the Withgott/Brennan Companion Website at **www.aw-bc.com/withgott**. The **GRAPHit!** tutorials allow you to plot your own data, and help you further expand your comprehension of graphs.

Appendix B   Metric System

Measurement	Unit and Abbreviation	Metric Equivalent	Metric to English Conversion Factor	English to Metric Conversion Factor
Length	1 kilometer (km)	= 1,000 (10^3) meters	1 km = 0.62 mile	1 mile = 1.61 km
	1 meter (m)	= 100 (10^2) centimeters	1 m = 1.09 yards	1 yard = 0.914 m
		= 1,000 millimeters	1 m = 3.28 feet	1 foot = 0.305 m
			1 m = 39.37 inches	
	1 centimeter (cm)	= 0.01 (10^{-2}) meter	1 cm = 0.394 inch	1 foot = 30.5 cm
				1 inch = 2.54 cm
	1 millimeter (mm)	= 0.001 (10^{-3}) meter	1 mm = 0.039 inch	
Area	1 square meter (m^2)	= 10,000 square centimeters	1 m^2 = 1.1960 square yards	1 square yard = 0.8361 m^2
			1 m^2 = 10.764 square feet	1 square foot = 0.0929 m^2
	1 square centimeter (cm^2)	= 100 square millimeters	1 cm^2 = 0.155 square inch	1 square inch = 6.4516 cm^2
Mass	1 metric ton (t)	= 1,000 kilograms	1 t = 1.103 ton	1 ton = 0.907 t
	1 kilogram (kg)	= 1,000 grams	1 kg = 2.205 pounds	1 pound = 0.4536 kg
	1 gram (g)	= 1,000 milligrams	1 g = 0.0353 ounce	1 ounce = 28.35 g
			1 g = 15.432 grains	
	1 milligram (mg)	= 0.001 gram	1 mg = approx. 0.015 grain	
Volume (solids)	1 cubic meter (m^3)	= 1,000,000 cubic centimeters	1 m^3 = 1.3080 cubic yards	1 cubic yard = 0.7646 m^3
			1 m^3 = 35.315 cubic feet	1 cubic foot = 0.0283 m^3
	1 cubic centimeter (cm^3 or cc)	= 0.000001 cubic meter	1 cm^3 = 0.0610 cubic inch	1 cubic inch = 16.387 cm^3
		= 1 milliliter		
	1 cubic millimeter (mm^3)	= 0.000000001 cubic meter		
Volume (liquids and gases)	1 kiloliter (kl or kL)	= 1,000 liters	1 kL = 264.17 gallons	1 gallon = 3.785 L
	1 liter (l or L)	= 1,000 milliliters	1 L = 0.264 gallons	1 quart = 0.946 L
			1 L = 1.057 quarts	
	1 milliliter (ml or mL)	= 0.001 liter	1 ml = 0.034 fluid ounce	1 quart = 946 ml
		= 1 cubic centimeter	1 ml = approximately $\frac{1}{4}$ teaspoon	1 pint = 473 ml
			1 ml = approx. 15–16 drops (gtt.)	1 fluid ounce = 29.57 ml
				1 teaspoon = approx. 5 ml
Time	1 second (s)	= $\frac{1}{60}$ minute		
	1 millisecond (ms)	= 0.001 second	1 second (s)	= $\frac{1}{60}$ minute
Temperature	Degrees Celsius (°C)		°C = $\frac{5}{9}$ (°F − 32)	°F = $\frac{9}{5}$ °C + 32
Energy and Power	1 kilowatt-hour	= 34,113 BTUs = 860,421 calories		
	1 watt	= 3.413 BTU/hr		
		= 14.34 calorie/min		
	1 calorie	= the amount of heat necessary to raise the temperature of 1 gram (1 cm^3) of water 1 degree Celsius		
	1 horsepower	= 7.457 × 102 watts		
	1 joule	= 9.481 × 10^{-4}) BTU		
		= 0.239 cal		
		= 2.778 × 10^{-7} kilowatt-hour		
Pressure	1 pound per square inch (psi)	= 6894.757 pascal (Pa)		
		= 0.068045961 atmosphere (atm)		
		= 51.71493 millimeters of mercury (mm hg = Torr)		
		= 68.94757 millibars (mbar)		
		= 68.94757 (hectopascal hPa)		
		= 6.894757 kilopascal (kPa)		
		= 0.06894757 bar (bar)		
	1 atmosphere (atm)	= 101.325 kilopascal (kPa)		

Appendix C   Periodic Table of the Elements

Representative (main group) elements

IA	IIA	IIIB	IVB	VB	VIB	VIIB	VIIIB	VIIIB	VIIIB	IB	IIB	IIIA	IVA	VA	VIA	VIIA	VIIIA
1 H 1.0079 Hydrogen																	2 He 4.003 Helium
3 Li 6.941 Lithium	4 Be 9.012 Beryllium											5 B 10.811 Boron	6 C 12.011 Carbon	7 N 14.007 Nitrogen	8 O 15.999 Oxygen	9 F 18.998 Fluorine	10 Ne 20.180 Neon
11 Na 22.990 Sodium	12 Mg 24.305 Magnesium											13 Al 26.982 Aluminum	14 Si 28.086 Silicon	15 P 30.974 Phosphorus	16 S 32.066 Sulfur	17 Cl 35.453 Chlorine	18 Ar 39.948 Argon
19 K 39.098 Potassium	20 Ca 40.078 Calcium	21 Sc 44.956 Scandium	22 Ti 47.88 Titanium	23 V 50.942 Vanadium	24 Cr 51.996 Chromium	25 Mn 54.938 Manganese	26 Fe 55.845 Iron	27 Co 58.933 Cobalt	28 Ni 58.69 Nickel	29 Cu 63.546 Copper	30 Zn 65.39 Zinc	31 Ga 69.723 Gallium	32 Ge 72.61 Germanium	33 As 74.922 Arsenic	34 Se 78.96 Selenium	35 Br 79.904 Bromine	36 Kr 83.8 Krypton
37 Rb 85.468 Rubidium	38 Sr 87.62 Strontium	39 Y 88.906 Yttrium	40 Zr 91.224 Zirconium	41 Nb 92.906 Niobium	42 Mo 95.94 Molybdenum	43 Tc 98 Technetium	44 Ru 101.07 Ruthenium	45 Rh 102.906 Rhodium	46 Pd 106.42 Palladium	47 Ag 107.868 Silver	48 Cd 112.411 Cadmium	49 In 114.82 Indium	50 Sn 118.71 Tin	51 Sb 121.76 Antimony	52 Te 127.60 Tellurium	53 I 126.905 Iodine	54 Xe 131.29 Xenon
55 Cs 132.905 Cesium	56 Ba 137.327 Barium	57 La 138.906 Lanthanum	72 Hf 178.49 Hafnium	73 Ta 180.948 Tantalum	74 W 183.84 Tungsten	75 Re 186.207 Rhenium	76 Os 190.23 Osmium	77 Ir 192.22 Iridium	78 Pt 195.08 Platinum	79 Au 196.967 Gold	80 Hg 200.59 Mercury	81 Tl 204.383 Thallium	82 Pb 207.2 Lead	83 Bi 208.980 Bismuth	84 Po 209 Polonium	85 At 210 Astatine	86 Rn 222 Radon
87 Fr 223 Francium	88 Ra 226.025 Radium	89 Ac 227.028 Actinium	104 Rf 261 Unniquadium	105 Db 262 Unnilpentium	106 Sg 263 Unnilhexium	107 Bh 262 Unnilseptium	108 Hs 265 Unniloctium	109 Mt 266 Unnilennium	110 Uun 269 Ununnilium	111 Uuu 272 Unununium	112 Uub 277 Ununbium		114		116		

— Transition metals — (groups IIIB through IIB)

Rare earth elements

Lanthanides

58 Ce 140.115 Cerium	59 Pr 140.908 Praseodymium	60 Nd 144.24 Neodymium	61 Pm 145 Promethium	62 Sm 150.36 Samarium	63 Eu 151.964 Europium	64 Gd 157.25 Gadolinium	65 Tb 158.925 Terbium	66 Dy 162.5 Dysprosium	67 Ho 164.93 Holmium	68 Er 167.26 Erbium	69 Tm 168.934 Thulium	70 Yb 173.04 Ytterbium	71 Lu 174.967 Lutetium

Actinides

90 Th 232.038 Thorium	91 Pa 231.036 Protactinium	92 U 238.029 Uranium	93 Np 237.048 Neptunium	94 Pu 244 Plutonium	95 Am 243 Americium	96 Cm 247 Curium	97 Bk 247 Berkelium	98 Cf 251 Californium	99 Es 252 Einsteinium	100 Fm 257 Fermium	101 Md 258 Mendelevium	102 No 259 Nobelium	103 Lr 262 Lawrencium

The periodic table arranges elements according to atomic number and atomic weight into horizontal rows called periods and vertical columns called groups.

Elements of each group in Class A have similar chemical and physical properties. This reflects the fact that members of a particular group have the same number of valence shell electrons, which is indicated by the group's number. For example, group IA elements have one valence shell electron, group IIA elements have two, and group VA elements have five. In contrast, as you progress across a period from left to right, properties of the elements change, varying from the very metallic properties of groups IA and IIA to the nonmetallic properties of group VIIA to the inert elements (noble gases) in group VIIIA. This reflects changes in the number of valence shell electrons.

Class B elements, or transition elements, are metals, and generally have one or two valence shell electrons. In these elements, some electrons occupy more distant electron shells before the deeper shells are filled.

In this periodic table, elements with symbols printed in black exist as solids under standard conditions (25 °C and 1 atmosphere of pressure), while elements in red exist as gases, and those in dark blue as liquids. Elements with symbols in green do not exist in nature and must be created by some type of nuclear reaction.

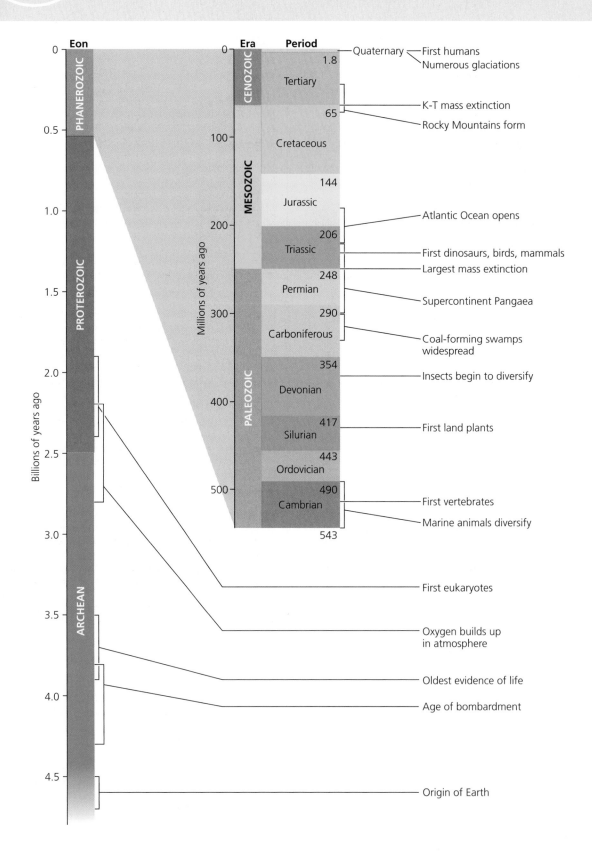

Glossary

abiotic factor Any nonliving component of the *environment*. Compare *biotic factor*.

acid drainage A process in which sulfide minerals in newly exposed rock surfaces react with oxygen and rainwater to produce sulfuric acid, which causes chemical *runoff* as it *leaches* metals from the rocks. Although acid drainage is a natural phenomenon, mining can greatly accelerate its rate by exposing many new rock surfaces at once.

acidic The property of a solution in which the concentration of hydrogen (H⁺) *ions* is greater than the concentration of hydroxide (OH⁻) ions. Compare *basic*.

acidic deposition The settling of *acidic* or acid-forming pollutants from the *atmosphere* onto Earth's surface. This can take place by *precipitation*, by fog, by gases, or by the deposition of dry particles. Compare *acid rain*.

acid rain *Acidic deposition* that takes place through rain. Compare *acidic deposition*.

active solar energy collection An approach in which technological devices are used to focus, move, or store solar energy. Compare *passive solar energy collection*.

acute exposure Exposure to a *toxicant* occurring in high amounts for short periods of time. Compare *chronic exposure*.

adaptive management The systematic testing of different management approaches to improve methods over time.

adaptive radiation A burst of *species* formation due to natural selection.

adaptive trait (adaptation) A trait that confers greater likelihood that an individual will reproduce.

aerobic Occurring in an *environment* where oxygen is present. For example, the decay of a rotting log proceeds by aerobic decomposition. Compare *anaerobic*.

age distribution The relative numbers of organisms of each age within a *population*. Age distributions can have a strong effect on rates of population growth or decline and are often expressed as a ratio of age classes, consisting of organisms (1) not yet mature enough to reproduce, (2) capable of reproduction, and (3) beyond their reproductive years.

age structure See *age distribution*.

age structure diagram (age pyramid) A diagram demographers use to show the *age distribution* of a population. The width of each horizontal bar represents the relative number of individuals in each age class.

agricultural revolution The shift around 10,000 years ago from a hunter-gatherer lifestyle to an agricultural way of life in which people began to grow their own crops and raise domestic animals. Compare *industrial revolution*.

agriculture The practice of cultivating *soil*, producing crops, and raising livestock for human use and consumption.

A horizon A layer of *soil* found in a typical *soil profile*. It forms the top layer or lies below the *O horizon* (if one exists). It consists of mostly inorganic mineral components such as *weathered* substrate, with some organic matter and humus from above mixed in. The A horizon is often referred to as *topsoil*. Compare *B horizon; C horizon; E horizon; R horizon*.

air pollutants Gases and particulate material added to the *atmosphere* that can affect *climate* or harm people or other organisms.

air pollution The act of polluting the air, or the condition of being polluted by *air pollutants*.

allergen A *toxicant* that overactivates the immune system, causing an immune response when one is not necessary.

allopatric speciation Species formation due to the physical separation of populations over some geographic distance.

ambient air pollution See *outdoor air pollution*.

amino acids Organic molecules that join in long chains to form *proteins*.

anaerobic Occurring in an *environment* that has little or no oxygen. The conversion of organic matter to *fossil fuels* (*crude oil, coal, natural gas*) at the bottom of a deep lake, swamp, or shallow sea is an example of anaerobic decomposition. Compare *aerobic*.

anthropocentrism A human-centered view of our relationship with the *environment*.

aquaculture The raising of aquatic organisms for food in controlled *environments*.

aquifer An underground water reservoir.

artificial selection *Natural selection* conducted under human direction. Examples include the *selective breeding* of crop plants, pets, and livestock.

asbestos Any of several types of mineral that form long, thin microscopic fibers—a structure that allows asbestos to insulate buildings for heat, muffle sound, and resist fire. When inhaled and lodged in lung tissue, asbestos scars the tissue and may eventually lead to lung cancer or *asbestosis*.

asbestosis A disorder resulting from lung tissue scarred by acid following prolonged inhalation of *asbestos*.

atmosphere The thin layer of gases surrounding planet Earth. Compare *biosphere; hydrosphere; lithosphere*.

atmospheric deposition The wet or dry deposition on land of a wide variety of pollutants, including mercury, nitrates, organochlorines, and others. *Acidic deposition* is one type of atmospheric deposition.

atom The smallest component of an *element* that maintains the chemical properties of that element.

autotroph (primary producer) An organism that can use the energy from sunlight to produce its own food. Includes green plants, algae, and cyanobacteria.

***Bacillus thuringiensis* (Bt)** A naturally occurring *soil* bacterium that produces a protein that kills many *pests*, including caterpillars and the larvae of some flies and beetles.

background rate of extinction The average rate of *extinction* that occurred before the appearance of humans. For example, the *fossil record* indicates that for both birds and mammals, one *species* in the world typically became extinct every 500–1,000 years. Compare *mass extinction event*.

baghouse A system of large filters that physically removes *particulate matter* from *incinerator* emissions.

basic The property of a solution in which the concentration of hydroxide (OH⁻) *ions* is greater than the concentration of hydrogen (H⁺) ions. Compare *acidic*.

bedrock The continuous mass of solid rock that makes up Earth's *crust*.

benthic Of, relating to, or living on the bottom of a water body. Compare *pelagic*.

benthic zone The bottom layer of water body. Compare *littoral zone; limnetic zone; profundal zone*.

B horizon The layer of *soil* that lies below the *E horizon* and above the *C horizon*. Minerals that leach out of the E horizon are carried down into the B horizon (or subsoil) and accumulate there. Sometimes called the "zone of accumulation" or "zone of deposition." Compare *A horizon; O horizon; R horizon*.

bioaccumulation The buildup of *toxicants* in the tissues of an animal.

biocentrism A philosophy that ascribes relative values to actions, entities, or properties on the basis of their effects on all living things or on the integrity of the *biotic* realm in general. The biocentrist evaluates an action in terms of its overall impact on living things, including—but not exclusively focusing on—human beings.

biodiesel Diesel fuel produced by mixing vegetable oil, used cooking grease, or animal fat with small amounts of *ethanol* or methanol (wood alcohol) in the presence of a chemical catalyst.

biodiversity (biological diversity) The sum total of all organisms in an area, taking into account the diversity of *species*, their *genes*, their *populations*, and their *communities*.

biodiversity hotspot An area that supports an especially great diversity of *species*, particularly species that are *endemic* to the area.

biofuel Fuel produced from *biomass energy* sources and used primarily to power automobiles.

biogenic Type of *natural gas* created at shallow depths by the anaerobic decomposition of organic matter by bacteria. Consists of nearly pure *methane*.

biogeochemical cycle See *nutrient cycle*.

biological control (biocontrol) The attempt to battle *pests* and weeds with organisms that prey on or parasitize them, rather than by using *pesticides*.

biological diversity See *biodiversity*.

biological hazard Human health hazards that result from ecological interactions among organisms. These include *parasitism* by viruses, bacteria, or other *pathogens*. Compare *infectious disease; chemical hazard; cultural hazard; physical hazard*.

biomagnification The magnification of the concentration of *toxicants* in an organism caused by its consumption of other organisms in which toxicants have *bioaccumulated*.

biomass Organic material that makes up living organisms.

biomass energy *Energy* harnessed from plant and animal matter, including wood from trees, charcoal from burned wood, and combustible animal waste products, such as cattle manure. *Fossil fuels* are not considered biomass energy sources because their organic matter has not been part of living organisms for millions of years and has undergone considerable chemical alteration since that time.

biome A major regional complex of similar plant *communities*; a large *ecological* unit defined by its dominant plant type and vegetation structure.

biophilia A phenomenon that E. O. Wilson has defined as "the connections that human beings subconsciously seek with the rest of life."

biopower The burning of *biomass energy* sources to generate electricity.

biosphere The sum total of all the planet's living organisms and the *abiotic* portions of the *environment* with which they interact.

biosphere reserve A tract of land with exceptional *biodiversity* that couples preservation with *sustainable development* to benefit local people. Biosphere reserves are designated by UNESCO (the *United Nations Educational, Scientific, and Cultural Organization*) following application by local stakeholders.

biotechnology The material application of biological *science* to create products derived from organisms. The creation of *transgenic* organisms is one type of biotechnology.

biotic factor Any living component of the *environment*. Compare *abiotic factor*.

biotic potential An organism's capacity to produce offspring.

bog A type of wetland in which a pond is thoroughly covered with a thick floating mat of vegetation. Compare *freshwater marsh; swamp*.

boreal forest A *biome* of northern coniferous forest that stretches in a broad band across much of Canada, Alaska, Russia, and Scandinavia. Also known as taiga, boreal forest consists of a limited number of *species* of evergreen trees, such as black spruce, that dominate large regions of forests interspersed with occasional bogs and lakes.

bottom-trawling Fishing practice that involves dragging weighted nets across the seafloor to catch benthic organisms. Trawling crushes many organisms in its path and leaves long swaths of damaged sea bottom.

breakdown products *Compounds* that result from the degradation of toxicants.

by-catch That portion of a commercial fishing catch consisting of animals caught unintentionally. By-catch kills many thousands of fish, sharks, marine mammals, and birds each year.

capitalist market economy An *economy* in which buyers and sellers interact to determine which *goods* and *services* to produce, how much of them to produce, and how to distribute them. Compare *centrally planned economy*.

captive breeding The practice of capturing members of threatened and endangered *species* so that their young can be bred and raised in controlled *environments* and subsequently reintroduced into the wild.

carbohydrate An *organic compound* consisting of *atoms* of carbon, hydrogen, and oxygen.

carbon An *element* with six protons, six *neutrons*, and six *electrons*. Carbon is the requisite element for *organic compounds*, and thus is abundant in organisms. See Appendix C, *Periodic table of the elements*.

carbon cycle A major *nutrient cycle* consisting of the routes that carbon *atoms* take through the nested networks of environmental *systems*.

carbon dioxide (CO_2) A colorless gas used by plants for *photosynthesis*, given off by *respiration*, and released by burning *fossil fuels*. Carbon dioxide is a primary *greenhouse gas* whose buildup contributes to *global climate change*.

carbon monoxide (CO) A colorless, odorless gas produced primarily by the incomplete combustion of fuel. Carbon monoxide is an EPA *criteria pollutant*.

carcinogen A chemical or type of radiation that causes cancer.

carrying capacity The maximum *population size* that a given *environment* can sustain.

cell The most basic organizational unit of organisms.

cellular respiration The process by which a *cell* uses the chemical reactivity of oxygen to split glucose into its constituent parts, water and carbon dioxide, and thereby release *chemical energy* that can be used to form chemical bonds or to perform other tasks within the cell. Compare *photosynthesis*.

centrally planned economy An *economy* in which a nation's government determines how to allocate resources in a top-down manner. Also called a "state socialist economy." Compare *capitalist market economy*.

chaparral A *biome* consisting mostly of densely thicketed evergreen shrubs occurring in limited small patches. Its "Mediterranean" *climate* of mild, wet winters and warm, dry summers is induced by oceanic influences. In addition to ringing the Mediterranean Sea, chaparral occurs along the coasts of California, Chile, and southern Australia.

chemical energy *Potential energy* held in the bonds between atoms.

chemical hazard Chemicals that pose human health hazards. These include *toxins* produced naturally, as well as many of the disinfectants, *pesticides*, and other synthetic chemicals that our society produces. Compare *biological hazard; cultural hazard; physical hazard*.

Chernobyl Site of a nuclear power plant in Ukraine (then part of the Soviet Union), where in 1986 an explosion caused the most severe *nuclear reactor* accident the world has yet seen. As with *Three Mile Island*, the term is often used to denote the accident itself.

chlorofluorocarbon (CFC) One of a group of human-made *organic compounds* derived from simple *hydrocarbons*, such as ethane and *methane*, in which hydrogen *atoms* are replaced by chlorine, bromine, or fluorine. CFCs deplete the protective *ozone layer* in the *stratosphere*.

chlorophyll The light-absorbing pigment that enables *photosynthesis* and makes plants green.

chloroplast A cell organelle containing *chlorophyll* in which *photosynthesis* occurs.

C horizon The layer of *soil* that lies below the *B horizon* and above the *R horizon*. It contains rock particles that are larger and less *weathered* than the layers above. It consists of *parent material* that has been altered only slightly or not at all by the process of *soil* formation. Compare *A horizon; E horizon; O horizon*.

chronic exposure Exposure for long periods of time to a *toxicant* occurring in low amounts. Compare *acute exposure*.

city planning The professional pursuit that attempts to design cities in such a way as to maximize their efficiency, functionality, and beauty.

classical economics Founded by *Adam Smith*, the study of the behavior of buyers and sellers in a free-market *economy*. Holds that individuals acting in their own self-interest may benefit society, provided that their behavior is constrained by the rule of law and by private property rights and operates within competitive markets. See also *neoclassical economics*.

Clean Air Act Congressional *legislation* that funds research into pollution control, sets standards for air quality, imposes limits on emissions from new stationary and mobile sources, enables citizens to sue parties that violate the standards, and introduced an emissions trading program for sulfur dioxide. First enacted in 1963 and amended multiple times since, particularly in 1970 and 1990.

clear-cutting The harvesting of timber by cutting all the trees in an area, leaving only stumps. Although it is the most cost-efficient method of harvesting timber, clear-cutting is also the most damaging to the *environment*.

climate The pattern of atmospheric conditions found across large geographic regions over long periods of time. Compare *weather*.

climate diagram (climatograph) A visual representation of a region's average monthly temperature and *precipitation*.

climax community In the traditional view of ecological *succession*, a *community* that remains in place with little modification until disturbance restarts the succession process. Today, ecologists recognize that community change is more variable and less predictable than originally thought, and that assemblages of species may instead form complex mosaics in space and time.

clumped distribution *Population distribution* pattern in which organisms arrange themselves in patches, generally according to the availability of the resources they need.

coal A *fossil fuel* composed of organic matter that was compressed under very high pressure to form a dense, solid carbon structure.

coevolution Process by which two or more species evolve in response to one another. Parasites and hosts may coevolve, as may flowers and their pollinators.

co-firing A process in which *biomass* is combined with *coal* in coal-fired power plants. Can be a relatively easy and inexpensive way for *fossil-fuel*-based utilities to expand their use of *renewable energy*.

cogeneration A practice in which the extra heat generated in the production of electricity is captured and put to use heating workplaces and homes, as well as producing other kinds of power.

command and control An approach to protecting the *environment* that sets strict legal limits and threatens punishment for violations of those limits.

communicable disease See *infectious disease*.

community A group of *populations* of organisms that live in the same place at the same time.

community-based conservation The practice of engaging local people to protect land and wildlife in their own region.

community ecology The study of the interactions among *species*, from one-to-one interactions to complex interrelationships involving entire *communities*.

competition A relationship in which multiple organisms seek the same limited resource.

compost a mixture produced when decomposers break down organic matter, including food and crop waste, in a controlled *environment*.

composting The conversion of organic *waste* into mulch or humus by encouraging, in a controlled manner, the natural biological processes of decomposition.

compound A *molecule* whose *atoms* are composed of two or more *elements*.

concentrated animal feeding operation (CAFO) See *feedlot*.

concession The right to extract a resource, granted by a government to a corporation.

conservation biology A scientific discipline devoted to understanding the factors, forces, and processes that influence the loss, protection, and restoration of *biological diversity* within and among *ecosystems*.

conservation ethic An *ethic* holding that humans should put *natural resources* to use but also have a responsibility to manage them wisely. Compare *preservation ethic*.

consumer See *heterotroph*.

consumptive use *Freshwater* use in which water is removed from a particular *aquifer* or surface water body and is not returned to it. *Irrigation* for *agriculture* is an example of consumptive use. Compare *nonconsumptive use*.

contour farming The practice of plowing furrows sideways across a hillside, perpendicular to its slope, to help prevent the formation of rills and gullies. The technique is so named because the furrows follow the natural contours of the land.

control The portion of an *experiment* in which a *variable* has been left unmanipulated, to serve as a point of comparison with the *treatment*.

control rods Rods made of a metallic alloy that absorbs *neutrons*, which are placed in a *nuclear reactor* among the water-bathed *fuel rods* of uranium. Engineers move these control rods into and out of the water to maintain the *fission* reaction at the desired rate.

controlled experiment An *experiment* in which the effects of all *variables* are held constant, except the one whose effect is being tested by comparison of *treatment* and *control* conditions.

convective circulation A circular *current* (of air, water, magma, etc.) driven by temperature differences. In the *atmosphere*, warm air rises into regions of lower atmospheric pressure, where it expands and cools and then descends and becomes denser, replacing warm air that is rising. The air picks up heat and moisture near ground level and prepares to rise again, continuing the process.

convention A *treaty* or binding agreement among national governments.

conventional law International law that arises from *conventions*, or treaties, that nations agree to enter into. Compare *customary law*.

Convention on Biological Diversity An international treaty that aims to conserve *biodiversity*, use biodiversity in a *sustainable* manner, and ensure the fair distribution of biodiversity's benefits. Although many nations have agreed to the treaty (as of 2005, 188 nations had become parties to it), several others, including the United States, have not.

Convention on International Trade in Endangered Species of Wild Fauna and Flora (CITES) A 1973 treaty facilitated by the *United Nations* that protects endangered *species* by banning the international transport of their body parts.

convergent plate boundary Area where tectonic plates collide. Can result in *subduction* or mountain range formation.

coral Tiny marine animals that build *coral reefs*. Corals attach to rock or existing reef and capture passing food with stinging tentacles. They also derive nourishment from photosynthetic symbiotic algae known as *zooxanthellae*.

coral reef A mass of calcium carbonate composed of the skeletons of tiny colonial marine organisms called *corals*.

core The innermost part of the Earth, made up mostly of iron, which lies beneath the *crust* and *mantle*.

Coriolis effect The apparent deflection of north-south air *currents* to a partly east-west direction, caused by the faster spin of regions near the equator than of regions near the poles as a result of Earth's rotation.

correlation A relationship among *variables*.

corridor A passageway of protected land established to allow animals to travel between islands of protected *habitat*.

corrosive Able to corrode metals. One criterion for defining *hazardous waste*.

cost-benefit analysis A method commonly used by *neoclassical economists*, in which estimated costs for a proposed action are totaled and then compared to the sum of benefits estimated to result from the action.

coupled general circulation models Computer programs that combine what is known about weather patterns, atmospheric circulation, atmosphere-ocean interactions, and feedback mechanisms to simulate *climate* processes.

covalent bond A chemical bond in which the uncharged *atoms* in a *molecule* share *electrons*. For example, the uncharged atoms of carbon and oxygen in carbon dioxide form a covalent bond. Compare *ionic bond*.

criteria pollutants Six *air pollutants*—carbon monoxide, sulfur dioxide, nitrogen dioxide, tropospheric ozone, particulate matter, and lead—for which the *Environmental Protection Agency* has established maximum allowable concentrations in ambient outdoor air because of the threats they pose to human health.

cropland Land that humans use to raise plants for food and fiber.

crop rotation The practice of alternating the kind of crop grown in a particular field from one season or year to the next.

crude oil (petroleum) A *fossil fuel* produced by the conversion of *organic compounds* by heat and pressure. Crude oil is a mixture of hundreds of different types of *hydrocarbon* molecules characterized by carbon chains of different length.

crust The lightweight outer layer of the Earth, consisting of rock that floats atop the malleable *mantle*, which in turn surrounds a mostly iron *core*.

cultural hazard Human health hazards that result from the place we live, our socioeconomic status, our occupation, or our behavioral choices. These include choosing to smoke cigarettes, or living or working with people who do. Compare *biological hazard*; *chemical hazard*; *physical hazard*.

current The flow of a liquid or gas in a certain direction.

customary law International law that arises from long-standing practices, or customs, held in common by most *cultures*. Compare *conventional law*.

dam Any obstruction placed in a river or stream to block the flow of water so that water can be stored in a reservoir. Dams are built to prevent floods, provide drinking water, facilitate *irrigation*, and generate electricity.

Darwin, Charles (1809–1882) English naturalist who proposed the concept of *natural selection* as a mechanism for *evolution* and as a way to explain the great variety of living things. See also *Wallace, Alfred Russell*.

data Information, generally quantitative information.

deciduous Term describing trees that lose their leaves each fall and remain dormant during winter, when hard freezes would endanger leaves.

decomposer An organism, such as a fungus or bacterium, that breaks down leaf litter and other nonliving matter into simple constituents that can be taken up and used by plants. Compare *detritivore*.

deep-well injection A *hazardous waste* disposal method in which a well is drilled deep beneath an area's *water table* into porous rock below an impervious *soil* layer. Wastes are then injected into the well, so that they will be absorbed into the porous rock and remain deep underground, isolated from *groundwater* and human contact. Compare *surface impoundment*.

deforestation The clearing and loss of forests.

demand The amount of a product people will buy at a given price if free to do so.

demographic transition A theoretical model of economic and cultural change that explains the declining death rates and birth rates that occurred in Western nations as they became industrialized. The model holds that industrialization caused these rates to fall naturally by decreasing mortality and by lessening the need for large families. Parents would thereafter choose to invest in quality of life rather than quantity of children.

demography A *social science* that applies the principles of *population ecology* to the study of statistical change in human *populations*.

denitrifying bacteria Bacteria that convert the nitrates in *soil* or water to gaseous nitrogen and release it back into the *atmosphere*.

density-dependent factor A *limiting factor* whose effects on a *population* increase or decrease depending on the *population density*. Compare *density-independent factor*.

density-independent factor A *limiting factor* whose effects on a *population* are constant regardless of *population density*. Compare *density-dependent factor*.

deoxyribonucleic acid See *DNA*.

dependent variable The *variable* that is affected by manipulation of the *independent variable*.

deposition The arrival of eroded *soil* at a new location. Compare *erosion*.

desalination (desalinization) The removal of salt from seawater.

desert The driest *biome* on Earth, with annual *precipitation* of less than 25 cm. Because deserts have relatively little vegetation to insulate them from temperature extremes, sunlight readily heats them in the daytime, but daytime heat is quickly lost at night, so temperatures vary widely from day to night and in different seasons.

desertification A loss of more than 10% of a land's productivity due to *erosion*, *soil* compaction, forest removal, *overgrazing*, drought, *salinization*, *climate* change, depletion of water sources, or other factors. Severe desertification can result in the actual expansion of desert areas or creation of new ones in areas that once supported fertile land.

detritivore An organism, such as a millipede or soil insect, that scavenges the waste products or dead bodies of other community members. Compare *decomposer*.

dike A long raised mound of earth erected along a river bank to protect against floods by holding rising water in the main channel.

divergent plate boundary Area where *magma* surging upward to the surface divides tectonic plates and pushes them apart, creating new *crust* as it cools and spreads. A prime example is the Mid-Atlantic ridge. Compare *transform plate boundary* and *convergent plate boundary*.

DNA (deoxyribonucleic acid) A double-stranded *nucleic acid* composed of four nucleotides, each of which contains a sugar (deoxyribose), a phosphate group, and a nitrogenous base. DNA carries the hereditary information for living organisms and is responsible for passing traits from parents to offspring. Compare *RNA*.

dose The amount of *toxicant* a test animal receives in a dose-response test.

dose-response curve A curve that plots the response of test animals to different doses of a *toxicant*. The response is generally quantified by measuring the proportion of animals exhibiting negative effects.

downwelling In the ocean, the flow of warm surface water toward the ocean floor. Downwelling occurs where surface *currents* converge. Compare *upwelling*.

driftnet Fishing net that captures everything in its path, including substantial numbers of dolphins, seals, sea turtles, and nontarget fish.

Dust Bowl An area that loses huge amounts of *topsoil* to wind *erosion* as a result of drought and/or human impact; first used to name the region in the North American Great Plains severely affected by drought and topsoil loss in the 1930s. The term is now also used to describe that historical event and others like it.

ecocentrism A philosophy that considers actions in terms of their damage or benefit to the integrity of whole ecological *systems*, including both *biotic* and *abiotic* elements. For an ecocentrist, the well-being of an individual organism—human or otherwise—is less important than the long-term well-being of a larger integrated ecological system.

ecolabeling The practice of designating on a product's label how the product was grown, harvested, or manufactured, so that consumers buying it are aware of the processes involved and can differentiate between brands that use processes believed to be *environmentally* beneficial (or less harmful than others) and those that do not.

ecological economics A developing school of *economics* that applies the principles of *ecology* and *systems* thinking to the description and analysis of *economies*. Compare *environmental economics*; *neoclassical economics*.

ecological footprint The cumulative amount of land and water required to provide the raw materials a person or *population* consumes and to dispose of or *recycle* the *waste* that is produced.

ecological restoration Efforts to reverse the effects of human disruption of *ecological systems* and to restore *communities* to their "natural" state.

ecology The *science* that deals with the distribution and abundance of organisms, the interactions among them, and the interactions between organisms and their *abiotic environments*.

economically recoverable Extractable such that income from a resource's sale exceeds the costs of extracting it. Applied to *fossil fuel* deposits.

economics The study of how we decide to use scarce resources to satisfy the demand for *goods* and *services*.

economy A social *system* that converts resources into *goods* and *services*.

ecosystem All organisms and nonliving entities that occur and interact in a particular area at the same time.

ecosystem-based management The attempt to manage the harvesting of resources in ways that minimize impact on the *ecosystems* and ecological processes that provide the resources.

ecosystem diversity The number and variety of ecosystems in a particular area. One way to express *biodiversity*. Related concepts consider the geographic arrangement of *habitats*, *communities*, or *ecosystems* at the landscape level, including the sizes, shapes, and interconnectedness of patches of these entities.

ecosystem service An essential service an *ecosystem* provides that supports life and makes *economic* activity possible. For example, ecosystems naturally purify air and water, cycle *nutrients*, provide for plants to be *pollinated* by animals, and serve as receptacles and *recycling* systems for the *waste* generated by our economic activity.

ecotone A transitional zone where *ecosystems* meet.

ecotourism Visitation of natural areas for tourism and recreation. Most often involves tourism by more-affluent people, which may generate *economic* benefits for less-affluent communities near natural areas and thus provide economic incentives for conservation of natural areas.

ED$_{50}$ (effective dose–50%) The amount of a *toxicant* it takes to affect 50% of a *population* of test animals. Compare *threshold dose; LD$_{50}$*.

E horizon The layer of *soil* that lies below the *A horizon* and above the *B horizon*. The letter "E" stands for "eluviation," meaning "loss," and the E horizon is characterized by the loss of certain minerals through *leaching*. It is sometimes called the "zone of leaching." Compare *C horizon; O horizon; R horizon.*

El Niño The exceptionally strong warming of the eastern Pacific Ocean that occurs every 2 to 7 years and depresses local fish and bird *populations* by altering the marine *food web* in the area. Originally, the name that Spanish-speaking fishermen gave to an unusually warm surface *current* that sometimes arrived near the Pacific coast of South America around Christmas time. Compare *La Niña.*

electricity A secondary form of energy that can be transferred over long distances and applied for a variety of uses.

electrolysis A process in which electrical current is passed through a *compound* to release *ions*. Electrolysis offers one way to produce hydrogen for use as fuel: Electrical current is passed through water, splitting the water *molecules* into hydrogen and oxygen *atoms*.

electron A negatively charged particle that surrounds the nucleus of an *atom*.

element A fundamental type of matter; a chemical substance with a given set of properties, which cannot be broken down into substances with other properties. Chemists currently recognize 92 elements that occur in nature, as well as more than 20 others that have been artificially created.

emigration The departure of individuals from a *population*.

Endangered Species Act (ESA) The primary *legislation*, enacted in 1973, for protecting *biodiversity* in the United States. It forbids the government and private citizens from taking actions (such as developing land) that would destroy endangered *species* or their *habitats*, and it prohibits trade in products made from endangered species.

endemic Native or restricted to a particular geographic region. An endemic species occurs in one area and nowhere else on Earth.

endocrine disruptor A *toxicant* that interferes with the endocrine (hormone) system.

energy conservation The practice of reducing *energy* use as a way of extending the lifetime of our *fossil fuel* supplies, of being less wasteful, and of reducing our impact on the *environment*.

energy An intangible phenomenon that can change the position, physical composition, or temperature of matter.

entropy The degree of disorder in a substance, *system*, or process. See *second law of thermodynamics*.

environment The sum total of our surroundings, including all of the living things and nonliving things with which we interact.

environmental economics A developing school of *economics* that modifies the principles of *neoclassical economics* to address environmental challenges. An environmental economist believes that we can attain *sustainability* within our current economic *systems*. Whereas ecological economists call for revolution, environmental economists call for reform. Compare *ecological economics; neoclassical economics.*

environmental ethics The application of *ethical standards* to environmental questions.

environmental health Environmental factors that influence human health and quality of life and the health of *ecological* systems essential to environmental quality and long-term human well-being.

environmental impact statement (EIS) A report of results from detailed studies that assess the potential effects on the *environment* that would likely result from development projects or other actions undertaken by the government.

environmental justice A movement based on a moral sense of fairness and equality that seeks to expand society's domain of ethical concern from rich to poor, and from majority races and ethnic groups to minority ones.

environmental policy *Public policy* that pertains to human interactions with the *environment*. It generally aims to regulate resource use or reduce *pollution* to promote human welfare and/or protect natural systems.

environmental resistance The collective force of limiting factors, which together stabilize a population size at its carrying capacity.

Environmental Protection Agency (EPA) An administrative agency created by executive order in 1970. The EPA is charged with conducting and evaluating research, monitoring environmental quality, setting standards, enforcing those standards, assisting the states in meeting standards and goals, and educating the public.

environmental science The study of how the natural world works and how humans and the *environment* interact.

environmental studies An academic *environmental science* program that heavily incorporates the social sciences as well as the natural sciences.

environmental toxicology The study of *toxicants* that come from or are discharged into the *environment*, including the study of health effects on humans, other animals, and *ecosystems*.

environmentalism A social movement dedicated to protecting the natural world.

enzyme A chemical that catalyzes a chemical reaction.

epidemiological study A study that involves large-scale comparisons among groups of people, usually contrasting a group known to have been exposed to some *toxicant* and a group that has not.

equilibrium theory of island biogeography A *theory* that was initially applied to oceanic islands to explain how *species* come to be distributed among them. Since its development, researchers have increasingly applied the theory to islands of *habitat* (patches of one type of habitat isolated within vast "seas" of others). Aspects of the theory include *immigration* and *extinction* rates, the effect of island size, and the effect of distance from the mainland.

erosion The removal of material from one place and its transport to another by the action of wind or water.

estuary An area where a river flows into the ocean, mixing *freshwater* with salt water.

ethanol The alcohol in beer, wine, and liquor, produced as a *biofuel* by fermenting biomass, generally from *carbohydrate*-rich crops such as corn.

ethical standards The criteria that help differentiate right from wrong.

ethics The study of good and bad, right and wrong. The term can also refer to a person's or group's set of moral principles or values.

eukaryote A multicellular organism. The *cells* of eukaryotic organisms consist of a membrane-enclosed nucleus that houses *DNA*, an outer membrane of lipids, and an inner fluid-filled chamber containing *organelles*. Compare *prokaryote*.

European Union (EU) Political and economic organization formed after World War II to promote Europe's economic and social progress. As of 2005, the EU consisted of 25 member nations.

eutrophic Term describing a water body that has high-*nutrient* and low-oxygen conditions. Compare *oligotrophic*.

eutrophication The process of *nutrient* enrichment, increased production of organic matter, and subsequent *ecosystem* degradation in a water body.

evaporation The conversion of a substance from a liquid to a gaseous form.

even-aged Condition of timber plantations—generally *monocultures* of a single *species*—in which all trees are of the same age. Most *ecologists* view plantations of even-aged stands more as crop *agriculture* than as ecologically functional forests. Compare *uneven-aged*.

evenness See *relative abundance*.

evolution Genetically based change in the appearance, functioning, and/or behavior of organisms across generations, often by the process of *natural selection.*

evolutionary arms race A duel of escalating adaptations between species. Like rival nations racing to stay ahead of one another in military technology, *host* and *parasite* may repeatedly evolve new responses to the other's latest advance.

e-waste Term for "electronic waste," which includes discarded computers, televisions, VCRs, cell phones, and other electronic devices.

experiment An activity designed to test the validity of a *hypothesis* by manipulating *variables*. See *manipulative experiment* and *natural experiment*.

exploitative interaction A species interaction in which one participant benefits while another is harmed; that is, one species exploits the other. Such interactions include *predation, parasitism,* and *herbivory.*

exploratory drilling Drilling that takes place after a *fossil fuel* deposit has been identified, in order to gauge how much of the fuel exists and whether extraction will prove worthwhile. Involves drilling small holes that descend to great depths.

exponential growth The increase of a *population* (or of anything) by a fixed percentage each year.

external cost A negative *externality;* a cost borne by someone not involved in an economic transaction. Examples include harm to citizens from water *pollution* or *air pollution* discharged by nearby factories.

extinction The disappearance of an entire *species* from the face of the Earth. Compare *extirpation.*

extirpation The disappearance of a particular *population* from a given area, but not the entire *species* globally. Compare *extinction.*

factory farm See *feedlot.*

feedback loop A circular process in which a *system*'s output serves as input to that same system. See *negative feedback loop; positive feedback loop.*

feedlot A huge barn or outdoor pen designed to deliver *energy*-rich food to animals living at extremely high densities. Also called a factory farm or concentrated animal feeding operation (CAFO).

Ferrel cell One of a pair of cells of *convective circulation* between 30° and 60° north and south latitude that influence global *climate* patterns. Compare *Hadley cell; polar cell.*

fertilizer A substance that promotes plant growth by supplying essential *nutrients* such as nitrogen or phosphorus.

first law of thermodynamics Physical law stating that *energy* can change from one form to another but cannot be created or lost. The total energy in the universe remains constant and is said to be conserved.

flagship species A *species* that has wide appeal with the public and that can be used to promote conservation efforts that also benefit other less charismatic species.

flat-plate solar collectors See *solar panels.*

flexible fuel vehicle A vehicle that runs on E-85, a mix of 85% *ethanol* and 15% gasoline.

floodplain The region of land over which a river has historically wandered and periodically floods.

flux The movement of nutrients among pools or reservoirs in a *nutrient cycle.*

food chain A linear series of feeding relationships. As organisms feed on one another, energy is transferred from lower to higher *trophic levels.*

food security An adequate, reliable, and available food supply to all people at all times.

food web A visual representation of feeding interactions within an *ecological community* that shows an array of relationships between organisms at different *trophic levels.*

forestry The professional management of forests.

fossil The remains, impression, or trace of an animal or plant of past geological ages that has been preserved in rock or *sediments.*

fossil fuel A *nonrenewable natural resource,* such as *crude oil, natural gas,* or *coal,* produced by the decomposition and compression of organic matter from ancient life.

fossil record The cumulative body of *fossils* worldwide, which paleontologists study to infer the history of past life on Earth.

Framework Convention on Climate Change International agreement to reduce *greenhouse gas* emissions to 1990 levels by the year 2000, signed by nations represented at the 1992 Earth Summit convened in Rio de Janeiro by the *United Nations.* The FCCC called for a voluntary, nation-by-nation approach, but by the late 1990s it had become apparent that it would not succeed. Its imminent failure sparked introduction of the *Kyoto Protocol.*

free rider A party that fails to invest in controlling *pollution* or carrying out other *environmentally* responsible activities and instead relies on the efforts of other parties to do so. For example, a factory that fails to control its emissions gets a "free ride" on the efforts of other factories that do make the sacrifices necessary to reduce emissions.

freshwater Water that is relatively pure, holding very few dissolved salts.

freshwater marsh A type of wetland in which shallow water allows plants such as cattails to grow above the water surface. Compare *swamp; bog.*

fuel rods Rods of uranium that supply the fuel for nuclear *fission,* and are kept bathed in a *moderator* in a *nuclear reactor.*

fungicide A type of chemical *pesticide* that kills fungi.

genera Plural of *genus.*

gasification A process in which *biomass* is vaporized at extremely high temperatures in the absence of oxygen, creating a gaseous mixture including hydrogen, *carbon monoxide,* and *methane,* in order to produce *biopower* or *biofuels.*

General Land Ordinances of 1785 and 1787 Laws that gave the U.S. government the right to manage Western lands and created a grid system for surveying them and readying them for private ownership.

generalist A *species* that can survive in a wide array of *habitats* or use a wide array of resources. Compare *specialist.*

genetically modified (GM) organism An organism that has been *genetically engineered* using a technique called *recombinant DNA* technology.

genetic diversity A measurement of the differences in *DNA* composition among individuals within a given *species.*

genetic engineering Any process scientists use to manipulate an organism's genetic material in the lab by adding, deleting, or changing segments of its *DNA.*

genus A taxonomic level in the Linnaean classification system that is above *species* and below family. A genus is made up of one or more closely related species.

geothermal energy Renewable *energy* that is generated deep within Earth. The radioactive decay of elements amid the extremely high pressures and temperatures at depth generate heat that rises to the surface in magma and through fissures and cracks. Where this energy heats *groundwater,* natural eruptions of heated water and steam are sent up from below.

geyser A natural spurt of heated groundwater and steam sent up from belowground that erupts through the surface.

global climate change Any change in aspects of Earth's *climate,* such as temperature, *precipitation,* and storm intensity. Generally refers today to the current warming trend in global temperatures and associated climatic changes. Compare *global warming.*

global warming An increase in Earth's average surface temperature. The term is most frequently used in reference to the pronounced warming trend of recent years and decades. Global warming is one aspect of *global climate change.*

good A material commodity manufactured for and bought by individuals and businesses.

greenhouse effect The warming of Earth's surface and *atmosphere* (especially the *troposphere*) caused by the *energy* emitted by *greenhouse gases.*

greenhouse gas A gas that absorbs infrared radiation released by Earth's surface and then warms the surface and *troposphere* by emitting *energy,* thus giving rise to the *greenhouse effect.* Greenhouse gases include carbon dioxide (CO_2), water vapor, ozone (O_3), nitrous oxide (N_2O), halocarbon gases, and methane (CH_4).

green manure Organic *fertilizer* comprised of freshly dead plant material.

green revolution An intensification of the industrialization of *agriculture* in the developing world in the latter half of the 20th century that has dramatically increased crop yields produced per unit area of farmland. Practices include devoting large areas to *monocultures* of crops specially bred for high yields and rapid growth; heavy use of *fertilizers, pesticides,* and *irrigation* water; and sowing and harvesting on the same piece of land more than once per year or per season.

green tax A levy on *environmentally* harmful activities and products aimed at providing a market-based incentive to correct for *market failure.* Compare *subsidy.*

gross primary production The *energy* that results when *autotrophs* convert solar energy (sunlight) to energy of chemical bonds in sugars through *photosynthesis*. Autotrophs use a portion of this production to power their own metabolism, which entails oxidizing *organic compounds* by *cellular respiration*. Compare *net primary production*.

ground source heat pump A pump that harnesses *geothermal energy* from near-surface sources of earth and water, and that can help heat residences.

groundwater Water held in aquifers underground.

growth rate The net change in a *population*'s size, per 1,000 individuals. Calculated by adding the crude birth rate to the *immigration* rate and then subtracting the crude death rate and the *emigration* rate, each expressed as the number per 1,000 individuals per year.

habitat The specific *environment* in which an organism lives, including both *biotic* and *abiotic factors*.

habitat conservation plan A cooperative agreement that allows landowners to harm threatened or endangered *species* in some ways if they voluntarily improve habitat for the species in others.

habitat selection The process by which organisms select *habitats* from among the range of options they encounter.

habitat use The process by which organisms use *habitats* from among the range of options they encounter.

Hadley cell One of a pair of cells of *convective circulation* between the equator and 30° north and south latitude that influence global *climate* patterns. Compare *Ferrel cell; polar cell*.

half-life The amount of time it takes for one-half the atoms of a *radioisotope* to emit radiation and decay. Different radioisotopes have different half-lives, ranging from fractions of a second to billions of years.

harmful algal bloom A *population* explosion of toxic algae caused by excessive *nutrient* concentrations.

hazardous waste *Waste* that is toxic, chemically reactive, flammable, or corrosive. Compare *industrial solid waste; municipal solid waste*.

herbicide A type of chemical *pesticide* that kills plants.

herbivory The consumption of plants by animals.

heterotroph (consumer) An organism that consumes other organisms. Includes most animals, as well as fungi and microbes that decompose organic matter.

horizon A distinct layer of *soil*. See *A horizon; B horizon; C horizon; E horizon; O horizon; R horizon*.

host The organism in a parasitic relationship that suffers harm while providing the *parasite* nourishment or some other benefit.

Hubbert's peak The peak in production of *crude oil* in the United States, which occurred in 1970 just as Shell Oil geologist M. King Hubbert had predicted in 1956.

hydrocarbon An *organic compound* consisting solely of hydrogen and carbon *atoms*.

hydroelectric power (hydropower) The generation of electricity using the *kinetic energy* of moving water.

hydrogen An *element* with one *proton*, one *neutron*, and one *electron*. Hydrogen is the simplest element in the universe, is abundant in living things, and can serve as a fuel for renewable energy systems. See Appendix C, *Periodic table of the elements*.

hydrogenase An enzyme that can trigger algae to stop producing oxygen as a metabolic by-product and start releasing hydrogen instead.

hydrologic cycle The flow of water—in liquid, gaseous, and solid forms— through our *biotic* and *abiotic environment*.

hydropower See *hydroelectric power*.

hydrothermal vent Location in the deep ocean where heated water spurts from the seafloor, carrying minerals that precipitate to form rocky structures. Unique and recently discovered *ecosystems* cluster around these vents; tubeworms, shrimp, and other creatures here use symbiotic bacteria to derive their energy from chemicals in the heated water rather than from sunlight.

hypothesis An educated guess that explains a phenomenon or answers a *scientific* question. Compare *theory*.

hypoxia The condition of extremely low dissolved oxygen concentrations in a body of water.

igneous rock One of the three main categories of rock. Formed from cooling *magma*. Granite and basalt are examples of igneous rock. Compare *metamorphic rock; sedimentary rock*.

ignitable Easily able to catch fire. One criterion for defining *hazardous waste*.

immigration The arrival of individuals from outside a *population*.

inbreeding depression A state that occurs in a *population* when genetically similar parents mate and produce weak or defective offspring as a result.

incineration A controlled process of burning solid waste for disposal in which mixed garbage is combusted at very high temperatures. Compare *sanitary landfill*.

independent variable The *variable* that the scientist manipulates in a *manipulative experiment*.

indoor air pollution *Air pollution* that occurs indoors.

industrial ecology A holistic approach to industry that integrates principles from engineering, chemistry, *ecology, economics*, and other disciplines and seeks to redesign industrial *systems* in order to reduce resource inputs and minimize inefficiency.

industrial revolution The shift in the mid-1700s from rural life, animal-powered agriculture, and manufacturing by craftsmen to an urban society powered by *fossil fuels* such as *coal* and *crude oil*. Compare *agricultural revolution*.

industrial smog Gray-air smog caused by the incomplete combustion of *coal* or oil when burned. Compare *photochemical smog*.

industrial solid waste Nonliquid *waste* that is not especially hazardous and that comes from production of consumer goods, mining, *petroleum* extraction and *refining*, and *agriculture*. Compare *hazardous waste; municipal solid waste*.

industrial stage The third stage of the *demographic transition* model, characterized by falling birth rates that close the gap with falling death rates and reduce the rate of *population* growth. Compare *pre-industrial stage; post-industrial stage; transitional stage*.

industrialized agriculture A form of *agriculture* that uses large-scale mechanization and *fossil fuel* combustion, enabling farmers to replace horses and oxen with faster and more powerful means of cultivating, harvesting, transporting, and processing crops. Other aspects include *irrigation* and the use of *inorganic fertilizers*. Use of chemical herbicides and *pesticides* reduces *competition* from weeds and *herbivory* by insects. Compare *traditional agriculture*.

infectious disease A disease in which a pathogen attacks a host.

inorganic fertilizer A *fertilizer* that consists of mined or synthetically manufactured mineral supplements. Inorganic fertilizers are generally more susceptible than *organic fertilizers* to *leaching* and *runoff* and may be more likely to cause unintended off-site impacts.

insecticide A type of chemical *pesticide* that kills insects.

integrated pest management (IPM) The use of multiple techniques in combination to achieve long-term suppression of *pests*, including *biocontrol*, use of *pesticides*, close monitoring of *populations*, *habitat* alteration, *crop rotation, transgenic* crops, alternative tillage methods, and mechanical pest removal.

intercropping Planting different types of crops in alternating bands or other spatially mixed arrangements.

interdisciplinary field A field that borrows techniques from several more traditional fields of study and brings together research results from these fields into a broad synthesis.

interest group A small group of people seeking private gain that may work against the larger public interest.

Intergovernmental Panel on Climate Change (IPCC) An international panel of *atmospheric* scientists, *climate* experts, and government officials established in 1988 by the *United Nations Environment Programme* and the World Meteorological Organization, whose mission is to assess information relevant to questions of human-induced *global climate change*. The IPCC's 2001 *Third Assessment Report* summarizes current and probable future global trends and represents the scientific consensus of climate experts around the world.

interspecific competition *Competition* that takes place among members of two or more different *species*. Compare *intraspecific competition*.

intertidal Of, relating to, or living along shorelines between the highest reach of the highest *tide* and the lowest reach of the lowest tide.

intraspecific competition *Competition* that takes place among members of the same *species*. Compare *interspecific competition*.

invasive species A *species* that spreads widely and rapidly becomes dominant in a *community*, interfering with the community's normal functioning.

inversion layer In a *temperature inversion*, the band of air in which temperature rises with altitude (instead of falling with altitude, as temperature does normally).

ion An electrically charged *atom* or combination of atoms.

ionic bond A chemical bond in which oppositely charged *ions* are held together by electrical attraction. Compare *covalent bond*.

ionic compound (salt) An association of *ions* that are bonded electrically in an *ionic bond*.

IPAT model A formula that represents how humans' total impact (I) on the *environment* results from the interaction among three factors: *population* (P), affluence (A), and technology (T).

irrigation The artificial provision of water to support *agriculture*.

isotope One of several forms of an *element* having differing numbers of *neutrons* in the nucleus of its *atoms*. Chemically, isotopes of an element behave almost identically, but they have different physical properties because they differ in mass.

kelp Large brown algae or seaweed that can form underwater "forests," providing habitat for marine organisms.

keystone species A *species* that has an especially far-reaching effect on a *community*.

kinetic energy *Energy* of motion. Compare *potential energy*.

K–selected Term denoting a *species* with low biotic potential whose members produce a small number of offspring and take a long time to gestate and raise each of their young, but invest heavily in promoting the survival and growth of these few offspring. *Populations* of K–selected species are generally regulated by *density-dependent factors*. Compare *r–selected*.

Kyoto Protocol An international agreement drafted in 1997 that calls for nations to reduce, by 2012, emissions of six *greenhouse gases* to levels lower than their levels in 1990. Although the United States has refused to ratify the protocol, it came into force in 2005 when Russia ratified it, the 127th nation to do so.

La Niña An exceptionally strong cooling of surface water in the equatorial Pacific Ocean that occurs every 2 to 7 years and has widespread climatic consequences. Compare *El Niño*.

landfill gas A mix of gases that consists of roughly half *methane* produced by anaerobic decomposition deep inside *landfills*.

landscape ecology An approach to the study of organisms and their *environments* at the landscape scale, focusing on geographical areas that include multiple *ecosystems*.

land trust Local or regional organization that preserves lands valued by its members. In most cases, land trusts purchase land outright with the aim of preserving it in its natural condition. The Nature Conservancy may be considered the world's largest land trust.

latitudinal gradient The increase in *species richness* as one approaches the equator. This pattern of variation with latitude has been one of the most obvious patterns in *ecology*, but one of the most difficult ones for scientists to explain.

lava *Magma* that is released from volcanic vents and flows or spatters across Earth's surface.

LD$_{50}$ (lethal dose–50%) The amount of a *toxicant* it takes to kill 50% of a *population* of test animals. Compare *ED$_{50}$*; *threshold dose*.

leachate Liquids that seep through liners of a *sanitary landfill* and leach into the *soil* underneath.

leaching The process by which solid materials such as minerals are dissolved in a liquid (usually water) and transported to another location.

lead A heavy metal that may be ingested through water or paint, or that may enter the *atmosphere* as a particulate pollutant through combustion of leaded gasoline or other processes. Atmospheric lead deposited on land and water can enter the *food chain*, accumulate within body tissues, and cause *lead poisoning* in animals and people. Lead is an EPA *criteria pollutant*.

lead poisoning Poisoning by ingestion or inhalation of the heavy metal *lead*, causing an array of maladies, including damage to the brain, liver, kidney, and stomach; learning problems and behavioral abnormalities; anemia; hearing loss; and even death. Lead poisoning can result from drinking water that passes through old lead pipes or ingesting dust or chips of old lead-based paint.

legislation Statutory law.

Leopold, Aldo (1887–1949) American scientist, scholar, philosopher, and author. His book *The Land Ethic* argued that humans should view themselves and the land itself as members of the same *community* and that humans are obligated to treat the land *ethically*.

levee See *dike*.

life-cycle analysis In *industrial ecology*, the examination of the entire life cycle of a given product—from its origins in raw materials, through its manufacturing, to its use, and finally its disposal—in an attempt to identify ways to make the process more *ecologically* efficient.

life expectancy The average number of years that individuals in particular age groups are likely to continue to live.

lifestyle hazard See *cultural hazard*.

light-dependent reactions Reactions in *photosynthesis* in which water molecules are split, forming hydrogen ions (H$^+$), molecular oxygen (O$_2$), and small, high-energy molecules used to fuel *light-independent reactions*.

light-independent reactions Reactions in *photosynthesis* in which carbon atoms from carbon dioxide link together to manufacture sugars.

limiting factor A physical, chemical, or biological characteristic of the *environment* that restrains *population* growth.

limnetic zone In a water body, the layer of open water through which sunlight penetrates. Compare *littoral zone*; *benthic zone*; *profundal zone*.

lipid One of a chemically diverse group of *macromolecules* that are classified together because they do not dissolve in water. Lipids include fats, phospholipids, waxes, pigments, and steroids.

littoral See *intertidal*.

littoral zone The region ringing the edge of a water body. Compare *benthic zone*; *limnetic zone*; *profundal zone*.

Living Planet Index A metric that summarizes trends in the *populations* of over 1,100 *species* that are well enough monitored to provide reliable data. Developed by scientists at the World Wildlife Fund and the *United Nations Environment Programme* to give an overall idea of how natural populations are faring. Between 1970 and 2000, this index fell by roughly 40%.

loam *Soil* with a relatively even mixture of *clay-*, *silt-*, and *sand-*sized particles.

lobbying The expenditure of time or money in an attempt to influence an elected official.

logistic growth curve A plot that shows how the initial *exponential growth* of a *population* is slowed and finally brought to a standstill by *limiting factors*.

longline fishing Fishing practice that involves dragging extremely long lines with baited hooks. Kills turtles, sharks, and an estimated 300,000 seabirds each year.

low-input agriculture *Agriculture* that uses smaller amounts of *pesticides*, *fertilizers*, growth hormones, water, and *fossil fuel* energy than are used in *industrial agriculture*.

macromolecule A very large molecule, such as a *protein*, *nucleic acid*, *carbohydrate*, or *lipid*.

magma Molten, liquid rock.

malnutrition The condition of lacking *nutrients* the body needs, including a complete complement of vitamins and minerals.

Malthus, Thomas (1766–1834) British economist who maintained that increasing human *population* would eventually deplete the available food supply until starvation, war, or disease arose and reduced the population.

mangrove A tree with a unique type of roots that curve upward to obtain oxygen, which is lacking in the mud in which they grow, and that serve as stilts to support the tree in changing water levels. Mangrove forests grow on the coastlines of the tropics and subtropics.

manipulative experiment An *experiment* in which the researcher actively chooses and manipulates the *independent variable*. Compare *natural experiment*.

mantle The malleable layer of rock that lies beneath Earth's *crust* and surrounds a mostly iron *core*.

marine protected area (MPA) An area of the ocean set aside to protect marine life from fishing pressures. An MPA may be protected from some human activities but be open to others. Compare *marine reserve*.

marine reserve An area of the ocean designated as a "no-fishing" zone, allowing no extractive activities. Compare *marine protected area*.

marketable emissions permit A permit issued to polluters that allows them to emit a certain fraction of the total amount of *pollution* the government will allow an entire industry to produce. Polluters are then allowed to buy, sell, and trade these permits with other polluters. See also *permit-trading*.

market failure The failure of markets to take into account the *environment*'s positive effects on *economies* (for example, *ecosystem services*) or to reflect the negative effects of economic activity on the environment and thereby on people (*external costs*).

mass extinction event The extinction of a large proportion of the world's *species* in a very short time period due to some extreme and rapid change or catastrophic event. Earth has seen five mass extinction events in the past half-billion years.

mass number The combined number of *protons* and *neutrons* in an *atom*.

materials recovery facility (MRF) A *recycling* facility where items are sorted, cleaned, shredded, and prepared for reprocessing into new items.

maximum sustainable yield The maximal harvest of a particular *renewable natural resource* that can be accomplished while still keeping the resource available for the future.

meltdown The accidental melting of the uranium fuel rods inside the core of a *nuclear reactor*, causing the release of radiation.

metamorphic rock One of the three main categories of rock. Formed by great heat and/or pressure that reshapes crystals within the rock and changes its appearance and physical properties. Common metamorphic rocks include marble and slate. Compare *igneous rock; sedimentary rock*.

methane (CH$_4$) A colorless gas produced primarily by *anaerobic* decomposition. The major constituent of *natural gas*, and a *greenhouse gas* that is molecule-for-molecule more potent than *carbon dioxide*.

methane hydrate An ice-like solid consisting of molecules of *methane* embedded in a crystal lattice of water molecules. Methane hydrates are being investigated as a potential new source of *energy* from *fossil fuels*.

Milankovitch cycle One of three types of variations in Earth's rotation and orbit around the sun that result in slight changes in the relative amount of solar radiation reaching Earth's surface at different latitudes. As the cycles proceed, they change the way solar radiation is distributed over Earth's surface and contribute to changes in *atmospheric* heating and circulation that have triggered the ice ages and other *climate* changes.

Millennium Ecosystem Assessment The most comprehensive scientific assessment of the present condition of the world's ecological systems and their ability to continue supporting our civilization. Prepared by over 2,000 of the world's leading environmental scientists from nearly 100 nations, and completed in 2005.

moderator Within a *nuclear reactor*, a substance, most often water or graphite, that slows the *neutrons* bombarding uranium so that *fission* can begin.

molecule A combination of two or more *atoms*.

monoculture The uniform planting of a single crop over a large area. Characterizes *industrialized agriculture*.

Montreal Protocol International treaty ratified in 1987 in which 180 signatory nations agreed to restrict production of *chlorofluorocarbons (CFCs)* in order to forestall stratospheric ozone depletion. Because of its effectiveness in decreasing global CFC emissions, the Montreal Protocol is considered the most successful effort to date in addressing a global *environmental* problem.

monumentalism The impulse to protect enormous, unusual, or beautiful natural features (as in national parks).

mortality Rate of death within a *population*.

mountaintop removal A large-scale form of *coal* mining in which entire mountaintops are leveled. The technique exerts extreme environmental impact on surrounding ecosystems and human residents.

Muir, John (1838–1914) Scottish immigrant to the United States who eventually settled in California and made the Yosemite Valley his wilderness home. Today, he is most strongly associated with the *preservation ethic*. He argued that nature deserved protection for its own inherent values (an *ecocentrist* argument) but also claimed that nature played a large role in human happiness and fulfillment (an *anthropocentrist* argument).

municipal solid waste Nonliquid *waste* that is not especially hazardous and that comes from homes, institutions, and small businesses. Compare *hazardous waste; industrial solid waste*.

mutagen A *toxicant* that causes *mutations* in the *DNA* of organisms.

mutation An accidental change in *DNA* that may range in magnitude from the deletion, substitution, or addition of a single nucleotide to a change affecting entire sets of chromosomes. Mutations provide the raw material for evolutionary change.

mutualism A relationship in which all participating organisms benefit from their interaction. Compare *parasitism*.

nacelle Compartment in a *wind turbine* containing machinery for generating power.

natality Rate of birth within a *population*.

National Environmental Policy Act (NEPA) A U.S. law enacted on January 1, 1970, that created an agency called the Council on Environmental Quality and required that an *environmental impact statement* be prepared for any major federal action.

national forest Public lands consisting of 191 million acres (more than 8% of the nation's land area) in many tracts spread across all but a few states.

National Forest Management Act *Legislation* passed by the U.S. Congress in 1976, mandating that plans for renewable resource management be drawn up for every national forest. These plans were to be explicitly based on the concepts of *multiple use* and *sustainable development* and be subject to broad public participation.

national park A scenic area set aside for recreation and enjoyment by the public. The national park system today numbers 388 sites totaling 78.8 million acres and includes national historic sites, national recreation areas, national wild and scenic rivers, and other types of areas.

national wildlife refuge An area set aside to serve as a haven for wildlife and also sometimes to encourage hunting, fishing, wildlife observation, photography, environmental education, and other public uses.

natural experiment An *experiment* in which the researcher cannot directly manipulate the *variables* and therefore must observe nature, comparing conditions in which variables differ, and interpret the results. Compare *manipulative experiment*.

natural gas A *fossil fuel* composed primarily of *methane* (CH$_4$), produced as a by-product when bacteria decompose organic material under *anaerobic* conditions.

natural rate of population change The rate of change in a *population*'s size resulting from birth and death rates alone, excluding migration.

natural resource Any of the various substances and *energy* sources we need in order to survive.

natural science An academic discipline that studies the natural world. Compare *social science*.

natural selection The process by which traits that enhance survival and reproduction are passed on more frequently to future generations of organisms than those that do not, thus altering the *genetic* makeup of populations through time. Natural selection acts on genetic variation and is a primary driver of evolution.

negative feedback loop A *feedback loop* in which output of one type acts as input that moves the *system* in the opposite direction. The input and output essentially neutralize each other's effects, stabilizing the system. Compare *positive feedback loop*.

neoclassical economics A *theory* of *economics* that explains market prices in terms of consumer preferences for units of particular commodities.

Buyers desire the lowest possible price, whereas sellers desire the highest possible price. This conflict between buyers and sellers results in a compromise price being reached and the "right" quantity of commodities being bought and sold. Compare *ecological economics; environmental economics.*

net primary production The *energy* or biomass that remains in an ecosystem after *autotrophs* have metabolized enough for their own maintenance through *cellular respiration.* Net primary production is the energy or biomass available for consumption by *heterotrophs.* Compare *gross primary production; secondary production.*

net primary productivity The rate at which *net primary production* is produced. See *productivity; gross primary production; net primary production; secondary production.*

neurotoxin A *toxicant* that assaults the nervous system. Neurotoxins include heavy metals, *pesticides*, and some chemical weapons developed for use in war.

neutron An electrically neutral (uncharged) particle in the nucleus of an *atom.*

new urbanism A school of thought among architects, planners, and developers that seeks to design neighborhoods in which homes, businesses, schools, and other amenities are within walking distance of one another. In a direct rebuttal to *sprawl*, proponents of new urbanism aim to create functional neighborhoods in which families can meet most of their needs close to home without the use of a car.

niche The functional role of a *species* in a *community.* See *fundamental niche; realized niche.*

nitrification The conversion by bacteria of ammonium *ions* (NO_4^+) first into nitrite ions (NO_2^-) and then into nitrate ions (NO_3^-).

nitrogen An *element* with seven *protons*, seven *neutrons*, and seven *electrons.* Nitrogen is a key *nutrient* for plant growth, and is abundant in organisms. See Appendix C, *Periodic table of the elements.*

nitrogen cycle A major *nutrient cycle* consisting of the routes that nitrogen *atoms* take through the nested networks of environmental *systems.*

nitrogen dioxide (NO_2) A foul-smelling reddish brown gas that contributes to *smog* and *acidic deposition.* It results when atmospheric nitrogen and oxygen react at the high temperatures created by combustion engines. An EPA *criteria pollutant.*

nitrogen fixation The process by which inert nitrogen gas combines with hydrogen to form ammonium *ions* NO_4^+, which are chemically and biologically active and can be taken up by plants.

nitrogen-fixing Term describing bacteria that live in a *mutualistic* relationship with many types of plants and provide *nutrients* to the plants by converting nitrogen to a usable form.

nonconsumptive use *Freshwater* use in which the water from a particular *aquifer* or surface water body either is not removed or is removed only temporarily and then returned. The use of water to generate electricity in hydroelectric *dams* is an example. Compare *consumptive use.*

nongovernmental organization (NGO) An organization unaffiliated with any government or corporation that exists to promote an issue or agenda. Many operate internationally, and some exert influence over environmental policy.

nonmarket value A value that is not usually included in the price of a *good* or *service.*

non-point source A diffuse source of *pollutants*, often consisting of many small sources. Compare *point source.*

nonrenewable natural resource A *natural resource* that is in limited supply and is formed much more slowly than we use it. Compare *renewable natural resource.*

no-till See *no-tillage.*

no-tillage *Agriculture* that does not involve tilling (plowing, disking, harrowing, or chiseling) the *soil.*

nuclear energy The *energy* that holds together *protons* and *neutrons* within the nucleus of an *atom.* Several processes, each of which involves transforming *isotopes* of one *element* into isotopes of other elements, can convert nuclear energy into thermal energy, which is then used to generate electricity. See also *nuclear fission; nuclear reactor.*

nuclear fission The conversion of the *energy* within an *atom*'s nucleus to usable thermal energy by splitting apart atomic nuclei. Compare *nuclear fusion.*

nucleic acid A *macromolecule* that directs the production of *proteins.* Includes *DNA* and *RNA.*

nutrient An *element* or *compound* that organisms consume and require for survival.

nutrient cycle The comprehensive set of cyclical pathways by which a given *nutrient* moves through the *environment.*

ocean thermal energy conversion (OTEC) A potential *energy* source that involves harnessing the solar radiation absorbed by tropical oceans in the tropics.

O horizon The top layer of *soil* in some *soil profiles*, made up of organic matter, such as decomposing branches, leaves, crop residue, and animal waste. Compare *A horizon; B horizon; C horizon; E horizon; R horizon.*

oil sands Dense, hard, oily substances that can be mined from the ground, and that some envision as a replacement for current *petroleum* as this resource declines.

oligotrophic Term describing a water body that has low- *nutrient* and high-oxygen conditions. Compare *eutrophic.*

organic agriculture *Agriculture* that uses no synthetic *fertilizers* or *pesticides* but instead relies on biological approaches such as *composting* and *biocontrol.*

organic compound A *compound* made up of carbon *atoms* (and, generally, hydrogen atoms) joined by *covalent bonds* and sometimes including other *elements*, such as nitrogen, oxygen, sulfur, or phosphorus. The unusual ability of carbon to build elaborate molecules has resulted in millions of different organic compounds showing various degrees of complexity.

organic fertilizer A *fertilizer* made up of natural materials (largely the remains or wastes of organisms), including animal manure, crop residues, fresh vegetation, and compost. Compare *inorganic fertilizer.*

Organization of Petroleum Exporting Countries (OPEC) Cartel of predominantly Arab nations that in 1973 embargoed oil shipments to the United States and other nations supporting Israel, setting off the nation's first oil shortage.

outdoor air pollution *Air pollution* that occurs outdoors.

overgrazing The consumption by too many animals of plant cover, impeding plant regrowth and the replacement of biomass. Overgrazing can exacerbate damage to *soils*, natural *communities*, and the land's productivity for further grazing.

overnutrition A condition of excessive food intake in which people receive more than their daily caloric needs.

oxygen An *element* with eight *protons*, eight *neutrons*, and eight *electrons.* Oxygen is abundant in organisms, is produced by *photosynthesis*, and is consumed in *respiration.* See Appendix C, *Periodic table of the elements.*

ozone A *molecule* consisting of three atoms of oxygen. Absorbs ultraviolet radiation in the *stratosphere.* Compare *ozone layer; tropospheric ozone.*

ozone layer A portion of the *stratosphere*, roughly 17–30 km (10–19 mi) above sea level, which contains most of the *ozone* in the *atmosphere.*

paradigm A dominant philosophical and theoretical framework within a scientific discipline.

parasite The organism in a parasitic relationship that extracts nourishment or some other benefit from the *host.*

parasitism A relationship in which one organism, the *parasite*, depends on another, the *host*, for nourishment or some other benefit while simultaneously doing the host harm. Compare *mutualism.*

parasitoid An insect that parasitizes other insects, generally causing eventual death of the *host.* Compare *parasite.*

parent material The base geological material in a particular location.

particulate matter Solid or liquid particles small enough to be suspended in the *atmosphere* and able to damage respiratory tissues when inhaled. Includes *primary pollutants* such as dust and soot as well as *secondary pollutants* such as sulfates and nitrates. An EPA *criteria pollutant.*

passive solar energy collection An approach in which buildings are designed and building materials are chosen to maximize their direct absorption of sunlight in winter, even as they keep the interior cool in the summer. Compare *active solar energy collection*.

peat A kind of precursor stage to *coal*, produced when organic material that is broken down by *anaerobic* decomposition remains wet, near the surface, and poorly compressed.

peer review The process by which a manuscript submitted for publication in an academic journal is examined by other specialists in the field, who provide comments and criticism (generally anonymously) and judge whether the work merits publication in the journal.

pelagic Of, relating to, or living between the surface and floor of the ocean. Compare *benthic*.

periodic table of the elements (see Appendix C) Standard table in chemistry that summarizes information on the *elements* in a comprehensive and elegant way.

permafrost In *tundra*, underground soil that remains more or less permanently frozen.

peroxyacyl nitrate A chemical created by the reaction of NO_2 with *hydrocarbons*, which can induce further reactions that damage living tissues in animals and plants.

pest A pejorative term for any organism that damages crops that are valuable to us. The term is subjective and defined by our own economic interest, and is not biologically meaningful. Compare *weed*.

pesticide An artificial chemical used to kill insects (insecticide), plants (herbicide), or fungi (fungicide).

pesticide drift airborne transport of *pesticides*.

petroleum See *crude oil*.

pH A measure of the concentration of hydrogen *ions* in a solution. The pH scale ranges from 0 to 14: A solution with a pH of 7 is neutral; solutions with a pH below 7 are *acidic*, and those with a pH higher than 7 are *basic*. Because the pH scale is logarithmic, each step on the scale represents a tenfold difference in hydrogen ion concentration.

phosphorus An *element* with 15 *protons*, 15 *neutrons*, and 15 *electrons*. Phosphorus is a key *nutrient* for plant growth. See Appendix C, *Periodic table of the elements*.

phosphorus cycle A major *nutrient cycle* consisting of the routes that phosphorus *atoms* take through the nested networks of environmental *systems*.

photobiological Term describing a process that produces fuel by utilizing light and living organisms.

photochemical smog Brown-air smog caused by light-driven reactions of *primary pollutants* with normal atmospheric *compounds* that produce a mix of over 100 different chemicals, ground-level *ozone* often being the most abundant among them. Compare *industrial smog*.

photoelectric effect Effect that occurs when light strikes one of a pair of metal plates in a *photovoltaic cell*, causing the release of *electrons*, which are attracted by electrostatic forces to the opposing plate. The flow of electrons from one plate to the other creates an electrical current.

photosynthesis The process by which *autotrophs* produce their own food. Sunlight powers a series of chemical reactions that convert carbon dioxide and water into sugar (glucose), thus transforming low-quality *energy* from the sun into high-quality energy the organism can use. Compare *cellular respiration*.

photovoltaic (PV) cell A device designed to collect sunlight and convert it to electrical *energy* directly by making use of the *photoelectric effect*.

photovoltaic effect See *photoelectric effect*.

phylogenetic tree A treelike diagram that represents the history of divergence of *species* or other taxonomic groups of organisms.

physical hazard Physical processes that occur naturally in our environment and pose human health hazards. These include discrete events such as earthquakes, volcanic eruptions, fires, floods, blizzards, landslides, hurricanes, and droughts, as well as ongoing natural phenomena such as ultraviolet radiation from sunlight. Compare *biological hazard; chemical hazard; cultural hazard*.

phytoplankton Microscopic photosynthetic algae, protists, and cyanobacteria that drift near the surface of water bodies and generally form the first *trophic level* in an aquatic *food chain*. Compare *zooplankton*.

Pinchot, Gifford (1865–1946) The first professionally trained American *forester*, Pinchot helped establish the U.S. Forest Service. Today, he is the person most closely associated with the *conservation ethic*.

pioneer species A *species* that arrives earliest, beginning the ecological process of *succession* in a terrestrial or aquatic *community*.

plate tectonics The process by which Earth's surface is shaped by the extremely slow movement of tectonic plates, or sections of *crust*. Earth's surface includes about 15 major tectonic plates. Their interaction gives rise to processes that build mountains, cause earthquakes, and otherwise influence the landscape.

point source A specific spot—such as a factory's smokestacks—where large quantities of *pollutants* are discharged. Compare *non-point source*.

polar A covalent bond, or a *molecule* with such a bond, in which electrons are shared unequally, with one *atom* exerting a greater pull.

polar cell One of a pair of cells of *convective circulation* between the poles and 60° north and south latitude that influence global *climate* patterns. Compare *Ferrel cell; Hadley cell*.

policy A rule or guideline that directs individual, organizational, or societal behavior.

pollination An interaction in which one organism (for example, bees) transfers pollen (male sex cells) from one flower to the ova (female cells) of another, fertilizing the female flower, which subsequently grows into a fruit.

polluter pays principle Principle in which the party that produces *pollution* pays the costs of cleaning up or mitigating the pollution.

pollution Any matter or *energy* released into the *environment* that causes undesirable impacts on the health and well-being of humans or other organisms. Pollution can be physical, chemical, or biological and can affect water, air, or soil.

polyculture The planting of multiple crops in a mixed arrangement or in close proximity. An example is some traditional Native American farming that mixed maize, beans, squash, and peppers. Compare *monoculture*.

polymer A chemical *compound* or mixture of compounds consisting of long chains of repeated *molecules*. Some polymers play key roles in the building blocks of life.

pool A location in which nutrients in a *biogeochemical cycle* remain for a period of time before moving to another pool. Can be living or nonliving entities. Compare *flux; residence time*.

population A group of organisms of the same *species* that live in the same area. Species are often composed of multiple populations.

population density The number of individuals within a *population* per unit area. Compare *population size*.

population dispersion See *population distribution*.

population distribution The spatial arrangement of organisms within a particular area.

population ecology Study of the quantitative dynamics of how individuals within a *species* interact with one another—in particular, why *populations* of some species decline while others increase.

population size The number of individual organisms present at a given time.

positive feedback loop A *feedback loop* in which output of one type acts as input that moves the *system* in the same direction. The input and output drive the system further toward one extreme or another. Compare *negative feedback loop*.

post-industrial stage The fourth and final stage of the *demographic transition* model, in which both birth and death rates have fallen to a low level and remain stable there, and *populations* may even decline slightly. Compare *industrial stage; pre-industrial stage; transition stage*.

potential energy Energy of position. Compare *kinetic energy*.

precautionary principle The idea that one should not undertake a new action until the ramifications of that action are well understood.

precipitation Water that condenses out of the *atmosphere* and falls to Earth in droplets or crystals.

predation The process in which one *species* (the *predator*) hunts, tracks, captures, and ultimately kills its *prey*.

predator An organism that hunts, captures, kills, and consumes individuals of another species, the *prey*.

prediction A specific statement, generally arising from a *hypothesis*, that can be tested directly and unequivocally.

pre-industrial stage The first stage of the *demographic transition* model, characterized by conditions that defined most of human history. In pre-industrial societies, both death rates and birth rates are high. Compare *industrial stage; post-industrial stage; transitional stage.*

prescribed (controlled) burns The practice of burning areas of forest or grassland under carefully controlled conditions to improve the health of *ecosystems*, return them to a more natural state, and help prevent uncontrolled catastrophic fires.

preservation ethic An ethic holding that we should protect the natural *environment* in a pristine, unaltered state. Compare *conservation ethic.*

prey An organism that is killed and consumed by a *predator*.

primary consumer An organism that consumes *producers* and feeds at the second *trophic level.*

primary pollutant A hazardous substance, such as soot or carbon monoxide, that is emitted into the *troposphere* in a form that is directly harmful. Compare *secondary pollutant.*

primary production The conversion of solar energy to the energy of chemical bonds in sugars during *photosynthesis*, performed by *autotrophs*.

primary succession A stereotypical series of changes as an *ecological community* develops over time, beginning with a lifeless substrate. In terrestrial *systems*, primary succession begins when a bare expanse of rock, *sand*, or *sediment* becomes newly exposed to the atmosphere and *pioneer species* arrive. Compare *secondary succession.*

primary treatment A stage of *wastewater* treatment in which contaminants are physically removed. Wastewater flows into tanks in which sewage solids, grit, and particulate matter settle to the bottom. Greases and oils float to the surface and can be skimmed off. Compare *secondary treatment.*

probability A quantitative description of the likelihood of a certain outcome.

producer See *autotroph.*

productivity The rate at which plants convert solar *energy* (sunlight) to biomass. *Ecosystems* whose plants convert solar energy to biomass rapidly are said to have high productivity. See *net primary productivity; gross primary production; net primary production.*

profundal zone In a water body, the volume of open water that sunlight does not reach. Compare *littoral zone; benthic zone; limnetic zone.*

prokaryote A typically unicellular organism. The *cells* of prokaryotic organisms lack *organelles* and a nucleus. All bacteria and archaeans are prokaryotes. Compare *eukaryote.*

protein A *macromolecule* made up of long chains of amino acids.

proton A positively charged particle in the nucleus of an *atom.*

proven recoverable reserve The amount of a given *fossil fuel* in a deposit that is technologically and economically feasible to remove under current conditions.

proxy indicator Indirect evidence, such as pollen from *sediment* cores and air bubbles from ice cores, of the *climate* of the past.

public policy *Policy* that is made by governments, including those at the local, state, federal, and international levels; it consists of *legislation, regulations*, orders, incentives, and practices intended to advance societal welfare. See also *environmental policy.*

pyrolysis A method of heating *biomass* in the absence of oxygen that produces a mix of solids, gases, and liquids, including a liquid fuel that can be burned to generate *electricity.*

radioactive Property of an element that spontaneously emits high-energy radiation by the decay of atomic nuclei.

radioisotope A radioactive *isotope* that emits subatomic particles and high-energy radiation as it "decays" into progressively lighter isotopes until becoming a stable isotope.

radon A highly *toxic*, radioactive, colorless gas that seeps up from the ground in areas with certain types of bedrock and can build up inside basements and homes with poor air circulation.

rangeland Land used for grazing livestock.

random distribution *Population distribution* pattern in which individuals are located haphazardly in space in no particular pattern (often when needed resources are spread throughout an area and other organisms do not strongly influence where individuals settle).

reactive Chemically unstable and readily able to react with other compounds, often explosively or by producing noxious fumes. One criterion for defining *hazardous waste.*

recombinant DNA *DNA* that has been patched together from the DNA of multiple organisms in an attempt to produce desirable traits (such as rapid growth, disease and *pest* resistance, or higher nutritional content) in organisms lacking those traits.

recombination Process by which organisms mix, or recombine, their genetic material (as during sexual reproduction), so that some of each parent's *genes* are included in the genes of the offspring. Recombination produces novel combinations of genes, generating variation among individuals.

recovery Waste management strategy composed of *recycling* and *composting.*

recycling The collection of materials that can be broken down and reprocessed to manufacture new items.

Red List An updated list of *species* facing unusually high risks of *extinction*. The list is maintained by the World Conservation Union.

red tide A *harmful algal bloom* consisting of algae that produce reddish pigments that discolor surface waters.

regional planning *City planning* done on broader geographic scales, generally involving multiple municipal governments.

regulation A specific rule issued by an administrative agency, based on the more broadly written statutory law passed by Congress and enacted by the president.

relative abundance The extent to which numbers of individuals of different species are equal or skewed. One way to express species diversity. See *evenness;* compare *species richness.*

relativist An ethicist who maintains that *ethics* do and should vary with social context. Compare *universalist.*

renewable natural resource A *natural resource* that is virtually unlimited or that is replenished by the *environment* over relatively short periods of hours to weeks to years. Compare *nonrenewable natural resource.*

replacement fertility The *total fertility rate (TFR)* that maintains a stable *population* size.

replicate (as verb): To stage multiple tests of the same experiment. (as noun): One of multiple tests of the same experiment.

reserves-to-production ratio (R/P ratio) The total remaining reserves of a *fossil fuel* divided by the annual rate of production (extraction and processing).

reservoir See *pool.*

residence time In a biogeochemical cycle, the amount of time a nutrient remains in a given pool or reservoir before moving to another. Compare *flux; pool.*

resilience The ability of an ecological *community* to change in response to disturbance but later return to its original state. Compare *resistance.*

resistance The ability of an ecological *community* to remain stable in the presence of a disturbance. Compare *resilience.*

Resource Conservation and Recovery Act (RCRA) Congressional *legislation* (enacted in 1976 and amended in 1984) that specifies, among other things, how to manage *sanitary landfills* to protect against environmental contamination.

resource management Strategic decision making about who should extract resources and in what ways, so that resources are used wisely and not wasted.

resource partitioning The process by which *species* adapt to *competition* by evolving to use slightly different resources, or to use their shared resources in different ways, thus minimizing interference with one another.

response The type or magnitude of negative effects an animal exhibits in response to a *dose* of *toxicant* in a dose-response test.

restoration ecology The study of the historical conditions of *ecological communities* as they existed before humans altered them.

R horizon The bottommost layer of *soil* in a typical *soil profile*. Also called *bedrock*. Compare *A horizon; B horizon; C horizon; E horizon; O horizon.*

ribonucleic acid See *RNA*.

risk The mathematical probability that some harmful outcome (for instance, injury, death, *environmental* damage, or *economic* loss) will result from a given action, event, or substance.

risk assessment The quantitative measurement of *risk*, together with the comparison of risks involved in different activities or substances.

risk management The process of considering information from scientific *risk assessment* in light of economic, social, and political needs and values, in order to make decisions and design strategies to minimize *risk*.

RNA (ribonucleic acid) A usually single-stranded *nucleic acid* composed of four nucleotides, each of which contains a sugar (ribose), a phosphate group, and a nitrogenous base. RNA carries the hereditary information for living organisms and is responsible for passing traits from parents to off-spring. Compare *DNA*.

rock cycle The very slow process in which rocks and the minerals that make them up are heated, melted, cooled, broken, and reassembled, form-ing *igneous*, *sedimentary*, and *metamorphic* rocks.

rotation time The number of years that pass between the time a forest stand is cut for timber and the next time it is cut.

r–selected Term denoting a *species* with high biotic potential whose mem-bers produce a large number of offspring in a relatively short time but do not care for their young after birth. *Populations* of r–selected species are generally regulated by *density-independent factors*. Compare *K–selected*.

runoff The water from *precipitation* that flows into streams, rivers, lakes, and ponds, and (in many cases) eventually to the ocean.

run-of-river Any of several methods used to generate *hydroelectric power* without greatly disrupting the flow of river water. Run-of-river approaches eliminate much of the *environmental* impact of large *dams*. Compare *storage*.

safe harbor agreement A cooperative agreement that allows landowners to harm threatened or endangered *species* in some ways if they voluntarily improve habitat for the species in others.

salinization The buildup of salts in surface *soil* layers.

salt See *ionic compound*.

salt marsh Flat land that is intermittently flooded by the ocean where the *tide* reaches inland. Salt marshes occur along temperate coastlines and are thickly vegetated with grasses, rushes, shrubs, and other herbaceous plants.

salvage logging The removal of dead trees following a natural disturbance. Although it may be economically beneficial, salvage logging can be ecologi-cally destructive, because the dead trees provide food and shelter for a vari-ety of insects and wildlife and because removing timber from recently burned land can cause severe *erosion* and damage to *soil*.

sanitary landfill A site at which solid waste is buried in the ground or piled up in large mounds for disposal, designed to prevent the waste from contaminating the *environment*. Compare *incineration*.

savanna A *biome* characterized by grassland interspersed with clusters of acacias and other trees. Savanna is found across parts of Africa (where it was the ancestral home of our *species*), South America, Australia, India, and other dry tropical regions.

science A systematic process for learning about the world and testing our understanding of it.

scientific method A formalized method for testing ideas with observations that involves several assumptions and a more or less consistent series of interrelated steps.

scrubber Technology to chemically treat gases produced in combustion to remove hazardous components and neutralize acidic gases, such as sulfur dioxide and hydrochloric acid, turning them into water and salt, in order to reduce smokestack emissions.

secondary consumer An organism that consumes *primary consumers* and feeds at the third *trophic level*.

secondary pollutant A hazardous substance produced through the reac-tion of substances added to the *atmosphere* with chemicals normally found in the atmosphere. Compare *primary pollutant*.

secondary succession A stereotypical series of changes as an *ecological community* develops over time, beginning when some event disrupts or dramatically alters an existing community. Compare *primary succession*.

secondary treatment A stage of *wastewater* treatment in which biological means are used to remove contaminants remaining after *primary treatment*. Wastewater is stirred up in the presence of *aerobic* bacteria, which degrade organic pollutants in the water. The wastewater then passes to another settling tank, where remaining solids drift to the bottom. Compare *primary treatment*.

second-growth Term describing trees that have sprouted and grown to partial maturity after virgin timber has been cut.

second law of thermodynamics Physical law stating that the nature of *energy* tends to change from a more-ordered state to a less-ordered state; that is, *entropy* increases.

sediment The eroded remains of rocks.

sedimentary rock One of the three main categories of rock. Formed when dissolved minerals seep through *sediment* layers and act as a kind of glue, crystallizing and binding sediment particles together. Sandstone and shale are examples of sedimentary rock. Compare *igneous rock*; *metamorphic rock*.

seed bank A storehouse for samples of the world's crop diversity.

selective breeding See *artificial selection*.

septic system A *wastewater* disposal method, common in rural areas, con-sisting of an underground tank and series of drainpipes. Wastewater runs from the house to the tank, where solids precipitate out. The water pro-ceeds downhill to a drain field of perforated pipes laid horizontally in gravel-filled trenches, where microbes decompose the remaining waste.

service Work done for others as a form of business.

sex ratio The proportion of males to females in a *population*.

shale oil Sedimentary rock filled with organic matter that was not buried deeply enough to form *oil*. Some envision this as a replacement for current *petroleum* as this resource declines.

shelterbelt A row of trees or other tall perennial plants that are planted along the edges of farm fields to break the wind and thereby minimize wind *erosion*.

sick-building syndrome An illness produced by indoor *pollution* in which the specific cause is not identifiable.

SLOSS (Single Large or Several Small) dilemma The debate over whether it is better to make reserves large in size and few in number or many in number but small in size.

smart growth A *city planning* concept in which a community's growth is managed in ways that limit *sprawl* and maintain or improve residents' quality of life. It involves guiding the rate, placement, and style of development such that it serves the *environment*, the *economy*, and the community.

Smith, Adam (1723–1790) Scottish philosopher known today as the father of *classical economics*. He believed that when people are free to pursue their own economic self-interest in a competitive marketplace, the marketplace will behave as if guided by "an invisible hand" that ensures that their actions will benefit society as a whole.

social science An academic discipline that studies human interactions and institutions. Compare *natural science*.

soil profile The cross-section of a *soil* as a whole, from the surface to the *bedrock*.

soil A complex plant-supporting *system* consisting of disintegrated rock, organic matter, air, water, *nutrients*, and microorganisms.

solar cooker A simple portable oven that uses reflectors to focus sunlight onto food and cook it.

solar energy Energy from the sun. It is perpetually renewable and may be harnessed in several ways.

solar panels Panels generally consisting of dark-colored, heat-absorbing metal plates mounted in flat boxes covered with glass panes, often installed on rooftops to harness *solar energy*.

source reduction The reduction of the amount of material that enters the *waste stream* to avoid the costs of disposal and *recycling*, help conserve resources, minimize *pollution*, and save consumers and businesses money.

specialist A *species* that can survive only in a narrow range of *habitats* that contain very specific resources. Compare *generalist*.

speciation The process by which new *species* are generated.

species A *population* or group of populations of a particular type of organism, whose members share certain characteristics and can breed freely with one another and produce fertile offspring. Different biologists may have different approaches to diagnosing species boundaries.

species-area curve A graph showing how number of *species* varies with the geographic area of a landmass or water body. *Species richness* commonly doubles as area increases tenfold.

species diversity The number and variety of *species* in the world or in a particular region.

species richness The number of species in a particular region. One way to express species diversity. Compare *evenness*; *relative abundance*.

sprawl The unrestrained spread of urban or *suburban* development outward from a city center and across the landscape.

steady-state economy An *economy* that does not grow or shrink but remains stable.

storage Technique used to generate *hydroelectric power*, in which large amounts of water are impounded in a reservoir behind a concrete *dam* and then passed through the dam to turn *turbine* s that generate electricity. Compare *run-of-river*.

stable isotopes *Isotopes* that are not radioactive.

stratosphere The layer of the *atmosphere* above the *troposphere* and below the mesosphere; it extends from 11 km (7 mi) to 50 km (31 mi) above sea level.

strip-mining The use of heavy machinery to remove huge amounts of earth to expose *coal* or minerals, which are mined out directly. Compare *subsurface mining*.

subduction The *plate tectonic* process by which denser ocean *crust* slides beneath lighter continental crust at a *convergent plate boundary*.

subsidy A government incentive (a giveaway of cash or publicly owned resources, or a tax break) intended to encourage a particular activity. Compare *green tax*.

subsistence economy A survival *economy*, one in which people meet most or all of their daily needs directly from nature and do not purchase or trade for most of life's necessities.

subspecies Populations of a *species* that occur in different geographic areas and vary from one another in some characteristics. Subspecies are formed by the same processes that drive *speciation* but result when divergence does not proceed far enough to create separate species.

subsurface mining Method of mining underground *coal* deposits, in which shafts are dug deeply into the ground and networks of tunnels are dug or blasted out to follow coal seams. Compare *strip-mining*.

suburb A smaller community that rings a city.

succession A stereotypical series of changes in the composition and structure of an *ecological community* through time. See *primary succession; secondary succession*.

sulfur dioxide (SO₂) A colorless gas resulting in part from the combustion of *coal*. In the *atmosphere*, it may react to form sulfur trioxide and sulfuric acid, which may return to Earth in *acidic deposition*. Sulfur dioxide is an EPA *criteria pollutant*.

Superfund A program administered by the *Environmental Protection Agency* in which experts identify sites polluted with hazardous chemicals, protect *groundwater* near these sites, and clean up the *pollution*.

supply The amount of a product offered for sale at a given price.

surface impoundment A *hazardous waste* disposal method in which a shallow depression is dug and lined with impervious material, such as *clay*. Water containing small amounts of hazardous waste is placed in the pond and allowed to evaporate, leaving a residue of solid hazardous waste on the bottom. Compare *deep-well injection*.

sustainability A guiding principle of *environmental science* that requires us to live in such a way as to maintain Earth's systems and its *natural resources* for the foreseeable future.

sustainable agriculture *Agriculture* that does not deplete *soils* faster than they form.

sustainable development Development that satisfies our current needs without compromising the future availability of *natural resources* or our future quality of life.

swamp A type of wetland consisting of shallow water rich with vegetation, occurring in a forested area. Compare *bog; freshwater marsh*.

symbiosis A *parasitic* or *mutualistic* relationship between different *species* of organisms that live in close physical proximity.

synergistic effect An interactive effect (as of *toxicants*) that is more than or different from the simple sum of their constituent effects.

system A network of relationships among a group of parts, elements, or components that interact with and influence one another through the exchange of *energy*, matter, and/or information.

taiga See *boreal forest*.

tar sands See *oil sands*.

tax break Governmental reduction or elimination of taxes required of a business or an individual, for the purpose of promoting industries or activities deemed desirable.

taxonomist A scientist who classifies species, using an organism's physical appearance and/or genetic makeup, and who groups species by their similarity into a hierarchy of categories meant to reflect evolutionary relationships.

technically recoverable Extractable using current technology. Applied to *fossil fuel* deposits.

temperate deciduous forest A *biome* consisting of midlatitude forests characterized by broad-leafed trees that lose their leaves each fall and remain dormant during winter. These forests occur in areas where *precipitation* is spread relatively evenly throughout the year: much of Europe, eastern China, and eastern North America.

temperate grassland A *biome* whose vegetation is dominated by grasses and features more extreme temperature differences between winter and summer and less *precipitation* than *temperate deciduous forests*.

temperate rainforest A *biome* consisting of tall coniferous trees, cooler and less species-rich than *tropical rainforest* and milder and wetter than *temperate deciduous forest*.

temperature (thermal) inversion A departure from the normal temperature distribution in the *atmosphere*, in which a pocket of relatively cold air occurs near the ground, with warmer air above it. The cold air, denser than the air above it, traps *pollutants* near the ground and causes a buildup of smog.

teratogen A *toxicant* that causes harm to the unborn, resulting in birth defects.

terracing The cutting of level platforms, sometimes with raised edges, into steep hillsides to contain water from *irrigation* and *precipitation*. Terracing transforms slopes into series of steps like a staircase, enabling farmers to cultivate hilly land while minimizing their loss of *soil* to water *erosion*.

tertiary consumer An organism that consumes *secondary consumers* and feeds at the fourth *trophic level*.

theory A widely accepted, well-tested explanation of one or more cause-and-effect relationships that has been extensively validated by a great amount of research. Compare *hypothesis*.

thermal inversion See *temperature inversion*.

thermal mass Construction materials that absorb heat, store it, and release it later, for use in *passive solar energy* approaches.

thermogenic Type of *natural gas* created by compression and heat deep underground. Contains *methane* and small amounts of other *hydrocarbon* gases.

Three Mile Island Nuclear power plant in Pennsylvania that in 1979 experienced a partial *meltdown*. The term is often using to denote the accident itself, the most serious *nuclear reactor* malfunction that the United States has thus far experienced.

threshold dose The amount of a *toxicant* at which it begins to affect a *population* of test animals. Compare ED_{50}; LD_{50}.

tide The periodic rise and fall of the ocean's height at a given location, caused by the gravitational pull of the moon and sun.

topsoil That portion of the *soil* that is most nutritive for plants and is thus of the most direct importance to *ecosystems* and to *agriculture*. Also known as the *A horizon*.

total fertility rate (TFR) The average number of children born per female member of a *population* during her lifetime.

toxic Poisonous; able to harm health of people or other organisms when a substance is inhaled, ingested, or touched. One criterion for defining *hazardous waste*.

toxic air pollutant *Air pollutant* that is known to cause cancer, reproductive defects, or neurological, developmental, immune system, or respiratory problems in humans, and/or to cause substantial *ecological* harm by affecting the health of nonhuman animals and plants. The *Clean Air Act of 1990* identifies 188 toxic air pollutants, ranging from the heavy metal mercury to *volatile organic compounds* such as benzene and methylene chloride.

toxicant A substance that acts as a poison to humans or wildlife.

toxicity The degree of harm a chemical substance can inflict.

toxicology The scientific field that examines the effects of poisonous chemicals and other agents on humans and other organisms.

toxin A *toxic* chemical stored or manufactured in the tissues of living organisms. For example, a chemical that plants use to ward off *herbivores* or that insects use to deter *predators*.

traditional agriculture Biologically powered *agriculture*, in which human and animal muscle power, along with hand tools and simple machines, perform the work of cultivating, harvesting, storing, and distributing crops. Compare *industrialized agriculture*.

transform plate boundary Area where two tectonic plates meet and slip and grind alongside one another. For example, the Pacific Plate and the North American Plate rub against each other along California's San Andreas Fault.

transgene A *gene* that has been extracted from the *DNA* of one organism and transferred into the DNA of an organism of another *species*.

transgenic Term describing an organism that contains *DNA* from another *species*.

transitional stage The second stage of the *demographic transition* model, which occurs during the transition from the *pre-industrial stage* to the *industrial stage*. It is characterized by declining death rates but continued high birth rates. See also *post-industrial stage*. Compare *industrial stage*; *post-industrial stage*; *pre-industrial stage*.

transmissible disease See *infectious disease*.

transpiration The release of water vapor by plants through their leaves.

treatment The portion of an *experiment* in which a *variable* has been manipulated in order to test its effect. Compare *control*.

treaty See *convention*.

tributary A smaller river that flows into a larger one.

trophic level Rank in the feeding hierarchy of a food chain. Organisms at higher trophic levels consume those at lower trophic levels.

tropical dry forest A *biome* that consists of deciduous trees and occurs at tropical and subtropical latitudes where wet and dry seasons each span about half the year. Widespread in India, Africa, South America, and northern Australia.

tropical rainforest A *biome* characterized by year-round rain and uniformly warm temperatures. Found in Central America, South America, southeast Asia, west Africa, and other tropical regions. Tropical rainforests have dark, damp interiors; lush vegetation; and highly diverse *biotic communities*.

troposphere The bottommost layer of the *atmosphere*; it extends to 11 km (7 mi) above sea level. See also *stratosphere*.

tropospheric ozone *Ozone* that occurs in the *troposphere*, where it is a *secondary pollutant* created by the interaction of sunlight, heat, nitrogen oxides, and volatile carbon-containing chemicals. A major component of *smog*, it can injure living tissues and cause respiratory problems. An EPA *criteria pollutant*.

tundra A *biome* that is nearly as dry as *desert* but is located at very high latitudes along the northern edges of Russia, Canada, and Scandinavia. Extremely cold winters with little daylight and moderately cool summers with lengthy days characterize this landscape of lichens and low, scrubby vegetation.

two-way metering Process by which owners of houses with photovoltaic systems can sell their excess *solar energy* to their local power utility.

umbrella species A *species* for which meeting its *habitat* needs automatically helps meet those of many other species. Umbrella species generally are species that require large areas of habitat.

undernutrition A condition of insufficient *nutrition* in which people receive less than 90% of their daily caloric needs.

uneven-aged Term describing stands of trees in timber plantations that are of different ages. Uneven-aged stands more closely approximate a natural forest than do *even-aged* stands.

uniform distribution *Population distribution* pattern in which individuals are evenly spaced (as when individuals hold territories or otherwise compete for space).

United Nations (U.N.) Organization founded in 1945 to promote international peace and to cooperate in solving international economic, social, cultural, and humanitarian problems.

United Nations Environment Programme (UNEP) Agency within the United Nations that deals with *environmental policy*. Created in 1972.

universalist An *ethicist* who maintains that there exist objective notions of right and wrong that hold across cultures and situations. Compare *relativist*.

upwelling In the ocean, the flow of cold, deep water toward the surface. Upwelling occurs in areas where surface *currents* diverge. Compare *downwelling*.

urban ecology A scientific field that views cities explicitly as *ecosystems*. Researchers in this field seek to apply the fundamentals of *ecosystem ecology* and *systems* science to urban areas.

urban growth boundary (UGB) In *city planning*, a geographic boundary intended to separate areas desired to be urban from areas desired to remain rural. Development for housing, commerce, and industry are encouraged within urban growth boundaries, but beyond them such development is severely restricted.

urbanization The shift from rural to city and *suburban* living.

variable In an *experiment*, a condition that can change. See *dependent variable* and *independent variable*.

vector An organism that transfers a *pathogen* to its *host*. An example is a mosquito that transfers the malaria pathogen to humans.

volatile organic compound (VOC) One of a large group of potentially harmful organic chemicals used in industrial processes.

Wallace, Alfred Russell (1823–1913) English naturalist who proposed, independently of *Charles Darwin*, the concept of *natural selection* as a mechanism for *evolution* and as a way to explain the great variety of living things.

waste Any unwanted product that results from a human activity or process.

waste management Strategic decision making to minimize the amount of *waste* generated and to dispose of waste safely and effectively.

waste stream The flow of *waste* as it moves from its sources toward disposal destinations.

waste-to-energy (WTE) facility An incinerator that uses heat from its furnace to boil water to create steam that drives electricity generation or that fuels heating systems.

wastewater Any water that is used in households, businesses, industries, or public facilities and is drained or flushed down pipes, as well as the polluted *runoff* from streets and storm drains.

waterlogging The saturation of *soil* by water, in which the *water table* is raised to the point that water bathes plant roots. Waterlogging deprives roots of access to gases, essentially suffocating them and eventually damaging or killing the plants.

watershed The entire area of land from which water drains into a given river.

water table The upper limit of *groundwater* held in an *aquifer*.

weather The local physical properties of the *troposphere*, such as temperature, pressure, humidity, cloudiness, and wind, over relatively short time periods. Compare *climate*.

weathering The physical, chemical, and biological processes that break down rocks and minerals, turning large particles into smaller particles.

weed A pejorative term for any plant that competes with our crops. The term is subjective and defined by our own economic interest, and is not biologically meaningful. Compare *pest*.

wetland A system that combines elements of freshwater and dry land. These biologically productive systems include *freshwater marshes, swamps*, and *bogs*.

wilderness area Federal land that is designated off-limits to development of any kind but is open to public recreation, such as hiking, nature study, and other activities that have minimal impact on the land.

windbreak See *shelterbelt*.

wind energy *Energy* from the motion of wind. In this source of renewable energy, the passage of wind through *wind turbines* is used to generate *electricity*.

wind farm A development involving a group of *wind turbines*.

wind turbine A mechanical assembly that converts the wind's *kinetic energy*, or energy of motion, into electrical energy.

wise-use movement A loose confederation of individuals and groups that coalesced in the 1980s and 1990s as a response to the increasing success of environmental advocacy. The movement favors extracting more resources from public lands, obtaining greater local control of lands, and obtaining greater motorized recreational access to public lands.

World Bank Institution founded in 1944 that serves as one of the globe's largest sources of funding for *economic* development, including such major projects as *dams*, *irrigation* infrastructure, and other undertakings.

world heritage site A location internationally designated by the United Nations for its cultural or natural value. There are over 700 such sites worldwide.

World Trade Organization (WTO) Organization based in Geneva, Switzerland, that represents multinational corporations and promotes free trade by reducing obstacles to international commerce and enforcing fairness among nations in trading practices.

worldview A way of looking at the world that reflects a person's (or a group's) beliefs about the meaning, purpose, operation, and essence of the world.

xeriscaping Landscaping using plants adapted to arid conditions.

zoning The practice of classifying areas for different types of development and land use.

zooplankton Tiny aquatic animals that feed on *phytoplankton* and generally comprise the second trophic level in an aquatic food chain. Compare *phytoplankton*.

zooxanthellae symbiotic algae that inhabit the bodies of corals and produce food through photosynthesis.

Photo Credits

Part Opening Photos:
    Part One Bill Hatcher/National Geographic Image Collection
    Part Two Bill Haber/AP Photo

Chapter 1 **Opening Photo** NASA/Johnson Space Center **1.2b** Charles O'Rear/CORBIS **1.3a** Art Resource **1.3b** Bettman/CORBIS **The Science behind the Story: The Lesson of Easter Island** Richard T. Nowitz/CORBIS **1.6** Joel W. Rogers/CORBIS **1.7a** Jay Withgott **1.7b** Reuters NewMedia Inc./CORBIS **1.11** Library of Congress **1.12** CORBIS **1.13** CORBIS **1.14** Bettmann/CORBIS

Chapter 2 **Opening Photo** Denis Poroy/Associated Press **2.3** Annie Griffiths Belt/CORBIS **2.4a** Konrad Wothe/Minden Pictures **2.4b** Bill Hatcher/National Geographic Image Collection **2.4c** Bruce Forster/The Image Bank/Getty Images **2.4d** CORBIS **2.4ef** Frans Lanting/Minden Pictures **2.4g** Charles O'Rear/CORBIS **2.14** Kristin Piljay **2.6a** Bettmann/CORBIS **2.6b** Museum of History & Industry/CORBIS **2.6c** University of Washington Libraries **2.7** Erich Hartmann/Magnum Photos **2.8** Bettmann/CORBIS **2.9b** AP/Wide World Photos **2.11** Don Boroughs/The Image Works

Chapter 3 **Opening Photo** Owen Franken/CORBIS **3.17a** Harry Taylor/Dorling Kindersley **3.17b** Dave King/Dorling Kindersley **3.17c** Harry Taylor/Dorling Kindersley

Chapter 4 **Opening Photo** Michael & Patricia Fogden/CORBIS **4.1a** Nicholas Athanas/Tropical Birding **4.1b** Michael Fogden/DRK Photo **4.1c** Michael & Patricia Fogden/CORBIS **4.1d** Michael Fogden/Photolibrary **4.2** Bishop Museum, Honolulu **4.7a** G. I. Bernard/Photo Researchers, Inc. **4.7b** Wisconsin Historical Society **4.8a** Kennan Ward/CORBIS **4.8b** Art Wolfe/The Image Bank/Getty Images **4.8c** Getty Images

Chapter 5 **Opening Photo** Wolfgang Polzer **5.3a** Peter Johnson/CORBIS **5.3b** Tom Brakefield/CORBIS **5.3c** Michael & Patricia Fogden/CORBIS **5.12a** Pat O'Hara/CORBIS **5.13a** Philip Gould/CORBIS **5.14a** Charles Mauzy/CORBIS **5.15a** David Samuel Robbins/CORBIS **5.16a** O. Alamany & E. Vicens/CORBIS **5.17a** Wolfgang Kaehler/CORBIS **5.18a** Joe McDonald/CORBIS **5.19a** Darrell Gulin/CORBIS **5.20a** Liz Hymans/CORBIS **5.21a** Charles Mauzy/CORBIS

Chapter 6 **Opening Photo** Louise Gubb/The Image Works **6.1** Reuters NewMedia Inc./CORBIS **6.9c** SETBOUN/CORBIS **The Science behind the Story: Causes of Fertility Decline in Bangladesh** Mark Edwards/Peter Arnold, Inc. **6.16a** Elie Bernager/Stone **6.16b** Ed Kashi/IPN/AURORA

Chapter 7 **Opening Photo** Joanna B. Pinneo/AURORA **7.5a** Sylvan Wittwer/Visuals Unlimited **7.5b** Kevin Horan/Stone **7.5c** Ron Giling/Peter Arnold **7.5d** Keren Su/Stone **7.5e** Yann Arthus-Bertrand/

CORBIS **7.5f** U.S. Department of Agriculture **7.6a** Getty Images **7.6b** Carol Cohen/CORBIS **7.9** Alexandra Avakian/Contact Press Images Inc. **7.10** Art Rickerby/Time Life Pictures/Getty Images **7.12ab** Department of Natural Resources, Queensland, Australia **7.13** Bob Rowan, Progressive Image/CORBIS **The Science behind the Story: Transgenic Contamination of Native Maize?** Peg Skorpinski **7.16b** Hal Fritts, Native Seeds/SEARCH **7.18** Arthur C. Smith III/Grant Heilman Photography **7.21** AP Photo **7.22** AP Wide World Photos

Chapter 8 **Opening Photo** Maurice Hornocker **The Science behind the Story: Testing and Applying Island Biogeography Theory** Stock Connection, Inc./Alamy **8.14** CORBIS Sygma **8.15** Tom & Pat Leeson/Photo Researchers, Inc.

Chapter 9 **Opening Photo** Phil Schermeister/CORBIS **9.2ab** U.S. EPA **9.3a** Lester Lefkowitz/CORBIS **9.3b** Bob Krist/CORBIS **9.3c** David R. Frazier Photolibrary, Inc. **9.3d** Aldo Torelli/Getty Images **The Science behind the Story: Measuring the Impacts of Sprawl** EditorialFotos/Alamy **9.4** Chicago Historical Society **9.5** Cooper Carry & Associates **9.7** Steve Semler/Alamy **9.8** Bettmann/CORBIS **9.9** Bohemian Nomad Picturemakers/CORBIS **9.15** Rob Badger/Getty Images **9.17** North Carolina Museum of Art/CORBIS **9.19a** Gary Braasch/CORBIS **9.19b** James Zipp/Photo Researchers, Inc.

Chapter 10 **Opening Photo** Howard K. Suzuki **10.1a** Ingram Publishing/Alamy **10.1b** Spencer Grant/Photo Edit **10.1c** Martin Dohrn/Photo Researchers, Inc. **10.1d** LWA-Dann Tardif/CORBIS **10.3** Bettmann/CORBIS **10.5** Peg Skorpinski **The Science behind the Story: Pesticides and Child Development in Mexico's Yaqui Valley,** all Elizabeth A. Guillette

Chapter 11 **Opening Photo** Lester Lefkowitz/CORBIS **11.2** William Manning/CORBIS **11.8a** AP Photo **11.8b** Ian Berry/Magnum Photos **11.9** High Country News **11.10a** Gilles Saussier/Liaison/Getty Images **11.10b** NASA Visible Earth **11.11** AP Photo **11.16a** Laguna Design/Science Photo Library/Photo Researchers, Inc. **11.16b** Bruce Robison/Minden Pictures **11.17a** Ralph A. Clevenger/CORBIS **11.17b** Jeff Hunter/Getty Images, Inc. **11.18** Mark A. Johnson/CORBIS

Chapter 12 **Opening Photo** Adam Woolfitt/CORBIS **12.5a** NASA Earth Observatory **12.5b** Bettmann/CORBIS **12.7** Pittsburgh Post-Gazette **12.8b** Allen Russell/Index Stock Imagery **12.9** NASA/Goddard Space Flight Center **The Science behind the Story: Acid Rain at Hubbard Brook Research Forest** Will & Deni McIntyre/CORBIS **12.17a** Ted Spiegel/CORBIS **12.17b** David Whillas, CSIRO Atmospheric Research **12.20** Ben Sklar/AP Photo **12.21** Lowell Georgia/CORBIS

Chapter 13 **Opening Photo** Steve Armbrust/AlaskaStock.com **13.7** CORBIS **13.10** Vivian Stockman **13.11** Steven Kazlowski/AlaskaStock **13.18a** Paul Fusco/Magnum Photos **13.18b** Reuters/

Selected Sources and References for Further Reading

Chapter 1

Bahn, Paul, and John Flenley. 1992. *Easter Island, Earth island*. Thames and Hudson, London.

Barbour, Ian G. 1992. *Ethics in an age of technology*. Harper Collins, San Francisco.

Bowler, Peter J. 1993. *The Norton history of the environmental sciences*. W. W. Norton, New York.

Cole, Luke W., and Sheila R. Foster. 2001. *From the ground up: Environmental racism and the rise of the environmental justice movement*. New York University Press, New York.

Diamond, Jared. 2005. *Collapse: How societies choose to fail or succeed*. Viking, New York.

Ehrlich, Paul. 1968. *The population bomb*. 1997 reprint, Buccaneer Books, Cutchogue, New York.

Elliot, Robert, and Arran Gare, eds. 1983. *Environmental philosophy: A collection of readings*. Pennsylvania State University Press, University Park.

Flenley, John, and Paul Bahn. 2003. *The enigmas of Easter Island*. Oxford University Press, New York.

Fox, Stephen. 1985. *The American conservation movement: John Muir and his legacy*. University of Wisconsin Press, Madison.

Goudie, Andrew. 2000. *The human impact on the natural environment*, 5th ed. MIT Press, Cambridge, Massachusetts.

Hardin, Garrett. 1968. The tragedy of the commons. *Science* 162: 1243–1248.

Katzner, Donald W. 2001. *Unmeasured information and the methodology of social scientific inquiry*. Kluwer, Boston.

Kuhn, Thomas S. 1962. *The structure of scientific revolutions*, 2nd ed., 1970. University of Chicago Press, Chicago.

Leopold, Aldo. 1949. *A Sand County almanac, and sketches here and there*. Oxford University Press, New York.

Malthus, Thomas R. *An essay on the principle of population*. 1983 ed. Penguin USA, New York.

Millennium Ecosystem Assessment. 2005. *Ecosystems and human well-being: General synthesis*. Millennium Ecosystem Assessment and World Resources Institute.

Musser, George. 2005. The climax of humanity. *Scientific American* 293(3): 44–47.

Nash, Roderick F. 1989. *The rights of nature*. University of Wisconsin Press, Madison.

Nash, Roderick F. 1990. *American environmentalism: Readings in conservation history*, 3rd ed. McGraw-Hill, New York.

O'Neill, John O., R. Kerry Turner, and Ian J. Bateman, eds. 2001. *Environmental ethics and philosophy*. Elgar, Cheltenham, U.K.

Ponting, Clive. 1991. *A green history of the world: The environment and the collapse of great civilizations*. Penguin Books, New York.

Popper, Karl R. 1959. *The logic of scientific discovery*. Hutchinson, London.

Porteous, Andrew. 2000. *Dictionary of environmental science and technology*, 3rd ed. John Wiley & Sons, Hoboken, New Jersey.

Redman, Charles R. 1999. *Human impact on ancient environments*. University of Arizona Press, Tucson.

Sagan, Carl. 1997. *The demon-haunted world: Science as a candle in the dark*. Ballantine Books, New York.

Schneiderman, Jill S., ed. 2003. *The Earth around us: Maintaining a livable planet*. Perseus Books, New York.

Siever, Raymond. 1968. Science: Observational, experimental, historical. *American Scientist* 56: 70–77.

Singer, Peter, ed. 1993. *A companion to ethics*. Blackwell Publishers, Oxford.

Sterba, James P., ed. 1995. *Earth ethics: Environmental ethics, animal rights, and practical applications*. Prentice Hall, Upper Saddle River, New Jersey.

Stone, Christopher D. 1972. Should trees have standing? Towards legal rights for natural objects. *Southern California Law Review* 1972: 450–501.

Valiela, Ivan. 2001. *Doing science: Design, analysis, and communication of scientific research*. Oxford University Press, Oxford.

Van Tilburg, Jo Anne. 1994. *Easter Island: Archaeology, ecology, and culture*. Smithsonian Institution Press, Washington, D.C.

Venetoulis, Jason, et al., 2004. *Ecological footprint of nations 2004*. Redefining Progress, Oakland, California.

Wackernagel, Mathis, and William Rees. 1996. *Our ecological footprint: Reducing human impact on the earth*. New Society Publishers, Gabriola Island, British Columbia, Canada.

Wackernagel, Mathis, Lillemor Lewan, and Carina Borgström-Hansson. 1999. Evaluating the use of natural capital with the ecological footprint. *Ambio* 28: 604.

Wenz, Peter S. 2001. *Environmental ethics today*. Oxford University Press, Oxford.

White, Lynn. 1967. The historic roots of our ecologic crisis. *Science* 155: 1203–1207.

Chapter 2

Balmford, Andrew, et al. 2002. Economic reasons for conserving wild nature. *Science* 297: 950–953.

Brown, Lester. 2001. *Eco-economy: Building an economy for the Earth*. Earth Policy Institute and W. W. Norton, New York.

Clark, Ray, and Larry Canter. 1997. *Environmental policy and NEPA: Past, present, and future*. St. Lucie Press, Boca Raton, Florida.

Costanza, Robert, et al. 1997. *An introduction to ecological economics*. St. Lucie Press, Boca Raton, Florida.

Costanza, Robert, et al. 1997. The value of the world's ecosystem services and natural capital. *Nature* 387: 253–260.

Daily, Gretchen. 1997. *Nature's services: Societal dependence on natural ecosystems*. Island Press, Washington, D.C.

Daly, Herman E. 1996. *Beyond growth*. Beacon Press, Boston.

Daly, Herman E. 2005. Economics in a full world. *Scientific American* 293(3): 100–107.

Dietz, Thomas, et al. 2003. The struggle to govern the global commons. *Science* 302: 1907–1912.

Field, Barry C., and Martha K. Field. 2001. *Environmental economics*, 3rd ed. McGraw-Hill, New York.

Fogleman, Valerie M. 1990. *Guide to the National Environmental Policy Act*. Quorum Books, New York.

French, Hilary. 2000. Environmental treaties gain ground. Pp. 134–135 in *Vital Signs 2000*. Worldwatch Institute and W. W. Norton, Washington D.C., and New York.

Gardner, Gary, et al. 2004. The state of consumption today. Pp. 3–23 in *State of the world 2004*. Worldwatch Institute and W. W. Norton, Washington, D.C., and New York.

Gardner, Gary, and Erik Assadourian. 2004. Rethinking the good life. Pp. 164–179 in *State of the world 2004*. Worldwatch Institute and W. W. Norton, Washington, D.C., and New York.

Goodstein, Eban. 1999. *The tradeoff myth: Fact and fiction about jobs and the environment*. Island Press, Washington, D.C.

Goodstein, Eban. 2005. *Economics and the environment*, 4th ed. John Wiley & Sons, Hoboken, New Jersey.

Green Scissors, 2004. *Green Scissors 2004: Cutting wasteful and environmentally harmful spending*. Friends of the Earth, Taxpayers for Common Sense, and U.S. Public Interest Research Group.

Hawken, Paul, Amory Lovins, and L. Hunter Lovins. 1999. *Natural capitalism*. Little, Brown, and Co., Boston.

Herzog, Lawrence A. 1990. *Where north meets south: Cities, space, and politics on the U.S.–Mexico border*. Center for Mexican-American Studies, University of Texas at Austin.

Houck, Oliver, 2003. Tales from a troubled marriage: Science and law in environmental policy. *Science* 302: 1926–1928.

Kolstad, Charles D. 2000. *Environmental economics*. Oxford University Press, Oxford.

Kraft, Michael E. 2003. *Environmental policy and politics*, 3rd ed. Longman, New York.

Kubasek, Nancy K., and Gary S. Silverman. 2004. *Environmental law*, 5th ed. Prentice Hall, Upper Saddle River, New Jersey.

Mastny, Lisa. 2002. Ecolabeling gains ground. Pp. 124–125 in *Vital signs 2002*. Worldwatch Institute and W. W. Norton, Washington D.C., and New York.

Millennium Ecosystem Assessment. 2005. *Ecosystems and human well-being: Opportunities and challenges for business and industry*. Millennium Ecosystem Assessment and World Resources Institute.

Myers, Norman, and Jennifer Kent. 2001. *Perverse subsidies: How misused tax dollars harm the environment and the economy*. Island Press, Washington, D.C.

National Environmental Policy Act of 1969, The, as amended (Pub. L. 91–190, 42 U.S.C. 4321–4347, January 1, 1970, as amended by Pub. L. 94–52, July 3, 1975, Pub. L. 94–83, August 9, 1975, and Pub. L. 97–258, § 4(b), Sept. 13, 1982). http://ceq.eh.doe.gov/nepa/regs/nepa/nepaeqia.htm

Pearson, Charles S. 2000. *Economics and the global environment*. Cambridge University Press, Cambridge.

Ricketts, Taylor, et al. 2004. Economic value of tropical forest to coffee production. *Proceedings of the National Academy of Sciences of the USA* 101: 12579–12582.

Sachs, Jeffrey. 2005. Can extreme poverty be eliminated? *Scientific American* 293(3): 56–65.

Shafritz, Jay M. 1993. *The HarperCollins dictionary of American government and politics*. HarperCollins, New York.

Smith, Adam. 1776. *An inquiry into the nature and causes of the wealth of nations*. 1993 ed., Oxford University Press, Oxford.

Southwest Center for Environmental Research and Policy, and San Diego State University. Tijuana River Watershed Atlas Project. http://geography.sdsu.edu/Research/Projects/TWRP/tjatlas.html

Steel, Brent S., Richard L. Clinton, and Nicholas P. Lovrich. 2002. *Environmental politics and policy*. McGraw-Hill, New York.

Tietenberg, Tom. 2003. *Environmental economics and policy*, 4th ed. Addison Wesley, Boston.

Turner, R. Kerry, David Pearce, and Ian Bateman. 1993. *Environmental economics: An elementary introduction*. Johns Hopkins University Press, Baltimore.

United States Congress. House. H.R. 3378. 2000. The Tijuana River Valley Estuary and Beach Sewage Cleanup Act of 2000.

Venetoulis, Jason, and Cliff Cobb. 2004. *The genuine progress indicator 1950–2002 (2004 update)*. Redefining Progress, Oakland, California.

Vig, Norman J., and Michael E. Kraft, eds. 2002. *Environmental policy: New directions for the twenty-first century*, 5th ed. CQ Press, Congressional Quarterly, Inc., Washington, D.C.

Wilkinson, Charles F. 1992. *Crossing the next meridian: Land, water, and the future of the West*. Island Press, Washington, D.C.

Chapter 3

Berry, R. Stephen. 1991. *Understanding energy: Energy, entropy and thermodynamics for every man*. World Scientific Publishing Co.

Campbell, Neil A., and Jane B. Reece. 2005. *Biology*, 7th ed. Benjamin Cummings, San Francisco.

Capra, Fritjof. 1996. *The web of life: A new scientific understanding of living systems*. Anchor Books Doubleday, New York.

Carpenter, Edward J., and Douglas G. Capone, eds. 1983. *Nitrogen in the marine environment*. Academic Press, New York.

Committee on Environment and Natural Resources, 2000. *An integrated assessment: Hypoxia in the northern Gulf of Mexico*. CENR, National Science and Technology Council, Washington, D.C.

Ferber, Dan. 2004. Dead zone fix not a dead issue. *Science* 305: 1557.

Field, Christopher B., et al., 1998. Primary production of the biosphere: Integrating terrestrial and oceanic components. *Science* 281: 237–240.

Hall, David O., and Krishna Rao. 1999. *Photosynthesis*, 6th ed. Cambridge University Press, Cambridge.

Jacobson, Michael, et al. 2000. *Earth system science from biogeochemical cycles to global changes*. Academic Press.

Keller, Edward A. 2004. *Introduction to environmental geology*, 3rd ed. Prentice Hall, Upper Saddle River, New Jersey.

Lancaster, M., 2002. *Green chemistry*. Royal Society of Chemistry, London.

Larsen, Janet. 2004. Dead zones increasing in world's coastal waters. *Eco-economy update #41*, 16 June 2004. Earth Policy Institute, www.earth-policy.org/Updates/Update41.htm

Manahan, Stanley E. 2004. *Environmental chemistry*, 8th ed. Lewis Publishers, CRC Press, Boca Raton, Florida.

McMurry, John E. 2003. *Organic chemistry*, 6th ed. Brooks/Cole, San Francisco.

Mississippi River/Gulf of Mexico Watershed Nutrient Task Force. 2001. *Action plan for reducing, mitigating, and controlling hypoxia in the northern Gulf of Mexico*. Washington, D.C.

Mitsch, William J., et al. 2001. Reducing nitrogen loading to the Gulf of Mexico from the Mississippi River Basin: Strategies to counter a persistent ecological problem. *BioScience* 51: 373–388.

Montgomery, Carla. 2005. *Environmental geology*, 7th ed. McGraw-Hill, New York.

National Oceanic and Atmospheric Administration: National Ocean Service. 2000. Hypoxia in the Gulf of Mexico: Progress toward the completion of an integrated assessment. www.nos.noaa.gov/products/pubs_hypox.html

National Science and Technology Council, Committee on Environment and Natural Resources. 2003. *An assessment of coastal hypoxia and eutrophication in U.S. waters*. National Science and Technology Council, Washington, D.C.

Rabalais, Nancy N., R. E. Turner, and D. Scavia. 2002. Beyond science into policy: Gulf of Mexico hypoxia and the Mississippi River. *BioScience* 52: 129–142.

Rabalais, Nancy N., R. E. Turner, and W. J. Wiseman, Jr. 2002. Hypoxia in the Gulf of Mexico, a.k.a. "The dead zone." *Annual Review of Ecology and Systematics* 33: 235–263.

Raloff, Janet. 2004. Dead waters: Massive oxygen-starved zones are developing along the world's coasts. *Science News* 165: 360–362. June 5, 2004.

Raloff, Janet. 2004. Limiting dead zones: How to curb river pollution and save the Gulf of Mexico. *Science News* 165: 378–380. June 12, 2004.

Ricklefs, Robert E., and Gary L. Miller. 2000. *Ecology*, 4th ed. W. H. Freeman and Co., New York.

Schlesinger, William H. 1997. *Biogeochemistry: An analysis of global change*, 2nd ed. Academic Press, London.

Skinner, Brian J., and Stephen C. Porter. 2003. *The dynamic earth: An introduction to physical geology*, 5th ed. John Wiley and Sons, Hoboken, New Jersey.

Smith, Robert L., and Thomas M. Smith. 2001. *Ecology and field biology*, 6th ed. Benjamin Cummings, San Francisco.

Stiling, Peter. 2002. *Ecology: Theories and applications*, 4th ed. Prentice Hall, Upper Saddle River, New Jersey.

Takahashi, Taro. 2004. The fate of industrial carbon dioxide. *Science* 305: 352–353.

Turner, R. Eugene, and Nancy N. Rabalais. 2003. Linking landscape and water quality in the Mississippi River Basin for 200 years. *BioScience* 53: 563–572.

Van Dover, Cindy Lee, 2000. *The ecology of deep-sea hydrothermal vents.* Princeton University Press, Princeton.

Van Ness, H.C. 1983. *Understanding thermodynamics.* Dover Publications, Mineola, New York.

Vitousek, Peter M., et al. 1997. Human alteration of the global nitrogen cycle: Sources and consequences. *Ecological Applications* 7: 737–750.

Whittaker, Robert H. 1975. *Communities and ecosystems*, 2nd ed. Macmillan, New York.

Chapter 4

Allen, K. C., and D. E. G. Briggs, eds. 1989. *Evolution and the fossil record.* John Wiley & Sons, Hoboken, New Jersey.

Alvarez, Luis W., et al. 1980. Extraterrestrial cause for the Cretaceous-Tertiary extinction. *Science* 208: 1095–1108.

Begon, Michael, Martin Mortimer, and David J. Thompson. 1996. *Population ecology: A unified study of animals and plants*, 3rd ed. Blackwell Scientific, Oxford.

Campbell, Neil A., and Jane B. Reece. 2005. *Biology*, 7th ed. Benjamin Cummings, San Francisco.

Clark K. L., et al. 1998. Cloud water and precipitation chemistry in a tropical montane forest, Monteverde, Costa Rica. *Atmospheric Environment* 32: 1595–1603.

Crump, L. Martha, et al. 1992. Apparent decline of the golden toad: Underground or extinct? *Copeia* 1992: 413–420.

Darwin, Charles. 1859. *The origin of species by means of natural selection.* John Murray, London.

Endler, John A. 1986. *Natural selection in the wild.* Monographs in Population Biology 21, Princeton University Press, Princeton.

Fenchel, Tom. 2003. *Origin and early evolution of life.* Oxford University Press, Oxford.

Fortey, Richard. 1998. *Life: A natural history of the first four billion years of life on Earth.* Alfred Knopf, New York.

Freeman, Scott, and Jon C. Herron. 2003. *Evolutionary analysis*, 3rd ed. Prentice Hall, Upper Saddle River, New Jersey.

Futuyma, Douglas J. 2005. *Evolution.* Sinauer Associates, Sunderland, Massachusetts.

Gee, Henry. 1999. *In search of deep time: Beyond the fossil record to a new history of life.* Free Press, New York.

Krebs, Charles J. 2001. *Ecology: The experimental analysis of distribution and abundance*, 5th ed. Benjamin Cummings, San Francisco.

Lawton, Robert O., et al. 2001. Climatic impact of tropical lowland deforestation on nearby montane cloud forests. *Science* 294: 584–587.

Molles, Manuel C., Jr. 2005. *Ecology: Concepts and applications*, 3rd ed. McGraw-Hill, Boston.

Nadkarni, Nalini M., and Nathaniel T. Wheelwright, eds. 2000. *Monteverde: Ecology and conservation of a tropical cloud forest.* Oxford University Press, New York.

Pounds, J. Alan. 2001. Climate and amphibian declines. *Nature* 410: 639.

Pounds, J. Alan, and Martha L. Crump. 1994. Amphibian declines and climate disturbance: The case of the golden toad and the harlequin frog. *Conservation Biology* 8: 72–85.

Pounds, J. Alan, Michael P. L. Fogden, and John H. Campbell. 1999. Biological response to climate change on a tropical mountain. *Nature* 398: 611–615.

Pounds, J. Alan, et al. 1997. Tests of null models for amphibian declines on a tropical mountain. *Conservation Biology* 11: 1307–1322.

Powell, James L. 1998. *Night comes to the Cretaceous: Dinosaur extinction and the transformation of modern geology.* W. H. Freeman, New York.

Raup, David M. 1991. *Extinction: Bad genes or bad luck?* W. W. Norton, New York.

Ricklefs, Robert E., and Gary L. Miller. 2000. *Ecology*, 4th ed. W. H. Freeman, New York.

Ridley, Mark. 2003. *Evolution*, 3rd ed. Blackwell Science, Cambridge, Massachusetts.

Savage, Jay M. 1966. An extraordinary new toad (*Bufo*) from Costa Rica. *Revista de Biologia Tropical* 14: 153–167.

Savage, Jay M. 1998. The "brilliant toad" was telling us something. *Christian Science Monitor*, 14 September 1998: 19.

Smith, Thomas M., and Robert L. Smith. 2006. *Elements of ecology*, 6th ed. Benjamin Cummings, San Francisco.

Ward, Peter D., and Donald Brownlee. 2000. *Rare Earth: Why complex life is uncommon in the universe.* Copernicus, New York.

Williams, George C. 1966. *Adaptation and natural selection.* Princeton University Press, Princeton.

Wilson, Edward O. 1992. *The diversity of life.* Harvard University Press, Cambridge, Massachusetts.

Chapter 5

Barbour, Michael G., et al. 1998. *Terrestrial plant ecology*, 3rd ed. Benjamin Cummings, Menlo Park, California.

Breckle, Siegmar-Walter. 1999. *Walter's vegetation of the Earth: The ecological systems of the geo-biosphere*, 4th ed. Springer-Verlag, Berlin, 1999.

Bronstein, Judith L. 1994. Our current understanding of mutualism. *Quarterly Journal of Biology* 69: 31–51.

Chase, Jonathan M., et al., 2002. The interaction between predation and competition: A review and synthesis. *Ecology Letters* 5: 302.

Connell, Joseph H., and Ralph O. Slatyer, 1977. Mechanisms of succession in natural communities. *American Naturalist* 111: 1119–1144.

Drake, John M., and Jonathan M. Bossenbroek. 2004. The potential distribution of zebra mussels in the United States. *BioScience* 54: 931–941.

Estes, J.A., et al. 1998. Killer whale predation on sea otters linking oceanic and nearshore ecosystems. *Science* 282: 473–476.

Ewald, Paul W., 1987. Transmission modes and evolution of the parasitism-mutualism continuum. *Annals of the New York Academy of Sciences* 503: 295–306.

Gurevitch, Jessica, and Dianna K. Padilla. 2004. Are invasive species a major cause of extinctions? *Trends in Ecology and Evolution* 19: 470–474.

Krebs, Charles J. 2001. *Ecology: The experimental analysis of distribution and abundance*, 5th ed. Benjamin Cummings, San Francisco.

Menge, Bruce A., et al. 1994. The keystone species concept: Variation in interaction strength in a rocky intertidal habitat. *Ecological Monographs* 64: 249–286.

Molles, Manuel C., Jr. 2005. *Ecology: Concepts and applications*, 3rd ed. McGraw-Hill, Boston.

Morin, Peter J. 1999. *Community ecology.* Blackwell, London.

Power, Mary E., et al., 1996. Challenges in the quest for keystones. *BioScience* 46: 609–620.

Ricklefs, Robert E., and Gary L. Miller. 2000. *Ecology*, 4th ed. W. H. Freeman, New York.

Ricklefs, Robert E., and Dolph Schluter, eds. 1993. *Species diversity in ecological communities.* University of Chicago Press, Chicago.

Shea, Katriona, and Peter Chesson, 2002. Community ecology theory as a framework for biological invasions. *Trends in Ecology and Evolutionary Biology* 17: 170–176.

Sih, Andrew, et al. 1985. Predation, competition, and prey communities: A review of field experiments. *Annual Review of Ecology and Systematics* 16: 269–311.

Smith, Robert L., and Thomas M. Smith. 2001. *Ecology and field biology*, 6th ed. Benjamin Cummings, San Francisco.

Springer, A.M., et al. 2003. Sequential megafaunal collapse in the North Pacific Ocean: An ongoing legacy of industrial whaling? *Proceedings of the National Academy of Sciences of the USA* 100: 12223–12228.

Strayer, David L., et al. 1999. Transformation of freshwater ecosystems by bivalves: A case study of zebra mussels in the Hudson River. *BioScience* 49: 19–27.

Strayer, David L., et al. 2004. Effects of an invasive bivalve (*Dreissena polymorpha*) on fish in the Hudson River estuary. *Canadian Journal of Fisheries and Aquatic Sciences* 61: 924–941.

Thompson, John N. 1999. The evolution of species interactions. *Science* 284: 2116–2118.

Weigel, Marlene, ed., 1999. *Encyclopedia of biomes*. UXL, Farmington Hills, Michigan.

Whittaker, Robert H., and William A. Niering. 1965. Vegetation of the Santa Catalina Mountains, Arizona: A gradient analysis of the south slope. *Ecology* 46: 429–452.

Woodward, Susan L., 2003. *Biomes of Earth: Terrestrial, aquatic, and human-dominated*. Greenwood Publishing, Westport, Connecticut.

Chapter 6

Cohen, Joel E. 1995. *How many people can the Earth support?* W. W. Norton, New York.

Cohen, Joel E. 2003. Human population: The next half century. *Science* 302: 1172–1175.

Cohen, Joel E. 2005. Human population grows up. *Scientific American* 293(3): 48–55.

De Souza, Roger-Mark, et. al., 2003. Critical links: Population, health, and the environment. *Population Bulletin* 58(3), 48 pp. Population Reference Bureau, Washington, D.C.

Eberstadt, Nicholas. 2000. China's population prospects: Problems ahead. *Problems of Post-Communism* 47: 28.

Ehrlich, Paul R., and Anne H. Ehrlich. 1990. *The population explosion*. Touchstone, New York.

Ehrlich, Paul R., and John P. Holdren. 1971. Impact of population growth: Complacency concerning this component of man's predicament is unjustified and counterproductive. *Science* 171: 1212–1217.

Engelman, Robert, Brian Halweil, and Danielle Nierenberg. 2002. Rethinking population, improving lives. Pp. 127–148 in *State of the world 2002*, Worldwatch Institute and W. W. Norton, Washington D.C., and New York.

Greenhalgh, Susan. 2001. Fresh winds in Beijing: Chinese feminists speak out on the one-child policy and women's lives. *Signs: Journal of Women in Culture & Society* 26: 847–887.

Harrison, Paul, and Fred Pearce, eds. 2000. *AAAS atlas of population & environment*. University of California Press, Berkeley.

Hesketh, Therese, and Wei Xing Zhu. 1997. Health in China: The one child family policy: The good, the bad, and the ugly. *British Medical Journal* 314: 1685.

Holdren, John P. and Ehrlich, Paul R. 1974. Human population and the global environment. *American Scientist* 62: 282–292.

Kane, Penny. 1987. *The second billion: Population and family planning in China*. Penguin Books, Australia, Ringwood, Victoria.

Kane, Penny, and Ching Y. Choi. 1999. China's one child family policy. *British Medical Journal* 319: 992.

Mastny, Lisa. 2005. HIV/AIDS crisis worsening worldwide. Pp. 68–69 in *Vital signs 2005*. Worldwatch Institute and W. W. Norton, Washington, D.C. and New York.

Mastny, Lisa, and Richard P. Cincotta. 2005. Examining the connections between population and security. Pp. 22–41 in *State of the world 2005*. Worldwatch Institute and W. W. Norton, Washington, D.C. and New York.

McDonald, Mia, with Danielle Nierenberg. 2003. Linking population, women, and biodiversity. Pp. 38–61 in *State of the world 2003*, Worldwatch Institute and W. W. Norton, Washington D.C., and New York.

Meadows, Donella, Jørgen Randers, and Dennis Meadows. 2004. *Limits to growth: The 30-year update*. Chelsea Green Publishing Co., White River Junction, Vermont.

Notestein, Frank. 1953. Economic problems of population change. Pp. 13–31 in *Proceedings of the Eighth International Conference of Agricultural Economists*. Oxford University Press, London.

Population Reference Bureau. 2005. *2005 World Population Data Sheet*. Population Reference Bureau, Washington, D.C., and John Wiley & Sons, Hoboken, New Jersey.

Redefining Progress. Programs: Sustainability indicators. www.rprogress.org/newprograms/sustIndi/index.shtml

Riley, Nancy E. 2004. *China's population: New trends and challenges*. Population Bulletin 59(2), 40 pp. Population Reference Bureau, Washington, D.C.

UNAIDS and World Health Organization. 2005. *AIDS epidemic update: December 2005*. UNAIDS and WHO, New York.

United Nations Economic and Social Commission for Asia and the Pacific. 2005. *2005 ESCAP population data sheet*. UNESCAP, New York.

United Nations Environment Programme. 2003. *Africa environment outlook: Past, present, and future perspectives*. UNEP, New York.

United Nations Population Division. 2004. *World population prospects: The 2004 revision*. UNPD, New York.

United Nations Population Fund. UNFPA, the 2005 World Summit and the millennium development goals. UNFPA. www.unfpa.org/icpd

United Nations Population Fund. *State of world population 2005*. UNFPA, New York.

United States Census Bureau. www.census.gov

Wackernagel, Mathis, and William Rees. 1996. *Our ecological footprint: Reducing human impact on the earth*. New Society Publishers, Gabriola Island, British Columbia, Canada.

World Bank. 2005. *World development indicators 2005*. World Bank, Washington, D.C.

Chapter 7

Ashman, Mark R., and Geeta Puri. 2002. *Essential soil science: A clear and concise introduction to soil science*. Blackwell Publishing, Malden, Massachusetts.

Bazzaz, Fakhri A. 2001. Plant biology in the future. *Proceedings of the National Academy of the United States of America* 98: 5441–5445.

Brown, Lester R. 2002. World's rangelands deteriorating under mounting pressure. *Eco-Economy Update #6*, 5 February 2002. Earth Policy Institute, www.earth-policy.org/Updates/Update6.htm

Brown, Lester R. 2004. *Outgrowing the Earth: The food security challenge in an age of falling water tables and rising temperatures*. Earth Policy Institute, Washington, D.C.

Buchmann, Stephen L., and Gary Paul Nabhan. 1996. *The forgotten pollinators*. Island Press/Shearwater Books, Washington, D.C./Covelo, California.

Charman, P. E. V., and Brian W. Murphy. 2000. *Soils: Their properties and management*, 2nd ed. Oxford University Press, South Melbourne, Australia.

Commission for Environmental Cooperation. 2004. *Maize and biodiversity: The effects of transgenic maize in Mexico*. CEC Secretariat.

[Correspondence to *Nature*, various authors]. 2002. *Nature* 416: 600–602, and 417: 897–898.

Farm Scale Evaluations of spring-sown genetically modified crops, The. 2003. A themed issue from *Philosophical Transactions of the Royal Society of London B: Biological Sciences* 358(1439), 29 November 2003.

Fedoroff, Nina, and Nancy Marie Brown, 2004. *Mendel in the kitchen: A scientist's view of genetically modified foods*. National Academies Press, Washington, D.C.

Food and Agriculture Organization of the United Nations. 2001. Conservation agriculture: Case studies in Latin America and Africa. *FAO Soils Bulletin No. 78*. FAO, Rome.

Food and Agriculture Organization of the United Nations. 2004. *The state of world fisheries and aquaculture, 2004*. FAO, Rome.

Gardner, Gary, and Brian Halweil. 2000. *Underfed and overfed: The global epidemic of malnutrition*. Worldwatch Paper #150. Worldwatch Institute, Washington, D.C.

Glanz, James. 1995. *Saving our soil: Solutions for sustaining Earth's vital resource*. Johnson Books, Boulder, Colorado.

Halweil, Brian. 2002. Farmland quality deteriorating. Pp. 102–103 in *Vital signs 2002*. Worldwatch Institute and W. W. Norton, Washington D.C., and New York.

Halweil, Brian. 2004. *Eat here: Reclaiming homegrown pleasures in a global supermarket.* Worldwatch Institute, Washington, D.C.

Halweil, Brian. 2005. Aquaculture pushes fish harvest higher. Pp. 26–27 in *Vital signs 2005.* Worldwatch Institute and W. W. Norton, Washington, D.C. and New York.

Halweil, Brian. 2005. Grain harvest and hunger both grow. Pp. 22–23 in *Vital signs 2005.* Worldwatch Institute and W. W. Norton, Washington, D.C., and New York.

Halweil, Brian, and Danielle Nierenberg. 2004. Watching what we eat. Pp. 68–95 in *State of the world 2004.* Worldwatch Institute and W. W. Norton, Washington, D.C., and New York.

Harrison, Paul, and Fred Pearce, eds. 2000. *AAAS atlas of population & environment.* University of California Press, Berkeley.

International Food Information Council. 2004. Food biotechnology. IFIC, Washington, D.C. www.ific.org/food/biotechnology/index.cfm

James, Clive. 2004. *Global status of GM crops, their contribution to sustainability, and future prospects.* International Service for the Acquisition of Agri-biotech Applications.

Jenny, Hans. 1941. *Factors of soil formation: A system of quantitative pedology.* McGraw-Hill, New York.

Kuiper, Harry A. 2000. Risks of the release of transgenic herbicide-resistant plants with respect to humans, animals, and the environment. *Crop Protection* 19: 773.

Larsen, Janet. 2003. Deserts advancing, civilization retreating. *Eco-Economy Update #23,* 27 March 2003. Earth Policy Institute, www.earth-policy.org/Updates/Update23.htm

Liebig, Mark A., and John W. Doran. 1999. Impact of organic production practices on soil quality indicators. *Journal of Environmental Quality* 28: 1601–1609.

Losey, John E., Linda S. Raynor, and Maureen E. Carter. 1999. Transgenic pollen harms monarch larvae. *Nature* 399: 214.

Maeder, Paul, et al. 2002. Soil fertility and biodiversity in organic farming. *Science* 296: 1694–1697.

Mann, Charles C. 2002. Transgene data deemed unconvincing. *Science* 296: 236–237.

Manning, Richard. 2000. *Food's frontier: The next green revolution.* North Point Press, New York.

Millennium Ecosystem Assessment. 2005. *Ecosystems and human well-being: Desertification synthesis.* Millennium Ecosystem Assessment and World Resources Institute.

Miller, Henry I., and Gregory Conko. 2004. *The frankenfood myth: How protest and politics threaten the biotech revolution.* Praeger Publishers, Westport, Connecticut.

Morgan, R. P. C. 2005. *Soil erosion and conservation,* 3rd ed. Blackwell, London.

Natural Resources Conservation Service. 2001. *National resources inventory 2001: Soil erosion.* NRCS, USDA, Washington, D.C.

Natural Resources Conservation Service. Soils. NRCS, USDA. http://soils.usda.gov

Nierenberg, Danielle. 2005. *Happier meals: Rethinking the global meat industry.* Worldwatch Paper #171. Worldwatch Institute, Washington, D.C.

Nierenberg, Danielle. 2005. Meat production and consumption rise. Pp. 24–25 in *Vital signs 2005.* Worldwatch Institute and W. W. Norton, Washington, D.C., and New York.

Nierenberg, Danielle, and Brian Halweil. 2005. Cultivating food security. Pp. 62–79 in *State of the world 2005.* Worldwatch Institute and W. W. Norton, Washington, D.C., and New York.

Nestle, Marion. 2002. *Food politics: How the food industry influences nutrition and health.* University of California–Berkeley Press, Berkeley.

Norris, Robert F., Edward P. Caswell-Chen, and Marcos Kogan. 2002. *Concepts in integrated pest management.* Prentice Hall, Upper Saddle River, New Jersey.

Paoletti, Maurizio G., and David Pimentel. 1996. Genetic engineering in agriculture and the environment: Assessing risks and benefits. *BioScience* 46: 665–673.

Pearce, Fred. 2002. The great Mexican maize scandal. *New Scientist* 174: 14 (15 June 2002).

Pedigo, Larry P., and Marlin E. Rice, 2006. *Entomology and pest management,* 5th ed. Prentice Hall, Upper Saddle River, New Jersey.

Pieri, Christian, et al. 2002. *No-till farming for sustainable rural development.* Agriculture & Rural Development Working Paper. International Bank for Reconstruction and Development, Washington, D.C.

Pierzynski, Gary M., et al. 2005. *Soils and environmental quality,* 3rd ed. CRC Press, Boca Raton, Florida.

Pimentel, David. 1999. Population growth, environmental resources, and the global availability of food. *Social Research,* Spring 1999.

Pinstrup-Andersen, Per, and Ebbe Schioler, 2001. *Seeds of contention: World hunger and the global controversy over GM (genetically modified) crops.* International Food Policy Research Institute, Washington, D.C.

Polak, Paul. 2005. The big potential of small farms. *Scientific American* 293(3): 84–91.

Pretty, Jules, and Rachel Hine. 2001. *Reducing food poverty with sustainable agriculture: A summary of new evidence.* Center for Environment and Society, University of Essex. *Occasional Paper 2001–2.*

Pringle, Peter. 2003. *Food, Inc.: Mendel to Monsanto—The promises and perils of the biotech harvest.* Simon and Schuster, New York.

Quist, David, and Ignacio H. Chapela. 2001. Transgenic DNA introgressed into traditional maize landraces in Oaxaca, Mexico. *Nature* 414: 541–543.

Richter, Daniel D. Jr., and Daniel Markewitz. 2001. *Understanding soil change: Soil sustainability over millennia, centuries, and decades.* Cambridge University Press, Cambridge.

Ruse, Michael, and David Castle, eds. 2002. *Genetically modified foods: Debating technology.* Prometheus Books, Amherst, New York.

Schmeiser, Percy. Monsanto vs. Schmeiser. www. percyschmeiser.com/

Shaxson, T. F. 1999. The roots of sustainability; concepts and practice: Zero tillage in Brazil. *ABLH Newsletter ENABLE; World Association for Soil and Water Conservation (WASWC) Newsletter.*

Shiva, Vandana. 2000. *Stolen harvest: The hijacking of the global food supply.* South End Press, Cambridge, Massachusetts.

Smil, Vaclav. 2001. *Feeding the world: A challenge for the twenty-first century.* MIT Press, Cambridge, Massachusetts.

Soil Science Society of America. 2001. Internet glossary of soil science terms. www.soils.org/sssagloss

Stewart, C. Neal. 2004. *Genetically modified planet: Environmental impacts of genetically engineered plants.* Oxford University Press, Oxford.

Stocking, M. A. 2003. Tropical soils and food security: The next 50 years. *Science* 302: 1356–1359.

Teitel, Martin, and Kimberly Wilson. 2001. *Genetically engineered food: Changing the nature of nature.* Park Street Press.

Trimble, Stanley W., and Pierre Crosson. 2000. U.S. soil erosion rates— myth and reality. *Science* 289: 248–250.

Troeh, Frederick R., and Louis M. Thompson. 2004. *Soil and soil fertility,* 6th ed. Blackwell Publishing, London.

Troeh, Frederick R., J. Arthur Hobbs, and Roy L. Donahue. 2004. *Soil and water conservation for productivity and environmental protection,* 4th ed. Prentice Hall, Upper Saddle River, New Jersey.

Tuxill, John. 1999. Appreciating the benefits of plant biodiversity. Pp. 96–114 in *State of the world 1999,* Worldwatch Institute and W. W. Norton, Washington D.C., and New York.

United Nations Convention to Combat Desertification. 2001. *Global alarm: Dust and sandstorms from the world's drylands.* UNCCD and others, Bangkok, Thailand.

United Nations Environment Programme. 2002. Land. Pp. 62–89 in *Global environment outlook 3 (GEO-3).* UNEP and Earthscan Publications, Nairobi and London.

Uri, Noel D. 2001. The environmental implications of soil erosion in the United States. *Environmental Monitoring and Assessment* 66: 293–312.

Westra, Lauren. 1998. Biotechnology and transgenics in agriculture and aquaculture: The perspective from ecosystem integrity. *Environmental Values* 7: 79.

Wilkinson, Bruce H. 2005. Humans as geologic agents: A deep-time perspective. *Geology* 33: 161–164.

Wolfenbarger, L. LaReesa 2000. The ecological risks and benefits of genetically engineered plants. *Science* 290: 2088.

Chapter 8

Balmford, Andrew, et al. 2002. Economic reasons for conserving wild nature. *Science* 297: 950–953.

Barnosky, Anthony D., et al. 2004. Assessing the causes of late Pleistocene extinctions on the continents. *Science* 306: 70–75.

Baskin, Yvonne. 1997. *The work of nature: How the diversity of life sustains us*. Island Press, Washington, D.C.

Bright, Chris. 1998. *Life out of bounds: Bioinvasion in a borderless world*. Worldwatch Institute and W. W. Norton, Washington D.C., and New York.

CITES Secretariat. Convention on International Trade in Endangered Species of Wild Fauna and Flora. www. cites.org/

Convention on Biological Diversity. www.biodiv.org/

Daily, Gretchen C., ed. 1997. *Nature's services: Societal dependence on natural ecosystems*. Island Press, Washington, D.C.

Ehrenfeld, David W. 1970. *Biological conservation*. International Thomson Publishing, London.

Gaston, Kevin J., and John I. Spicer. 2004. *Biodiversity: An introduction*, 2nd ed. Blackwell, London.

Groom, Martha J., et al. 2005. *Principles of conservation biology*, 3rd ed. Sinauer Associates, Sunderland, Massachusetts.

Groombridge, Brian, and Martin D. Jenkins. 2002. *Global biodiversity: Earth's living resources in the 21st century*. UNEP, World Conservation Monitoring Centre, and Aventis Foundation; World Conservation Press, Cambridge, U.K.

Groombridge, Brian, and Martin D. Jenkins. 2002. *World atlas of biodiversity: Earth's living resources in the 21st century*. University of California Press, Berkeley.

Hanken, James. 1999. Why are there so many new amphibian species when amphibians are declining? *Trends in Ecology and Evolution* 14: 7–8.

Harris, Larry D. 1984. *The fragmented forest: Island biogeography theory and the preservation of biotic diversity*. University of Chicago Press, Chicago.

Harrison, Paul, and Fred Pearce, eds. 2000. *AAAS atlas of population & environment*. University of California Press, Berkeley.

Jenkins, Martin, 2003. Prospects for biodiversity. *Science* 302: 1175–1177.

Louv, Richard. 2005. *Last child in the woods: Saving our children from nature-deficit disorder*. Algonquin Books, Chapel Hill, North Carolina.

MacArthur, Robert H., and Edward O. Wilson. 1967. *The theory of island biogeography*. Princeton University Press, Princeton.

Mackay, Richard. 2002. *The Penguin atlas of endangered species: A worldwide guide to plants and animals*. Penguin, New York.

Maehr, David S., Reed F. Noss, and Jeffrey Larkin, eds. 2001. *Large mammal restoration: Ecological and sociological challenges in the 21st century*. Island Press, Washington, D.C.

Matthiessen, Peter. 2000. *Tigers in the snow*. North Point Press, New York.

Millennium Ecosystem Assessment. 2005. *Ecosystems and human well-being: Biodiversity synthesis*. Millennium Ecosystem Assessment and World Resources Institute.

Miquelle, Dale, Howard Quigley, and Maurice Hornocker. 1999. A habitat protection plan for Amur Tiger conservation: A proposal outlining habitat protection measures for the Amur Tiger. Hornocker Wildlife Institute.

Mooney, Harold A., and Richard J. Hobbs, eds. 2000. *Invasive species in a changing world*. Island Press, Washington, D.C.

Newmark, William D. 1987. A land-bridge perspective on mammal extinctions in western North American parks. *Nature* 325: 430.

Pimm, Stuart L., and Clinton Jenkins. 2005. Sustaining the variety of life. *Scientific American* 293(3): 66–73.

Primack, Richard B. 2004. *Essentials of conservation biology*, 3rd ed. Sinauer Associates, Sunderland, Massachusetts.

Quammen, David. 1996. *The song of the dodo: Island biogeography in an age of extinction*. Touchstone, New York.

Rosenzweig, Michael L. 1995. *Species diversity in space and time*. Cambridge University Press, Cambridge.

Simberloff, Daniel S. 1969. Experimental zoogeography of islands: A model for insular colonization. *Ecology* 50: 296–314.

Simberloff, Daniel. 1998. Flagships, umbrellas, and keystones: Is single-species management passé in the landscape era? *Biological Conservation* 83: 247–257.

Simberloff, Daniel S., and Edward O. Wilson. 1969. Experimental zoogeography of islands: The colonization of empty islands. *Ecology* 50: 278–296.

Simberloff, Daniel S., and Edward O. Wilson. 1970. Experimental zoogeography of islands: A two-year record of colonization. *Ecology* 51: 934–937.

Soulé, Michael E. 1986. *Conservation biology: The science of scarcity and diversity*. Sinauer Associates, Sunderland, Massachusetts.

Takacs, David. 1996. *The idea of biodiversity: Philosophies of paradise*. Johns Hopkins University Press, Baltimore.

United Nations Environment Programme. 2002. Biodiversity. Pp. 120–149 in *Global environment outlook 3 (GEO-3)*. UNEP and Earthscan Publications, Nairobi and London.

United Nations Environment Programme. 2003. Sustaining life on Earth: How the Convention on Biological Diversity promotes nature and human well-being. www.biodiv. org/doc/publications/guide.asp

United States Fish and Wildlife Service. The endangered species act of 1973. Accessible online at http://endangered.fws.gov/esa.html

Wilson, Edward O. 1984. *Biophilia*. Harvard University Press, Cambridge, Massachusetts.

Wilson, Edward O. 1992. *The diversity of life*. Harvard University Press, Cambridge, Massachusetts.

Wilson, Edward O. 1994. *Naturalist*. Island Press, Shearwater Books, Washington, D.C.

Wilson, Edward O. 2002. *The future of life*. Alfred A. Knopf, New York.

Wilson, Edward O., and Daniel S. Simberloff. 1969. Experimental zoogeography of islands: Defaunation and monitoring techniques. *Ecology* 50: 267–278.

World Conservation Union. 2006. IUCN Red List. www.iucnredlist.org

Chapter 9

Abbott, Carl. 2001. *Greater Portland: Urban life and landscape in the Pacific Northwest*. University of Pennsylvania Press.

Abbott, Carl. 2002. Planning a sustainable city. Pp. 207–235 in Squires, Gregory D., ed. *Urban sprawl: Causes, consequences, and policy responses*. Urban Institute Press, Washington, D.C.

Beck, Roy, et. al., 2003. *Outsmarting smart growth: Population growth, immigration, and the problem of sprawl*. Center for Immigration Studies, Washington, D.C.

Breuste, Jurgen, et al. 1998. *Urban ecology*. Springer-Verlag.

British Columbia Ministry of Forests. Introduction to Silvicultural Systems. www.for.gov.bc.ca/hfd/pubs/SSIntroworkbook/index.htm British Columbia Ministry of Forests, Victoria, B.C.

Clary, David. 1986. *Timber and the Forest Service*. University Press of Kansas, Lawrence.

Cronon, William. 1991. *Nature's metropolis: Chicago and the great West*. W. W. Norton, New York.

Duany, Andres, et al. 2001. *Suburban nation: The rise of sprawl and the decline of the American dream*. North Point Press, New York.

Ewing, Reid, et al. 2002. *Measuring sprawl and its impact*. Smart Growth America.

Ewing, Reid, et al. 2003. Measuring sprawl and its transportation impacts. *Transportation Research Record* 1831: 175–183.

Food and Agriculture Organization of the United Nations. 2005. *Global forest resources assessment*. FAO Forestry, Rome.

Foster, Bryan C., and Peggy Foster. 2002. *Wild logging: A guide to environmentally and economically sustainable forestry.* Mountain Press, Missoula, Montana.

Gardner, Gary. 2005. Forest loss continues. Pp. 92–93 in *Vital signs 2005.* Worldwatch Institute and W. W. Norton, Washington, D.C., and New York.

Girardet, Herbert. 2004. *Cities people planet: Livable cities for a sustainable world.* Academy Press.

Hall, Kenneth B., and Gerald A. Porterfield. 2001. *Community by design: New urbanism for suburbs and small communities.* McGraw-Hill, New York.

Harrison, Paul, and Fred Pearce, eds. 2000. *AAAS atlas of population & environment.* University of California Press, Berkeley.

Haynes, Richard W., and Gloria E. Perez, tech. eds. 2001. Northwest Forest Plan research synthesis. *Gen. Tech. Rep. PNW-GTR-498.* USDA Forest Service, Pacific Northwest Research Station, Portland, Oregon.

Horizon International. 2003. Efficient transportation for successful urban planning in Curitiba. www.solutions-site.org/artman/publish/printer_62.shtml

Jacobs, Jane. 1992. *The death and life of great American cities.* Vintage.

Jacobs, Lynn. 1991. *Waste of the West: Public lands ranching.* Lynn, Jacobs, Tucson, Arizona.

Kalnay, Eugenia, and Ming Cai. 2003. Impact of urbanization and land-use change on climate. *Nature* 423: 528–531.

Kirdar, Uner, ed. 1997. *Cities fit for people.* United Nations, New York.

Litman, Todd. 2004. *Rail transit in America: A comprehensive evaluation of benefits.* Victoria Transport Policy Institute and American Public Transportation Association.

Logan, Michael F. 1995. *Fighting sprawl and city hall.* University of Arizona Press, Tucson.

Metro. www.metro-region.org

Myers, Norman, and Jennifer Kent. 2001. *Perverse subsidies: How misused tax dollars harm the environment and the economy.* Island Press, Washington, D.C.

National Forest Management Act of 1976. October 22, 1976 (P.O. 94–588, 90 Stat. 2949, as amended; 16 U.S.C.)

Natural Resources Canada. 2005. *The state of Canada's forests, 2004-2005.* Natural Resources Canada, Ottawa.

New Urbanism. www.newurbanism.org

Northwest Environment Watch. 2004. *The Portland exception: A comparison of sprawl, smart growth, and rural land loss in 15 U.S. cities.* Northwest Environment Watch, Seattle.

Portney, Kent. E. 2003. *Taking sustainable cities seriously: Economic development, the environment, and quality of life in American cities (American and comparative environmental policy).* MIT Press, Cambridge, Massachusetts.

Pugh, Cedric, ed. 1996. *Sustainability, the environment, and urbanization.* Earthscan Publications, London.

Runte, Alfred. 1979. *National parks and the American experience.* University of Nebraska Press, Lincoln.

Sedjo, Robert A. 2000. *A vision for the US Forest Service.* Resources for the Future, Washington, D.C.

Sheehan, Molly O'Meara. 2001. *City limits: Putting the brakes on sprawl.* Worldwatch Paper #156. Worldwatch Institute, Washington, D.C.

Sheehan, Molly O'Meara. 2002. What will it take to halt sprawl? *Worldwatch* (Jan/Feb 2002): 12–23.

Singh, Ashbindu, et al. 2001. An assessment of the status of the world's remaining closed forests. United Nations Environmental Program, UNEP/DEWA/TR 01–2l, August 2001.

Smith, David M., et al. 1996. *The practice of silviculture: Applied forest ecology,* 9th ed. Wiley, New York.

Smith, W. Brad, et al. 2004. Forest resources of the United States, 2002. *Gen. Tech. Rep.* NC-241, North Central Research Station, USDA Forest Service, St. Paul, Minnesota.

Soulé, Michael E., and John Terborgh, eds. 1999. *Continental conservation.* Island Press, Washington, D.C.

Sprawl City. www.sprawlcity.org

Stren, R., et al. 1992. *Sustainable cities: Urbanization and the environment in international perspective.* Westview Press, Boulder, Colorado, and San Francisco.

United Nations Environment Programme. 2002. Forests. Pp. 90–119 in *Global environment outlook 3 (GEO-3).* UNEP and Earthscan Publications, Nairobi and London.

United Nations Environment Programme. 2002. Urban areas. Pp. 240–269 in *Global environment outlook 3 (GEO-3).* UNEP and Earthscan Publications, Nairobi and London.

United States Census Bureau. www.census.gov/

United States Environmental Protection Agency. Smart growth. www.epa.gov/smartgrowth

United States National Park Service. 2002. *National Park Service statistical abstract 2002.* NPS Public Use Statistics Office, United States Department of the Interior, Denver, Colorado.

United States Department of Agriculture Forest Service. 2001. *U.S. forest facts and historical trends.* FS-696, March 2001.

Wiewel, Wim, and Jospeh J. Persky., eds. 2002. *Suburban sprawl: Private decisions and public policy.* M. E. Sharpe, Armond, New York.

Chapter 10

Ames, Bruce N., Margie Profet, and Lois Swirsky Gold. 1990. Nature's chemicals and synthetic chemicals: Comparative toxicology. *Proceedings of the National Academy of the USA* 87: 7782–7786.

Bloom, Barry. 2005. Public health in transition. *Scientific American* 293(3): 92–99.

Carlsen, Elisabeth, et al. 1992. Evidence for decreasing quality of semen during past 50 years. *British Medical Journal* 305: 609–613.

Carson, Rachel. 1962. *Silent spring.* Houghton Mifflin, Boston.

Colburn, Theo, Dianne Dumanoski, and John P. Myers. 1996. *Our stolen future.* Penguin USA, New York.

Crain, D. Andrew, and Louis J. Guillette Jr. 1998. Reptiles as models of contaminant-induced endocrine disruption. *Animal Reproduction Science* 53: 77–86.

Guillette, Elizabeth A., et al. 1998. An anthropological approach to the evaluation of preschool children exposed to pesti-cides in Mexico. *Environmental Health Perspectives* 106: 347–353.

Guillette, Louis J. Jr., et al. 1999. Plasma steroid concentrations and male phallus size in juvenile alligators from seven Florida lakes. *General and Comparative Endocrinology* 116: 356–372.

Guillette, Louis J. Jr., et al. 2000. Alligators and endocrine disrupting contaminants: A current perspective. *American Zoologist* 40: 438–452.

Halweil, Brian. 1999. Sperm counts dropping. Pp. 148–149 in *Vital signs 1999.* Worldwatch Institute and W. W. Norton, Washington, D.C. and New York.

Hayes, Tyrone, et al. 2003. Atrazine-induced hermaphroditism at 0.1 PPB in American leopard frogs (*Rana pipiens*): Laboratory and field evidence. *Environmental Health Perspectives* 111: 568–575.

Hunt, Patricia A., et al. 2003. Bisphenol A exposure causes meiotic aneuploidy in the female mouse. *Current Biology* 13: 546–553.

Kolpin, Dana W., et al. 2002. Pharmaceuticals, hormones, and other organic wastewater contaminants in U.S. streams, 1999–2000: A national reconnaissance. *Environmental Science and Technology* 36: 1202–1211.

Landis, Wayne G., and Ming-Ho Yu. 2004. *Introduction to environmental toxicology,* 3rd ed. Lewis Press, Boca Raton, Florida.

Loewenberg, Samuel. 2003. E.U. starts a chemical reaction. *Science* 300: 405.

Millennium Ecosystem Assessment. 2005. *Ecosystems and human well-being: Health synthesis.* World Health Organization.

Manahan, Stanley E. 2000. *Environmental chemistry,* 7th ed. Lewis Publishers, CRC Press, Boca Raton, Florida.

McGinn, Anne Platt. 2000. *Why poison ourselves? A precautionary approach to synthetic chemicals.* Worldwatch Paper #153. Worldwatch Institute, Washington, D.C.

McGinn, Anne Platt. 2002. Reducing our toxic burden. Pp. 75–100 in *State of the world 2002*. Worldwatch Institute and W. W. Norton, Washington, D.C. and New York.

McGinn, Anne Platt. 2003. Combating malaria. Pp. 62–84 in *State of the world 2003*. Worldwatch Institute and W. W. Norton, Washington, D.C. and New York.

Moeller, Dade. 2004. *Environmental health*, 3rd ed. Harvard University Press, 2004.

National Center for Environmental Health; U.S. Centers for Disease Control and Prevention. 2005. *Third national report on human exposure to environmental chemicals*. NCEH Pub. No. 05-0570, Atlanta.

National Center for Health Statistics, 2004. *Health, United States, 2004, with chartbook on trends in the health of Americans*. Hyattsville, Maryland.

Pirages, Dennis. 2005. Containing infectious disease. Pp. 42–61 in *State of the world 2005*. Worldwatch Institute and W. W. Norton, Washington, D.C., and New York.

Renner, Rebecca. 2002. Conflict brewing over herbicide's link to frog deformities. *Science* 298: 938–939.

Rodricks, Joseph V. 1994. *Calculated risks: Understanding the toxicity of chemicals in our environment*. Cambridge University Press, Cambridge.

Salem, Harry, and Eugene Olajos. 1999. *Toxicology in risk assessment*. CRC Press, Boca Raton, Florida.

Spiteri, I. Daniel, Louis J. Guillette Jr., and D. Andrew Crain. 1999. The functional and structural observations of the neonatal reproductive system of alligators exposed *in ozo* to atrazine, 2,4-D, or estradiol. *Toxicology and Industrial Health* 15: 181–186.

Stancel, George, et al. 2001. "Report of the bisphenol A sub-panel." Chapter 1 in *National Toxicology Program's report of the endocrine disruptors low-dose peer review*. U.S. EPA and NIEHS, NIH.

Stockholm Convention on Persistant Organic Pollutants. www.pops.int/

United States Environmental Protection Agency. 2003. Pesticide registration program. www.epa.gov/pesticides/factsheets/registration.htm

United States Environmental Protection Agency. 2003. Toxic Substances Control Act. www.epa.gov/region5/defs/html/tsca.htm

United States Environmental Protection Agency. 2003. *EPA's draft report on the environment*. EPA 600-R-03-050. EPA, Washington, D.C.

Williams, Phillip L., Robert C. James, and Stephen M. Roberts, eds. 2000. *The principles of toxicology: Environmental and industrial applications*, 2nd ed. Wiley-Interscience, New York.

World Health Organization, 2004. *World health report 2004: Changing history*. WHO, Geneva, Switzerland.

Yu, Ming-Ho, 2004. *Environmental toxicology: Biological and health effects of pollutants*, 2nd ed. CRC Press, Boca Raton, Florida.

Chapter 11

American Rivers. 2002. *The ecology of dam removal: A summary of benefits and impacts*. American Rivers, Washington D.C., February 2002.

Bellwood, David R., et al. 2004. Confronting the coral reef crisis. *Nature* 429: 827–833.

British Geographical Society and Bangladesh Department of Public Health Engineering. 2001. *Arsenic contamination of groundwater in Bangladesh*. Technical Report WC/00/19, Volume 1: Summary.

[Correspondence to *Science*, various authors]. 2001. *Science* 295: 1233–1235.

De Villiers, Marq, 2000. *Water: The fate of our most precious resource*. Mariner Books.

Food and Agriculture Organization of the United Nations. 2004. *The state of world fisheries and aquaculture, 2004*. FAO, Rome.

Garrison, Tom. 2005. *Oceanography: An invitation to marine science*, 5th ed. Brooks/Cole, San Francisco.

Gell, Fiona R., and Callum M. Roberts. 2003. Benefits beyond boundaries: The fishery effects of marine reserves. *Trends in Ecology and Evolution* 18: 448–455.

Gleick, Peter. H. 2003. Global freshwater resources: Soft-path solutions for the 21st century. *Science* 302: 1524–1527.

Gleick, Peter. H., et al. 2004. *The world's water 2004–2005: The biennial report on freshwater resources*. Island Press, Washington D.C.

Halpern, Benjamin S., and Robert R. Warner. 2002. Marine reserves have rapid and lasting effects. *Ecology Letters* 5: 361–366.

Halpern, Benjamin S., and Robert R. Warner. 2003. Matching marine reserve design to reserve objectives. *Proceedings of the Royal Society of London B*: 270: 1871–1878.

Harrison, Paul, and Fred Pearce, eds. 2000. *AAAS atlas of population & environment*. University of California Press, Berkeley.

Institute of Governmental Studies, University of California, Berkeley. Imperial Valley-San Diego water transfer controversy. www.igs.berkeley.edu/library/htImperialWaterTransfer2003.html

Jackson, Jeremy B. C., et al. 2001. Historical overfishing and the recent collapse of coastal ecosystems. *Science* 293: 629–638.

Jenkins, Matt. 2002. The royal squeeze. *High Country News* 35(1), January 20, 2003.

Jenkins, Matt. 2003. California's water binge skids to a halt. *High Country News* 34(17), September 16, 2002.

Larsen, Janet. 2005. Wild fish catch hits limits—Oceanic decline offset by increased fish farming *Eco-economy indicators*. Earth Policy Institute, www.earth-policy.org/Indicators/Fish/2005.htm

Marston, Ed. 2001. Quenching the big thirst. *High Country News* 33(10), May 21, 2001.

Millennium Ecosystem Assessment. 2005. *Ecosystems and human well-being: Wetlands and water synthesis*. Millennium Ecosystem Assessment and World Resources Institute.

Myers, Ransom A., and Boris Worm. 2003. Rapid worldwide depletion of predatory fish communities. *Nature* 423: 280–283.

National Academy of Public Administration. 1999. *Protecting Our National Marine Sanctuaries*. Center for the Economy and the Environment, NAPA, Washington, D.C.

National Center for Ecological Analysis and Synthesis (NCEAS) and Communication Partnership for Science and the Sea (COMPASS), sponsors. 2001. *Scientific consensus statement on marine reserves and marine protected areas*. Available online at www.nceas.ucsb.edu/consensus

National Research Council. 2003. *Oil in the sea III: Inputs, fates, and effects*. National Academies Press, Washington, D.C.

Norse, Elliott, and Larry B. Crowder, eds. 2005. *Marine conservation biology: The science of maintaining the sea's biodiversity*. Island Press, Washington, D.C.

Nybakken, James W., and Mark D. Bertness. 2004. *Marine biology: An ecological approach*, 6th ed. Benjamin Cummings, San Francisco.

Palumbi, Stephen. 2003. *Marine reserves: A tool for ecosystem management and conservation*. Pew Oceans Commission.

Pauly, Daniel, et al. 2002. Towards sustainability in world fisheries. *Nature* 418: 689–695.

Pauly, Daniel, et al. 2003. The future for fisheries. *Science* 302: 1359–1361.

Pew Oceans Commission. 2003. *America's living oceans: Charting a course for sea change*. A report to the nation. May 2003. Pew Oceans Commission, Arlington, Virginia.

Pinet, Paul R. 1999. *Invitation to oceanography*, 2nd ed. Jones & Bartlett, Boston.

Postel, Sandra. 1999. *Pillar of sand: Can the irrigation miracle last?* W. W. Norton, New York.

Postel, Sandra. 2005. *Liquid assets: The critical need to safeguard freshwater ecosystems*. Worldwatch Paper #170. Worldwatch Institute, Washington, D.C.

Postel, Sandra, and Amy Vickers. 2004. Boosting water productivity. Pp. 46–67 in *State of the world 2004*. Worldwatch Institute and W. W. Norton, Washington, D.C. and New York.

Reisner, Marc. 1986. *Cadillac desert: The American West and its disappearing water*. Viking Penguin, New York.

Roberts, Callum M., et al. 2001. Effects of marine reserves on adjacent fisheries. *Science* 294: 1920–1923.

Sampat, Payal. 2001. Uncovering groundwater pollution. Pp. 21–42 in *State of the world 2001*. Worldwatch Institute and W. W. Norton, Washington, D.C., and New York.

Sibley, George. 1997. A tale of two rivers: The desert empire and the mountain. *High Country News* 29(21), November 10, 1997.

Stone, Richard. 1999. Coming to grips with the Aral Sea's grim legacy. *Science* 284: 30–33.

Sumich, James L., and John F. Morrissey, 2004. *Introduction to the biology of marine life*, 8th ed. Jones & Bartlett, Boston.

Thurman, Harold V., and Alan P. Trujillo, 2004. *Introductory oceanography*, 10th ed. Prentice Hall, Upper Saddle River, New Jersey.

United Nations Environment Programme. 2002. Coastal and marine areas. Pp. 180–209 in *Global environment outlook 3 (GEO-3)*. UNEP and Earthscan Publications, Nairobi and London.

United Nations Environment Programme. 2002. Freshwater. Pp. 150–179 in *Global environment outlook 3 (GEO-3)*. UNEP and Earthscan Publications, Nairobi and London.

United Nations World Water Assessment Programme. 2003. *U.N. world water development report: Water for people, water for life*. Paris, New York, and Oxford, UNESCO and Berghahn Books.

United States Bureau of Reclamation, Lower Colorado Regional Office. www.usbr.gov/lc/region

United States Commission on Ocean Policy. 2004. *An ocean blueprint for the 21st century*. Final Report. Washington, D.C.

United States Department of Commerce and United States Department of the Interior. Marine protected areas of the United States. www.mpa.gov.

United States Environmental Protection Agency. 1998. *Wastewater primer*. EPA 833-K-98-001, Office of Wastewater Management, May 1998.

United States Environmental Protection Agency. 2003. *EPA's draft report on the environment*. EPA 600-R-03-050. EPA, Washington, D.C.

United States Environmental Protection Agency. 2003. *Water on tap: What you need to know*. EPA 816-K-03-007. Office of Water, EPA, Washington, D.C.

Weber, Michael L. 2001. *From abundance to scarcity: A history of U.S. marine fisheries policy*. Island Press, Washington, D.C.

Wolf, Aaron T., et al. 2005. Managing water conflict and cooperation. Pp. 80–99 in *State of the world 2005*. Worldwatch Institute and W. W. Norton, Washington, D.C., and New York.

World Health Organization. 2000. *Global water supply and sanitation assessment 2000 report*. WHO, Geneva, Switzerland.

Youth, Howard. 2005. Wetlands drying up. Pp. 90–91 in *Vital signs 2005*. Worldwatch Institute and W. W. Norton, Washington, D.C., and New York.

Chapter 12

Ahrens, C. Donald. 2003. *Meteorology today*, 7th ed. Brooks/Cole, San Francisco.

Akimoto, Hajime. 2003. Global air quality and pollution. *Science* 302: 1716–1719.

Alley, Richard B. 2000. *The two-mile time machine: Ice cores, abrupt climate change, and our future*. Princeton University Press, Princeton, New Jersey.

Bell, Michelle L., and Devra L. Davis. 2001. Reassessment of the lethal London fog of 1952: Novel indicators of acute and chronic consequences of acute exposure to air pollution. *Environmental Health Perspectives* 109(Suppl 3): 389–394.

Bernard, Susan M., et al. 2001. The potential impacts of climate variability and change on air pollution-related health effects in the United States. *Environmental Health Perspectives* 109(Suppl 2): 199–209.

Biscaye, Pierre E., et al. 2000. Eurasian air pollution reaches eastern North America. *Science* 290: 2258–2259.

Boubel, Richard W., et al., eds. 1994. *Fundamentals of air pollution*, 3rd ed. Academic Press, San Diego, California.

Bruce, Nigel, Rogelio Perez-Padilla, and Rachel Albalak. 2000. Indoor air pollution in developing countries: A major environmental and public health challenge. *Bulletin of the World Health Organization* 78: 1078–1092.

Burroughs, William James. 2001. *Climate change: A multidisciplinary approach*. Cambridge University Press, Cambridge.

Cooper, C. David, and F. C. Alley. 2002. *Air pollution control*, 3rd ed. Waveland Press.

Davis, Devra. 2002. *When smoke ran like water: Tales of environmental deception and the battle against pollution*. Basic Books, New York.

Drake, Frances. 2000. *Global warming: The science of climate change*. Oxford University Press, Oxford.

Driscoll, Charles T., et al. 2001. *Acid rain revisited: Advances in scientific understanding since the passage of the 1970 and 1990 Clean Air Act Amendments*. Hubbard Brook Research Foundation. Science Links™ Publication. Vol. 1, no.1.

Dunn, Seth. 2001. Decarbonizing the energy economy. Pp. 83–102 in *State of the world 2001*. Worldwatch Institute and W. W. Norton, Washington, D.C., and New York.

Dunn, Seth, and Christopher Flavin. 2002. Moving the climate change agenda forward. Pp. 24–50 in *State of the world 2002*. Worldwatch Institute and W. W. Norton, Washington, D.C., and New York.

Ezzati, Majid, and Daniel M. Kammen. 2001. Quantifying the effects of exposure to indoor air pollution from biomass combustion on acute respiratory infections in developing countries. *Environmental Health Perspectives* 109: 481–488.

Gardner, Gary. 2005. Air pollution still a problem. Pp. 94–95 in *Vital signs 2005*. Worldwatch Institute and W. W. Norton, Washington, D.C., and New York.

Gelbspan, Ross. 1997. *The heat is on: The climate crisis, the cover-up, the prescription*. Perseus Books, New York.

Gelbspan, Ross. 2004. *Boiling point: How politicians, big oil and coal, journalists, and activists are fueling the climate crisis—and what we can do to avert disaster*. Basic Books, New York.

Godish, Thad. 2003. *Air quality*, 4th ed. CRC Press, Boca Raton, Florida.

Hoffman, Matthew J. 2005. *Ozone depletion and climate change: Constructing a global response*. SUNY Press, New York.

Intergovernmental Panel on Climate Change. 2001. *IPCC third assessment report—Climate change 2001: The scientific basis*. World Meteorological Organization and United Nations Environment Programme.

Intergovernmental Panel on Climate Change. 2001. *IPCC third assessment report—Climate change 2001: Impacts, adaptations, and vulnerability*. World Meteorological Organization and United Nations Environment Programme.

Intergovernmental Panel on Climate Change. 2001. *IPCC third assessment report—Climate change 2001: Mitigation*. World Meteorological Organization and United Nations Environment Programme.

Intergovernmental Panel on Climate Change. 2001. *IPCC third assessment report—Climate change 2001: Synthesis report*. World Meteorological Organization and United Nations Environment Programme.

Intergovernmental Panel on Climate Change. 2001. *Technical Summary of the Working Group 1 report*.

Intergovernmental Panel on Climate Change. www.ipcc.ch

Jacobson, Mark Z. 2002. *Atmospheric pollution: History, science, and regulation*. Cambridge University Press, New York.

Kareiva, Peter M., Joel G. Kingsolver, and Raymond B. Huey, eds. 1993. *Biotic interactions and global change*. Sinauer Associates, Sunderland, Massachusetts.

Karl, Thomas R., and Kevin E. Trenberth, 2003. Modern global climate change. *Science* 302: 1719–1723.

Kunzli, Nino, et al. 2000. Public-health impact of outdoor and traffic-related air pollution: A European assessment. *Lancet* 356: 795–801.

Lelieveld, Jos, et al. 2001. The Indian Ocean experiment: Widespread air pollution from South and Southeast Asia. *Science* 291: 1031–1036.

Likens, Gene E. 2004. Some perspectives on long-term biogeochemical research from the Hubbard Brook ecosystem study. *Ecology* 85: 2355–2362.

Lomborg, Bjorn. 2001. *The skeptical environmentalist: Measuring the real state of the world*. Cambridge University Press, Cambridge.

London, Stephanie J., and Isabelle Romieu. 2000. Health costs due to outdoor air pollution by traffic. *Lancet* 356: 782–783.

Mastny, Lisa. 2005. Global ice melting accelerating. Pp. 88–89 in *Vital signs 2005*. Worldwatch Institute and W. W. Norton, Washington, D.C., and New York.

Mayewski, Paul A., and Frank White. 2002. *The ice chronicles: The quest to understand global climate change*. University Press of New England, Hanover, New Hampshire.

Molina, Mario J., and F. Sherwood Rowland. 1974. Stratospheric sink for chlorofluoromethanes: Chlorine atom catalyzed destruction of ozone. *Nature* 249: 810–812.

National Assessment Synthesis Team. 2000. *Climate change impacts on the United States: The potential consequences of climate variability and change*. U.S. Global Change Research Program. Cambridge University Press, Cambridge.

National Research Council, Board on Atmospheric Sciences and Climate, Commission on Geosciences, Environment, and Resources. 1998. *The atmospheric sciences: Entering the twenty-first century*. National Academies Press, Washington, D.C.

National Research Council, Committee on the Science of Climate Change, Division of Earth and Life Studies. 2001. *Climate change science: An analysis of some key questions*. National Academies Press, Washington, D.C.

Nordhaus, William D. 1998. Assessing the economics of climate change: An introduction. In Nordhaus, William D., ed., *Economic and policy issues in climate change*. Resources for the Future Press, Washington D.C.

Pal Arya, S. 1998. *Air pollution: Meteorology and dispersion*. Oxford University Press, Oxford.

Parmesan, Camille, and Gary Yohe. 2003. A globally coherent fingerprint of climate change impacts across natural systems. *Nature* 421: 37–42.

Parson, Edward A. 2003. *Protecting the ozone layer: Science and strategy*. Oxford University Press, Oxford.

Real Climate. www.realclimate.org

Root, Terry L., et al. 2003. Fingerprints of global warming on wild animals and plants. *Nature* 421: 57–60.

Sawin, Janet L. 2005. Climate change indicators on the rise. Pp. 40–41 in *Vital signs 2005*. Worldwatch Institute and W. W. Norton, Washington, D.C., and New York.

Schneider, Stephen H., and Terry L. Root, eds. 2002. *Wildlife responses to climate change: North American case studies*. Island Press, Washington, D.C.

Seinfeld, John H., and Spyros N. Pandis. 2006. *Atmospheric chemistry and physics*, 2nd ed. Wiley-Interscience, New York.

Shapiro, Robert J., Kevin A. Hassett, and Frank S. Arnold. 2002. *Conserving energy and preserving the environment: The role of public transportation*. American Public Transportation Association, July 2002.

Speth, James Gustave. 2004. *Red sky at morning: America and the crisis of the global environment*. Yale University Press, New Haven, Connecticut.

Stevens, William K. 1999. *The change in the weather: People, weather and the science of climate*. Delta Trade Paperbacks, New York.

Taylor, David. 2003. Small islands threatened by sea level rise. Pp. 84–85 in *Vital signs 2003*. Worldwatch Institute and W. W. Norton, Washington D.C. and New York.

Victor, David G. 2004. *Climate change: Debating America's policy options*. U.S. Council on Foreign Relations Press, Washington, D.C.

United Nations. United Nations Framework Convention on Climate Change. http://unfccc.int/2860.php

United Nations. Kyoto Protocol. http://unfccc.int/resource/docs/convkp/kpeng.html.

United Nations Environment Programme. Montreal Protocol. http://hq.unep.org/ozone/Treaties_and_Ratification/2B_montreal_protocol.asp

United Nations Environment Programme. 2002. Atmosphere. Pp. 210–239 in *Global environment outlook 3 (GEO-3)*. UNEP and Earthscan Publications, Nairobi and London.

United States Congress. House Committee on Science. 2001. Climate change: The state of the science. Hearing before the Committee on Science, House of Representatives, One Hundred Seventh Congress, first session, 14 March 2001.

United States Environmental Protection Agency. 2003. *EPA's draft report on the environment*. EPA 600-R-03-050. Washington, D.C.

United States Environmental Protection Agency. 2003. *Latest findings on national air quality: 2002 status and trends*. EPA 454/K-03-001. Washington, D.C.

Wark, Kenneth, et al., 1997. *Air pollution: Its origin and control*, 3rd ed. Prentice Hall, Upper Saddle River, New Jersey.

World Health Organization. Indoor air pollution. WHO, Geneva, Switzerland. www.who.int/indoorair/en/index.html

Chapter 13

Association for the Study of Peak Oil and Gas. www.peakoil.net/

British Petroleum. 2006. *BP statistical review of world energy*. London, June 2006.

Campbell, Colin J. 1997. *The coming oil crisis*. Multi-Science Publishing Co., Essex, U.K.

Chandler, David. 2003. America steels itself to take the nuclear plunge. *New Scientist* (August 9, 2003): 10–13.

Deffeyes, Kenneth S. 2001. *Hubbert's peak: The impending world oil shortage*. Princeton University Press, Princeton, New Jersey.

Deffeyes, Kenneth S. 2005. *Beyond oil: The view from Hubbert's peak*. Farrar, Straus, and Giroux, New York.

Douglas, D. C., P.E. Reynolds, and E. B. Rhode, eds. 2002. *Arctic Refuge coastal plain terrestrial wildlife research summaries. Biological science report*. USGS/BRD/BSR-2002-0001. United States Geological Survey, Washington, D.C.

Dunn, Seth. 2001. Decarbonizing the energy economy. Pp. 83–102 in *State of the world 2001*, Worldwatch Institute and W. W. Norton, Washington, D.C., and New York.

Energy Information Administration, United States Department of Energy. www.eia.doe.gov

Energy Information Administration, United States Department of Energy. 1999. *Petroleum: An energy profile, 1999*. DOE/EIA-0545(99).

Energy Information Administration, United States Department of Energy. 2005. *Annual energy review 2004*. DOE/EIA-0384(2004). Washington, D.C.

Energy Information Administration, United States Department of Energy. 2005. *International energy annual 2003*. Washington, D.C.

European Commission/International Atomic Energy Agency/World Health Organization. 1996. One decade after Chernobyl: Summing up the consequences of the accident. Summary of the conference results. Vienna, Austria, 8–12 April, 1996. EC/IAEA/WHO.

Flavin, Christopher. 2005. Fossil fuel use surges. Pp. 30–31 in *Vital signs 2005*. Worldwatch Institute and W. W. Norton, Washington, D.C., and New York.

Freese, Barbara. 2003. *Coal: A human history*. Perseus Books, New York.

Goodstein, David. 2004. *Out of gas*. W. W. Norton, New York.

Holmes, Bob, and Nicola Jones. 2003. Brace yourself for the end of cheap oil. *New Scientist* (August 2, 2003): 9–11.

International Atomic Energy Agency. 2004. *Annual report 2003*. GC(48)/3.IAEA, Vienna, Austria.

International Atomic Energy Agency. *Nuclear power and sustainable development*. IAEA Information Series 02-01574/FS Series 3/01/E/Rev.1. Vienna, Austria.

International Energy Agency. 2005. *Key world energy statistics 2005*. IEA Publications, Paris.

International Energy Agency. 2005. *World energy outlook 2005*. IEA Publications, Paris.

International Energy Agency. 2005. *Resources to reserves: Oil and gas technologies for the energy markets of the future*. IEA Publications, Paris.

Kunstler, James Howard. 2005. *The long emergency: Surviving the converging catastrophes of the twenty-first century*. Atlantic Monthly Press, New York.

Lenssen, Nicholas. 2005. Nuclear power rises once more. Pp. 32–33 in *Vital signs 2005*. Worldwatch Institute and W. W. Norton, Washington, D.C., and New York.

Lovins, Amory B., et al. 2004. *Winning the oil endgame: Innovation for profits, jobs, and security*. Rocky Mountain Institute, Snowmass, Colorado.

Lovins, Amory B. 2005. More profit with less carbon. *Scientific American* 293(3): 74–83.

Nellemann, Christian, and Raymond D. Cameron. 1998. Cumulative impacts of an evolving oil-field complex on the distribution of calving caribou. *Canadian Journal of Zoology* 76: 1425–1430.

Nuclear Energy Agency, OECD. 2002. *Chernobyl: Assessment of radiological and health impacts. (2002 Update of Chernobyl: Ten Years On)*. OECD, Paris.

Nuclear Energy Agency. 2005. *NEA annual report 2004*. NEA, Organisation for Economic Co-operation and Development. OECD, Paris.

Organisation for Economic Co-operation and Development. 2000. *Business as usual and nuclear power*. OECD Publications, Paris.

Pelley, Janet. 2001. Will drilling for oil disrupt the Arctic National Wildlife Refuge? *Environmental Science and Technology* 35: 240–247.

Powell, Stephen G. 1990. Arctic National Wildlife Refuge: How much oil can we expect? *Resources Policy* Sept. 1990: 225–240.

Prugh, Tom, et al. 2005. Changing the oil economy. Pp. 100–121 in *State of the world 2005*. Worldwatch Institute and W. W. Norton, Washington, D.C., and New York.

Ristinen, Robert A., and Jack J. Kraushaar, 1998. *Energy and the environment*. John Wiley and Sons, New York.

Russell, D. E., and P. McNeil. 2005. *Summer ecology of the Porcupine caribou herd*. Porcupine Caribou Management Board, Whitehorse, Yukon.

Sawin, Janet L. 2004. Making better energy choices. Pp. 24–45 in *State of the world 2004*. Worldwatch Institute and W. W. Norton, Washington, D.C. and New York.

Skinner, Brian J., and Stephen C. Porter. 2003. *The dynamic earth: An introduction to physical geology*, 5th ed. John Wiley and Sons, Hoboken, New Jersey.

Spadaro, Joseph V., Lucille Langlois, and Bruce Hamilton. 2000. Greenhouse gas emissions of electricity generation chains: Assessing the difference. *IAEA Bulletin* 42(2).

United States Environmental Protection Agency. 2005. *Light-duty automotive technology and fuel economy trends: 1975 through 2005*. EPA420-R-05-001. EPA Office of Transportation and Air Quality, Washington, D.C.

United States Fish and Wildlife Service. 2001. Potential impacts of proposed oil and gas development on the Arctic Refuge's coastal plain: Historical overview and issues of concern. Web page of the Arctic National Wildlife Refuge, Fairbanks, Alaska. http://arctic.fws.gov/issues1.html.

United States Geological Survey. 2001. *The National Petroleum Reserve-Alaska (NPRA) data archive*. USGS Fact Sheet FS-024-01, March 2001.

United States Geological Survey. 2001. *Arctic National Wildlife Refuge, 1002 Area, petroleum assessment, 1998, including economic analysis*. USGS Fact Sheet FS-028-01, April 2001.

Walker, Donald A. 1997. Arctic Alaskan vegetation disturbance and recovery. Pp. 457–479 in *Disturbance and recovery in Arctic lands*, R.M.M. Crawford, ed. Kluwer Academic Publishers, Dordrecht, Netherlands.

Chapter 14

Aeck, Molly. 2005. Biofuel use growing rapidly. Pp. 38–39 in *Vital signs 2005*. Worldwatch Institute and W. W. Norton, Washington, D.C., and New York.

American Wind Energy Association. 2005. *Global wind energy market report*. AWEA, Washington, D.C.

Ananthaswamy, Anil. 2003. Reality bites for the dream of a hydrogen economy. *New Scientist*, (November 15, 2003): 6–7.

Arnason, Bragi, and and Thorsteinn I. Sigfusson. 2000. Iceland—a future hydrogen economy. *International Journal of Hydrogen Energy* 25: 389–394.

Ásmundsson, Jón Knútur. 2002. Will fuel cells make Iceland the 'Kuwait of the North?' *World Press Review*, 15 February 2002.

British Petroleum. 2005. *BP statistical review of world energy*. London, June 2005.

Burkett, Elinor. 2003. A mighty wind. *New York Times magazine*. June 15, 2003.

Chow, Jeffrey, et al. 2003. Energy resources and global development. *Science* 302: 1528–1531.

DaimlerChrysler. 2003. *360 DEGREES/DaimlerChrysler Environmental Report 2003*. DaimlerChrysler AG, Stuttgart, Germany.

Dunn, Seth. 2000. The hydrogen experiment. *Worldwatch* 13: 14–25.

Energy Information Administration, United States Department of Energy. www.eia.doe.gov

Energy Information Administration. 2005. *Annual energy outlook 2005*. Washington, D.C.

Energy Information Administration, United States Department of Energy. 2005. *Annual energy review 2004*. DOE/EIA-0384(2004). Washington, D.C.

Energy Information Administration, United States Department of Energy. 2005. *International energy annual 2003*. Washington, D.C.

Flavin, Christopher, and Seth Dunn. 1999. A new energy paradigm for the 21st century. *Journal of International Affairs* 53: 167–190.

Hirsch, Tim. 2001. Iceland launches energy revolution. *British Broadcasting Corporation News*, 24 December 2001.

Hydrogen & Fuel Cell Letter. 2003. World's first commercial hydrogen station opens in Iceland. *Hydrogen & Fuel Cell Letter* May 2003.

International Energy Agency. 2002. *Renewables in global energy supply: An IEA fact sheet*. IEA Publications, Paris.

International Energy Agency. 2005. *Key world energy statistics 2005*. IEA Publications, Paris.

International Energy Agency. 2005. *Renewables information 2005*. IEA Publications, Paris.

International Energy Agency. 2005. *World energy outlook 2005*. IEA Publications, Paris.

International Energy Agency Renewable Energy Working Party. 2002. *Renewable energy . . . into the mainstream*. SITTARD, The Netherlands.

Klass, Donald L. 2004. Biomass for Renewable Energy and Fuels. In *The Encyclopedia of Energy*, Elsevier.

Lovins, Amory B., et al. 2004. *Winning the oil endgame: Innovation for profits, jobs, and security*. Rocky Mountain Institute, Snowmass, Colorado.

Martinot, Eric, et al. 2002. Renewable energy markets in developing countries. *Annual Review of Energy and the Environment* 27: 309–48.

Martinot, Eric, Ryan Wiser, and Jan Hamrin. 2005. *Renewable energy markets and policies in the United States*. Center for Resource Solutions, San Francisco. www.martinot.info/Martinot_et_al_CRS.pdf

Melis, Anastasios, et al. 2000. Sustained photobiological hydrogen gas production upon reversible inactivation of oxygen evolution in the green alga *Chlamydomonas reinhardtii*. *Plant Physiology* 122: 127–135.

Murray, Danielle. 2005. Ethanol's potential: Looking beyond corn. *Eco-economy Update #49*, 5 June 2005. Earth Policy Institute, http://www.earth-policy.org/Updates/2005/Update49.htm

National Renewable Energy Lab, United States Department of Energy. www.nrel.gov

Office of Energy Efficiency and Renewable Energy, United States Department of Energy www.eere.energy.gov

Office of Energy Efficiency and Renewable Energy, United States Department of Energy. 2005. *Wind power today: Federal wind program highlights*. DOE/GO-102005-2115. Washington, D.C.

Randerson, James. 2003. The clean green energy dream. *New Scientist* (August 16, 2003): 8–11.

Reeves, Ari, with Fredric Beck. 2003. *Wind energy for electric power: A REPP issue brief*. Renewable Energy Policy Project, Washington, D.C.

REN21 Renewable Energy Policy Network. 2005. *Renewables 2005 global status report*. Worldwatch Institute, Washington, D.C.

Ristinen, Robert A., and Jack J. Kraushaar. 1998. *Energy and the environment.* John Wiley and Sons, New York.

Rocky Mountain Institute webpage. Energy. RMI, Snowmass, Colorado. www.rmi.org/sitepages/pid17.php

Sawin, Janet. 2004. *Mainstreaming renewable energy in the 21st century.* Worldwatch Paper 169. Worldwatch Institute, Washington, D.C.

Sawin, Janet L. 2005. Global wind growth continues. Pp. 34–35 in *Vital signs 2005.* Worldwatch Institute and W. W. Norton, Washington, D.C. and New York.

Sawin, Janet L. 2005. Solar energy markets booming. Pp. 36–37 in *Vital signs 2005.* Worldwatch Institute and W. W. Norton, Washington, D.C. and New York.

United Nations Food and Agriculture Organization. *Biomass energy in ASEAN member countries.* FAO/ASEAN/EC. FAO Regional Wood Energy Development Programme in Asia, Bangkok, Thailand.

United States Department of Energy National Laboratory directors.1997. *Technology opportunities to reduce U.S. greenhouse gas emissions.* DOE, Washington, D.C.

United States Environmental Protection Agency. Alternative fuels website. www.epa.gov/otaq/consumer/fuels/altfuels/altfuels.htm

Weisman, Alan. 1998. *Gaviotas: A village to reinvent the world.* Chelsea Green Publishing Co., White River Junction, Vermont.

World Alliance for Decentralized Energy. 2005. *World survey of decentralized energy 2005.* WADE, Edinburgh, Scotland.

Chapter 15

Ayres, Robert U., and Leslie W. Ayres. 1996. *Industrial ecology: Towards closing the materials cycle.* Edward Elgar Press, Cheltenham, U.K.

Beede, David N., and David E. Bloom. 1995. The economics of municipal solid waste. *World Bank Research Observer* 10: 113–150.

Diesendorf, Mark, and Clive Hamilton. 1997. *Human ecology, human economy.* Allen and Unwin, St. Leonards.

Edmonton, Alberta, City of. 2003. Waste management. www.edmonton.ca/portal/server.pt/gateway/PTARGS_0_2_104_0_0_35/http%3B/cmsserver/COEWeb/environment+waste+and+recycling/waste

Energy Information Administration. 2005. Municipal solid waste. EIA, U.S. Department of Energy, Washington, D.C. www.eia.doe.gov/cneaf/solar.renewables/page/mswaste/msw.html

Gitlitz, Jenny, and Pat Franklin. 2004. *The 10-cent incentive to recycle,* 3rd ed. Container Recycling Institute, Arlington, Virginia.

Graedel, Thomas E., and Braden R. Allenby. 2002. *Industrial ecology,* 2nd ed. Prentice Hall, Upper Saddle River, New Jersey.

Integrated Waste Services Association. WTE: About waste-to-energy. IWSA, Washington, D.C. www.wte.org/waste.html

Kaufman, Scott, et al. 2004. The state of garbage in America. *Biocycle* 45: 31–41.

Lilienfeld, Robert, and William Rathje. 1998. *Use less stuff: Environmental solutions for who we really are.* Ballantine, New York.

Manahan, Stanley E. 1999. *Industrial ecology: Environmental chemistry and hazardous waste.* Lewis Publishers, CRC Press, Boca Raton, Florida.

McDonough, William, and Michael Braungart. 2002. *Cradle to cradle: Remaking the way we make things.* North Point Press, New York.

McGinn, Anne Platt. 2002. Toxic waste largely unseen. Pp. 112–113 in *Vital signs 2002.* Worldwatch Institute and W. W. Norton, Washington D.C., and New York.

New York City Department of Planning. Fresh Kills: Landfill to landscape. www.nyc.gov/html/dcp/html/fkl/ada/about/1_0.html

New York City Department of Planning. Fresh Kills lifescape. www.nyc.gov/html/dcp/html/fkl/fkl_index.shtml

New York City Department of Sanitation. 2000. Closing the Fresh Kills landfill. *The DOS Report,* February 2000.

Rathje, William, and Colleen Murphy. 2001. *Rubbish! The archeology of garbage.* University of Arizona Press, March 2001.

Smith, Ronald S. 1998. *Profit centers in industrial ecology.* Quorum Books, Westport.

Socolow, Robert H., et al., 1994. *Industrial ecology and global change.* Cambridge University Press, Cambridge.

United Nations Environment Programme. 2000. *International source book on environmentally sound technologies (ESTs) for municipal solid waste management (MSWM).* UNEP IETC, Osaka, Japan.

United States Environmental Protection Agency. 2005. *Municipal solid waste generation, recycling, and disposal in the United States: Facts and figures for 2003.* EPA530-F-05-003, EPA Office of Solid Waste and Emergency Response.

United States Environmental Protection Agency. Municipal solid waste. www.epa.gov/epaoswer/non-hw/muncpl

Epilogue

Bartlett, Peggy, and Geoffrey W. Chase, eds. 2004. *Sustainability on campus: Stories and strategies for change.* MIT Press, Cambridge, Massachusetts.

Brower, Michael, and Warren Leon. 1999. *The consumer's guide to effective environmental choices: Practical advice from the Union of Concerned Scientists.* Three Rivers Press, New York.

Brown, Lester. 2001. *Eco-economy: Building an economy for the Earth.* Earth Policy Institute and W. W. Norton, New York.

Brown, Lester. 2006. *Plan B 2.0: Rescuing a planet under stress and a civilization in trouble.* Earth Policy Institute and W. W. Norton, New York.

Creighton, Sarah Hammond. 1998. *Greening the ivory tower: Improving the environmental track record of universities, colleges, and other institutions.* MIT Press, Cambridge, Massachusetts.

Daly, Herman E. 1996. *Beyond growth.* Beacon Press, Boston.

Dasgupta, Partha, Simon Levin, and Jane Lubchenco. 2000. Economic pathways to ecological sustainability. *BioScience* 50: 339–345.

De Graaf, John, David Wann, and Thomas Naylor. 2002. *Affluenza: The all-consuming epidemic.* Berrett-Koehler Publishers, San Francisco.

Durning, Alan. 1992. *How much is enough? The consumer society and the future of the Earth.* Worldwatch Institute, Washington, D.C.

Erickson, Jon D., and John M. Gowdy. 2002. The strange economics of sustainability. *BioScience* 52: 212.

French, Hilary. 2004. Linking globalization, consumption, and governance. Pp. 144–163 in *State of the world 2004.* Worldwatch Institute and W. W. Norton, Washington, D.C., and New York.

Gibbs, W. Wayt. 2005. How should we set priorities? *Scientific American* 293(3): 108–115.

Hawken, Paul. 1994. *The ecology of commerce: A declaration of sustainability.* HarperBusiness, New York.

Keniry, Julian. 1995. *Ecodemia: Campus environmental stewardship at the turn of the 21st century.* National Wildlife Federation, Washington, D.C.

McMichael, A. J., et al. 2003. New visions for addressing sustainability. *Science* 302: 1919–1921.

Millennium Ecosystem Assessment. 2005. *Ecosystems and human well-being: General synthesis.* Millennium Ecosystem Assessment and World Resources Institute.

McIntosh, Mary, et al. 2001. *State of the campus environment: A national report card on environmental performance and sustainability in higher education.* National Wildlife Federation Campus Ecology.

Meadows, Donella, Jørgen Randers, and Dennis Meadows. 2004. *Limits to growth: The 30-year update.* Chelsea Green Publ. Co., White River Junction, Vermont.

National Research Council, Board on Sustainable Development. 1999. *Our common journey: A transition toward sustainability.* National Academies Press, Washington, D.C.

National Wildlife Federation. Campus Ecology. www.nwf.org/campusecology.

Sanderson, Eric W., et al. 2002. The human footprint and the last of the wild. *BioScience* 52: 891–904.

Schor, Juliet B., and Betsy Taylor, eds. 2002. *Sustainable planet: Solutions for the twenty-first century.* The Center for a New American Dream. Beacon Press, Boston.

United Nations. 2002. *Report of the World Summit on Sustainable Development, Johannesburg, South Africa, 26 August–4 September 2002.* United Nations, New York.

United Nations. 2002. *The road from Johannesburg: What was achieved and the way forward*. United Nations, New York.

United Nations Division for Sustainable Development. 1990. *Agenda 21*. Accessible online at www.un.org/esa/sustdev/documents/agenda21/index.htm

United Nations Environment Programme. 2002. Outlook: 2002-2032. Pp. 319–400 in *Global environment outlook 3 (GEO-3)*. UNEP and Earthscan Publications, Nairobi and London.

University Leaders for a Sustainable Future. www.ulsf.org.

Wilson, Edward O. 1998. *Consilience: The unity of knowledge*. Alfred A. Knopf, New York.

World Commission on Environment and Development. 1987. *Our common future*. Oxford University Press, Oxford.

Index

Note to the reader: Italicized page numbers indicate highlighted text terms; tables are identified with a *t* following the page number and figures by an *f*.